Out of Africa I

Vertebrate Paleobiology and Paleoanthropology Series

Edited by

Eric Delson
Vertebrate Paleontology, American Museum of Natural History,
New York, NY 10024, USA
delson@amnh.org

Eric J. Sargis
Anthropology, Yale University
New Haven, CT 06520, USA
eric.sargis@yale.edu

Focal topics for volumes in the series will include systematic paleontology of all vertebrates (from agnathans to humans), phylogeny reconstruction, functional morphology, Paleolithic archaeology, taphonomy, geochronology, historical biogeography, and biostratigraphy. Other fields (e.g., paleoclimatology, paleoecology, ancient DNA, total organismal community structure) may be considered if the volume theme emphasizes paleobiology (or archaeology). Fields such as modeling of physical processes, genetic methodology, nonvertebrates or neontology are out of our scope.

Volumes in the series may either be monographic treatments (including unpublished but fully revised dissertations) or edited collections, especially those focusing on problem-oriented issues, with multidisciplinary coverage where possible.

For other titles published in this series, go to
www.springer.com/series/6978

Out of Africa I

The First Hominin Colonization of Eurasia

Contributions from the Second Stony Brook
Human Evolution Symposium and Workshop, September 27–30, 2005

Edited by

John G. Fleagle

Department of Anatomical Sciences and Turkana Basin Institute, Stony Brook University, Stony Brook, NY, USA

John J. Shea

Department of Anthropology and Turkana Basin Institute, Stony Brook University, Stony Brook, NY, USA

Frederick E. Grine

Departments of Anthropology and Anatomical Sciences, Stony Brook University, Stony Brook, NY, USA

Andrea L. Baden

Interdepartmental Doctoral Program in Anthropological Sciences, Stony Brook University, Stony Brook, NY, USA

Richard E. Leakey

Department of Anthropology and Turkana Basin Institute, Stony Brook University, Stony Brook, NY, USA

Springer

Editors
John G. Fleagle
Department of Anatomical Sciences
and Turkana Basin Institute
Stony Brook University
Stony Brook, NY 11794-8081
USA
john.fleagle@stonybrook.edu

Frederick E. Grine
Departments of Anthropology
and Anatomical Sciences
Stony Brook University
Stony Brook, NY 11794-4364
USA
frederick.grine@stonybrook.edu

Richard E. Leakey
Department of Anthropology
and Turkana Basin Institute
Stony Brook University
Stony Brook, NY 11794-4364
USA
richard.leakey@stonybrook.edu

John J. Shea
Department of Anthropology
and Turkana Basin Institute
Stony Brook University
Stony Brook, NY 11794-4364
USA
john.shea@sunysb.edu

Andrea L. Baden
Interdepartmental Doctoral Program
in Anthropological Sciences
Stony Brook University
Stony Brook, NY 11794-4364
USA
abaden@ic.sunysb.edu

ISBN 978-90-481-9035-5 e-ISBN 978-90-481-9036-2
DOI 10.1007/978-90-481-9036-2
Springer Dordrecht Heidelberg London New York

Library of Congress Control Number: 2010932000

Cover illustration: Clockwise from bottom left, photographs of SK 847 by Frederick E. Grine (© and courtesy of Frederick E. Grine and the Transvaal Museum), KNM-ER 3733 by Robert Campbell (© and courtesy of Robert Campbell and the Kenya National Museum), Dmanisi by David Lordkipanidze (© and courtesy of David Lordkipanidze), Yuanmou incisors by Richard Potts (© and courtesy of Richard Potts), and Modjokerto by Karen Baab (© and courtesy of Karen Baab). Image design by Luci Betti-Nash.

Printed on acid-free paper

Springer is part of Springer Science+Business Media (www.springer.com)

Preface

Today, humans are the most cosmopolitan of all primates, and one of the most widespread of all living mammals. Our lineage has not always been so geographically widespread. For over two-thirds of our evolutionary history, from our divergence from chimpanzees more than 6 million years ago, until as recently as 2 million years ago, hominins were an endemic African group. Sometime near the beginning of the Pleistocene, just under 2 million years ago, this all changed, and archeological and paleontological evidence of early hominins appears in many parts of Eurasia. The papers in this volume address the many facets of the first hominin range expansion from Africa into Eurasia.

These include not just the obvious issues of when hominins first appear in Eurasia and the type of evidence indicating their presence, but a whole series of broader questions about the nature of this initial hominin intercontinental dispersal. Why did hominins first leave Africa in the Early Pleistocene and no earlier? What do we know about the adaptations of the hominins that dispersed – their diet, locomotor abilities, cultural abilities? Was there a single dispersal event or several? Did this initial dispersal lead to successful colonization of Eurasia, or were they just temporary excursions? Was the hominin dispersal part of a broader faunal expansion of African mammals northward, or is there something unique about our lineage? What route or routes did dispersing populations take? Why do we have evidence of Early Pleistocene hominins in western and eastern Asia, but nothing in between?

In order to address these and many other issues regarding the initial dispersal of hominins out of Africa, we organized the Second Stony Brook Human Evolution Symposium and Workshop in 2005. An international group of scholars assembled in Stony Brook for 5 days of presentations and discussions to explore our current understanding of this aspect of human evolution and to identify priorities for future research. This volume is the result of that meeting. Since the mid-1980s, the lower boundary of the Pleistocene Epoch (i.e., the beginning of the Quaternary Period) has been regarded as corresponding with the base of the Calabrian stratotype, at 1.81 Ma. Recently, however, the International Union of Geological Sciences has recognized the base of the Gelasian stratotype, which corresponds to the Matuyama (C2r) chronozone, or the Gauss-Matuyama boundary, as defining the Pliocene-Pleistocene boundary at 2.588 Ma. Because many of the papers were submitted before that change, in this volume, the older convention has been followed with the base of the Pleistocene at 1.8 Ma.

The workshop was sponsored by Stony Brook University and the Turkana Basin Institute, with important contributions by Kay Harrigan Woods, Bill and Kathy Cleary, Jim and Marilyn Simons, The Richard and Rhoda Goldman Fund, The Leakey Foundation and the Office of the Provost, Stony Brook University. The President of Stony Brook, Dr. Shirley Strum Kenny, kindly offered her home at Sunwood for the workshop. In addition to the contributors to this volume, Susan Antón, Karen Baab, Amanuel Beyin, Paul Bingham, Ari Grossman, Lawrence Martin, James Rossie, Danielle Royer, Matthew Sisk, Fred Spoor, Kathryn Twiss, and Frank Brown. Elizabeth Wilson contributed to the success of the symposium and workshop in many ways.

The chapters in this volume underwent extensive peer-review. We thank our many colleagues who contributed their time, effort, and wisdom to the review process. Luci Betti-Nash, Amanuel Beyin, Chris Gilbert, Jessica Lodwick, and Brandon Wheeler provided extensive assistance in the editing of this volume. We thank Eric Delson and Eric Sargis, senior co-editors of the Vertebrate Paleontology and Paleoanthropology Series for their endless support, encouragement, and especially their patience in seeing this volume through to publication.

Stony Brook University

John G. Fleagle
John J. Shea
Frederick E. Grine
Andrea L. Baden
Richard E. Leakey

Contents

Contributors

Miriam Belmaker
Department of Anthropology, Peabody Museum, Harvard University, 11 Divinity Avenue, Cambridge, MA 02138, USA
belmaker@fas.harvard.edu

Parth R. Chauhan
The Stone Age Institute & CRAFT Research Center, 1392 W. Dittemore Road, Bloomington (Gosport), IN 47433, USA
pchauhan@indiana.edu

Russell L. Ciochon
Department of Anthropology, University of Iowa, 114 Macbride Hall, Iowa City, IA 52242-1322, USA
russell-ciochon@uiowa.edu

Robin Dennell
Department of Archaeology, University of Sheffield, S1 4 ET, UK
r.dennell@sheffield.ac.uk

John G. Fleagle
Department of Anatomical Sciences, Stony Brook University Medical Center, Stony Brook, NY 11794-8081, USA
john.fleagle@stonybrook.edu

Ya-Mei Hou
Institute of Vertebrate Paleontology and Paleoanthropology, Chinese Academy of Sciences, No. 142, Xizhimenwai Dajie, Beijing 100044, PR China
houyamei@ivpp.ac.cn

Marta Mirazón Lahr
Leverhulme Centre for Human Evolutionary Studies, University of Cambridge, Cambridge, CB2 1QH, UK
mbml1@cam.ac.uk

Meave Leakey
Turkana Basin Institute, Department of Anthropology, Stony Brook University, Stony Brook, NY 11794-4364, USA
meaveleakey@uuplus.net

Margaret E. Lewis
School of Mathematical and Natural Sciences (Biology), The Richard Stockton College of New Jersey, PO Box 195, Pomona, NJ 08240-0195, USA
margaret.lewis@stockton.edu

David Lordkipanidze
Georgian National Museum, 3, Rustaveli Avenue, 0105 Tbilisi, Georgia
dlordkipanidze@museum.ge

Bienvenido Martínez-Navarro
Institut Català de Paleoecologia Humana i Evolució Social-IPHES,
Universitat Rovira i Virgili, Pl Imperial Tarraco, 143005 Tarragona, Spain
bienvenido@icrea.es

Avinash C. Nanda
Wadia Institute of Himalayan Geology, Dehra Dun 248 001, India
nandaac@rediffmail.com

Rajeev Patnaik
Center of Advanced Study in Geology, Punjab University, Chandigarh 160014, India
rajeevpatnaik@gmail.com

Michael D. Petraglia
Research Laboratory for Archaeology and the History of Art, Dyson Perrins Building,
University of Oxford, South Parks Road, Oxford OX1 3QY
michael.petraglia@rlaha.ox.ac.uk

Richard Potts
Human Origins Program, National Museum of Natural History, Smithsonian Institution,
Washington DC 20013-7012, USA
pottsr@si.edu

G. Philip Rightmire
Department of Anthropology, Peabody Museum, Harvard University,
Cambridge MA 02138, USA
Department of Anthropology, Binghamton University, NY 13902, USA
gprightm@fas.harvard.edu

John J. Shea
Anthropology Department & Turkana Basin Institute, Stony Brook University,
Stony Brook, NY 11794-4364, USA
John.Shea@sunysb.edu

Robin Teague
Center for Advanced Study of Hominid Paleobiology, George Washington University,
Washington DC 20013, USA
Human Origins Program, National Museum of Natural History,
Smithsonian Institution, Washington DC 20013-7012, USA
rteague@gwu.edu

Lars Werdelin
Department of Palaeozoology, Swedish Museum of Natural History,
Box 50007S, 104 05, Stockholm, Sweden
werdelin@nrm.se

Yahdi Zaim
Department of Geology, Institut Teknologi Bandung, Jl. Ganesa no.10
Bandung – 40132, Indonesia
zaim@gc.itb.ac.id

Ling-Xia Zhao
Institute of Vertebrate Paleontology and Paleoanthropology, Chinese Academy of Sciences,
No. 142, Xizhimenwai Dajie, Beijing 100044, PR China
zhaolingxia@ivpp.ac.cn

Part I
The African Background

Introduction

The papers in this first part discuss the aspects of geography, climate, faunal composition and hominin culture that enabled or led to the initial dispersal of hominins into Eurasia.

In "Early Pleistocene Mammals of Africa: Background to Dispersal", Meave Leakey and Lars Werdelin place the evolution and dispersal of hominins in the broader context of the evolution of anthropoid primates in Africa and their patterns of dispersal over the past 20 million years. In particular, they compare the ecology of early *Homo* with that of *Theropithecus,* another widely dispersing catarrhine from the Plio-Pleistocene. They also review patterns of African carnivore evolution and dispersal from the Miocene through the Pleistocene. They argue that with the advent of culture, in the form of stone tools, hominins moved into a new feeding niche and became part of the carnivore guild, with the added ability to also exploit plant resources. This would have affected many aspects of their ecology, including home range size and dispersal abilities, and also affected the ecology and survival within the diverse carnivore communities of the Pliocene and Early Pleistocene. They also review the herbivore communities of the Pliocene and Pleistocene, noting that the hominin shift to a carnivore feeding niche took place in the context of a diverse assemblage of large herbivores that were potential prey. Cut marks on a broad size range of taxa support the view that Early Pleistocene hominins did indeed exploit a diversity of taxa for meat.

Much of the discussion regarding the initial dispersal of hominins from Africa to Eurasia in the earliest Pleistocene centers around the role of synchronic carnivores, especially the genus *Megantereon*, both as a co-disperser and as a potential competitor. In "Carnivoran Dispersal Out of Africa During the Early Pleistocene: Relevance for Hominins", Margaret Lewis and Lars Werdelin review the diverse carnivores in the Pliocene and Pleistocene of Africa. They then summarize the complex taxonomic history of the genus *Megantereon* and questions surrounding the intercontinental dispersal of various species within that genus, including the identity of the species of *Megantereon* from the site of Dmanisi. Finally, they review the nature of the likely ecological relationships and possible interactions between hominins and *Megantereon.*

In "Saharan Corridors and Their Role in the Evolutionary Geography of 'Out of Africa I'", Marta Mirazón Lahr observes that dispersal of hominins out of Africa must have first involved dispersal within Africa since the fossil record of early hominin evolution is almost exclusively from sub-Saharan Africa. Accordingly, she points out that questions of dispersal can be examined at very different scales, such as species or populations and continents or ecological zones. It is at the latter scales that evolution takes place. In a review of the paleontological and archeological record of North Africa, she finds that evidence of a hominin presence is very rare before the Early and Middle Pleistocene, and the earliest record, at Ain Hanech in Algeria, corresponds to the timing of the colonization of Eurasia. Thus, trans-Saharan dispersal was very rare, and hominin presence in North Africa was probably ephemeral prior to the Middle Pleistocene. She argues that it is most likely that the initial hominin dispersal into Eurasia was across the Bab-el-Mandeb Strait from the Horn of

Africa to the Arabian Peninsula. Ironically, Early Pleistocene hominin dispersals into Eurasia were apparently more successful and more long-lived than those into North Africa.

It is generally accepted that the development of culture represented by the manufacture and use of stone tools was one of the most important factors that enabled hominins to successfully disperse into Eurasia and the remains of these artifacts are often the primary evidence we have of early hominin presence. In "Stone Age Visiting Cards Revisited: A Strategic Perspective on the Lithic Technology of Early Hominin Dispersal", John Shea investigates what differences in stone tool technology tell us about the identity and behavior of the hominins that made them. Shea notes that efforts to link specific lithic technologies uniquely with one taxon of hominin have repeatedly failed, and the lack of correspondence is increasing with increased diversity in hominin phylogeny. Most notably, pebble core tools such as those that characterize the Oldowan industries persist from the late Pliocene to the present. He suggests that a more profitable approach might be to look at different technologies from a strategic perspective. Following an examination of the likely costs and benefits of the production and use of pebble cores such as those characterizing Oldowan industries, and Large Cutting Tool (LCT) core technology, he argues that the former are likely characteristic of low density, mobile, colonizing populations, while LCT technologies are more characteristic of larger, stable populations.

The Editors

Chapter 1
Early Pleistocene Mammals of Africa: Background to Dispersal

Meave Leakey and Lars Werdelin

Abstract The initial dispersal of humans out of Africa was a significant event in human evolution raising many questions. Why did this happen at this particular time? Was it part of a major migration of mammals out of Africa and did any species move into Africa at the same time? Were climate and habitat changes taking place that might have been contributing factors? With the advent of culture at 2.6 Ma, hominins moved from the primate to the carnivore feeding niche, thus avoiding constraints that had previously determined their distribution. Here we look at fossil carnivores and cercopithecids for factors that provide a background to this significant event in our evolutionary history and we also look at herbivore diversity as a potential source of prey for meat-eating hominins.

Keywords Carnivore • Dispersal • Catarrhine • *Homo erectus*

Introduction

Until the first evidence of culture in the hominin record, the geographic distribution of human ancestors was subjected to life history constraints common to all higher primates. Catarrhines today are largely restricted to tropical and subtropical forests and woodlands that provide relatively high quality foods that are largely available year round. The few extant African catarrhines found outside these habitats, such as baboons (*Papio* sp.) and vervet monkeys (*Chlorocebus aethiops*), are semi-terrestrial eclectic feeders tolerant of extreme seasonal variations in their food supply. The ability to utilize and manufacture stone tools was the first in a series of adaptations that enabled hominins to shift feeding strategy and compete with carnivores with increasing success. In the absence of constraints that had previously limited hominin distribution, hominins migrated into habitats from which other catarrhines were excluded, and ultimately to latitudes far removed from the tropics.

To fully understand the relative significance of the factors influencing this event, it is essential to assess whether the dispersal of *Homo erectus* (*sensu lato*; see Antón 2003 for discussion; and Spoor et al. 2007 for use of *Homo erectus* for African specimens previously referred to as *Homo ergaster*) out of Africa 1.8 million years ago was part of a major migration event affecting many species, was restricted to a very few species with similar life histories and diets, or was unique to *H. erectus*. To explore possible scenarios we first look at anthropoid evolutionary history in the Miocene, Pliocene and Pleistocene as a background to hominin evolution prior to the advent of culture, when early hominins were an integral part of the fossil catarrhine assemblages and subject to similar constraints on their distribution. We then examine patterns of cercopithecid and carnivore dispersal in the critical time interval, and particularly at or near 1.8 Ma to assess whether other species moved out of Africa at this time. We then discuss the Early Pleistocene carnivore guilds and how these might have influenced meat eating hominin food procurement strategies. Finally, we look at Early Pleistocene herbivore faunal assemblages as a potential food resource for the newly emerged meat eating hominins.

The Turkana Basin, Kenya, is a source for much of this discussion since it documents a particularly rich record over the appropriate time interval. The many archeological sites at Olduvai Gorge, Tanzania, and the fossil assemblages of the Transvaal caves in South Africa are also important sources of evidence.

Anthropoid Faunal Assemblages

With the notable exception of *Homo*, the distribution of anthropoids today is largely constrained to the tropics; apes and most monkeys are found in tropical and subtropical forests and woodlands where food resources are relatively

M. Leakey (✉)
Turkana Basin Institute, Department of Anthropology, Stony Brook University, Stony Brook, NY 11794-4364
e-mail: meaveleakey@uuplus.net

L. Werdelin
Department of Palaeozoology, Swedish Museum of Natural History, Box 50007, S-104 05, Stockholm, Sweden
e-mail: werdelin@nrm.se

J.G. Fleagle et al. (eds.), *Out of Africa I: The First Hominin Colonization of Eurasia*, Vertebrate Paleobiology and Paleoanthropology,
DOI 10.1007/978-90-481-9036-2_1, © Springer Science+Business Media B.V. 2010

reliable and year round. Whereas in Africa today apes are rare and represented by few species inhabiting limited areas of tropical forest, monkeys in contrast are more common, widely dispersed and diverse, largely inhabiting low latitude tropical and riparian forests and closed woodlands across central Africa. The difference in ape and monkey distributions may reflect the preference of apes for higher quality foods, especially fruits, in less seasonal environments, and their more protracted reproductive schedules (Jablonski et al. 2000). Alternatively the switch from ape dominance and diversity in the Late Miocene to monkey dominance today may have been related to climate and habitat changes during the Late Miocene of Africa and Europe (Fleagle 1999).

Although Pliocene and Pleistocene fossil apes are virtually unknown (McBrearty and Jablonski 2005), the cercopithecid fossil record at this time is good and monkeys were relatively common in the faunal assemblages (Table 1.1). Surprisingly, in the past, African cercopithecids were less restricted in their distribution than today, occurring widely in eastern, southern and northern Africa. The majority of modern African cercopithecid species are not representative either of their fossil precursors or of the majority of their Asian cousins in either distribution or diet.

Cercopithecid diversity stems from small bodied semi-terrestrial papionins and colobines common in woodland and well watered habitats in the Late Miocene (Frost 2001, 2002; Leakey et al. 2003). The morphology of the dentition of these early cercopithecid monkeys suggests that they had similar diets with a high proportion of seeds (Leakey et al. 2003) and their locomotor adaptations indicate that they were more terrestrial than their modern counterparts. Their success may relate to an ability to adapt to the seasonality that increasingly came to dominate open country habitats.

Monkeys underwent a significant radiation between 3.5 and 2.0 Ma, most clearly seen in the colobines. In contrast to the relatively small-bodied, arboreal, leaf-eating modern colobines found in forests and closed woodlands, the Plio-Pleistocene colobines are represented by three genera (*Rhinocolobus, Paracolobus* and *Cercopithecoides*, including several species) of large-bodied, comparatively terrestrial monkeys inhabiting open woodlands and savannas. In many aspects of their morphology, these extinct African colobines most closely resemble the larger extant colobines from Asia; the "temple langurs" (*Semnopithecus entellus*) and the "odd-nosed colobines" (*Rhinopithecus*, *Pygathrix*, and *Nasalis*), which today are widely distributed in more open habitats (Jablonski and Leakey 2008). The general perception that colobines are and have always been arboreal species confined to forests is erroneous.

Table 1.1 The cercopithecoid faunal assemblage from the Upper Burgi and KBS Members of the Koobi Fora Formation, Omo Group deposits, Turkana Basin Kenya. *Theropithecus oswaldi* (in bold) far outnumbers any other species, being represented by over 200 specimens in contrast to all other species which are represented by less than 25 specimens

Parapapio sp. indet. A	*Colobus freedmani*
Parapapio sp. indet. B	*Rhinocolobus turkanaensis*
Lophocebus cf. *L. albigena*	*Cercopithecoides kimeui*
Theropithecus oswaldi	*Cercopithecoides williamsi*
Cercopithecus sp. indet. A	*Paracolobus mutiwa*

The most common genus of monkey in the Plio-Pleistocene was *Theropithecus*. Although today the single remaining species of *Theropithecus, T. gelada*, is confined to the Ethiopian highlands, where it maintains a tenuous hold on this small locality in Africa, in the past this genus was widespread and the most common cercopithecid in the Plio-Pleistocene. First appearing approximately 3.5 Ma ago (Leakey et al. 2001), it derives from a papionin ancestor that diverged to specialize in a graminivorous diet. The evolution of *Theropithecus*, with its increasingly marked dental adaptations towards exclusive graminivory, is well documented (Jolly 1985; Jablonski 1993b; Leakey 1993). *Theropithecus brumpti*, the common gelada in the Late Pliocene, inhabited more closed woodland habitats than its successor *Theropithecus oswaldi*, which replaced *T. brumpti* about 2.5 Ma ago (Eck 1987a; Eck and Jablonski 1987; Leakey 1993; Jablonski and Leakey 2008). This unusual, highly specialized grass eating monkey was extremely successful, surviving for 2 million years, from 2.5 Ma until its extinction in the last 500,000 years. Its known occurrences are widespread; fossil *T. oswaldi* have been recorded from sites in eastern, southern, and northern Africa (Jablonski 1993a), the Levant (Belmaker 2002, 2009), Italy (Rook et al. 2004; but see Patel et al. 2007), and Spain (Gibert et al. 1995) (the phalanx CV-O, ascribed to *Homo* by Gibert and co-workers has recently been reclassified as *Theropithecus* in Martínez-Navarro et al. 2005), in Europe, and the Siwaliks in India (Gupta and Sahni 1981; Delson 1993). *Theropithecus* was the most widespread cercopithecid known in the Late Pliocene and Early to Middle Pleistocene.

Competition from *Theropithecus* and the large bodied colobines between 3.0 and 1.5 Ma appears to have led to a reduction in the earlier papionin diversity in eastern Africa where, at this time papionins are rather uncommon. It was only in the last 1.0 Ma that *Papio* became more common and today this genus is the most widespread cercopithecid in Africa. In the Plio-Pleistocene Transvaal cave deposits of South Africa, where there is less diversity among fossil colobines, with only *Cercopithecoides williamsi* (the most terrestrial of the African fossil colobines) represented (Freedman 1957), papionins were common throughout this time (Freedman 1976; Freedman and Brain 1977). Fossil guenons (*Cercopithecus* sp.) were rare in the East African fossil record; few specimens are known from the Pliocene and few have been recovered from the Early Pleistocene of the Turkana Basin (Leakey 1988; Eck 1987b; Jablonski and Leakey 2008). Although molecular studies indicate that

Cercopithecus originated in the Late Miocene and that the major subdivisions of the genus are Pliocene in age (Tosi et al. 2005), it is not until the last 1.5 Ma that there is fossil evidence that might suggest the beginning of the radiation that led to their modern status as the most common and diverse African genus of monkeys, largely inhabiting tropical forests in central Africa. The modern, rather restricted distribution of African monkeys is thus in direct contrast to that in the Plio-Pleistocene, when cercopithecid species were taxonomically diverse and geographically widespread, frequenting diverse habitats ranging from riparian woodlands, grasslands and relatively open country. Although the most common monkeys today are the guenons, the majority of which inhabit tropical forests (Gautier-Hion et al. 1988), the most widespread species are the baboons (species of *Papio*) and vervets (*Chlorocebus aethiops*), eclectic feeders that are able to utilize fall-back foods in times of drought and food shortages (Kingdon 1997).

Hominins too have a well documented fossil record and their early occurrences frequently coincide with those of cercopithecids. Hominins were almost as widely dispersed geographically in the Pliocene and Early Pleistocene as the most widespread monkeys, although they were less common. Although the earliest hominins are known from the Late Miocene (Brunet et al. 2002; Haile-Selassie 2001; Senut et al. 2001), it is not until 3.5 Ma that there is evidence of diversity in the fossil record (Leakey et al. 2001). Between 2.5 and 1.8 Ma, when hominins shared their habitat with a diverse cercopithecid assemblage including three genera of colobines and one species of *Theropithecus*, they came to increasingly externalize their food procurement functions in the form of stone tools.

Patterns of Catarrhine Dispersal

What is it that determines the distribution of catarrhines and, as the closest relatives of hominins, what can the distribution of past cercopithecid and hominoid assemblages tell us about the dispersal of hominins? Cercopithecid migrations in the past appear to have been exclusively out of Africa (Fleagle and Gilbert 2006). The earliest evidence for the divergence of the modern subfamilies is in Africa with the presence of colobines at 11 Ma in the Baringo Basin in Kenya (Benefit and Pickford 1986). Colobines are first found in Europe soon after this with the appearance of *Mesopithecus* (Delson 1973, 1975a, b, 1994), they first appear in Asia (northern India and Pakistan) by the Late Miocene between 7 and 5 Ma (Barry 1987), and in the latest Miocene of China (Delson 1994). The earliest known cercopithecines are from the latest Miocene, dated to approximately 7 Ma (Delson 1973, 1975a, b; Szalay and Delson 1979; Leakey et al. 1996, 2003). At about this time, cercopithecines are recognized in Europe; *Procynocephalus* appears in the latest Miocene, *Dolichopithecus* appears in the early Middle Pliocene and *Paradolichopithecus* appears in the Middle Pliocene. There is no evidence of anthropoid faunal exchange in the Pliocene, and in the Pleistocene, the only cercopithecid known to have spread out of Africa is *Theropithecus*. The earliest recorded *Theropithecus* out of Africa is from Pirro Nord in Italy, which has an estimated biochronological date of 1.6–1.3 Ma (Rook et al. 2004). However, the ocurrence at Pirro Nord has been questioned by Patel et al. (2007) who also note the absence of *Theropithecus* from any other Plio-Pleistocene sites outside of Africa and the Levant (including Dmanisi). It is thus possible that *Theropithecus* migrated out of Africa at a similar time to *Homo erectus*, even though the genus is not present at localities such as Dmanisi that include *H. erectus* or Venta Micena and Appolonia-1 that (along with Dmanisi) include *Megantereon*, another putative African migrant (see below).

The earliest evidence of fossil "apes" outside Africa is between 16.5 and 17 Ma from southern Germany and they are also recorded at Paşalar in Turkey (Mourtzou and Andrews 2008). Hominoids are widespread in Eurasia in the Middle Miocene in contrast to Africa where they are very rare. Hominoids disappeared from Europe in the Late Miocene. The earliest fossil hominins appeared in the Late Miocene of Africa, but the evidence is inconclusive as to whether extant African apes and humans originated in Africa or from the Middle Miocene hominoid diversity of Eurasia; two contrasting schools of thought currently pertain to this problem (Kordos and Begun 2002). Fossil hominins are relatively abundant in eastern and southern Africa in the Pliocene, but the earliest evidence of migration of hominins out of Africa is that of *Homo erectus* at ~1.7 Ma (Antón et al. 2002).

In order to assess parameters determining the distribution of the Catarrhini, Jablonski et al. (2000) examined the impact of environmental change in the Late Pliocene and Pleistocene in China. During this time there were decreases in the extent of tropical and subtropical ecozones as a result of southward latitudinal shifts. The responses of the major catarrhine genera to these habitat shifts were diverse, but dietary selectivity and life history parameters were noted to be strong predictors of the type and magnitude of responses of individual taxa. Relative to apes, monkeys can survive on a wider variety of vegetation in extreme seasonal habitats and have shorter gestation times, weaning periods and interbirth intervals. Apes have a preference for high quality foods, especially ripe fruits, in less seasonal environments and more protracted reproductive schedules and lower intrinsic rates of population increase. Apes, including gibbons, orang-utans and *Gigantopithecus*, were found to be more sensitive to environmental change than monkeys, including macaques and langurs (Jablonski et al. 2000).

Hominins, like apes, have an advanced age for onset of reproduction, long gestation and weaning periods and long interbirth intervals. They are an extreme example of "K-selection" reproductive strategy in which high parental investment in low numbers of offspring per lifetime yield a low intrinsic rate of population increase (Pianka 1978). They are therefore expected to conform to the ape pattern. Jablonski et al. (2000) found this to be generally true, with the exception of early *Homo sapiens*, because advanced culture enabled this species to colonize highly seasonal habitats including tundra, which were unavailable to other catarrhines.

Antón et al. (2002) note that rate of dispersal, like foraging strategy, is influenced by home range size and body size. These factors likely also contributed to the wide Pleistocene dispersal of *Theropithecus oswaldi*. In contrast to the relatively small home range sizes of modern forest dwelling monkeys, extant *Theropithecus gelada* has a relatively extensive home range in which the exceptionally large troops forage for limited food resources. This was presumably also true for the large bodied *Theropithecus oswaldi* in the Plio-Pleistocene. Carnivores have increased home range sizes relative to herbivores, so that the home range size of early *Homo* would have increased with the shift to meat eating. Indeed archeological evidence indicates large home range sizes for early hominins; the distribution of archeological sites within Africa at this time shows increasing complexity of the archeological record and increasing distance from lithic raw material sources (Cachel and Harris 1995; Delagnes and Roche 2005). The rapid rates of dispersal of *Homo erectus* appear to have been promoted by changes in foraging strategy that led to increases in home range size and body size facilitated by changes in ecosystem structure during the Plio-Pleistocene (Antón et al. 2002). But what were the factors that may have affected hominin success as they changed foraging strategy?

Patterns of Carnivore Dispersal

The dispersal of mammalian species into new geographical areas requires corridors of appropriate habitats with access to water and suitable food. Species that require specific but limited feeding niches are less likely to be widely distributed than those more tolerant of variations in their food supply. Carnivores are thus generally more widely distributed than herbivores, since suitable food is widely distributed.

In contrast to catarrhines, carnivore dispersal in the Miocene was almost exclusively from Eurasia into Africa (Werdelin 2006, unpublished data). It was not until the very end of the Miocene that this pattern changed. Between 5.5 and 4.0 Ma, carnivores continued to move into Africa and the first migrations out of Africa took place. After this time, carnivore migration has been almost exclusively out of Africa, involving taxa of medium to large (>10 kg) body size.

Out of 31 carnivoran species that were present in East Africa between 2.1 and 1.8 Ma (of which 18–19 including *Mellivora*, were large and terrestrial), only one (*Megantereon whitei*) possibly migrated out of Africa at this time. *M. whitei* can be parsimoniously regarded as present in some southern European sites (e.g., Venta Micena, Dmanisi, Appolonia-1) dated 1.8–1.5 Ma, though this identification is not unproblematic (Werdelin and Lewis 2002; Lewis and Werdelin 2007). *Crocuta* left Africa some time prior to 2 Ma, as the current FAD for this genus in China is ~2.2 Ma (Qiu et al. 2003), while the FAD for Europe is <0.8 Ma.

In contrast, diverse carnivoran taxa appear to migrate out of Africa at various times before and after 2.1–1.8 Ma. Both *Megantereon* and *Homotherium* may have migrated from Africa at ca. 3.5 Ma or earlier. *Panthera* and *Acinonyx* also have FADs in Europe at ca. 3 Ma, which is later than in Africa. *P. leo* reaches Eurasia from Africa ca. 0.5 Ma. With all these data, there is no evidence among carnivore taxa for a peak of migrations at a time corresponding to the timing of the first appearance of *Homo* outside Africa.

The composition of the carnivore faunal assemblages changes with time, which may be partly a result of the influence of hominins. Up to the end of the Pliocene, carnivore assemblages include numerous large-bodied specialists, some of which were undoubtedly preying on early australopithecines. The time interval 2–1.5 Ma is the time of extinction of these large, specialist carnivores, while after this the majority of remaining carnivores in Africa can be considered ecological generalists (Table 1.2; Lewis and Werdelin 2007;

Table 1.2 The carnivoran faunal assemblage from eastern and southern Africa dated between 2 and 1.5 Ma ago. Large-bodied specialist taxa are in bold; extinct species not replaced by closely related taxa are starred

Caracal caracal	*Ichneumia albicauda*
Chasmaporthetes nitidula*	***Lycyaenops silberbergi****
Crocuta dietrichi	***Megantereon whitei****
Crocuta ultra	*Mellivora capensis*
Crossarchus transvaalensis	*Mungos dietrichi*
Cynictis penicillata	*Mungos minutus*
Dinofelis aronoki*	*Nyctereutes terblanchei**
Dinofelis barlowi*	*Pachycrocuta brevirostris**
Dinofelis piveteaui*	*Panthera leo*
Dinofelis* sp. (Olduvai)	*Panthera pardus*
Felis sp.	*Parahyaena brunnea*
Galerella debilis	*Proteles amplidentus*
Galerella primitivus	*Prototocyon recki*
Genetta genetta	***Pseudocivetta ingens****
Genetta tigrina	***Sivaonyx* sp.***
Helogale hirtula	*Suricata suricatta*
Herpestes ichneumon	*Torolutra ougandensis*
Homotherium* sp.	*Vulpes chama*
Hyaena hyaena	*Vulpes pulcher*
Hyaena makapani	

Peters et al. 2008). This sequence of events is clearly correlated with the time when hominins evolved new dietary strategies and began directly competing with carnivorans for prey (Lewis and Werdelin 2007). Increased competition from hominins, as well as increased seasonality may have led to times of acute food shortages that were unlikely to have been present earlier and may have affected the survival of specialist species.

The Early Pleistocene Carnivore Guild

Werdelin and Lewis (2005) examined carnivore species diversity through the African Plio-Pleistocene. Carnivore species richness in the last 4 Ma reaches a maximum between 3.9 and 3.6 Ma, with a further peak between 2.1 and 1.8 Ma. This corresponds to a high rate of originations and extinctions of taxa at 4.0–3.5 Ma and another peak of origination between 2.0–1.5 Ma. This latter peak was, however, outstripped by a higher rate of extinction that continues to the present day, resulting in the modern relatively depauperate carnivore assemblages. It is likely that a large proportion of these extinctions were the direct result of increasing competition from hominins.

Lewis (1997) examined the functional anatomy of extant and extinct fossil carnivorans to assess their impact on carcass availability for early meat eating hominins. Her analysis shows that Plio-Pleistocene carnivorans as a group engaged in a wider range of behaviors than their modern counterparts, largely due to the presence of sabertooth cats, which are hypothesized to have been providers of large carcasses for hominin scavenging (Blumenschine 1987; Marean 1989; Van Valkenburgh et al. 1990). The largest carcasses would have been provided by the sabertooth *Homotherium*, while *Dinofelis* and *Megantereon* probably fed from carcasses of similar size to those killed by lions and leopards today (Lewis 1997).

The advent of culture placed hominins in a new feeding niche. Whereas previously human ancestors were carnivore prey (Brain 1981) they now had a way of accessing meat and became part of the carnivore guild. Cut-marked bones, and bones smashed to extract marrow attest to the adoption of meat eating strategies as early as 2.6 Ma (Semaw et al. 1997). The method of procurement of meat is not known, but it is likely that hominins initially scavenged carcasses left by other predators and only later improved their skills to become active hunters. Early butchery sites are not common but at Olduvai Gorge, an elephant skeleton in Bed 1 at FLK North and a *Deinotherium* skeleton just above the base of Bed II were both found with artifacts (Leakey 1971). Although it is not clear how the animals died, these occurrences show that, between 1.8 and 1.6 Ma hominids butchered the carcasses of large mammals. Early hominins would have directly competed with other large carnivores, especially hyaenas, for their meat and marrow. Subsequent carnivoran extinction events between 2 and 1 million years ago would have had a substantial effect on carcass availability for hominins. But it may have been the hunting skills perfected by the hominins that precipitated these extinctions. Certainly, as hominins became proficient hunters, they would have increasingly occupied the niches of the large predators, particularly *Homotherium*. This is reflected in the elevated extinction rate of carnivores in eastern Africa 2–1.5 Ma (Lewis and Werdelin 2007).

It should not be forgotten that hominins, like apes, are not exclusively carnivorous and a wide variety of edible plant resources would have offered a rich source of alternative foods. Because no evidence of plant remains is preserved in archeological sites, it is not possible to assess the importance of plant foods to these early hominins, and indeed this aspect of their diet is often ignored. However, plant resources would certainly have added to their dietary flexibility and ability to move widely through diverse habitats. The combination of access to high calorie meat and marrow together with the ability to utilize plant resources when meat became scarce would have made these early hominins opportunistic flexible feeders with a wide range of dietary options, including many large herbivorous mammals.

Early Pleistocene Herbivores: Proboscideans, Perissodactyls and Artiodactyls

As with the large bodied carnivores, the large bodied herbivores were more diverse in the eastern African Pliocene and Pleistocene than today, even when the faunal assemblages from a wide range of modern habitats are combined. In the Early Pleistocene, the large herbivore species in a single geographic area in the Turkana basin included three proboscideans, three hippos, and three giraffids, where only one species of each of these taxa is known in the wider region of East Africa today; six suids and six equids where only two and three respectively are known today; and a great diversity of large bovids which included two species of *Pelorovis* and two of *Megalotragus*, all now extinct (Table 1.3) (Harris 1983, 1991; Harris et al. 1988; Bobe et al. 2007).

The composition of the herbivore assemblages varied through time reflecting increased climatic variability and seasonality as the environments became more open and more arid through the Late Pliocene and Early Pleistocene. Between 2.3 and 2.1 Ma, bovid abundance and diversity increased in the Shungura Formation, lower Omo Valley, in the Turkana Basin (Bobe and Eck 2001); this was apparently related to greater environmental heterogeneity at the initiation of glacial cycles in the north and to a drier climate in the tropics of Africa. An analysis that included all the major

Table 1.3 The Bovidae faunal assemblage from the Omo Group deposits, Turkana Basin, Kenya, including the Upper Burgi and KBS Members of the Koobi Fora Formation, the Kalachoro and Kaitio Members of the Nachukui Formation and Member G unit 24 through to Member J of the Shungura Formation (from Bobe et al. 2007). The dominant species (represented by 50 or more specimens) are in bold

Aepyceros shungurae-melampus	*Hippotragus gigas*
Beatragus antiquus	*Oryx* sp.
Connochaetes gentryi	*Madoqua* sp.
Damaliscus (Parmularius) eppsi	*Raphicerus* sp.
Megalotragus isaaci	***Kobus ancystrocera***
Parmularius altidens	*Kobus ellipsiprymnus*
Antidorcas recki	*Kobus kob*
Gazella cf. *granti*	*Kobus* aff. *leche*
Gazella janenschi	***Kobus sigmoidalis***
Gazella praethomsoni	***Menelikia lyrocera***
Pelorovis oldowayensis	Tragelaphus gaudryi
Pelorovis turkanensis	Tragelaphus nakuae
Syncerus acoelotus	***Tragelaphus strepsiceros***
Cephalophus sp.	

families of large to medium sized mammals collected from the Shungura Formation (Bobe et al. 2007) shows changes in dominance of three of the most common families, the Suidae, Cercopithecidae and Bovidae and increased faunal variability after 2.5 Ma, with the relative numbers of species favoring closed woodland and forested environments higher at some intervals and those associated with open grassland environments dominating at others. These short term fluctuations are superimposed on a long term trend from more forested to more open arid woodland and bushland environments.

Analyses of a more comprehensive data set from the Omo-Turkana Basin, that includes the Shungura, Nachukui and Koobi Fora Formations indicates three peaks of bovid diversity in the time interval between 4 and 2 Ma that occurred at about 3.8–3.4 Ma, 2.8–2.4 Ma and 2.0–1.4 Ma (Bobe et al. 2007). The last two peaks correspond to previously identified periods of high faunal turnover in the Turkana Basin at 2.4–2.2 and 2.0–1.8 Ma (Bobe and Behrensmeyer 2002, 2004), and the last peak was also a time of diversification of grazing bovids inhabiting the expanding grasslands which included the species *Pelorovis oldowayensis, Megalotragus isaaci* and *Beatragus antiquus* (Bobe and Behrensmeyer 2004). Three tribes (Alcelaphini, Antilopini and Hippotragini) that are strongly indicative of open seasonally arid grasslands and bushlands increase in relative abundance during this time. These patterns in faunal distribution and abundance are consistent with the known record of climate change derived from marine sediments (deMenocal 1995; deMenocal and Bloemendal 1995; Dupont and Leroy 1995; Denton 1999).

It appears that at the time that early *Homo* was adapting to an increasingly carnivorous life style, the bovid assemblages were diverse, rich in open country bushland species and

Table 1.4 Species identified in archeological sites in the KBS and Okote Members of the Koobi Fora Formation, East Turkana (From Bunn 1997)

Rodentia	*Thryonomys* sp.
	Hystrix sp.
	Rodentia indet.
Primates	*Papio* sp
	Cercopithecus sp.
	Theropithecus oswaldi
	Colobus sp.
	Cercocebus sp.
	Australopithecus boisei
	Homo sp.
Carnivora	*Viverridae* indet.
	Felidae indet.
	Canis mesomelas
Proboscidea	*Elephas recki*
Equidae	*Equus tabeti*
	Hipparion aethiopicus
Rhinocerotidae	*Ceratotherium simum*
	Diceros bicornis
Hippopotamidae	*Hexaprotodon karumensis*
	Hippopotamus gorgops
Suidae	*Metridiochoerus andrewsi*
	Kolpochoerus limnetes
Giraffidae	*Giraffa jumae*
Bovidae	*Aepyceros* sp.
	Antidorcus recki
	Gazella granti
	Kobus ellipsyprymnus
	Megalotragus isaaci
	Parmularius altidens
	Pelorovis sp.
	Sivatherium maurusium
	Tragelaphus strepsiceros
Pisces	*Polypterus*
	Clarius sp.
	Cichlidae indet.
Chelonia	*Trionyx* sp.
	Chelonia indet.
Squamata	Squamata indet.
Aves	Aves indet.
Crocodylidae	*Euthecodon* sp.
	Crocodylidae indet.

variable in species composition reflecting the climatic variability that increasingly came to dominate the opening bushlands and grasslands. The diverse assemblage of large herbivores thus provided a plentiful and diverse potential meat resource for early hominins which they appear to have fully utilized. Vertebrate taxa identified from archeological sites in the KBS and Okote Members at East Turkana (Bunn 1997), some showing cut marks, are listed in Table 1.4. Similarly an extraordinary diversity of vertebrate taxa ranging in size from micromammals and frogs to elephants, rhinos and hippos has been recovered from the many Bed 1 and Lower Bed II sites at Olduvai; these too were utilized by early hominins (see Appendix B by Margaret Leakey in Leakey 1971).

These faunal assemblages excavated from living floors show diversity in both the taxa and the body size of the animals represented. Furthermore, the composition and diversity of the taxa represented as well as the body parts preserved (with a high representation of limb bone elements), indicate that these sites are unlikely to be kill sites but rather that the hominins were transporting selected meaty parts of carcasses to central locations and there systematically processing these parts for marrow and meat. Cut marks have been found on bones ranging in size from small gazelles to large hippos, giraffes and even elephants, and the disposition of the cut marks indicates systematic butchery procedures for carcass skinning, joint disarticulation and meat removal as well as extensive breakage for marrow (Bunn 1997).

Conclusions

The almost simultaneous appearance of *Homo erectus* in geographically widespread sites, including Dmanisi in Georgia (Rightmire and Lordkipanidze 2010) and Perning (Mojokerto) in Java (Antón 2003), at approximately the same time that this species is first recognized in Africa, raises many questions. Evidence from Pliocene and Pleistocene faunal assemblages from eastern and southern Africa provide a background to this event. In this paper we have looked at African anthropoid, carnivoran and herbivore fossil faunal assemblages in order to explore possible influences that may have led to the initial dispersal of hominins out of Africa ~1.8 Ma ago.

Prior to the appearance of stone tools 2.6 Ma ago, hominins were subject to constraints on their dispersal and distribution that are common to all higher primates and relate to life history parameters and dietary selectivity; hominins were an integral component of primate faunal assemblages and were as widely dispersed as the most widespread cercopithecids. The most successful cercopithecids in the long term were those that could adapt to the increasingly dry open conditions and flexibility in dietary choice seems to have been the key adaptation that led to the success of early *Homo* at this time. The ability to combine plant and meat resources gave the Early Pleistocene hominins considerable dietary flexibility and the potential to exploit a wide variety of food sources in seasonal habitats. With this dietary shift, the constraints on the distribution of pre-meat eating hominins were largely removed and hominins were no longer confined to relatively tropical and subtropical habitats.

With the advent of culture, and this shift to meat eating, hominins came into direct competition with a diversity of large carnivores that shared their habitats. The carnivore guild, of which they became part, was more diverse than that in eastern Africa today and would have provided early hominins with many opportunities for scavenging meat from carcasses. The large herbivores show a corresponding high taxonomic diversity particularly among the largest taxa. Although it is not known whether hominins hunted their own prey or scavenged meat from carnivore kills, evidence from archeological sites at East Turkana and Olduvai dated between 1.8 and 1.5 Ma show that hominins were taking meat from herbivores of all body sizes, from small gazelles and rodents to rhinos, elephants and giraffe, and were transporting selected meaty parts of carcasses found elsewhere to central areas for consumption (Table 1.4).

To fully understand the relative significance of the factors discussed above, it is essential to assess whether the dispersal of *Homo erectus* out of Africa 1.8 million years ago was part of a major migration affecting many species, was restricted to very few species with similar life history and diets, or was unique to *H. erectus*. In order to attain a full understanding of this question, a number of aspects have to be approached in detail. The Plio-Pleistocene fauna of Africa, especially eastern Africa, which is the probable source area for *H. erectus* must be fully investigated so that possible migrants can be identified. The Plio-Pleistocene fauna of Eurasia must be similarly studied. This will allow possible African immigrants to be detected, as well as possible ancestors of such immigrants, should allopatric speciation have occurred rapidly. The pattern of migrations throughout the Plio-Pleistocene or even further back should be investigated, to determine whether there are any features unique to the 2.0–1.5 Ma time interval when *H. erectus* reached Eurasia. Finally, the ecological characteristics of the identified migrants must be understood, so that the causes of migration of each species can (ideally) be determined and compared with those inferred for *H. erectus*, to show whether possible coincident migrations were due to similar factors or were entirely fortuitous. At present, our assessment based on fossil carnivore and cercopithecid assemblages suggests that there were few if any migration events out of Africa contemporaneous with that of *H. erectus* and that those that may have occurred (*Megantereon, Hippopotamus, Theropithecus*; Rook et al. 2004, Martínez-Navarro 2004) could have been due to factors distinct from those that led to the dispersal of *H. erectus*.

References

Antón, S. C. (2003). Natural history of *Homo erectus*. *Yearbook of Physical Anthropology, 46*, 126–170.

Antón, S. C., Leonard, W. R., & Robertson, M. L. (2002). An ecomorphological model of the initial hominid dispersal from Africa. *Journal of Human Evolution, 43*, 773–785.

Barry, J. C. (1987). The history and chronology of Siwalik cercopithecids. *Human Evolution, 2*, 47–58.

Belmaker, M. (2002). First evidence of *Theropithecus* sp. in the Southern Levant. *Israel Journal of Zoology, 48*, 165.

Belmaker, M. (2010). The presence of a large cercopithecine (cf. *Theropithecus* sp.) in the 'Ubeidiya formation (Early Pleistocene, Israel). *Journal of Human Evolution, 58*, 79–89.

Benefit, B. R., & Pickford, M. (1986). Miocene fossil cercopithecoids from Kenya. *American Journal of Physical Anthropology, 69*, 441–464.

Blumenschine, R. (1987). Characteristics of an early hominid scavenging niche. *Current Anthropology, 28*, 383–407.

Bobe, R., & Behrensmeyer, A. K. (2002). Faunal change, environmental variability and late Pliocene hominin evolution. *Journal of Human Evolution, 42*, 475–497.

Bobe, R., & Behrensmeyer, A. K. (2004). The expansion of grassland ecosystems in Africa in relation to mammalian evolution and the origin of the genus *Homo*. *Palaeogeography, Palaeoclimatology, Palaeoecology, 207*, 399–420.

Bobe, R., & Eck, G. G. (2001). Responses of African bovids to Pliocene climate change. *Paleobiology Memoirs, 2*, 1–48.

Bobe, R., Behrensmeyer, A. K., Eck, G. G., & Harris, J. M. (2007). Patterns of abundance and diversity in late Cenozoic bovids from the Turkana and Hadar basins, Kenya and Ethiopia. In R. Bobe, Z. Alemseged, & A. K. Behrensmeyer (Eds.), *Hominin environments in the East African Pliocene: An assessment of the faunal evidence* (Vertebrate Paleobiology and Paleoanthropology Series). Dordrecht: Springer.

Brain, C. (1981). *The hunters or the hunted? An introduction to African cave taphonomy*. Chicago: University of Chicago Press.

Brunet, M., Guy, F., Pilbeam, D., Mackaye, H. T., Likius, A., Ahounta, D., et al. (2002). A new hominid from the Upper Miocene of Chad, Central Africa. *Nature, 418*, 145–151.

Bunn, H. T. (1997). The bone assemblages from the excavated sites. In G. L. Isaac & B. Isaac (Eds.), *Koobi Fora research project* (Plio-Pleistocene Archaeology, Vol. 5, pp. 402–444). Oxford: Clarendon.

Cachel, S., & Harris, J. W. K. (1995). Ranging patterns, land use and subsistence in *Homo erectus* from the perspective of evolutionary ecology. In J. R. F. Bower & S. Sartono (Eds.), *Evolution and ecology of* Homo erectus (pp. 51–66). Leiden: Pithecanthropus Centennial Foundation.

Delagnes, A., & Roche, H. (2005). Late Pliocene hominid knapping skills. The case of Lokalalei 2C, West Turkana, Kenya. *Journal of Human Evolution, 48*, 435–472.

Delson, E. (1973). Fossil colobine monkeys of the circum-Mediterranean region and the evolutionary history of the Cercopithecidae (Primates, Mammalia). Ph.D. dissertation, Columbia University, Columbia.

Delson, E. (1975a). Evolutionary history of the Cercopithecidae. *Contributions to Primatology, 5*, 167–217.

Delson, E. (1975b). Palaeoecology and zoogeography of the old world monkeys. In R. H. Tuttle (Ed.), *Primate functional morphology and evolution* (pp. 37–64). Paris: Mouton.

Delson, E. (1993). *Theropithecus* fossils from Africa and India and the taxonomy of the genus. In N. G. Jablonski (Ed.), Theropithecus: *the rise and fall of a primate genus* (pp. 157–189). Cambridge: Cambridge University Press.

Delson, E. (1994). Evolutionary history of the colobine monkeys in palaeoenvironmental perspective. In A. G. Davies & J. F. Oates (Eds.), *Colobine monkeys: their ecology, behaviour and evolution* (pp. 11–43). Cambridge: Cambridge University Press.

deMenocal, P. B. (1995). Plio-Pleistocene African climate. *Science, 270*, 53–59.

deMenocal, P. B., & Bloemendal, J. (1995). Plio-Pleistocene climatic variability in subtropical Africa and the paleoenvironment of hominid evolution: A combined data-model approach. In E. S. Vrba, G. H. Denton, T. C. Partridge, & L. H. Burckle (Eds.), *Paleoclimate and evolution with emphasis on human origins* (pp. 262–288). New Haven, CT: Yale University Press.

Denton, G. H. (1999). Cenozoic climate change. In T. G. Bromage & F. Schrenk (Eds.), *African biogeography, climate change and human evolution* (pp. 94–114). New York: Oxford University Press.

Dupont, L. M., & Leroy, S. A. G. (1995). Steps toward drier climatic conditions in northwestern Africa during the Upper Pliocene. In E. S. Vrba, G. H. Denton, T. C. Partridge, & L. H. Burckle (Eds.), *Paleoclimate and evolution with emphasis on human origins* (pp. 289–298). New Haven, CT: Yale University Press.

Eck, G. G. (1987a). *Theropithecus oswaldi* from the Shungura Formation, lower Omo Basin, southwestern Ethiopia. Les Faunes Plio-Pléistocènes de la Basse Vallée de l'Omo (Éthiopie). Tome 3. Cercopithecidae de la Formation de Shungura. Paris: Centre National de Recherche Scientifique, pp. 123–141.

Eck, G. G. (1987b). Plio-Pleistocene specimens of *Cercopithecus* from the Shungura Formation, southwestern Ethiopia. Les Faunes Plio-Pléistocènes de la Basse Vallée de l'Omo (Éthiopie). Tome 3. Cercopithecidae de la Formation de Shungura. Paris, Centre National de Recherche Scientifique (pp. 140–147).

Eck, G. G., & Jablonski, N. G. (1987). The skull of *Theropithecus brumpti* compared with those of other species of the genus *Theropithecus*. Les Faunes Plio-Pléistocènes de la Basse Vallée de l'Omo (Éthiopie). Tome 3. Cercopithecidae de la Formation de Shungura. Paris, Centre National de Recherche Scientifique (pp. 12–122).

Fleagle, J. G. (1999). *Primate adaptation and evolution*. San Diego, CA: Academic.

Fleagle, J. G., & Gilbert, C. C. (2006). The biogeography of primate evolution: the role of plate tectonics, climate, and chance. In S. M. Lehman & J. G. Fleagle (Eds.), *Primate biogeography*. Dordrecht: Springer.

Freedman, L. (1957). The fossil Cercopithecoidea of South Africa. *Annals of the Transvaal Museum, 23*, 121–257.

Freedman, L. (1976). South African fossil Cercopithecoidea: a reassessment including a description of new material from Makapansgat, Sterkfontein and Taung. *Journal of Human Evolution, 5*, 297–315.

Freedman, L., & Brain, C. K. (1977). A reexamination of the cercopithecoid fossils from Swartkrans (Mammalia: Cercopithecidae). *Annals of the Transvaal Museum, 30*, 211–218.

Frost, S. R. (2001). New early Pliocene Cercopithecidae (Mammalia: Primates) from Aramis, Middle Awash valley, Ethiopia. *American Museum Novitates, 3350*, 1–36.

Frost, S. R. (2002). Fossil Cercopithecidae from the Hadar Formation and surrounding areas of the Afar Depression, Ethiopia. *Journal of Human Evolution, 43*, 687–748.

Gautier-Hion, A., Bourlière, F., Gautier, J.-P., & Kingdon, J. (Eds.). (1988). *A primate radiation: evolutionary biology of the African Guenons*. Cambridge: Cambridge University Press.

Gibert, J., Ribot, F., Gibert, L., Leakey, M., Arribas, A., & Martínez-Navarro, B. (1995). Presence of the Cercopithecid genus *Theropithecus* in Cueva Victoria (Murcia, Spain). *Journal of Human Evolution, 28*, 487–493.

Gupta, V. L., & Sahni, A. (1981). *Theropithecus delsoni*, a new cercopithecine species from the Upper Siwaliks of India. *Bulletin of the Indian Geological Association, 14*, 69–71.

Haile-Selassie, Y. (2001). Late Miocene hominids from the Middle Awash, Ethiopia. *Nature, 412*, 178–181.

Harris, J. M. (Ed.). (1983). *Koobi Fora research project, volume 2: The Fossil Ungulates: Proboscidea, Perissodactyla, and Suidae*. Oxford: Clarendon.

Harris, J. M. (Ed.). (1991). *Koobi Fora research project, volume 3: The Fossil Ungulates: Geology, fossil Artiodactyls, and palaeoenvironments*. Oxford: Clarendon.

Harris, J. M., Brown, F. H., & Leakey, M. G. (1988). Stratigraphy and paleontology of the Nachukui Formation, Lake Turkana Region, Kenya. *Contributions in Science Los Angeles County Museum of Natural History, 399*, 1–128.

Jablonski, N. G. (Ed.). (1993a). Theropithecus: *The rise and fall of a Primate Genus*. Cambridge: Cambridge University Press.

Jablonski, N. G. (1993b). Evolution of the masticatory apparatus in *Theropithecus*. In N. G. Jablonski (Ed.), Theropithecus: *The rise and fall of a Primate Genus* (pp. 299–319). Cambridge: Cambridge University Press.

Jablonski, N. G., & Leakey, M. G. (Eds.). (2008). *Koobi Fora research project vol. 6. The fossil monkeys*. San Francisco: California Academy of Sciences.

Jablonski, N. G., Whitfort, M. J., Roberts-Smith, N., & Qinqi, X. (2000). The influence of life history and diet on the distribution of catarrhine primates during the Pleistocene in eastern Asia. *Journal of Human Evolution, 39*, 131–157.

Jolly, C. J. (1985). The seed-eaters: a new model of hominid differentiation based on a baboon analogy. In R. L. Ciochon & J. G. Fleagle (Eds.), *Primate and Human Origins* (pp. 323–332). Menlo Park: Benjamin/Cummings.

Kingdon, J. (1997). *The Kingdon field guide to African mammals*. London: Academic.

Kordos, L., & Begun, D. R. (2002). Rudabánya: A Late Miocene subtropical swamp deposit with evidence of the origin of the African Apes and humans. *Evolutionary Anthropology, 11*, 45–57.

Leakey, M. (1988). Fossil evidence of the evolution of the guenons. In A. Gautier-Hion, F. Bourlière, J.-P. Gautier, & J. Kingdon (Eds.), *A primate radiation: evolutionary biology of the African Guenons* (pp. 7–12). Cambridge: Cambridge University Press.

Leakey, M. D. (1971). *Olduvai Gorge Vol. 3. Excavations in beds I and II, 1960–1963*. Cambridge: Cambridge University Press.

Leakey, M. G. (1993). Evolution of *Theropithecus* in the Turkana Basin. In N. G. Jablonski (Ed.), Theropithecus: *The rise and fall of a primate genus* (pp. 85–123). Cambridge: Cambridge University Press.

Leakey, M. G., Feibel, C. S., Bernor, R. L., Cerling, T. E., Harris, J. M., Stewart, K. M., et al. (1996). Lothagam: A record of faunal change in the late Miocene of East Africa. *Journal of Vertebrate Palaeontology, 16*, 556–570.

Leakey, M. G., Spoor, F., Brown, F. H., Gathogo, P. N., Kiarie, C., Leakey, L. N., et al. (2001). New hominin genus from eastern Africa shows diverse middle Pliocene lineages. *Nature, 410*, 433–440.

Leakey, M. G., Teaford, M. F., & Ward, C. V. (2003). Cercopithecidae from Lothagam. In M. G. Leakey & J. M. H. Harris (Eds.), *Lothagam the dawn of humanity in Eastern Africa* (pp. 201–248). New York: Columbia University Press.

Lewis, M. E. (1997). Carnivoran paleoguilds of Africa: Implications for hominid food procurement strategies. *Journal of Human Evolution, 32*, 257–288.

Lewis, M. E., & Werdelin, L. (2007). Patterns of change in the Plio-Pleistocene carnivorans of eastern Africa: Implications for hominin evolution. In R. Bobe, Z. Alemseged, & A. K. Behrensmeyer (Eds.), *Hominin environments in the East African Pliocene: An assessment of the faunal evidence* (Vertebrate Paleobiology and Paleoanthropology Series, pp. 77–105). Dordrecht: Springer.

Marean, C. W. (1989). Sabretooth cats and their relevance for early hominid diet and evolution. *Journal of Human Evolution, 18*, 559–582.

Martínez-Navarro, B. (2004). Hippos, pigs, bovids, saber-toothed tigers, monkeys, and hominids: Dispersals through the Levantine Corridor during Late Pliocene and Early Pleistocene times. In N. Goren-Inbar & J. D. Speth (Eds.), *Human paleoecology in the Levantine Corridor* (pp. 37–51). Oxford: Oxbow Books.

Martínez-Navarro, B., Claret, A., Shabel, A. B., Pérez-Claros, J. A., Lorenzo, C. & Palmqvist, P. (2005). Early Pleistocene "hominid remains" from southern Spain and the taxonomic assignment of the Cueva Victoria phalanx. *Journal of Human Evolution, 48*, 517–523.

McBrearty, S., & Jablonski, N. G. (2005). First fossil chimpanzee. *Nature, 437*, 105–108.

Mourtzou, G., & Andrews, P. (2008). The deciduous dentition of *Griphopithecus alpani* from Paşalar, Turkey. *Journal of Human Evolution, 54*, 494–502.

Patel, B. A., Gilbert, C. C., & Ericson, K. E. (2007). Cercopithecoid cervical vertebral morphology and implications for the presence of *Theropithecus* in early Pleistocene Europe. *Journal of Human Evolution, 52*, 113–129.

Peters, C. R., Blumenschine, R. J., Hay, R. L., Livingstone, D. A., Marean, C. W., Harrison, T., et al. (2008). Paleoecology of the Serengeti-Mara ecosystem. In A. R. E. Sinclair, C. Packer, S. A. R. Mduma, & J. M. Fryxell (Eds.), *Serengeti III: Human impacts on ecosystem dynamics*. Chicago: University of Chicago Press.

Pianka, E. R. (1978). *Evolutionary ecology*. New York: Harper & Row.

Qiu, Z.-X., Deng, T., & Wang, B. (2003). Early Pleistocene mammalian fauna from Longdan, Dongxiang, Gansu, China. *Palaeontologica Sinica, 191*, 1–198.

Rightmire, G. P., & Lordkipanidze, D. (2010). Fossil skulls from Dmanisi: A paleodeme representing earliest *Homo* in Eurasia. In J. G. Fleagle, J. J. Shea, F. E. Grine, A. L. Baden, & R. E. Leakey (Eds.), *Out of Africa I: The first hominin colonization of Eurasia* (pp. 225–243). Dordrecht, The Netherlands: Springer.

Rook, L., Martínez-Navarro, B., & Howell, F. C. (2004). Occurrence of *Theropithecus* sp. in the Late Villafranchian of southern Italy and implications for Early Pleistocene "Out of Africa" dispersals. *Journal of Human Evolution, 47*, 267–277.

Semaw, S., Renne, P., Harris, J. W. K., Feibel, C. S., Bernor, R. L., Fesseha, N., et al. (1997). 2.5-million-year-old stone tools from Gona, Ethiopia. *Nature, 385*, 333–336.

Senut, B., Pickford, M., Gommery, D., Mein, P., Cheboi, K., & Coppens, Y. (2001). First hominid from the Miocene (Lukeino Formation, Kenya). *Comptes Rendus de l'Academie des Sciences, Sciences de la Terre et des Planètes, 332*, 137–144.

Spoor, F., Leakey, M. G., Gathogo, P. N., Brown, F. H., Antón, S. C., McDougall, I., et al. (2007). Implications of new early *Homo* fossils from Ileret, east of Lake Turkana, Kenya. *Nature, 448*, 688–691.

Szalay, F. S., & Delson, E. (1979). *Evolutionary history of the primates*. New York: Academic.

Tosi, A. J., Detwiler, K. M., & Disotell, T. R. (2005). X-chromosomal window into the evolutionary history of the guenons (Primates: Cercopithecini). *Molecular Phylogenetics and Evolution, 36*, 58–66.

Van Valkenburgh, B., Teaford, M. F., & Walker, A. (1990). Molar microwear and diet in large carnivores: Inferences concerning diet in the sabretooth cat *Smilodon fatalis*. *Journal of Zoology, 222*, 319–340.

Werdelin, L. (2006). *Twenty million years of migrations: is there anything special about leaving Africa 2 million years ago?* INQUA symposium: Towards understanding palaeoenvironments during the first out of Africa, Nairobi, Kenya, pp. 15–16.

Werdelin, L., & Lewis, M. E. (2002). Species identification in *Megantereon*: A reply to Palmqvist. *Journal of Paleontology, 76*, 931–933.

Werdelin, L., & Lewis, M. E. (2005). Plio-Pleistocene Carnivora of eastern Africa: Species richness and turnover patterns. *Zoological Journal of the Linnean Society, 144*, 121–144.

Chapter 2
Carnivoran Dispersal Out of Africa During the Early Pleistocene: Relevance for Hominins?

Margaret E. Lewis and Lars Werdelin

Abstract Carnivorans and hominins share a long history of interactions. This paper examines some of the evidence for carnivoran migration out of Africa at the same time as the earliest hominin dispersals. Of the two relevant taxa, *Crocuta* and *Megantereon*, *Megantereon* is the focus of this paper due to increased interest in this taxon in recent years and to the nature of the earliest records of dispersal of these two taxa, raising several questions related to *Megantereon* and its possible influence on hominins. To answer these questions, a brief summary of the literature on *Megantereon* in Eurasia and Africa is provided. While researchers do not agree on the number of species of *Megantereon* or the evolutionary relationships among those species, most would agree that *Megantereon* is a hypercarnivorous predator capable of grappling with relatively large prey for its body size. Despite the fact that carcasses generated by *Megantereon* were probably of value to hominins, the hypotheses that these carcasses were a major source of food or that they were a major force enabling hominins to migrate out of Africa are rejected. As indicated in the literature on extant carnivorans, kleptoparasitism (= food theft) by dominant members of a carnivore guild exacts a heavy price on lower ranking carnivores. In addition, there is nothing in the African fossil record to suggest a special relationship between *Megantereon* and hominins that did not exist between hominins and other large-bodied carnivorans. The hypothesis that a species of *Megantereon* migrated out of Africa at roughly the same time as early hominins is also considered. While this hypothesis cannot be rejected, alternative hypotheses to explain similarities between later African and Eurasian forms of *Megantereon* are proposed (e.g., shared characters are due to convergence or are symplesiomorphies). In the end, the small number of diverse African species (including hominins) who disperse into Eurasia at the Plio-Pleistocene transition may have been part of a sweepstakes dispersal where the factors permitting (or driving) dispersal may have differed from species to species.

M.E. Lewis (✉)
School of Mathematical and Natural Sciences (Biology), The Richard Stockton College of New Jersey, PO Box 195, Pomona, NJ 08240-0195, USA
e-mail: margaret.lewis@stockton.edu

L. Werdelin
Department of Palaeozoology, Swedish Museum of Natural History, Box 50007, S-104 05 Stockholm, Sweden
e-mail: werdelin@nrm.se

Keywords *Crocuta* • *Megantereon* • *Pachycrocuta* • Guild • Kleptoparasitism • Machairodont • Sabertooth • Scavenging

Introduction

The image of the first hominins dispersing from Africa into Eurasia is a compelling one. While the questions surrounding this event can be addressed in numerous ways, it is important to consider species that shared similar adaptations with the dispersing hominins. One group that probably overlapped significantly in diet and habitat with these hominins is the larger members of the Order Carnivora.

The relationship between carnivorans and hominins has changed through time. Early hominins fell prey to large-bodied carnivorans, as numerous lines of evidence attest (e.g., Brain 1981). At some point, hominins encroached upon the carnivore guild within Africa and entered into competitive relationships with large-bodied carnivorans (e.g., Lewis and Werdelin 2007, and all references therein, as well as Turner 1988; Lewis 1997; Brantingham 1998). Since it has been shown that carnivore guilds are tightly constrained in ecological space and that changes in part of this guild affect its entirety (Dayan and Simberloff 1996, 2005; Woodroffe and Ginsberg 2005), understanding the adaptations of any large-bodied carnivore is crucial for reconstructing the potential niche space for all other large-bodied carnivores (including hominins) present at that particular time and place. Changes in the adaptations of larger carnivorans and their dispersal events may yield critical information about factors affecting evolutionary events and dispersal patterns in hominins.

This paper uses the African fossil record to identify carnivoran taxa of relevance to the question of initial hominin dispersal to Eurasia. A literature survey and critical analysis of those taxa is then presented, with reference to the question of hominin dispersal.

J.G. Fleagle et al. (eds.), *Out of Africa I: The First Hominin Colonization of Eurasia*, Vertebrate Paleobiology and Paleoanthropology, DOI 10.1007/978-90-481-9036-2_2, © Springer Science+Business Media B.V. 2010

The Plio-Pleistocene Carnivoran Guilds of Africa

The carnivoran guilds of Plio-Pleistocene Africa included a higher diversity of genera and species than present today (see Table 2.1). Like *Panthera* today, some genera had more than one species present in a given area at a given time (e.g., *Dinofelis*, *Crocuta*; Werdelin and Lewis 2005; Lewis and Werdelin 2007). In addition, the carnivoran guilds of eastern and southern Africa were composed of different taxa (e.g., Turner 1990; Lewis 1995b, 1997). Within each region, detectable ecomorphological differences occur among congeners (e.g., differences in *Dinofelis* from Olduvai Gorge versus other eastern Africa localities of similar age, Werdelin and Lewis 2001).

Of the taxa listed in Table 2.1, some did not disperse to Eurasia (e.g., *Parahyaena*, but see Arribas et al. 2001). Others may have migrated significantly earlier than hominins (e.g., *Homotherium* and *Acinonyx*) or may be of New World origin (e.g., *Acinonyx*). Members of only two genera may have crossed into Eurasia at the same time as hominins: *Crocuta* and *Megantereon*.

The dietary adaptations and abilities of spotted hyenas (*Crocuta*) make this a very attractive species to study in comparison with tool-using, group-living, hunting hominins. *Crocuta* appears in Europe after 0.8 Ma, but is present in Asia much earlier. Although the dating is not exact, the current best estimate is that *Crocuta* must have entered Eurasia well before 2 Ma. This is attested to primarily by its presence in the Longdan Basin of China, in levels that are dated to ca. 2.2 Ma or even older (Qiu et al. 2004) and possibly in the Pinjor Formation of Indo-Pakistan (see Patnaik and Nanda 2010). Interestingly, and perhaps significantly, *Crocuta* is not recorded from Dmanisi (Vekua 1995). This may be an indication that it used a different dispersal route, possibly via the Indian Subcontinent, than did hominins at ca. 1.8 Ma. Unfortunately, the lack of Asian specimens around the crucial hominin dispersal period makes dispersals of *Crocuta* difficult to evaluate.

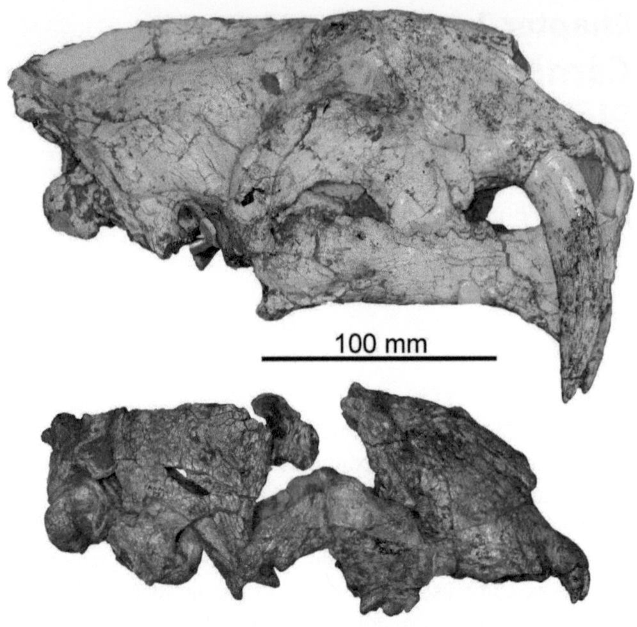

Fig. 2.1 Skulls of *Megantereon*. *Top*: *M. nihowanensis*, unnumbered skull, Hezheng Museum, Gansu, China from the Longdan Basin, Gansu. *Bottom*: *M. whitei*, KNM-ER 793A, Okote Mb., Koobi Fora Fm., Turkana Basin, Kenya. Note that despite the significantly smaller teeth of the latter specimen, the skull is only very slightly anteroposteriorly shorter (cf. Werdelin and Lewis 2002)

Table 2.1 Large-bodied carnivoran genera present in Plio-Pleistocene Africa

Family	Genus	Modern survivor
Canidae	*Canis*	*C. pictus* – African wild dog
Felidae	*Acinonyx*	*A. jubatus* – cheetah
Felidae	*Panthera*	*P. leo* (lion) & *P. pardus* (leopard)
Felidae	*Dinofelis*	None
Felidae	*Homotherium*	None
Felidae	*Megantereon*	None
Hyaenidae	*Chasmaporthetes*	None
Hyaenidae	*Lycyaenops*	None
Hyaenidae	*Crocuta*	*C. crocuta* – spotted hyena
Hyaenidae	*Hyaena*	*H. hyaena* – striped hyena
Hyaenidae	*Pachycrocuta*	None
Hyaenidae	*Parahyaena*	*P. brunnea* – brown hyena

Individual species are not listed due to the sheer number present (see Werdelin and Lewis 2005 for a complete listing). Some genera have multiple species present in the Plio-Pleistocene while others are not well known enough to assess taxonomic diversity. Note that modern survivors are not necessarily equivalent in behavior and ecology to their extinct congeners.

Among the African machairodont lineages present during the Plio-Pleistocene of Africa, *Megantereon* (Fig. 2.1) has been identified as being of crucial importance to the understanding of dispersals into Eurasia from Africa at the Plio-Pleistocene transition (e.g., Martínez-Navarro and Palmqvist 1996; Palmqvist et al. 1996; Arribas and Palmqvist 1999; Palmqvist et al. 2007; Martínez-Navarro 2010). Unfortunately, *Megantereon*, as a genus, is the most poorly known Plio-Pleistocene machairodont of Africa. In contrast to *Crocuta*, however, the few specimens of *Megantereon* that have been found are from crucial time periods and sites (see below). The rest of this paper will be confined to evaluating the evidence provided by *Megantereon*.

Questions Surrounding the Dispersal of *Megantereon*

Megantereon has been hypothesized to have migrated from Africa to Eurasia at roughly the same time as the earliest hominin migration (e.g., Martínez-Navarro 2010). This hypothesis is based on the assignment of specimens from 'Ubeidiya (Israel), Dmanisi (Georgia), and other Eurasian sites to the African species *M. whitei* rather than to a new species or to *M. cultridens*, which is found at older Eurasian localities (Martínez-Navarro and Palmqvist 1995, 1996; Palmqvist et al. 1996; Rook et al. 2004; Palmqvist et al. 2007; Martínez-Navarro et al. 2009; Martínez-Navarro 2010). The identification of this later Eurasian form and the Levantine material as being African in origin led to the suggestion that *Megantereon* made the first migration of hominins into Eurasia possible by providing carcasses for them to scavenge (Martínez-Navarro and Palmqvist 1996; Palmqvist et al. 1996, 2007). Such a food source has been suggested to have been sufficient for hominin subsistence, even in the presence of *Pachycrocuta*, which is reconstructed as a "strict scavenger" (Martínez-Navarro and Palmqvist 1996; Palmqvist et al. 1996).

Several questions must therefore be asked:

1. Do the specimens of *Megantereon* found at Venta Micena (Spain), Dmanisi (Georgia), Pirro Nord (Italy), Appolonia-1 (Greece), Untermassfeld (Germany), Argentario (Italy), Urkút (Hungary), Bugiulesti (Romania), and Java (Indonesia) (collectively referred to herein as late Eurasian *Megantereon*) that have been placed in the African species *M. whitei* (Martínez-Navarro and Palmqvist 1995, 1996; Palmqvist et al. 1996; Rook et al. 2004; Palmqvist et al. 2007; Martínez-Navarro et al. 2009; Martínez-Navarro 2010) truly differ from older European specimens of *Megantereon* (referred to herein as *Megantereon cultridens sensu stricto* or early Eurasian *Megantereon*)?
2. If the above specimens are different from *M. cultridens sensu stricto*, do they show similarities to the African lineage of *Megantereon* in general or to a specific African species (e.g., *M. whitei* or *M. ekidoit*)?
3. If they do show similarities to the African lineage or a specific African species, what is the nature of that similarity (i.e., is it due to dispersal from Africa to Eurasia or to convergence)?
4. Where do the affinities of the Levantine *Megantereon* from 'Ubeidiya (Israel) lie and what implications does this material have for understanding the biogeography of *Megantereon*?
5. Regardless of its affinities, could late Eurasian *Megantereon* have been a significant source of carcasses for scavenging by the earliest hominins in Europe even in the presence of the hyaenid *Pachycrocuta*?

Before these questions can be addressed, a discussion of the history of the study of *Megantereon* must be undertaken.

Brief History of the Taxonomy of *Megantereon*

Controversy over the attribution of specimens within the genus *Megantereon* has a long history. Summaries of the early history of this genus and its numerous species can be found elsewhere (e.g., Ficcarelli 1979; Sardella 1998; Palmqvist et al. 2007). Ficcarelli (1979) was the first to bring order to the taxonomic chaos that reigned within this genus. His revision identified one Eurasian species of *Megantereon*, *M. cultridens*, which was diagnosed as "small machairodonts with non-crenulated upper canines from both European and Asiatic Villafranchian …" (1979:18). Although Ficcarelli removed a large number of Asian forms from the genus, he considered the rest to be within an acceptable range of variation for the single species *M. cultridens*. Ficcarelli summarized the literature on African and North American specimens of *Megantereon*, but refrained from commenting on the taxonomic validity of the various species proposed for these specimens.

The next researcher to tackle the task of sorting out Eurasian and African *Megantereon* was Turner (1987). Turner undertook an exhaustive review of the published diagnoses of all *Megantereon* species known at the time to determine the number of valid taxa in Africa and to evaluate all valid taxa and comment on possible origination and dispersal events. Turner's review identified numerous diagnostic characteristics that were found in more than one African species suggesting to him that all African material then known should be placed within a single taxon. Turner also questioned the validity of diagnoses of Eurasian and North American species. Differences in size were suggested to be due to sexual dimorphism and geographic variation. Based on the problems that he uncovered in the published diagnoses and descriptions, Turner then went a step further than Ficcarelli and proposed that there was a single species, *Megantereon cultridens*, to which all North American, African, and Eurasian specimens belonged. Turner has since revised this viewpoint (Palmqvist et al. 2007; see below).

In a study published at roughly the same time as Turner's, Pons-Moya (1987) separated the European and Asian forms into separate subspecies (*M. c. cultridens* and *M. c. adroveri* in the European Villafranchian and Lower Pleistocene, respectively, and *M. c. nihowanensis*, in Asia). Although Pons-Moya reached conclusions that were superficially similar to those of Turner, he did distinguish between early and late forms of European *Megantereon*. More recently, Hemmer (2001) has followed Pons-Moya in using *M. c. adroveri* for the *Megantereon* found at the Early Pleistocene site of Untermassfeld in Germany.

In 1995, Martínez-Navarro and Palmqvist reinstated some of the species of *Megantereon*. Based on analyses of dental measurements, Martínez-Navarro and Palmqvist proposed that there were three species within the genus:

1. *M. cultridens* (Cuvier 1824) found in the North American Lower Pliocene, the Asian Upper Pliocene and Lower and Middle Pleistocene, and the European Upper Pliocene (Villafranchian). See Berta and Galiano (1983) and Turner (1987) for alternate views on the taxonomy of the North American specimens.
2. *M. whitei* (Broom 1937) found in the African Upper Pliocene and Lower Pleistocene and the European and Middle Eastern Lower Pleistocene. Later papers (Rook et al. 2004; Martínez-Navarro et al. 2009) expand the list of non-African sites to include Java and 'Ubeidiya.
3. *M. falconeri* (Pomel 1853) found in the Upper Pliocene of India. This species had been revised previously (Petter and Howell 1982).

This scheme was repeated in subsequent papers (e.g., Arribas and Palmqvist 1999; Martínez-Navarro and Palmqvist 1996; Palmqvist et al. 2007; Martínez-Navarro et al. 2009). These authors suggested that *Megantereon cultridens* arose in the New World (as originally proposed by Berta and Galiano 1983) and dispersed to the Old World approximately 3.5 Ma. *M. cultridens* then evolved into *M. falconeri* on the Indian subcontinent and *M. whitei* in Africa. *M. whitei* later dispersed from Africa to Eurasia. Later papers do not mention *M. falconeri* and only discuss Asian material that they have assigned to *M. cultridens* (Palmqvist et al. 2007). Alternate origins for *Megantereon* in Africa (Turner 1987) and Asia (Sotnikova 1989) have also been proposed.

The assignation of specimens to taxa by Martínez-Navarro, Palmqvist and colleagues was based solely on dental metrics and did not consider non-dental and non-metric characters. In their analyses, these authors assumed that if specimens were not statistically significantly different in dental metrics, then those specimens belonged to the same species. Differences in other measurements or in non-metric characters were ignored as was the potential confounding factor of disparate body sizes amongst species of *Megantereon*.

We performed a simple analysis of dental measurements relative to skull length to test whether the size of the teeth relative to skull size differs even if their absolute length and/or breadth does not (Fig. 2.2). Comparison of the relative proportions of upper carnassial (P4) length to skull condylobasal length demonstrates that the Koobi Fora skull (KNM-ER 793; African *M. whitei*) is considerably different from all of the specimens. The much smaller Dmanisi skull (Nr. 1341; considered to be *M. whitei* by Martínez-Navarro, Palmqvist and colleagues) has the longest upper carnassial relative to skull length, while the Koobi Fora specimen has the relatively shortest carnassial. When the range of variation in an extant felid (leopard) is considered, the Koobi Fora specimen can be seen to lie far outside the potential range of the other taxa. While this is just a rough analysis of one difference between the Dmanisi specimen and African *M. whitei*, it demonstrates that, at least in this feature, Dmanisi is substantially different from the African taxon.

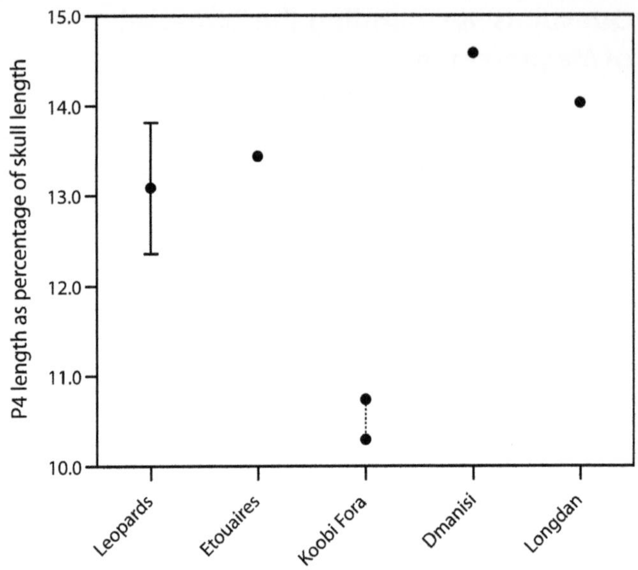

Fig. 2.2 Length of upper carnassial (P4) as a percentage of skull condylobasal length (CBL) for a sample of extant leopards, *Panthera pardus*, from Africa and Asia (left, $N = 15$, with 99.9% confidence interval) and some specimens of *Megantereon* species. P4 measurements are from Palmqvist et al. 2007. The lower point for the Koobi Fora specimen uses CBL as determined from Fig. 2.1 in Palmqvist et al. (2007), while the upper point uses the actual CBL as measured by us on the skull. The difference between the Koobi Fora specimen and all other *Megantereon* specimens is much greater than the variation within the sample of leopards

In contrast to Martínez-Navarro, Palmqvist and colleagues, Sardella (1998) further subdivided *Megantereon* based on morphometric analyses. Recognizing the incomplete nature of most of the specimens, his classification included both morphotypes and species and lumped all of the following into what he referred to as *Megantereon* ex. gr. *cultridens*:

1. *Megantereon* sp. 1 (primitive form)
   Locality/Age: Baode, China (late Miocene?).
   Characters: "P^3 is laterally compressed with an anterior and a posterior cusplet and is, on the whole, more developed than in all the other studied specimens of *Megantereon*. P^4 shows no preparastyle and a strong deuterocone [= protocone]. This tooth is morphologically very similar to that of the more advanced forms of *Megantereon*." (1998:6) (However, further investigation

strongly suggests that the Baode specimen belongs to a species of *Paramachairodus*, a genus already known from the Baode faunas).

2. *Megantereon* sp. 2 (primitive form)
   Locality/Age: Odessa Catacombs, Ukraine (Early Pliocene; Ruscinian).
   Characters: moderate reduction of P^3 and P_3. More recent study, however, has demonstrated that these specimens belong to a species of *Dinofelis* (Sotnikova in litt. to LW 07/02/2004).
3. *M. cultridens* (primitive form)
   Locality/Age: Perrier-Les Etouaires (Early Villafranchian).
   Characters: "low degree of development of the upper incisors, similar in shape to those of the living felids, while upper canines are well developed like in other dirktoothed cats" (1998:7).
4. *M. cultridens* (typical form)
   Locality/Age: Europe (Pardines, Puebla de Valverde, St. Vallier, Senèze, Fontana Acetosa, Olivola, Upper Valdarno) (Early-Late Villafranchian).
   Characters: "Machairodont the size of a panther; the skull shows a shortened muzzle; teeth are not crenulated; the incisors are stronger than in modern felids, but are not so developed than in other sabertoothed cats as *Homotherium* and *Machairodus*; the upper canines are not serrated with a very high and curved crown, P^3 and P^4 are very reduced with deuterocone variable in size; on the whole, the structure of the upper carnassial is close to modern felids. Very developed mandibular flange. C_1 is weak, P_3 reduced. The neck is long and limb bones are strong, with straight shortened diaphysis" (1998:7–8).
5. *M. cultridens* (advanced form)
   Locality/Age: Europe (Pirro Nord, Argentario, Urkút, Venta Micena, Apollonia 1, Dmanisi) (Late Villafranchian).
   More derived characters: "(1) very strong incisors; (2) upper canines greatly developed in size; (3) upper carnassial moderately reduced; (4) reduced P_4" (1998:9).
6. *Megantereon falconeri* (Pomel)
   Locality/Age: Asia (Late Pliocene-Middle Pleistocene)
   Late Pliocene forms: very "strong" upper canines and moderately reduced premolars.
   Early Pleistocene forms: large-sized specimens with moderately reduced P_4.
   Middle Pleistocene form: large.
7. *Megantereon whitei* (Broom)
   Locality/Age: Africa (Plio-Pleistocene).
   Characters: reduced P^4 and P_4 and very "strong" upper canines.

Sardella viewed the European morphotypes of *M. cultridens* as part of a single evolutionary lineage through time. Like Turner (1987), Sardella removed the North American specimens from the Bone Valley Formation (4.5 Ma) from *Megantereon* and thus concluded that *Megantereon* migrated from the Old World into North America. Most interestingly, Sardella concluded that his *M. cultridens* (advanced form) morphotype is related to the African *M. whitei* morphotype, but that the two forms are distinct. However, Sardella has since begun referring to his *M. cultridens* (advanced form) as *M. whitei* and referred material from Monte Argentario, Italy to this species (Sardella et al. 2008). This change in nomenclature was based on the hypothesis of Martínez-Navarro and Palmqvist reaching a "larger consensus" (Sardella et al. 2008:603), which means, presumably, the recent support for this hypothesis by Turner (i.e., Palmqvist et al. 2007). Sardella and colleagues note the anatomical differences between Pliocene *M. cultridens* and Early Pleistocene European *Megantereon*, but do not discuss the morphological justification for combining African *M. whitei* and Early Pleistocene European *Megantereon* into a single species.

In a recent contribution to the taxonomy of *Megantereon*, Liu (2005) made a distinction between two European forms (typified by the material from St. Vallier and Senèze, respectively) listing a series of craniodental characteristics said to distinguish the two. He then resurrected the name *Megantereon megantereon* for the St. Vallier form, and placed some Chinese material (and implicitly also *M. falconeri*) in this taxon. Evaluation of this perspective must be left for the future, but it is of significance that Liu also acknowledges the specific status of the African *M. whitei*. Younger European material was not included in the analysis.

Recent work by Palmqvist and colleagues (including Martínez-Navarro and Turner; Palmqvist et al. 2007) indicate that size differences among Eurasian and African specimens are not due to sexual dimorphism. This study expanded the number and geographic extent of specimens included in their previous morphometric analyses. No specimens from the Indian subcontinent are included, nor is the validity of *M. falconeri* discussed. In these analyses, specimens were classified a priori as either *M. whitei* (all Africa, European Lower Pleistocene, and 'Ubeidiya) or *M. cultridens* (European Upper Pliocene, Asia, and North America). Within Asia, material is included from China and Tajikistan, but not Java. Analyses of two (discriminant analysis: P_4 length and M_1 breadth) to four variables (principal components analysis: log length and breadth of P_4 and M_1) support these a priori classifications, although one might question, in particular, the use of discriminant analysis with only two variables. While measurements of the upper and lower canines, premolars, and molars are presented, only P_4 and M_1 measurements appear to be useful in discriminating these groups. These authors suggest that proportional changes throughout the dentition and concomitant changes in the rest of the skull led to *M. whitei* being able to hunt "more efficiently" (p. 173) than *M. cultridens*. The removal of Java from the list of sites with *M. whitei* present means that this taxon in their scenario did not penetrate very far into Asia. This study did not search for features that might distinguish sub-groups within these a priori groups or categorize the specimens in a different manner.

Each of the above researchers or research groups used different means of determining the taxonomic status of the various species of *Megantereon* and, not surprisingly, came to different conclusions (see Table 2.2). While there is little consensus among these researchers, there are some points on which most recent studies agree:

1. African and at least some Asian forms differ morphologically from what was originally described as European *M. cultridens* (but see Liu 2005, with respect to Asian forms).
2. There are two forms within Europe (or three, in the case of Liu 2005): the larger, more robust early form (which all agree is *M. cultridens*) and a later form characterized by dental reduction.
3. The latter form within Europe may share some affinity with African forms (although the nature of this affinity is disputed).
4. Differences in size within African *M. whitei*, particularly those in southern Africa, are most likely due to sexual dimorphism (a point made by Turner that has gone unchallenged by all subsequent researchers).

Table 2.2 Summary of changes in the taxonomy of *Megantereon* through time

Old World Species of *Megantereon* (valid and invalid)		
Europe	Asia	Africa
M. cultridens	*M. falconeri*	*M. ekidoit*
M. c. adroveri	*M. inexpectatus*	*M. eurynodon*
M. megantereon	*M. lantianensis*	*M. gracile*
	M. nihowanensis	*M. whitei*
Ficcarelli 1979		
M. cultridens	*M. cultridens*	No comment
Turner 1987		
M. cultridens	*M. cultridens*	*M. cultridens*
Martínez-Navarro and Palmqvist 1995		
M. cultridens (early form)	*M. falconeri*	*M. whitei*
M. whitei (late form)		
Sardella 1998		
M. ex gr. cultridens	*M. ex gr. cultridens*	*M. ex gr. cultridens*
(*M. cultridens* primitive form)	(*M. falconeri*)	(*M. whitei*)
M. ex gr. cultridens		
(*M. cultridens* typical form)		
M. ex gr. cultridens		
(*M. cultridens* advanced form)		
Liu 2005		
M. megantereon (St. Vallier)	*M. cf. megantereon*[a]	*M. whitei*
M. cultridens (Senèze)		Did not include later form
Did not include later form		
Palmqvist et al. 2007		
M. cultridens (early form)	*M. cultridens* (widespread)	*M. whitei*
M. whitei (late form)	*M. whitei* (limited)	
Current Paper (after Werdelin and Lewis 2000, 2002)		
M. cultridens (early form)	One or more taxa[b]	*M. ekidoit* (early form)
M. adroveri (late form; new rank)		*M. whitei* (late form)

[a] This form belongs to the genus named and has characters that may be compared usefully to the species-level taxon, though it may not actually belong to this species.

[b] While a discussion of Asian *Megantereon* taxonomy is beyond the purview of this paper, we believe the following may be valid species within Asia: *M. falconeri*, *M. inexpectatus*, or *M. nihowanensis*. *M. falconeri* has priority if there is only a single species of *Megantereon* within Asia.

Although numerous sites throughout Eurasia and Africa list *Megantereon* as present, the actual material is often fairly incomplete. As such, it is compelling that different research groups have come to some of the same conclusions, even if they dispute how these conclusions should be interpreted taxonomically.

A New Species of African *Megantereon*: Significance for Dmanisi

In 2000, a new species of *Megantereon* (*M. ekidoit*) was described from the Kenyan site of South Turkwel (3.5–3.2 Ma) (Werdelin and Lewis 2000). While only a single mandible of this species is known (Fig. 2.3), this specimen clearly belongs to *Megantereon* but differs from known members of the genus. *M. ekidoit* was diagnosed as "a *Megantereon* with a slender mandibular ramus, large salivary gland pit on the anteromedial face of the ramus, small masseteric and mental foramina, and well developed, hookshaped coronoid process" (2000:1173). The individual mandible upon which the description was based lacks the P_3, a feature that the authors excluded from the diagnosis due to the possibility that it was an individual variation.

The significance of this specimen is that it is the oldest described specimen of this genus from Africa and improves our understanding of evolution within the African lineage of *Megantereon*. Older material has been reported from Aramis at 4.4 Ma (WoldeGabriel et al. 1994), but has not yet been described. Specimens from the Lukeino Formation in Kenya are most likely to be *Paramachairodus* or a related taxon rather than *Megantereon* as they resemble the Baode material (LW, personal observation). Given the exclusion of the Lukeino material from *Megantereon*, along with the Baode and Odessa material as discussed above, the mandible of *M. ekidoit* is the oldest described specimen of *Megantereon* worldwide.

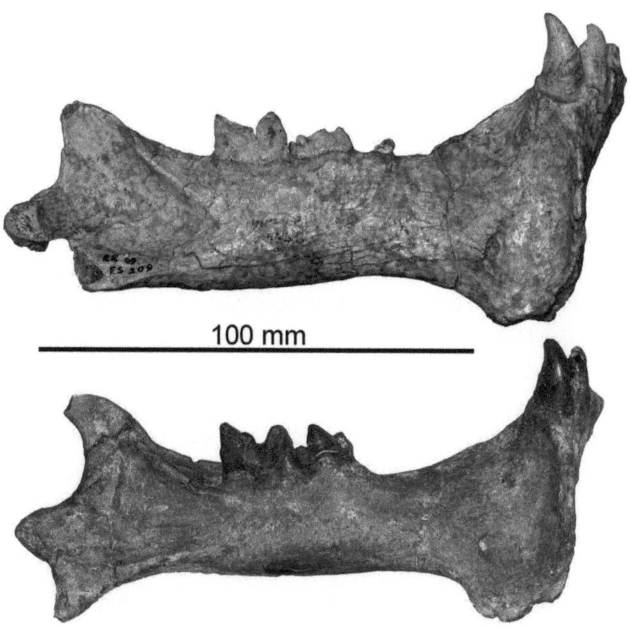

Fig. 2.3 Right mandibular rami of *Megantereon*. *Top*: *M. whitei*, KNM-ER 793B, Okote Mb., Koobi Fora Fm., Turkana Basin, Kenya. *Bottom*: *M. ekidoit*, KNM-ER ST 23812, South Turkwell, West Turkana, Turkana Basin, Kenya. Note that the latter is considerably more slender despite being ontogenetically older (as judged by tooth wear), indicating that *M. whitei* was a craniodentally more robust animal

Not everyone immediately accepted the new species. Palmqvist (2002) attempted to show that the new species fit comfortably within the existing African species, *M. whitei* based on a quantitative analysis of the mandibular dentition and a list of characters shared between the two.

Werdelin and Lewis issued a rebuttal (2002) noting that Palmqvist was correct that *M. ekidoit* and *M. whitei* were similar in the dental proportions mentioned (though not in all dental proportions) and that the diagnosis of the species was based on non-dental characters. In short, some aspects of the dentition within the African lineage of *Megantereon* remained the same while other characters evolved. In our experience, carnivoran teeth, particularly among felids, tend to be fairly conservative in comparison to the rest of the body.

Palmqvist (2002) listed seven characters that he believed synonymized *M. whitei* (including the Dmanisi and other European material) and *M. ekidoit*. Werdelin and Lewis (2002) countered by noting that five of the seven were features shared by *Megantereon* as a genus. Werdelin & Lewis dismissed some of the other characters as misunderstandings (e.g., misreading of the lack of P_3 as separating *M. ekidoit* from *M. whitei* rather than the possibility of individual variation).

Finally, the last characters that Palmqvist stated were shared between the two species are, in fact, shared between *M. ekidoit* and the Dmanisi *Megantereon* (as figured in Vekua 1995), but were not shared with *M. whitei sensu stricto* (i.e., the other African specimens). A feature shared by *M. ekidoit* and the Dmanisi form (but not *M. whitei sensu stricto*) is the presence of a long, shallow masseteric fossa that is developed well anterior to the posterior end of M_1. Palmqvist's last character, that of the hook-shaped coronoid process for which *M. ekidoit* was named, was not shared by all three taxa: the coronoid process of *M. whitei* is not hook-shaped, that of the South Turkwel specimen is, and the Dmanisi specimen is intermediate.

The significance of this debate is that like the analysis presented in Fig. 2.2, it casts doubt on the assignation of the Dmanisi material (and by extension other late Eurasian *Megantereon*) material to *M. whitei*. However, the possibility of an African origin of the Dmanisi form cannot be discarded due to the similarities between that form and *M. ekidoit*.

Ecomorphology of African *Megantereon*

Sabertoothed felids (subfamily Machairodontinae) in the African Plio-Pleistocene include representatives of three different tribes: the Metailurini (e.g., *Dinofelis*), the Homotheriini (e.g., *Homotherium*), and the Smilodontini (e.g., *Megantereon*). Representatives of these tribes are quite different in morphology and presumably behavior.

In comparison to other large-bodied carnivorans found in the African Plio-Pleistocene (e.g., *Dinofelis*, *Crocuta*, and *Homotherium*), *Megantereon* is very poorly represented. Craniodental and postcranial specimens are known from both eastern and southern Africa. However, partial skeletons are rare. One partial skeleton has been described from Kromdraai B (Vrba 1981). Associated bits and pieces of postcrania occur at Koobi Fora, but are extremely fragmentary (Lewis 1997; Werdelin and Lewis, in preparation).

The older species, *M. ekidoit*, is currently known only from eastern Africa from approximately 3.5–3.2 Ma. If the Aramis material belongs to this species, this extends its first appearance datum back to 4.4 Ma. Unfortunately, other specimens of *Megantereon* from the Pliocene of eastern Africa (e.g., Shungura Fm. Mbs. B-G) are isolated teeth, making taxonomic identifications below the genus level impossible (Werdelin and Lewis 2005). No postcrania have as yet been assigned to this taxon.

The younger species, *M. whitei*, is present in both eastern and southern Africa. Within eastern Africa, the only definite record of this taxon is in the Okote Mb. of the Koobi Fora Formation (Werdelin and Lewis 2005; Lewis and Werdelin 2007). *Megantereon whitei* is better represented in South Africa than at eastern African sites, with records from Kromdraai Mb. A, Swartkrans Mb. 3, Sterkfontein, Mbs 2, 3, 4, and Coopers (Broom and Schepers 1946; Broom 1948; Ewer 1955; Hendey 1973, 1974; Vrba 1981; Turner 1987, 1993; Lewis 1995a, b, 1997; Hartstone-Rose et al. 2007).

Like their close relative *Smilodon*, members of the genus *Megantereon* in both Europe and Africa have been shown to have extreme strength in the forelimb (Lewis 1995a, b, 1997; Martínez-Navarro and Palmqvist 1996) (see Fig. 2.4). Specimens from Kromdraai, South Africa possess a limb morphology that is more similar to that of extant jaguars than to any of the modern African felid taxa or other African sabertooths, although they were much more heavily muscled than jaguars (Lewis 1995a, b, 1997). As a result, African and European *Megantereon* have been identified as potential providers of large carcasses for hominins (Lewis 1995b, 1997; Martínez-Navarro and Palmqvist 1996; Arribas and Palmqvist 1999). However, based on body size and morphology, Lewis (1995b, 1997) concluded that African *Megantereon* could not have generated carcasses much larger than those generated by extant carnivorans. Thus, it is unclear whether *Megantereon* would have been as important a scavenging resource (or, conversely, as much of a threat as a kleptoparasite) as other sabertoothed felids.

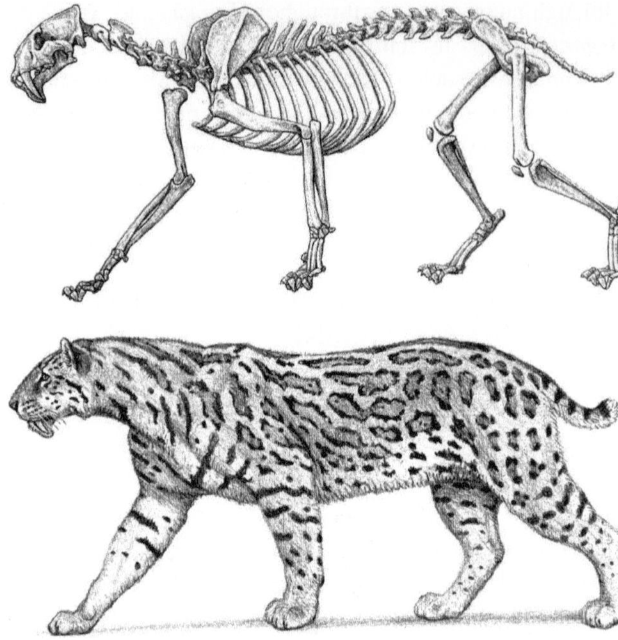

Fig. 2.4 Skeletal and life reconstructions of *Megantereon*, showing the long, low body, robust and heavily muscled forequarters, and short tail. Illustration by Mauricio Antón. Reprinted with permission from the artist

One should note that throughout much of the Plio-Pleistocene, *Megantereon* was not the only sabertooth present. In addition to at least one species of *Homotherium*, there were two species of *Dinofelis* living at any given time in eastern Africa, although not necessarily at the same location. One species of *Dinofelis* tended to be relatively larger (e.g., *D. aronoki*) and one tended to be a little smaller with a more crouched posture (e.g., *D. petteri* or *D. piveteaui*) (Werdelin and Lewis 2001; Lewis and Werdelin 2007). Species of *Megantereon*, however, were the smallest of the African machairodonts during this time.

Like *Dinofelis*, *Megantereon* has been suggested to have inhabited mixed/closed habitats (Lewis 1995a, b, 1997) or even dense forest (Marean 1989; Palmqvist et al. 2008) in contrast to *Homotherium*, which has been reconstructed as more open-habitat adapted in both Europe and Africa (e.g., Lewis 1995b, 1997; Palmqvist et al. 2003; Antón et al. 2005). Habitat preference does not mean that a species is limited to that habitat, however, as narrow categorizations of habitat preference cannot be made from carnivoran postcranial morphology (Van Valkenburgh 1987; Taylor 1989). Large, extant carnivorans in Africa may be found in a variety of habitats despite what their postcranial morphology might predict (e.g., lions, leopards, spotted hyenas; see review in Van Valkenburgh 2001). Of course, it is possible that the ability of many extant African carnivorans to inhabit a variety of habitats successfully is a key component of the suite of adaptations that ensured their survival to the present (Lewis and Werdelin 2007). While the crouched posture of *Megantereon* spp. is indicative of an ambush predator and their size and limb morphology suggest an ability to climb trees (Lewis 1995a, b, 1997; Martínez-Navarro and Palmqvist 1996), this does not mean that they were tied to specific habitats (nor does it mean that they were "partially arboreal" as reported by Hartstone-Rose et al. 2007 in a mis-citation of Lewis 1997). Their forelimb morphology may reflect prey grappling more than scansorial ability regardless of their habitat preference (Lewis 1997). However, carbon- and nitrogen-stable isotope analyses of *Megantereon* from Venta Micena (Spain), have suggested that at least this population focused on browsing and mixed feeding cervids in a closed habitat (Palmqvist et al. 2003, 2008).

Despite being the smallest of the sabertooths known from this time period, even if *Megantereon* spp. did climb trees, they would have been more likely to steal already cached carcasses and feed on them in the trees than to cache carcasses (Lewis 1997; Lewis and Werdelin 2007). Tree-caching a shifting carcass would have been a risky behavior with high potential for damage to the canines (Lewis 1997; Turner and Antón 1997; Lewis and Werdelin 2007).

In sum, the studies cited above have suggested that the various species of *Megantereon* were ambush predators that may have utilized mixed/closed habitats predominantly, although they may have been present in a variety of habitats. Despite their size, all studies have agreed that they could have taken down prey of a large enough size to be of interest to larger scavengers, including hominins. Carcasses generated by *Megantereon* likely had intact within-bone nutrients and varying amounts of flesh present due to its specialized dentition (e.g., Ewer 1973; Marean 1989; Lewis 1995a, b, 1997; Marean and Ehrhardt 1995; Turner 1988; Palmqvist et al. 2007). Nothing in the fossil record of *Megantereon* has suggested the possibility of group hunting, a behavior that would have strongly discouraged kleptoparasitism. However, the robust musculature in combination with the utilization of cover within mixed/closed habitats would have made even a solitary individual of *Megantereon* a formidable foe.

Megantereon and Hominin Behavior

Given the morphology of *Megantereon* spp., one can assume that a single individual of this taxon would have been more difficult to dislodge from a carcass or defend oneself from than a single modern leopard or lion. Of course, weapons and grouping behavior would have increased the relative rank of hominins within the carnivore guilds. Successful aggressive behaviors by hominins would also have conferred status.

What could confrontationally scavenging hominins have gained from *Megantereon* kills? If a group of hominins were drawn to a kill site soon after the kill occurred and were able to scare away the cat, there could have been a great benefit. If hominins were passively scavenging (i.e., waiting until the predator abandoned the carcass) or came upon a kill after the cat had finished with it, the story might be quite different. While Martínez-Navarro and Palmqvist have suggested that *Megantereon* would have exploited carcasses to a "small degree" (1996:871) such that there would be enough for hyaenids (e.g., *Pachycrocuta*), behaving as "strict scavengers" and scavenging hominins, not everyone would agree. Based on both an analogy to North American *Smilodon*, which has a large amount of tooth breakage, and the fact that modern big cats use their tongues as files to rasp flesh off bones, Van Valkenburgh (2001) has suggested that African sabertooths were probably quite capable of dismembering the carcass and engaging in bone-cracking. However, despite the fact that *Smilodon* and *Megantereon* are sister taxa, there is no evidence (e.g., broken teeth showing wear) in Africa, at least, to support the idea that *Megantereon* engaged in these behaviors at the level hypothesized for *Smilodon* (Lewis and Werdelin 2007). Given the dental morphology and reduced tooth row in *Megantereon*, and especially *M. whitei*, bone-cracking is highly improbable.

What is clear is that *Megantereon*, like all felids, was hypercarnivorous and probably could quickly deflesh a carcass if it needed to (i.e., if it was living in an area of high competition with marauding groups of hominins and large-bodied hyaenids). Despite debate over bone-cracking, *Megantereon* could not in all likelihood access larger chambers of the skull or bone marrow cavities in larger bones. Thus, if hominins did not arrive early on the scene or were not confrontational scavengers, the remains would still have been useful, but not bountiful. In addition, *Pachycrocuta* could access a wider range of carcass-based resources than *Megantereon*. If this large-bodied hyaenid arrived at a *Megantereon* kill prior to hominins, there might be even less left for hominins. [Note that at the time hominins initially dispersed to Eurasia, African *Pachycrocuta* was rare (southern Africa) or extinct (eastern Africa).]

If hominins scavenged regularly from one resource species, that species would have experienced a great deal of stress and would either have had to adopt new strategies to protect or hide their food or migrate to a hominin-free area to prevent at least local extinction (see Lewis 1997 for similar arguments against regular stealing of tree-cached carcasses by hominins). Kleptoparasitism by high ranking carnivores has been shown to drive populations of lower ranking taxa into suboptimal habitats (Woodroffe and Ginsberg 2005) or even to local extinction (Linnell and Strand 2000; Creel 2001). Given that *Megantereon* continued to be associated with hominins for some time even after hominins dispersed to Europe, one would surmise that by the time of migration hominins were not stealing carcasses from *Megantereon* at a rate that would cause severe stress. Of course, it is certainly possible that *Megantereon* fled Africa in an attempt to escape hominins with hominins in hot pursuit and that hominins were eventually successful in driving *Megantereon* extinct in both Eurasia and Africa. Given the timing of migration and co-occurrence of the two taxa at multiple sites, this scenario is highly unlikely. Even if hominins could be established as being primarily responsible for the eventual disappearance of *Megantereon*, it is probably not possible to determine whether they out competed *Megantereon* through hunting or through confrontational scavenging or by some combination thereof.

All of this, of course, raises the interesting question of why *Megantereon* kills might be favored over those of other large felids. Could *Megantereon* be the only non-pack living carnivoran large enough to take down prey of a size usable by hominins? This scenario is unlikely as *Dinofelis* would also fall into this category (see Marean and Ehrhardt 1995; Lewis 1997; Van Valkenburgh 2001; Antón et al. 2005, for discussions of pack living and/or hunting in various Eurasian and African carnivorans). Could the smaller body size of *Megantereon whitei* individuals make them more susceptible to hominin kleptoparasitism than other machairodonts? What, then, would prevent other carnivorans from engaging in kleptoparasitism against *M. whitei*? One must note that there is nothing in the African fossil record that suggests a special relationship between *Megantereon* and *Homo* to the exclusion of other large-bodied carnivorans.

While hominins may have benefited from occasional scavenging of *Megantereon* kills, it is unlikely that they could have relied on *Megantereon* as their sole source of meat. Assuming that *Megantereon* dispersed from Africa at the same time (or even slightly before) hominins, the presence of *Megantereon* was probably not the primary motivating factor in hominin dispersal. Ability to scavenge from *Megantereon* would have been useful and may have helped hominins establish themselves in Eurasia, but it seems likely that there were additional factors driving hominin dispersal. While it is possible that they were interested in the same prey species, it is also possible that they were interested in different prey species that happened to be dispersing out of Africa for the same reasons at roughly the same time. Both hypotheses are equally untestable at present.

Reiteration of Questions Posed Earlier

At this point it is probably useful to return to the five questions posed at the beginning of this paper and make some attempt to answer them. Not all of the questions can be answered here. Some hypotheses may simply be untestable while others may necessitate the discovery of more fossils.

1. Do the specimens of *Megantereon* found at Venta Micena (Spain), Dmanisi (Georgia), Pirro Nord (Italy), Appolonia-1 (Greece), Untermassfeld (Germany), Argentario (Italy), Urkút (Hungary), Bugiulesti (Romania), and Java (Indonesia) (collectively referred to herein as late Eurasian *Megantereon*) that have been referred to the African species *M. whitei* (Martínez-Navarro and Palmqvist 1995, 1996; Palmqvist et al. 1996; Rook et al. 2004; Palmqvist et al. 2007; Martínez-Navarro 2010) truly differ from older European specimens of *Megantereon* (referred to herein as *Megantereon cultridens sensu stricto* or early Eurasian *Megantereon*)?

Most researchers agree that there are two different morphotypes present in the fossil record of Eurasia (but see Ficcarelli 1979; Turner 1987). Whether one chooses to see them as two subspecies within a larger *M. cultridens* or as two species within *Megantereon*, this works out functionally to the same conclusion: there is a detectable difference in morphology between early and late specimens of *Megantereon*.

One should note, however, that the morphometric analyses do not include all material listed above as being a part of *M. whitei*. In many cases, the material does not preserve the necessary areas of the body. The Javan *Megantereon*, for example, consists exclusively of isolated upper canines, a portion of the skeleton that is not diagnostic at the species level in *Megantereon* (for example, see Fig. 4 and Table 2 in Palmqvist et al. 2007). However, this does not invalidate the argument that there are two species (or morphotypes) present in Eurasia. Eurasian sites stated to have *M. whitei* that are included in various multivariate analyses (Martínez-Navarro and Palmqvist 1996; Palmqvist et al. 2007) are Venta Micena, Dmanisi, Apollonia-1, Argentario, Pirro Nord, Untermassfeld, Urkút, and Bugiulesti. A clear difference can be seen between specimens from those sites and older European material. As such, we support a species-level distinction between early and late Eurasian *Megantereon*, although we do not concur that the later species is *M. whitei* (see next two items). Our understanding of the geographical and temporal extent of the later species will only be enhanced with the discovery and description of new fossils.

2. If the above specimens are different from *M. cultridens sensu stricto*, do they show affinities with the African lineage of *Megantereon* or with a specific African species (e.g., *M. whitei* or *M. ekidoit*)?

Martínez-Navarro and Palmqvist (1995, 1996; Palmqvist et al. 2007) have demonstrated morphometrically that specimens of late Eurasian *Megantereon* fall within the range of the dental proportions of African *M. whitei* rather than *M. cultridens*. Palmqvist (2002) noted that those particular dental proportions do not change between *M. whitei* and *M. ekidoit*. On the other hand, Werdelin and Lewis (2002) indicate that *M. ekidoit* and at least the Dmanisi specimen share some features to the exclusion of *M. whitei*. These shared features may be primitive (a view favored by the authors of this paper), which raises the interesting question of the timing of origin of the possibly more derived *M. whitei*. Another character of the Dmanisi specimen, coronoid process shape, was intermediate between the morphology of *M. whitei* and *M. ekidoit*. In sum, while the exact relationship between these specimens and the two African species is currently unknown, there is a general consensus that there is an affinity between these specimens and the African forms.

3. If they do show affinities with the African lineage or specific African species, what is the nature of that affinity (i.e., is it due to dispersal from Africa to Eurasia or to convergence)?

Unfortunately, not enough is known about the later group of Eurasian *Megantereon* to rule out convergence with the African forms. More specimens of *M. ekidoit* would also be useful. It is certainly possible that ecological changes enabling hominin migration and/or the appearance of hominins drove Eurasian *Megantereon* to converge on African forms. However, this is not currently a testable hypothesis. The hypothesis that later Eurasian *Megantereon* is derived from *M. ekidoit* also cannot be disproved. Based on the dissimilarity between the Dmanisi specimen and *M. whitei* in some features, we believe that late Eurasian *Megantereon* cannot be referred to *M. whitei*. It may instead be related to the Eurasian *M. cultridens* and, if so, could be placed in *Megantereon adroveri* Pons-Moya 1987 (new rank).

4. Where do the affinities of the Levantine *Megantereon* from 'Ubeidiya (Israel) lie and what implications does this material have for understanding the biogeography of *Megantereon*?

The site of 'Ubeidiya is critical in many ways to the understanding of the dispersal of African taxa into Eurasia (see Belmaker 2006, 2010a,b). The first specimen of *Megantereon* to be described from 'Ubeidiya was a well-preserved upper canine (UB 80) (Haas 1968; Ballesio 1986). While Ballesio (1986) assigned this tooth to *M.* cf. *cultridens*, he believed that the material was not sufficient to determine its taxonomic and geographic affinities. Martínez-Navarro, Palmqvist and colleagues (e.g., Martínez-Navarro and Palmqvist 1995; Palmqvist et al. 2007) refer the specimen to *M. whitei*, thus supporting their hypothesis that *M. whitei* dispersed from Africa and eventually replaced the larger Eurasian *M. cultridens*.

Two additional 'Ubeidiya specimens have now been assigned to *Megantereon* cf. *M. whitei* along with UB 80: a lower canine (UB 14) and a middle phalanx (UB 307) (Martínez-Navarro et al. 2009). Martínez-Navarro and colleagues note that precise identifications of all of the 'Ubeidiya material cannot currently be made due to the nature of the material. Their tentative assignation of the phalanx to *Megantereon* cf. *M. whitei*, however, is based on its small size and its similarity to an unpublished phalanx from Venta Micena believed to be *M. whitei* (presumably based on the assignation of more diagnostic portions of the skeleton at Venta Micena to this taxon). Measurements of both phalanges are presented in their paper and suggest that they are similar in size and proportion. These authors rule out the possibility that this is *Panthera*, particularly *P. leo* and *P. pardus*, based on the relative elongation of UB 307. However, no data is provided to support this statement. *Panthera* cf. *P. gombaszoegensis* is found at this site, but no mention is made of what distinguishes UB 307 from this species of *Panthera*.

Interestingly, Palmqvist et al. (2007) list measurements for an unpublished lower canine from 'Ubeidiya (presumably UB 14) and include width and breadth measurements that are larger than some of their *M. cultridens* measurements (thus suggesting their assignments of canines to species are based on something other than size). While they do not include measurements of the upper canine from 'Ubeidiya, their measurements of upper canines do not distinguish *M. cultridens* from *M. whitei* (as they define these two taxa). Measurements of lower canines are not included in the analyses and no mention is made of this specimen within descriptions of their statistical results.

Assignation of the published 'Ubeidiya material was tentative (Martínez-Navarro et al. 2009). We suggest that the 'Ubeidiya material is not complete enough to assign to a specific species. In addition, no new diagnostic material of *Megantereon* was found in the post Ballesio excavations from 1989–1994 and 1997–1999 (Belmaker M., personal communication, 2008). Martínez-Navarro and colleagues have suggested a similarity between the unpublished Venta Micena *Megantereon* phalanx and that from 'Ubeidiya. While it is certainly possible that the 'Ubeidiya material belongs to *M. whitei*, it is also possible that it is *M. cultridens*, *M. ekidoit*, or a completely new species. Until more diagnostic material is recovered, the taxonomic status of the 'Ubeidiya *Megantereon* remains unclear as are the biogeographic implications of this material.

5. Regardless of its affinities, could late Eurasian *Megantereon* have been a significant source of carcasses for scavenging by the earliest hominins in Europe even in the presence of the hyaenid *Pachycrocuta*?

The key word here is "significant". Given the behavior of modern large-bodied carnivorans, one would expect hominins at this time to attempt to take carcasses from *Megantereon*. How important that resource was to migrating hominins is unknown. It seems doubtful that this would be the only factor or even the dominant factor enabling hominin migration out of Africa. For reasons mentioned above, regular, confrontational use of this source of carcasses might cause the resource to disappear. While there may have been few species actively dispersing from Africa to Eurasia at this time, there were indigenous species that hominins would have encountered as they moved northwards. While stealing carcasses from *Megantereon* probably occurred occasionally, it is likely that hominins encountered other sources of meat and marrow along their journey.

Conclusions

The evolution of *Megantereon* is not well enough understood to fully comprehend the nature of dispersals within this genus (see also Leakey and Werdelin 2010). While there are interesting hypotheses linking *Megantereon* and hominins, none are currently testable. Perhaps *Megantereon* dispersed from Africa at roughly the same time as hominins. Perhaps the appearance of hominins in Eurasia and/or ecological changes enabling hominin migration to this region drove later Eurasian *Megantereon* to converge on African forms. Perhaps the later form of *Megantereon* shared primitive features with African forms implying no dispersal and no convergence. We just do not know at present.

We can, however, set the scene in Africa for hominin dispersal. After 1.8 Ma, the carnivore guilds of Africa were decreasing in taxonomic diversity (Werdelin and Lewis 2005) and hominins were becoming increasingly dominant. Effective kleptoparasitic strategies, such as confrontational scavenging, by *Homo* could have destabilized the carnivore guilds, although it is probably not a sufficient explanation for all the species that become extinct during the Early Pleistocene (Lewis and Werdelin 2007). Most importantly, while some hominins migrate to Eurasia, others remain in Africa. These African hominins do not go extinct. What would cause some hominins to disperse while others remain (a question outside the purview of this paper)?

The point is that there are many more issues at play here than the relationship between hominins and carnivorans. While we can continue to ask what *Theropithecus oswaldi*, *Hippopotamus antiquus*, *Megantereon* and *Homo erectus* might have in common at the time of dispersal (Martínez-Navarro 2004; Rook et al. 2004; Various papers in this volume, 2010), this may also be the wrong question. Dispersal to Eurasia at this point may have been a Simpsonian sweepstakes event where the factors affecting dispersal may have been different for each taxon and dispersal may not have occurred all at once. The search then becomes much more difficult: a search for the stochastic needle in the paleoenvironmental haystack.

Acknowledgments We would like to thank the symposium and workshop organizers and book editors for the invitation to participate. We would also like to thank all of the workshop participants, and especially B. Martínez-Navarro and M. Belmaker, for thoughtful discussions and critical comments on the workshop manuscript. Special thanks to J. Fleagle for editorial comments and B. Martínez-Navarro and two anonymous reviewers for their comments on this manuscript. Thanks also to M. Lague for comments. This research was supported by a grant from the LSB Leakey Foundation to ML and K.E. Reed, NSF grant to ML, Stockton Distinguished Faculty Fellowships to ML, and Swedish Research Council Grants to LW.

References

Antón, M., Galobart, A., & Turner, A. (2005). Co-existence of scimitar-toothed cats, lions and hominins in the European Pleistocene. Implications of the post-cranial anatomy of *Homotherium latidens* (Owen) for comparative palaeoecology. *Quaternary Science Reviews, 24*, 1287–1301.

Arribas, A., & Palmqvist, P. (1999). On the ecological connection between sabre-tooths and hominids: Fauna dispersal events in the Lower Pleistocene and a review of the evidence for the first human arrival in Europe. *Journal of Archaeological Science, 26*, 571–585.

Arribas, A., Riquelme, J. A., Palmqvist, P., Garrido, G., Hernández, R., Laplana, C., et al. (2001). Un nuevo yacimiento de grandes mamíferos villafranquienses en la Cuenca de Guadix-Baza (Granada): Fonelas P-1, primer registro de una fauna próxima al límite Plio-Pleistoceno en la Península Ibérica. *Boletín Geológico y Minero, 112*, 3–34.

Ballesio, R. (1986). Les Carnivores du gisement Pléistocène d'Oubeidiyeh (Israël). *Mémoires et Travaux du Centre de Recherche Français de Jerusalem, 5*, 63–92.

Belmaker, M. (2006). *Community structure through time: 'Ubeidiya, a Lower Pleistocene site as a case study*. Ph.D. dissertation, Hebrew University, Jerusalem.

Belmaker, M. (2010a). The presence of a large cercopithecine (cf. *Theropithecus* sp.) in the 'Ubeidiya formation (Early Pleistocene, Israel). *Journal of Human Evolution* 58, 79–89.

Belmaker, M. (2010b). Early Pleistocene faunal connections between Africa and Eurasia: an ecological perspective. In J. G. Fleagle, J. J. Shea, F. E. Grine, A. L. Baden, R. E. Leakey (Eds.), *Out of Africa I: the first hominin colonization of Eurasia* (pp. 183–205). Dordrecht: Springer.

Berta, A., & Galiano, H. (1983). *Megantereon hesperus* from the late Hemphillian of Florida with remarks on the phylogenetic relationships of machairodonts (Mammalia, Felidae, Machairodontinae). *Jurnal of Paleontology, 57*, 892–899.

Brain, C. K. (1981). *The hunters or the hunted? An introduction to African cave taphonomy*. Chicago: University of Chicago Press.

Brantingham, P. J. (1998). Hominid-carnivore coevolution and invasion of the predatory guild. *Journal of Anthropological Archaeology, 17*, 327–353.

Broom, R. (1937). On some new Pleistocene mammals from limestone caves of the Transvaal. *South African Journal of Science, 33*, 750–768.

Broom, R. (1948). Some South African Pliocene and Pleistocene mammals. *Annals of the Transvaal Museum, 21*, 1–38.

Broom, R., & Schepers, G. (1946). The South African fossil ape men: The Australopithecinae. *Transvaal Museum Memoirs, 2*, 1–272.

Creel, S. (2001). Four factors modifying the effect of competition on carnivore population dynamics as illustrated by African wild dogs. *Conservation Biology, 15*, 271–274.

Cuvier, G. (1824). Recherches sur les ossements fossiles, où l'on rétablit les caractères de plusieurs animaux dont les révolutions du globe ont détruit les espèces. Paris.

Dayan, T., & Simberloff, D. (1996). Patterns of size separation in carnivore communities. In J. L. Gittleman (Ed.), *Carnivore behavior, ecology, and evolution* (Vol. 2, pp. 243–266). Ithaca, NY: Cornell University Press.

Dayan, T., & Simberloff, D. (2005). Ecological and community-wide character displacement: the next generation. *Ecology Letters, 8*, 875–894.

Ewer, R. (1955). The fossil carnivores of the Transvaal caves: Machairodontinae. *Proceedings of the Zoological Society, London, 125*, 587–615.

Ewer, R. (1973). *The carnivores*. New York: Cornell University Press.

Ficcarelli, G. (1979). The Villafranchian machairodonts of Tuscany. *Palaeontographia Italica, 79*, 17–26.

Haas, G. (1968). *On the fauna of `Ubeidiya*. Jerusalem: The Israel Academy of Science and Humanities.

Hartstone-Rose, A., De Ruiter, D. J., Berger, L. R., & Churchill, S. E. (2007). A sabre-tooth felid from Coopers cave (Gauteng, South Africa) and its implications for *Megantereon* (Felidae: Machairodontinae) taxonomy. *Palaeontologia Africana, 42*, 99–108.

Hemmer, H. (2001). Die Felidae aus dem Epivillafranchium von Untermassfeld. *Monographien des Römisch-Germanischen Zentralmuseum, 40*, 699–782.

Hendey, Q. (1973). Carnivore remains from the Kromdraai australopithecine site (Mammalia: Carnivora). *Annals of the Transvaal Museum, 28*, 99–112.

Hendey, Q. (1974). The late Cenozoic Carnivora of the south-western Cape Province. *Annals of the South African Museum, 63*, 1–369.

Leakey, M. G. & Werdelin, L. (2010). Early Pleistocene mammals of Africa: Background to dispersal. In J. G. Fleagle, J. J. Shea, F. E. Grine, A. L. Baden, & R. E. Leakey (Eds.) *Out of Africa I: The first hominin colonization of Eurasia* (pp. 3–11). Dordrecht, The Netherlands: Springer.

Lewis, M. E. (1995a). Functional morphology of the sabertoothed felid *Megantereon* from Kromdraai, South Africa. *Journal of Vertebrate Paleontology, 15*, 40A.

Lewis, M. E. (1995b). *Plio-Pleistocene carnivoran guilds: Implications for hominid paleoecology*. Ph.D. thesis, State University of New York, Stony Brook.

Lewis, M. E. (1997). Carnivoran paleoguilds of Africa: Implications for hominid food procurement strategies. *Journal of Human Evolution, 32*, 257–288.

Lewis, M. E., & Werdelin, L. (2007). Patterns of change in the Plio-Pleistocene carnivorans of eastern Africa: Implications for hominin evolution. In R. Bobe, Z. Alemseged, & A. K. Behrensmeyer (Eds.), *Hominin environments in the East African Pliocene: An assessment of the faunal evidence* (pp. 77–106). Dordrecht, The Netherlands: Springer.

Linnell, J. D. C., & Strand, O. (2000). Interference interactions, co-existence and conservation of mammalian carnivores. *Diversity and Distributions, 6*, 169–176.

Liu, J.-Y. (2005). *Megantereon* fossil remains from Renzidong Cave, Fanchang County, Anhui Province, China. *Vertebrata PalAsiatica, 43*, 122–134.

Marean, C. W. (1989). Sabertooth cats and their relevance for early hominid diet and evolution. *Journal of Human Evolution, 18*, 559–582.

Marean, C. W., & Ehrhardt, C. L. (1995). Paleoanthropological and paleoecological implications of the taphonomy of a sabertooth's den. *Journal of Human Evolution, 29*, 515–547.

Martínez-Navarro, B. (2004). Hippos, pigs, bovids, sabertoothed tigers, monkeys and hominids dispersals during late Pliocene and early Pleistocene times through the Levantine corridor. In: N.L. Goren-Inbar, & J. D. Speth (Eds.), *Proceedings of the Conference entitled "Human Paleoecology in the Levantine Corridor," Jerusalem July 2002* (pp. 37–51). London: Oxbow Books.

Martínez-Navarro, B. (2010). Early Pleistocene faunas of Eurasia and hominin dispersals. In J. G. Fleagle, J. J. Shea, F. E. Grine, A. L. Baden, R. E. Leakey (Eds.), *Out of Africa I: the first hominin colonization of Eurasia* (pp. 207–224). Dordrecht: Springer.

Martínez-Navarro, B., & Palmqvist, P. (1995). Presence of the African machairodont *Megantereon whitei* (Broom, 1937) (Felidae, Carnivora, Mammalia) in the Lower Pleistocene site of Venta Micena (Orce, Granada, Spain), with some considerations on the origin, evolution and dispersal of the genus. *Journal of Archaeological Science, 22*, 569–582.

Martínez-Navarro, B., & Palmqvist, P. (1996). Presence of the African saber-toothed felid *Megantereon whitei* (Broom, 1937) (Mammalia, Carnivora, Machairodontinae) in Apollonia-1 (Mygdonia Basin, Macedonia, Greece). *Journal of Archaeological Science, 23*, 869–872.

Martínez-Navarro, B., Belmaker, M., & Bar-Yosef, O. (2009). The large carnivores from 'Ubeidiya (early Pleistocene, Israel): Biochronological and biogeographical implications. *Journal of Human Evolution, 56*, 514–524.

Palmqvist, P. (2002). On the presence of *Megantereon whitei* at the South Turkwel Hominid Site, northern Kenya. *Journal of Paleontology, 76*, 928–930.

Palmqvist, P., Martínez-Navarro, B., & Arribas, A. (1996). Prey selection by terrestrial carnivores in a Lower Pleistocene community. *Paleobiology, 22*, 514–534.

Palmqvist, P., Gröcke, D. R., Arribas, A., & Fariña, R. A. (2003). Paleoecological reconstruction of a lower Pleistocene large mammal community using biogeochemical ($\delta^{13}C$, $\delta^{15}N$, $\delta^{18}O$, Sr:Zn) and ecomorphological approaches. *Paleobiology, 29*, 205–229.

Palmqvist, P., Torregrosa, V., Pérez-Claros, J. A., Martínez-Navarro, B., & Turner, A. (2007). A re-evaluation of the diversity of *Megantereon* (Mammalia, Carnivora, Machairodontinae) and the problem of species identification in extinct carnivores. *Journal of Vertebrate Paleontology, 27*, 160–175.

Palmqvist, P., Pérez-Claros, J. A., Janis, C. M., Figueirido, B., Torregrosa, V., & Gröcke, D. R. (2008). Biogeochemical and ecomorphological inferences on prey selection and resource partitioning among mammalian carnivores in an Early Pleistocene community. *Palaios, 23*, 724–737.

Patnaik, R., & Nanda, A. C. (2010). Early Pleistocene mammalian faunas of India and evidence of connections with other parts of the world. In J. G. Fleagle, J. J. Shea, F. E. Grine, A. L. Baden, R. E. Leakey (Eds.), *Out of Africa I: the first hominin colonization of Eurasia* (pp. 129–143). Dordrecht: Springer.

Petter, G., & Howell, F. C. (1982). Un Félidé machairodonte des formations plio-pleistocènes des Siwaliks: *Megantereon falconeri* Pomel/ =*M. sivalensis* (F. et C.)/. (Mammalia. Carnivora. Felidae). *Comptes Rendus de l'Académie des Sciences*, Paris, Série II 295, 281–284.

Pomel, M. (1853). Catalogue méthodique et descriptif des Vertébrés fossiles. Paris.

Pons-Moya, J. (1987). Los carnivoros (Mammalia) de Venta Micena (Granada, Espana). *Paleontologia i Evolució, Mem. Esp., 1*, 109–128.

Qiu, Z., Deng, T., & Wang, B. (2004). Early Pleistocene mammalian fauna from Longdan, Dongxiang, Gansu, China. *Palaeontologia Sinica, 191*, 1–198.

Rook, L., Martínez Navarro, B., & Howell, F. C. (2004). Occurrence of *Theropithecus* sp. in the Late Villafranchian of Southern Italy and implications for Early Pleistocene "out of Africa" dispersals. *Journal of Human Evolution, 47*, 267–277.

Sardella, R. (1998). The Plio-Pleistocene Old World dirk toothed cat *Megantereon* ex gr. *cultridens* (Mammalia, Felidae, Machairodontinae), with comments on taxonomy, origin and evolution. *Neues Jahrbuch für Geologie und Paläontologie, Abhandlungen, 207*, 1–36.

Sardella, R., Petrucci, M., & Rook, L. (2008). The African species *Megantereon whitei* from the Early Pleistocene of Monte Argentario (South Tuscany, Central Italy). *Comptes Rendus Palevol, 7*, 601–606.

Sotnikova, M. V. (1989). Late Pliocene-Early Pleistocene Carnivora: Stratigraphic significance. *Trudy, Akademia Nauk, SSSR, 440*, 1–123 (in Russian).

Taylor, M. E. (1989). Locomotor adaptations by carnivores. In J. L. Gittleman (Ed.), *Carnivore behavior, ecology, and evolution* (pp. 382–409). New York: Cornell University Press.

Turner, A. (1987). *Megantereon cultridens* (Cuvier) (Mammalia, Felidae, Machairodontinae) from Plio-Pleistocene deposits in Africa and Eurasia, with comments on the possibility of a New World origin. *Journal of Paleontology, 61*, 1256–1268.

Turner, A. (1988). Relative scavenging opportunities for east and south African Plio-Pleistocene hominids. *Journal of Archaeological Science, 15*, 327–341.

Turner, A. (1990). The evolution of the guild of larger terrestrial carnivores during the Plio-Pleistocene in Africa. *Geobios, 23*, 349–368.

Turner, A. (1993). New fossil carnivore remains from Swartkrans. In: C. K. Brain (Ed.), *Swartkrans: A cave's chronicle of early man* (pp. 161–165). Transvaal Museum Monograph No. 8. Pretoria: Transvaal Museum.

Turner, A., & Antón, M. (1997). *The big cats and their fossil relatives: An illustrated guide to their evolution and natural history*. New York: Columbia University Press. 234 pp.

Van Valkenburgh, B. (1987). Skeletal indicators of locomotor behavior in living and extinct carnivores. *Journal of Vertebrate Paleontology, 7*, 162–182.

Van Valkenburgh, B. (2001). The dog-eat-dog world of carnivores: A review of past and present carnivore community dynamics. In C. B. Stanford & H. T. Bunn (Eds.), *Meat-eating and human evolution* (pp. 101–121). Oxford: Oxford University Press.

Vekua, A. K. (1995). Die Wirbeltierfauna des Villafranchium von Dmanisi und ihre biostratigraphische Bedeutung. *Jahrbuch des Römisch-Germanischen Zentralmuseums, Mainz, 42*, 77–180.

Vrba, E. S. (1981). The Kromdraai australopithecine site revisited in 1980; Recent investigations and results. *Annals of the Transvaal Museum, 33*, 17–59.

Werdelin, L., & Lewis, M. E. (2000). Carnivora from the South Turkwel hominid site, northern Kenya. *Journal of Paleontology, 74*, 1173–1180.

Werdelin, L., & Lewis, M. E. (2001). A revision of the genus *Dinofelis* (Mammalia, Felidae). *Zoological Journal of the Linnean Society, 132*, 147–258.

Werdelin, L., & Lewis, M. E. (2002). Species identification in *Megantereon*: A reply to Palmqvist. *Journal of Paleontology, 76*, 931–933.

Werdelin, L., & Lewis, M. E. (2005). Plio-Pleistocene Carnivora of eastern Africa: Species richness and turnover patterns. *Zoological Journal of the Linnean Society, 144*, 121–144.

Werdelin, L., & Lewis, M. E. (in prep). *Koobi Fora Research Project, volume 6: The fossil carnivora*. San Francisco: California Academy of Sciences.

WoldeGabriel, G., White, T. D., Suwa, G., Renne, P., de Heinzelin, J., Hart, W. K., et al. (1994). Ecological and temporal placement of early Pliocene hominids at Aramis, Ethiopia. *Nature, 371*, 330–333.

Woodroffe, R., & Ginsberg, J. R. (2005). King of the beasts? Evidence for guild redundancy among large mammalian carnivores. In J. C. Ray, K. H. Redford, R. S. Steneck, & J. Berger (Eds.), *Large carnivores and the conservation of biodiversity* (pp. 154–178). Washington, DC: Island Press.

Chapter 3
Saharan Corridors and Their Role in the Evolutionary Geography of 'Out of Africa I'

Marta Mirazón Lahr

Abstract In any discussion of hominin dispersal it is possible, and important, to examine the event at many different scales. This paper examines the initial dispersal out of Africa at the scale of populations rather than species, looks at dispersal between ecological zones rather than continents, and considers dispersal within Africa prior to any dispersal out of Africa. Before hominins could disperse out of Africa they needed to disperse out of their likely area of endemism in sub-Saharan Africa to North Africa, the most likely departure point for Eurasia. Prior to the Middle Pleistocene, successful long term colonizations of North Africa by hominins were very rare, and apparently less successful than their colonizations of Eurasia. The Early Pleistocene hominin dispersal into Eurasia was most probably along the western coast of the Red Sea. The ability of hominins to successfully disperse into Eurasia and successfully colonize northern continents was made possible by the ecological and climatic diversity within Africa.

Keywords Climate • Dispersal • Ecology • North Africa • Population

Introduction

Most discussions of the first hominin dispersals out of Africa are, understandably, driven by the discovery of fossil and archeological remains in Eurasia. However, dispersal events are a problem in evolutionary geography (Lahr and Foley 1998), and they are shaped by the interaction through time of the species under study and the community ecology of the source and recipient areas of the dispersal event. Particularly when dealing with inter-continental movements, understanding the interaction of a target group and its environment requires that the dispersal process is analyzed in terms of fluctuating corridors and barriers, so as to obtain insights into the phylogeographic structuring of any species and the evolution of its adaptive pattern(s).

In the particular case of hominin dispersals into Eurasia, one of the problems lies, in my view, in the scale at which we analyze the dispersal event. In this paper, I want to experiment with an attempt at changing the scale at which the problem can be examined, from species to populations, from continents to ecological zones, and explore what this new look at what has become known as 'out of Africa I' may offer. The starting point for the scalar adjustment is that before hominins disperse out of Africa, they must first disperse out of their region of endemism – sub-Saharan Africa, and more specifically, eastern Africa. This implies that the starting point of what we usually consider 'out of Africa dispersal' is more accurately described as 'within Africa dispersal'. To put more strongly, Eurasian colonisation could be seen as a by-product of dispersals within Africa.

Hominins, then, did not disperse out of Africa, but out of sub-Saharan Africa, and in particular, out of East Africa.[1] This is not just a statement of fact. Only by reducing the question to the ecological scale at which the process actually operated (the scale at which populations share a particular set of competitors and resources), can we hope to find the causes and conditions for demographic fluctuations through time, as well as the evidence for interpreting the dispersal response in terms of niche expansion or exclusion. Furthermore, the identification of geographical corridors, barriers and refugia should set the parameters of the dispersing process in terms of speed of movement, degree of localized selective pressures (and thus progressive adaptive differentiation), spatial constraints on home range size (and thus extent of demographic bottlenecking and genetic drift effects), as well as asynchronicities in evolutionary trends.

M.M. Lahr (✉)
Leverhulme Centre for Human Evolutionary Studies,
University of Cambridge, Cambridge, UK
e-mail: mbml1@cam.ac.uk

[1] This does not negate the existence of other directions of hominin movement (particularly between South and East Africa, or between East and West Eurasia), but focuses on a set of dispersal events (the 'out of Africa' set) that significantly shaped the evolutionary history of the genus *Homo*.

J.G. Fleagle et al. (eds.), *Out of Africa I: The First Hominin Colonization of Eurasia*, Vertebrate Paleobiology and Paleoanthropology,
DOI 10.1007/978-90-481-9036-2_3,

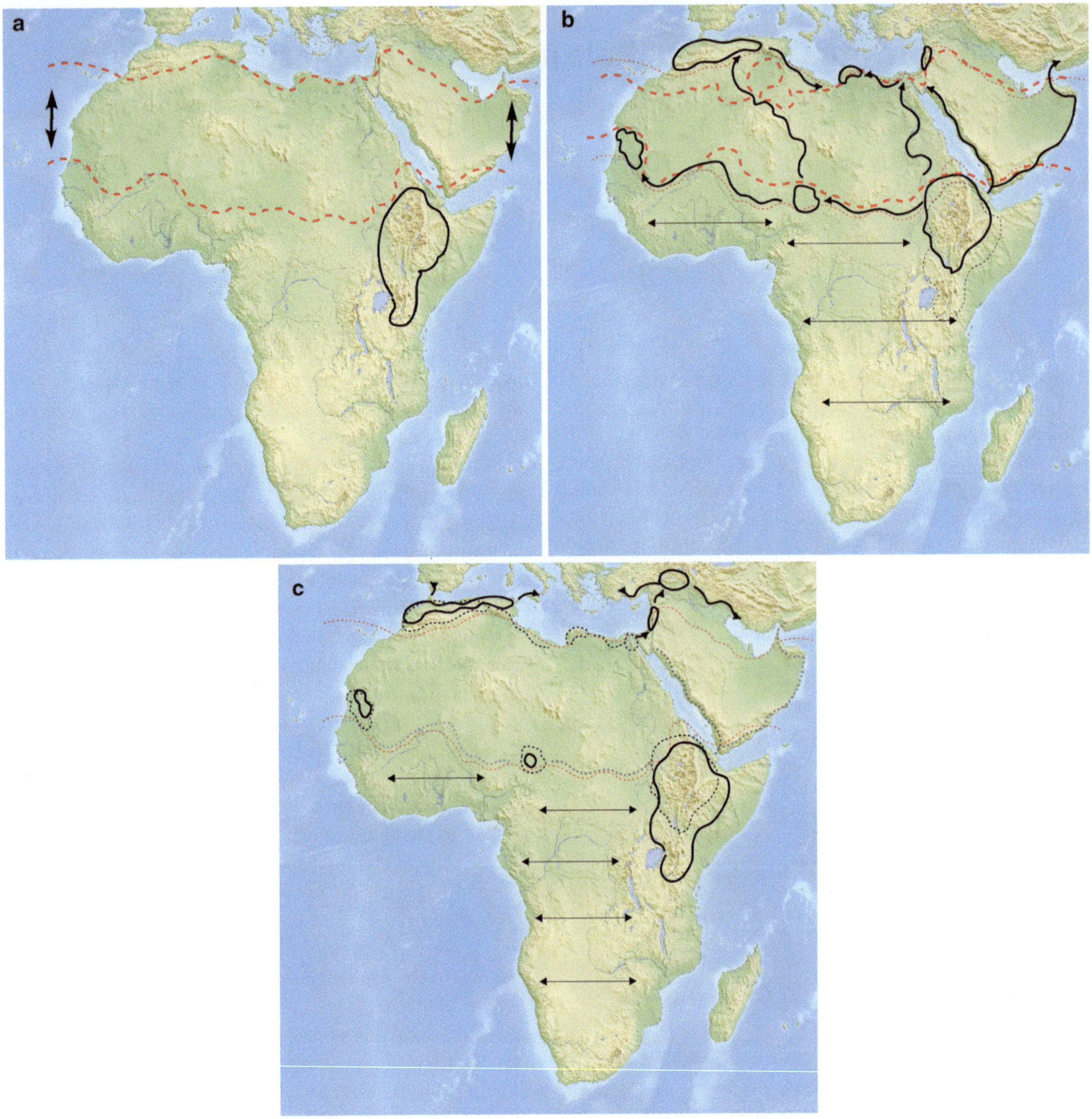

Fig. 3.1 A simplified model of the temporal sequence of early hominin dispersals out of East Africa given changing environmental conditions driven by climatic change after the formation of the Saharo-Arabian belt in the Late Pliocene. (**a**) Phase I – Late Pliocene formation of Saharo-Arabian desert belt (*red dotted line*) resulting in the separation of Africa into two zones – sub-Saharan and North Africa, of very unequal size and variability. Drawn is the hypothetical ecological range of a given hominin population in East Africa. Note that the ancestral range is characterized by moderate altitude, and the latitudinal extent of desert (and thus ecological zonation to be traversed) is much greater on the western (Atlantic) than the eastern (Strait of Hormuz) end of the Saharo-Arabian desert belt. (**b**) Phase II – a hypothetical 'wet episode' in which the Saharan barrier is latitudinally compressed (*thick red lines*) in relation to its maximum aridity extent (*thin red dotted lines*), and paleodrainage systems are formed (McCauley et al. 1997). The E-W expansion of forests in equatorial sub-Saharan Africa (Foley 1999; indicated by the *thin bilateral horizontal arrows*), together with the N-S expansion of savannahs on the northern and southern forest edges alters the range of the East African population and promotes a northward expansion that leads to colonization of new ecological zones by the use of temporary available corridors to the East, West and North. These 'jump dispersals', as defined by Tchernov (1992a) and Lahr and Foley (1994), may be followed by subsequent jump dispersals through corridors leading towards high carrying capacity environments, and thus to the establishment of 'daughter' populations in new ecological zones, normally discontiguous from the parental source. Note that during the last 2 million years, 'wet episodes' have varied markedly in frequency and extent. During the intense and stable arid period between 2 and 1 Ma, two periods of significantly increased precipitation have been identified. This differs markedly from the climatic pattern of Middle Pleistocene glacial cycles, during which the amplitude and periodicity of change increased, leading to shorter, more frequent and more pronounced high and low atmospheric water periods (major arid and wet episodes)

All these issues are prominent in the evolution of the genus *Homo*, with its recurrent pattern of 'out-of-Africa' movements, apparent simultaneous large-scale expansions (Plio-Pleistocene and Late Pleistocene dispersals), geographically differentiated contemporaneous lineages (Middle Pleistocene *Homo heidelbergensis* and *erectus*, Late Pleistocene *Homo sapiens* and *neanderthalensis*), pronounced localized differentiation (*Homo floresiensis*), unequal population levels of genetic variance (*Homo sapiens*), and so on. Dispersing behaviour can be seen as one of the major characteristics of the genus, one which arguably shaped its evolution to a greater extent than any other.

In this paper, I want to explore these ideas using the record of the colonization of North Africa by hominins in the Pleistocene. At its heart lies the simple observation that, compared to eastern Africa, the evidence for early hominin presence in northern Africa is remarkably scant. Two views can be taken on that observation. One, that the sparse early evidence is a research taphonomy, and that greater fieldwork efforts should change the picture dramatically. The other, that the scarcity of early sites is at least partly real, and thus part of the solution. Undoubtedly, more field research will improve the record. Nevertheless, given that paleoanthropological and paleontological work in North Africa has not been unsubstantial, I would argue that the available information can form the basis of a working model for analyzing the parameters shaping early dispersals of hominins from East Africa.

A Biogeographical Model for Dispersals Out of East Africa

Whole Cenozoic records of dust transport show that the most significant change in the last 65 million years – an order of magnitude increase in dust generation – occurred as a consequence of the onset of northern hemisphere glaciation in the Late Pliocene, reflecting major continental drying (Rea 1994). There are three main sources of dust today, eastern and central Asia, northwest Africa and Arabia. The Late Pliocene major dust episode reflects the establishment of two of these, the formation of the Sahara and Arabian desert belt. Therefore, since the Late Pliocene, Africa has been sharply divided by a major biogeographical barrier.

The degree of separation between sub-Saharan Africa and both the Mediterranean Basin and the South Asian landmass varied throughout the Pleistocene according to the interaction of precipitation levels and altitudinal profiles. That interaction, in turn, shaped corridors and refugia which promoted population isolation and movement. Furthermore, that same interaction between altitude and varying precipitation levels also acted on the source populations, by either expanding or contracting their ecological niche. Figure 3.1 shows a model of a simplified possible temporal sequence in the biogeography of populations dispersing from East to North Africa and Eurasia following climatic change.

The critical aspect of the model[2] is the identification of potential corridors and refugia. Although these are based on Late Pleistocene data, their definition as such derives from the interaction of fluctuating aridity and the Saharan landscape, and depending on the development of temporally specific barriers, should have had a similar ecological role during earlier periods. At this point, the presence of barriers beyond the Saharan-Arabian desert belt itself is not drawn, as these would be specific to geomorphological processes at different times. It is expected that, if these existed in such a way as to affect the extent or indeed existence of a given corridor, they would be identified through the examination of the Early Pleistocene record itself or from specific paleoenvironmental studies.

With the aim of testing whether this model fits the record of Early Pleistocene dispersals of hominins, and thus throws light onto their phylogeography and adaptive trajectories, the next part of this paper will consider the paleoanthropological record of North Africa, as well as the evidence for dispersals from North Africa into Eurasia. Finally, I will use the 'fitted' model to explore how an evolutionary geography theoretical framework might contribute to our understanding of hominin evolution at the time.

[2] This model was built on the basis of paleoenvironmental reconstructions of the Sahara during the second half of the last glacial cycle.

Fig. 3.1 (continued) Because of the sharpness and rapid succession of arid-wet episodes in the Middle Pleistocene, the expansion of East African populations early during wet episodes would have taken place after periods of major environmental fragmentation, with the potential for increased inter-population variance through drift. The greater potential for demographic fluctuations and drift during the Middle Pleistocene probably accelerated the rate of evolutionary and cultural change in relation to earlier periods. (**c**) Phase III – the re-establishment of arid conditions, resulting in the contraction of equatorial forests, development of barriers and disappearance of the ecological corridors that had allowed for jump dispersals to take place in the previous phase. The East African population would expand southwards at this point, recolonizing niches vacated by the retracting forest belt. Populations in the high carrying capacity areas colonized during the dispersal phase may survive for different lengths of time depending on the size, competitive community and resource structure of the refugia. Depending on the length of time until another wet episode opens the dispersal corridors, surviving populations in the refugia will differentiate from the parental population through both selection and drift. Only in this case, the dispersal event would have become a vicariant process of evolutionary consequence, leading to the establishment of an evolutionary lineage in the Simpsonian sense (Simpson 1961; Foley and Lahr 2007). Map, public domain (http://commons.wikimedia.org/wiki/File:Africa_topography_map.png)

Plio-Pleistocene North Africa

Any discussions of dispersals from eastern Africa to North Africa and Eurasia have to assume that such movements took place when paleoenvironmental conditions differed dramatically from those today, so as to eliminate, completely or partially, the Saharan desert barrier.

As briefly alluded to above, global climatic events characterize the period around 2.5 Ma. These led to major global cooling that resulted in significantly increased aridity in Africa and eustatic sea-level changes (Brunner and Maniscalco 1998; Shackleton et al. 1990; DeMenocal 1995). The effects on northern African environments were very pronounced and widespread (Sarnthein et al. 1982; Stein 1985; Tiedemann et al. 1989; LeHouerou 1992; Morel 1992). The establishment of the Saharo-Arabian desert belt (Rea 1994) was followed by a period of very dry conditions between 2 and 1 Ma, interrupted by two wet episodes. These pronounced climatic and consequently environmental changes led to faunal turnovers, the two most significant of which in Africa occurred ~2.5 Ma and 1.8 Ma (Vrba 1995).

During the Pliocene, hominins expanded at least once into Central Africa from the East, as shown by the mid-Pliocene remains of *Australopithecus bahrelghazali* (Brunet et al. 1995). The fauna suggests an open environment, drier than that of East African sites at the time, and a degree of endemism. However, no evidence exists that they expanded further north at the time. Stone-tools are conspicuously absent from the rich Maghrebian paleontological localities of Aïn Boucherit (~2.3 Ma) and Ahl al Oughlam (~2.4 Ma) (Raynal et al. 2001). Indeed, signs of hominin presence in the Early Pleistocene of northern Africa are extremely rare altogether. For the period between 1.8 and 1.0 Ma, several archeological sites are known across Eurasia, distributed from Spain to Java. In contrast, the North African record of this time is almost non-existent (Fig. 3.2). Sites with Mode 1 assemblages

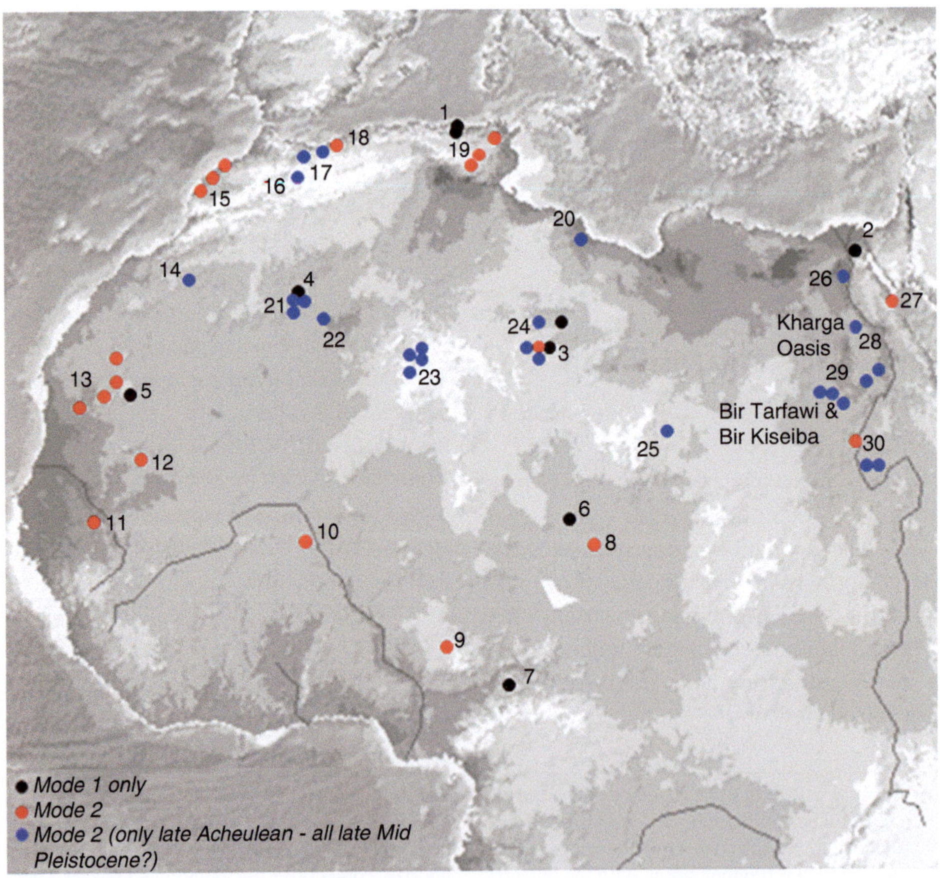

Fig. 3.2 Distribution of main Early Paleolithic sites in northern Africa (from various sources as discussed in the text). Sites include: (1) Aïn Hanech & El-Kherba, (2) Abassieh, (3) Shatti, BJJ6, (4) Saoura valley, (5) Richat Massif, Ouadane, Aftassa-Amzeili, (6) Sherda, Enneri group, (7) Beli, (8) Yayo, (9) Jos Plateau, (10) Mekrou group, (11) Faleme group (Djita, Sandande, Sare, Kare, Kidira), (12) Dhar Tichitt, (13) Richat group (Khatt Takfoil, Adrar, Tazazmout, Aderg Motleh, Hammami), (14) ACH1 Erqueiz Lahmar, (15) Sidi Abderrahman, Rabat, Khebibat, Sale, (16) Aïn Fritissa, (17) Tighenif, Ouzidane, (18) Lake Karar, (19) Sidi Zin, Sidi Mansour group, (20) Wadi Merdum, (21) Soura group, (22) Tachengit, (23) Tihodane group, (24) Fazzan group (Wadi al-Ajjal, Tadrart Acacus, Messak Settafet, NUS, BJJ, Murzuq), (25) Ounianga group, (26) Fayum group, (27) Saffagah, (28) Kharga Oasis, (29) Bir Tarfawi, Bir Kiseiba group, (30) Arkin eight group

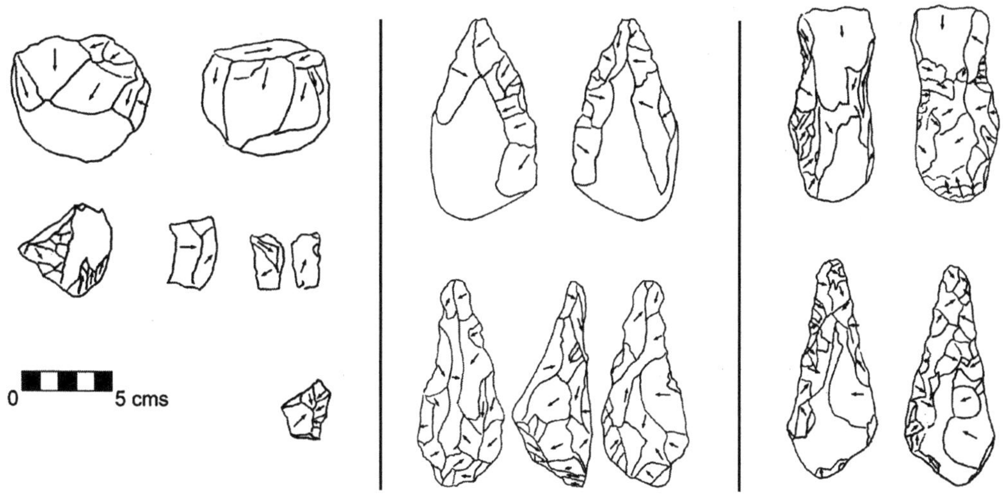

Fig. 3.3 Comparison of the Early Pleistocene lithic technology (Mode 1 and Mode 2) from North Africa. Aïn Hanech redrawn from Sahnouni et al. 2002; Thomas Quarry 1 Unit L and the Oulad Hamid 1 Quarry redrawn from Raynal et al. 2001 by R.A. Foley

are few, and none are uncontroversial in relation to their chronology.[3] The best contextualized of these is the site of Aïn Hanech, near Sétif in northern Algeria, and the recently described site of El-Kherba in its vicinity (Sahnouni et al. 2002). Aïn Hanech was discovered in 1947 by Camille Arambourg, who described an archeological assemblage from the site as Oldowan, with affinities towards those found at Olduvai Bed I/lower Bed II.

The site has been re-excavated in recent years by Sahnouni and colleagues, who have identified another Oldowan locality in the same wadi (El-Kherba). Sahnouni and colleagues have clarified a number of outstanding issues about the archeology of this site (notably that argued by Clark [1975] as not in primary context), especially the confirmation of the stratigraphy and the source of the Acheulean tools found (Sahnouni et al. 1996; 2002). The site has yielded 2,156 artefacts (Fig. 3.3), including some retouched pieces in which a microwear analysis identified evidence of meat processing. The controversial aspect of the recent studies has been the dating proposed by Sahnouni and colleagues. On the basis of normal paleomagnetism and the associated fauna (which includes the species *Mammuthus meridionalis, Equus tabeti* [but see comments below], *Sivatherium maurusium, Kolpochoerus phacochoeroides*, and the absence of *Anancus* [present at Aïn Boucherit]), Sahnouni and colleagues have proposed a date for the site within the Olduvai subchron (1.95–1.78 Ma) (Sahnouni et al. 1996, 2002). This dating is strongly disputed by Raynal, Geraads and colleagues, who believe the site to be closer to 1.2 Ma (Raynal et al. 2001; Geraads et al. 2004). Whether the earlier or later Early Pleistocene date of Aïn Hanech is confirmed, the site still represents the oldest stratigraphically contextualized hominin occupation of North Africa.

Besides the numerous surface finds of pebble tools and flaked stones, very few other Oldowan sites have been described in some detail for this vast region – of particular interest those of Abassieh, Cairo, and in the Richat Massif of Mauritania. The former was described in 1925 by Bovier-Lapierre (1925), who identified a number of archeological occurrences, some in stratigraphic sequence, on the plains to the East of the city of Cairo ('Plaine de l'Abassieh') and the surrounding highlands. Exciting as the finding of an Oldowan site at the 'door' of the Levantine corridor is, unfortunately no work followed these early observations, and the area has now been covered by the suburbs of Cairo. In the Massif des Richât, East of Ouadane, Mauritania, a number of Oldowan sites have been recorded, and some described (Vernet and Naffé 2003). These sites are located on the edge of paleolakes, such as those of El Beyedh and Guelb Er Richât (Baouba M., personal communication, 2004). The main sites among these are Wadi Akerdeil, the artefacts of which are described by Monod as showing affinities towards those of Olduvai

[3]It should be noted that a very conservative list of Mode 1-Oldowan sites is presented here. Another thirty or so localities where pebble tools and Oldowan cores and flakes have been observed could be added to this list (see Aumassip 2004). However, the majority of these are early observations without study or quantification, and as is the case in so many North African areas, part of mixed surface assemblages on the deflated landscapes.

Bed I, and Aftassa-Amzeili, to the SE of Zouerate (Monod 1976). Also important from a biogeographical perspective are the surface finds of Oldowan artifacts in Chad (Sherda, Enneri group) and, in the last two years, in Fazzan, Libya (Shatti, Barjuj 6). None of these findings have chronological control.

Clearly, the Early Pleistocene occupation of northern Africa by Mode 1 stone-tool making hominins was ephemeral at best, particularly in the eastern Sahara. As succinctly put by Veermersch "... if, during the Early Pleistocene, hominids followed the Nile Valley on their journey out of Africa, no clear traces of their presence have yet been found" (Veermersch 2001).

This rarity of early sites in North Africa can also be extended to the early Acheulean (early Mode 2). At present, the best Quaternary stratigraphic sequence is that of the complex of sites at Casablanca, Morocco, which comprises a series of deposits and terraces, ranging from 180 m a.s.l. to the coast, and variously cover the last 5.5 million years (Raynal et al. 2001). The oldest archeological occurrence at Casablanca is that of Thomas Quarry 1, Unit L, dated to ~1 Ma. The archeology is Mode 2, consisting of chopping-tools, polyhedrons, and some cleavers, trihedrons and bifaces. The bifaces are not symmetrical or carefully finished (Fig. 3.3).

The remaining sites are all likely to be Middle Pleistocene in age, ranging from early to very late. The earliest Middle Pleistocene sites in the Casablanca sequence are those of Oulad Hamida 1 Quarry, the Grotte de Rhinoceros at Thomas Quarry 1 (OH1-GDR), and the Hominid Cave. These deposits are currently dated as early Middle Pleistocene, between 0.7–0.6 Ma (Rhodes et al. 1994), of similar age and faunal associations as Tighenif. At OH1-GDR, a rich Acheulean assemblage, in which cleavers are rare and bifaces are large, was found (Fig. 3.3). The Thomas Quarry mandible probably derives from a pink breccia from the earliest levels of OH1-Th1, in which three new *Homo* teeth have been discovered (Raynal et al. 2003).

At both Tighenif and the neighbouring site of Lake Karar (a small spring) in Oran Province, northern Algeria, Acheulean tools vary from simple quartzite forms to carefully finished lanceolate bifaces. The assemblages contain cleavers, thought to show similarities to those of Olduvai Bed III by McBurney (1960), and evidence of the use of the Kombewa technique, thought to have developed in East Africa ~1 Ma (as evidenced at Olorgesailie; Potts 1989). The site of Sidi Zin, at the margin of a paleo-spring or water hole at the extreme eastern foothills of the Atlas Mountains in northern Tunisia, contains four stratigraphic units (Layers I to IV) (Clark 1975; McBurney 1960; Gobert 1950). In Layer I, beautiful narrow lanceolate handaxes were found, like the best at Lake Karar, together with flake tools and flattened river pebbles coarsely trimmed into massive scraping or chopping edges (which according to McBurney are similar to other such in sub-Saharan African industries); Layer II probably represents a short occupation, and contains elliptical, carefully finished cleavers and small flake tools, lacking pebble scrapers and choppers as well as lanceolate handaxes; Layer III contains an industry similar to that of Layer I (Layer IV is MSA, with Levallois prepared cores) (Freeman 1975). The site of Arkin 8, on the West margin of the Nile at the Sudanese-Egyptian border (Chmielewski 1968), is undated. However, the character of the stone-tool industry (dominated by heavy duty tools, cordiform, ovate and lanceolate handaxes, as well as trihedrons; see Sahnouni et al. 1996) tentatively aligns it with this early Middle Pleistocene group of sites. Similarly undated, early bifaces and trihedral pieces have recently been recorded for the Fazzan, Libya.

Of younger Middle Pleistocene age (~400 ka?) a number of sites have been found, including the localities at Sidi Abderrahman – Cap Chatelier, Grotte d'Ours, and Grotte des Littorines (locality of the Sidi Abderrahman *Homo* mandible) (Raynal et al. 2001). McBurney (1960) described two technological phases within the Sidi Abderrahman complex; an early one containing beautifully finished pear-shaped handaxes, and cleavers made on very large flakes; and a later one, which would correspond to the pink breccia horizons at the Grotte d'Ours and Grotte des Littorines. The site of Tihodaïne at the edge of the Tassili Massif in southern Algeria, contains a late Acheulean industry, with carefully made handaxes and cleavers, in association with a water dependent fauna, and some indications of more open environments (*Elephas recki, Rhinoceros simus, Equus zebra, Hippopotamus amphibious, Bubalus antiquus?, Bos primigenius, Gazella dorca,* and several antelopes including *Kobus*). The fauna and archeology at Tihodaïne has been argued to show correlations to those of Olduvai Bed IV (McBurney 1960). At Bir Kiseiba, Egypt, the assemblage has also been likened to that of Olduvai Bed IV. It contains cleavers and handaxes made on large flakes (Kombewa technique) (Haynes et al. 1997), similar to those at Bed IV at Olduvai (Leakey 1975) and other East African sites (Clark 1975), as well as to those of Gesher Benot Ya'aqov in the Dead Sea Rift (Goren-Inbar and Saragusti 1996), dated to 780 Ka (Verosub et al. 1998). The undated site of Ouzidane, near Tighenif, has been referred to as middle or late Acheulean in character (Vaufrey 1955).

Among the possibly later Acheulean sites, we find those at the Fayum depression (associated with the 40 m lake), the Wadi Midauwara in the Kharga depression, and Bir Tarfawi (Hill 2001). At Bir Tarfawi and Bir Sahara East, geochronometric dating of the Acheulean deposits suggest a minimum age of 350 Ka (Wendorf et al. 1994), while recent work on the geochronology of the fossil-spring tufas of the Kharga Oasis have provided U-series minimum ages of 300 ka (Smith et al. 2004).

The numerous Acheulean sites recorded (but minimally studied) in Mauritania, as well as those along southern tributaries of the Senegal River, have no chronological control, and the fauna of only one of these has been reported (Coppens et al. 1972). Similarly, the recently reported Acheulean deposits in the Tadrart Acacus and Messak Settafet in the westernmost Fazzan, Libya, are so far undated (Garcea 1997), as are the various Acheulean localities within the Edeyen Awbari and Murzuq (Lahr et al. 2007, 2008, 2009).

The character of the assemblages found at Tachengit, another central Saharan site, and that at the site of ACH1 Erqueiz Lahmar, on the edge of a paleoriver in the Sahrawi Arab Democratic Republic (Polisario-controlled western Sahara) is unclear, and it may represent a very early Middle Stone Age site in which bifaces are found together with flakes produced through Levallois methods (as in the second occupation levels of the Kharga Oasis, or the 9 m terraces of the Nile, or the uppermost levels at Cap Chatelier). Similar assemblages of small, thin bifaces of possible early MSA context have recently been recorded in Fazzan (Lahr et al. 2007, 2008, 2009).

The main aspects of the early North African record can be summarised as follows:

1. North Africa was not occupied by hominins in the Late Pliocene – at the richest Northwest African paleontological site, that of Ahl al Oughla (~2.4 Ma), as well as at the somewhat later site of Aïn Boucherit, no evidence of hominin occupation exists.
2. The earliest evidence for hominin dispersals into North Africa comes from the site of Aïn Hanech (northern Algeria), with Oldowan stone tools and a controversial date of 1.8 Ma, although they are certainly older than 1 Ma (the conservative estimate is 1.2 Ma). Therefore, current evidence suggests that the colonization of North Africa is synchronous with that of Eurasia.
3. The pre-Acheulean occupation of the region was minimal, and most probably extremely ephemeral.
4. The early Acheulean (of Early Pleistocene age) occupation was also sparse, and currently only documented with chronological certainty in the Casablanca sequence (Thomas Quarry 1 Unit L).
5. Of the few described Early Pleistocene sites (or localized groups of sites) – three Oldowan and one Acheulean, three are located along the Atlantic border of North Africa, while the fourth, on the edge of the Nile delta, must be considered as only tentative given that it is not available for new studies that could confirm its stratigraphic integrity or dating.
6. A later Acheulean tradition, of early to middle Middle Pleistocene age (700–400 kya?), is more widespread, represented at sites in the eastern, as well as the western Sahara, extending also into the Central Sahara (Bir Kiseiba, Sidi Zin, Tighenif, Lake Karar, the Oulad-Thomas Quarry deposits, the Sidi Abderrahman deposits, Tihodaïne, Wadi el-Ajal, Fazzan). In some of these sites, technological affinities with eastern African industries (through the Kombewa technique) are apparent, also pointing towards possible links with the Levant (Gesher Benot Ya'aqov). A somewhat later group of sites (400–200 kya?), or perhaps part of a northeastern African continuous Middle Pleistocene Acheulean tradition, is found between the Nile and the Libyan desert and plateau (Fayum, Kharga, Bir Tarfawi, Bir Sahara East). None of these sites show evidence of use of the Kombewa technique.
7. Later sites, of very late Acheulean or early MSA character are equally or more widespread as the early Middle Pleistocene Acheulean.

Across Deserts and Mountains, Seas and Deltas

The brief review of the Early Stone Age/Early Paleolithic record of North Africa clearly indicates that the Saharan desert remained a very major barrier to hominins until the Middle Pleistocene, when the establishment of glacial-interglacial cycles changed again, albeit intermittently and temporarily, the environmental conditions of the region. Clearly, getting out of sub-Saharan Africa was as much of a challenge as getting out of Africa.

However ephemeral the occupation, hominins did cross the Sahara during the Early Pleistocene and their presence along the southern Mediterranean coast, at least along its western portion, before 0.8 Ma is not disputed. These North African hominin populations have been generally assumed to have been part of the dispersal into Eurasia.

Nevertheless, once on the Mediterranean coast, getting out of Africa is also not uncontroversial. There is presently only one land connection between Africa and Eurasia, namely the Sinai Peninsula, and although this land bridge has been in existence since the Miocene (Tchernov 1992b) it involved the crossing of the Nile Delta, which paleontological evidence suggests acted as a partial barrier to faunal movements, reflected in the paucity of Eurasian elements in North African faunas (Tchernov 1992b). Besides the Sinai route, the crossings of both the Sicily-Tunisian Strait and the Strait of Gibraltar have also been proposed (Alimen 1975; Martínez-Navarro and Palmqvist 1995, 1996; Strauss 2001). Figures 3.4 and 3.5 show the bathymetric outline of the Gibraltar and Tunisian/Sicilian straits respectively. As discussed by several authors before, both areas currently reach depths of 200 m or more, and would have, therefore, not been fully exposed during the currently estimated low sea-level stands of the later Pleistocene. Nevertheless, because of

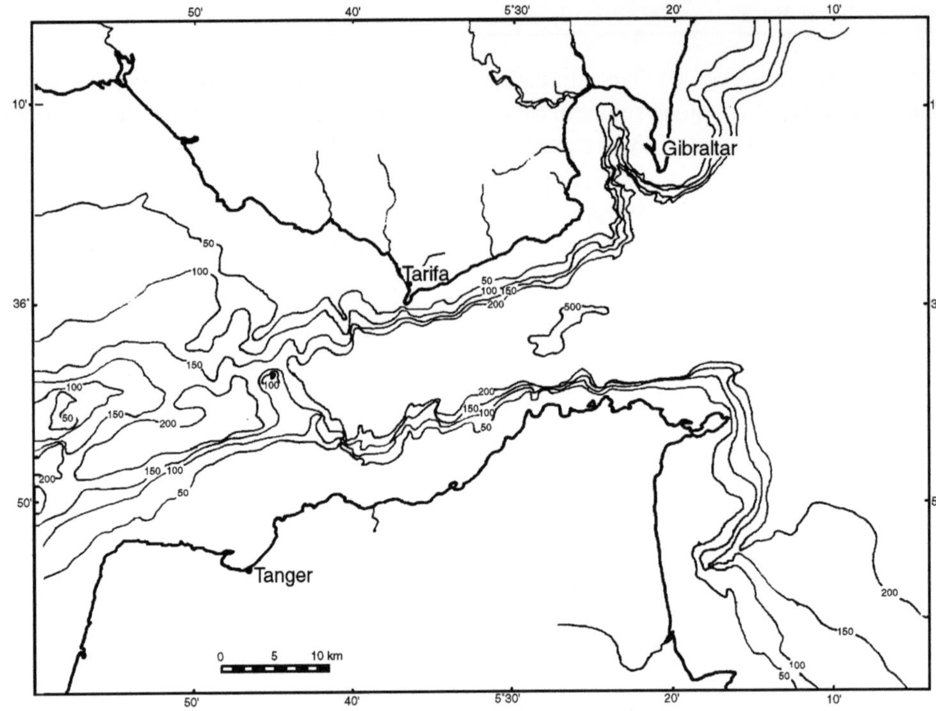

Fig. 3.4 Bathymetric outline of the Gibraltar Strait (From Lahr 1996)

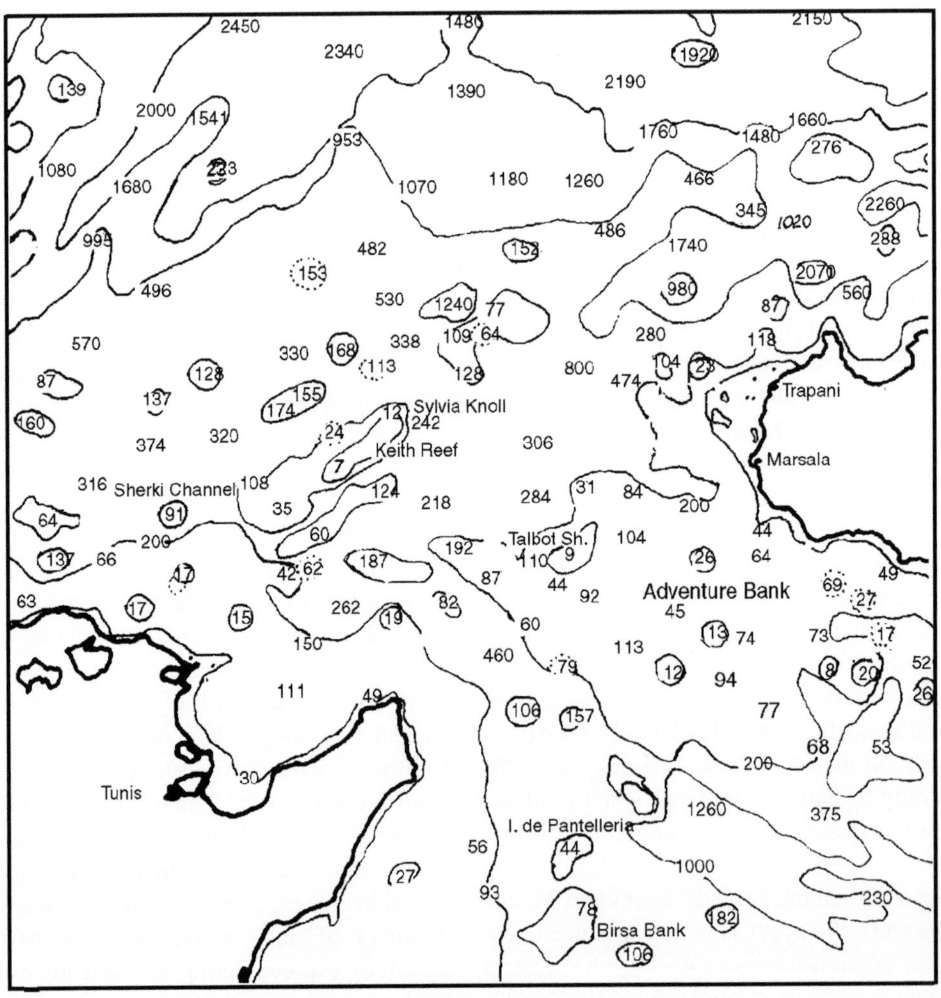

Fig. 3.5 Bathymetric outline of the Sicily-Tunisia Strait (From Lahr 1996)

uncertainties associated with the degree of tectonic movements, and the fact that the Sicilian Channel has been subsiding since the Middle Pleistocene (Alexander 1988; Bonfiglio and Berdar 1970), the question has not been settled.

From the Late Pliocene to ca. 1 Ma, three major drops in sea-level associated with important changes in the European paleoenvironment are known to have taken place – the 'Aquatraversan' or 'Elephant-*Equus* Event'; the 'Aullan' (the Italian 'Wolf Event'); and the 'Cassian' or 'end-Villafranchian' Event (Azzaroli 1995; Arribas and Palmqvist 1999). These have been dated to 2.6–2.4 Ma, 1.8–1.6 Ma, and 1.2–0.9 Ma respectively. The first of these had very pronounced consequences at both marine and terrestrial levels, and clearly correlates with the major global cooling at this time. In Europe it is associated with the arrival of the Asian elephants and monodactyl horses (Arribas and Palmqvist 1999), thus marking a period of E-W trans-Eurasian dispersals. However, the arrival in Eurasia of African faunas during the 'Aullan' Event (at the Neogene-Quaternary boundary) and their particular distributions across Europe have been argued to suggest possible trans-Mediterranean exchanges (Martínez-Navarro and Palmqvist 1995; 1996). These are considered in more detail below.

Alimen's (1975) argument for a crossing between Tunisia and Sicily during the Early Pleistocene was strongly influenced by the description of Acheulean findings in Sicily (Bianchini 1973). Several Mode 1 tools and assemblages have been reported from Sicily. These include both surface finds and in association with marine terraces. However, in the absence of radiometric dates, stratigraphic context and faunal associations, later Middle Pleistocene ages cannot be ruled out (Villa 2001). Contrastingly, the Quaternary paleontological record of Sicily strongly suggests that a land bridge between Tunisia and Sicily did not exist. The island has a rich Pleistocene paleontological record (Agnesi et al. 2004), usually divided into four stratigraphic units. The earliest of these is found in the northwestern half of the island (which was separated from the southeastern half until the Middle Pleistocene; see Bonfiglio and Piperno 1996). The small mammal component shows signs of endemism, while the large one is European in origin (Villa 2001). The exception is the large-sized ctenodactylid rodent *Pellegrinia panormensis*, a form of African affinities that has never been observed elsewhere in Europe (Thaler 1972). The second unit has been dated to the earliest Middle Pleistocene (Bonfiglio et al. 1997). It shows even more endemic features, including the dwarf elephant *E. falconeri* (Roth 1990). The subsequent assemblages are of certain Middle Pleistocene age, and are evidence of the uplift of the Hyblean Plateau and fusion of the two Early Pleistocene Sicilian landmasses (Bonfiglio and Piperno 1996). Therefore, the hypothesis of a connection between North Africa and Europe across the Sicilian Strait during the Early Pleistocene is not supported by existing data.

The Strait of Gibraltar is currently ~14.5 km wide (Fig. 3.4). It would require a sea-level drop of approximately 300 m to virtually close it, while a 200 m drop would narrow it to ~6.5 km (see Arribas and Palmqvist 1999; although several islands would become exposed whenever the sea-level dropped more than 100 m, making for several short crossings, see Martinet and Searight 1994). At present, positive evidence of such low sea-level stands during the Plio-Pleistocene does not exist. However, the argument for multiple Mediterranean crossings is based on the Early Pleistocene *circum*-Mediterranean paleontological record (Martínez-Navarro and Palmqvist 1996; Arribas and Palmqvist 1999). Palmqvist and colleagues have argued that the colonization of the Northern Mediterranean zones and western Asia through multiple routes from northern Africa would explain the chronologically close appearance of the same African mammals in Spain (Orce), Italy (Pirro Nord), Greece (Apollonia) and Georgia (Dmanisi), as well as their absence from Central Europe; such multiple dispersals would also be consistent with the apparent latitudinal constraint of *Megantereon whitei* to 40°N.

While the Sicilian faunal evidence strongly suggests that the area was not connected to North Africa in the Early Pleistocene, the argument proposed by Martínez-Navarro, Palmsqvist and colleagues for trans-Gibraltar crossings cannot be ruled out. However, an outstanding issue remains. The critical African species that dispersed into Eurasia between 1.8 and 1.6 Ma are the carnivores *Megantereon whitei* (a sabre-toothed cat) and *Pachycrocuta brevirostris*, the horse *Equus numidicus* (= *E. altidens*), the primate *Theropithecus oswaldi*, and the hippo *Hippopotamus antiquus*. At present, a particular link between these forms in northwestern Africa and southwestern Europe is not strong.

Equus numidicus is found in East Africa (Omo Shungura, Olduvai, East Turkana) and in the Late Pliocene Maghrebian site of Aïn Boucherit (? ~2 Ma; Sahnouni et al. 2002), but its presence at Aïn Hanech (possibly dated to the critical Plio-Pleistocene boundary; Sahnouni et al. 2002), instead of the derived form *Equus tabeti* (as reported by Sahnouni et al. 1996, 2002) or even *E.* cf *mauritanicus*, is controversial (Geraads et al. 2004). Although they are both species of the "*simplicidens* type" (Guerrero-Alba and Palmqvist 1997), the one dispersing into Eurasia in the Early Pleistocene (found at Orce, Cúllar de Baza, Cueva Victoria, Huéscar 1 in Spain; Pirro Nord and Selvella in Italy; Sainzelles in France, and Süssenborn in Germany) has been clearly identified as *E. numidicus*, a form that may have not been present in northwest Africa at ~1.8 Ma.[4]

Theropithecus oswaldi is not recorded in the Late Pliocene faunas of Ahl al Oughlam (Morocco, ~2.4 Ma), or at Aïn

[4]However, if the short chronology of Geraads, Raynal and colleagues (in which Aïn Boucherit would not be older than 2 Ma, and thus closer to the Plio-Pleistocene boundary, and Aïn Hanech not older than 1.2 Ma) is considered, then *E. numidicus* would be present in northwest Africa at the critical time.

Boucherit, where the species *Theropithecus atlanticus* was found (Alemseged and Geraads 1998; Raynal et al. 2001). According to Raynal and colleagues, *T. oswaldi* is first recorded in the Maghrebian stratigraphy at Tighenif, currently estimated as very late Early Pleistocene or earliest Middle Pleistocene in age. Although for some time the only European site in which this African primate had been found was Cueva Victoria, Spain (Arribas and Palmqvist 1999), thus suggesting a Maghrebian-Iberian exchange, remains of *T. oswaldi* have now been identified at Pirro Nord (Rook et al. 2004) and 'Ubeidiya (Belmaker 2002), as well as its previously known presence at Mirzapur, India (Delson 1993). Altogether, the distribution points to a western Asian point of entry followed by trans-Eurasian dispersals.

Similarly, the record of the two main large carnivores that dispersed into Eurasia in the Early Pleistocene (*Pachycrocuta* and *Megantereon*), as well as *Hippopotamus antiquus*, are clear evidence of an African dispersal at this time, but not particularly across the Mediterranean straits. The oldest record of *Pachycrocuta*, although still debated, is now considered to be in East Africa (Werdelin 1999). Its appearance in Europe pre-dates the beginning of the Pleistocene, having been found at Olivola, Italy, dated to ~2 Ma (Napoleone et al. 2003) and a generally earlier Late Pliocene date (from a time when hominins are absent from North Africa) cannot be ruled out. The identification of the *Megantereon* Eurasian form which replaced *M. cultridens* in the Early Pleistocene as the African *Megantereon whitei* has been strongly argued by Martínez- Navarro and colleagues (Martínez-Navarro and Palmqvist 1995, 1996); this species is thought to have entered Eurasia through the Levantine corridor around 1.8 Ma (Palmqvist and Arribas, 2001).

A possibly stronger case for a dispersal across the Strait of Gibraltar can be made for the Late Pliocene, when an African-Eurasian faunal exchange took place. As part of this exchange, a small number of African species (*Giraffa*, *Struthio* and *Kuabebihyrax*) are recorded at the sites of Bethlehem, Kuabebi in the Caucasus (2.6–2.5 Ma), and possible derivative forms in Europe (see Martínez-Navarro 2010), while *Equus* disperses to Africa, possibly through the Bab el Mandab (Tchernov, 1992b). Most interesting is the presence of Eurasian forms at Ahl al Oughlam (Geraads, 1997; Raynal et al. 2001), namely *Ursus cf. etruscus* and *Nyctereutes abdeslami*. Although they may have dispersed from either the Iberian Peninsula (both are present, see Kurtén and Crusafont 1977) or western Asia (*Nyctereutes* is found at Bethlehem, see Hooijer 1958), the absence of *Ursus* from other African sites is noticeable. Martínez-Navarro further points to the presence of the Eurasian *Capra primaeva* in the Late Pliocene Tunisian site of Aïn Brimba. Together with the presence (survival?) of the African rodent *Pellegrinia* in Sicily amidst an endemic Early Pleistocene fauna, the presence of these Eurasian forms in Late Pliocene Maghrebian sites may indicate a time when the Mediterranean straits were exposed. However, if those trans-Mediterranean exchanges occurred, they preceded the earliest hominin colonization of the Mediterranean basin.

If the Mediterranean straits were not exposed since the Late Pliocene, all potential Pleistocene hominin (and faunal) dispersals out of and into North Africa had to take place via the Sinai Peninsula. In this scenario, northeastern (the eastern Sahara) and northwestern Africa (the Maghreb) acquire extremely different biogeographical roles – one as a corridor, the other as a *cul-de-sac*.

Out of Africa Across the Sahara

Integrating the archeological information above with the lack of evidence for trans-Mediterranean crossings during the Pleistocene, it is a strong hypothesis that northwestern Africa acted as a *cul-de-sac* throughout the period, receiving intermittent faunal (and hominin) dispersals from Central and West Africa when climatic conditions allowed the formation of bodies of water (in the form of wadis, paleolakes, springs and water holes – all of which form the geomorphological context of all Quaternary sites in the region), and more rarely from Eurasia along the southern Mediterranean coast. This would make the Maghreb, or in particular the Atlantic and Mediterranean plains of the Atlas Mountains, a faunal refugium throughout the period. This may also have promoted the recurrent extinction, as well as extended survivorship, of species (including hominins and their cultural traditions) at times when these had already become extinct or become technologically different elsewhere. In a recent simulation study of extinction rates of large European carnivores, O'Regan and colleagues show the strict correlation between the spatial extent of a refugium and the likelihood of extinction, largely driven by inbreeding (O'Regan et al. 2002). Similar models could be built for the Maghreb, and tested again the paleontological record to assess the extent to which the area acted as a refugium during the Pleistocene.

In this context, it is interesting to note that the Maghrebian hominin fossil remains, traditionally attributed to *Homo erectus*, are all of Middle Pleistocene age (Oulad-Thomas Quarry, Tighenif, Sidi Abderrahman, Sale), when *Homo heidelbergensis* is found in sub-Saharan Africa (Rightmire 1996). If these affinities are correctly identified, that would be evidence for the relatively late survivorship and eventual extinction of *Homo erectus* in this area. On the other hand, these remains were studied a long time ago, within a different theoretical, comparative and chronological framework. A new assessment of their affinities would be extremely useful.

The outstanding question is whether northeast Africa was a corridor for dispersals between East Africa and the Levant during the Pleistocene. Geomorphological and paleontological evidence indicates that the Nile linked East African water sources with the Mediterranean in the Middle Pliocene, also supported by the molecular phylogenies of the single haplochromine cichlid fish found out of Africa, the Levantine species *Astatotilapia flaviijosephi* (Werner and Mokady 2004). During the Late Pliocene, the Paleo-Nile cut through Egypt (along a different course), but probably had its sources within Egypt itself (Said 1981), i.e., there was not a water corridor connecting the East African highlands to the Mediterranean at the time. With the onset of major aridity at the Plio-Pleistocene boundary, the traversal of the Nile along the Egyptian desert stopped, and it is not thought to have been formed again until the early Middle Pleistocene ~0.7 Ma (Said 1981; Rzóska 1976). Indeed, paleo-ichtyological studies show that several hydrological systems in Africa lost their pan-African faunas in the Early Pleistocene, reflecting the disappearance of trans-regional waterways due to increased aridity and geomorphological change (Stewart 2001). The formation of the Proto-Nile, and subsequent Pre-Nile, in the Middle Pleistocene is characterized by the renewed connection to sub-Saharan African watersheds.

The lack of Early Pleistocene sites (with the exception of the already discussed poorly contextualized and undated site of Abassieh), together with the evidence for (a) the establishment of the Saharan desert barrier during the Plio-Pleistocene (Tchernov 1988), (b) the lack of a water corridor connecting East Africa to the Mediterranean shores until the Middle Pleistocene, and (c) the strikingly different faunistic composition of the sites of 'Ubeidiya and Maghrebian sites such as Irhoud Ocre and Sis Abdalla (Tchernov 1992b), suggests that northeastern Africa was not a dispersal corridor between East Africa and the Levant during the Early Pleistocene. Significantly, the African component of Early Pleistocene Eurasian faunas has been argued to be of East, rather than North African affinity ("the East African stamp upon the fauna of 'Ubeidiya", see Tchernov 1992b).

If this assessment was correct, the few Early Pleistocene northwest African sites may have been the result of short-term northward expansions of hominins and fauna from Central Africa, using either the Central Saharan (through Chad, Tibesti, Hoggar-Tassili, southern Atlas wadis) or more probably the Western Saharan (Senegal, Mauritania) waterways (a complex series of paleolakes, paleodrainage systems and highland springs), and be thus unrelated to the events taking place in western Asia and eventually to the East, West and North of the Levant.

The available data suggest very different conditions at and after the beginning of the Middle Pleistocene. Not only the restoration of the Nile re-established a waterway between East Africa and the Levantine corridor, but the increased amplitude and frequency of climatic change initiated at the time (when not only the known ~100 and ~41 ka orbital precession-scale mechanisms were operating, but a marked increased in sub-Milankovtich variability on a millennial scale, probably related to monsoon dynamics and the size of northern hemisphere ice sheets; Larrasoana et al. 2003) led to markedly greater levels of precipitation for short periods of time. Two examples illustrate the unique temporary paleoenvironmental conditions in the Sahara during the Middle and Late Pleistocene. In the eastern Sahara, sedimentary evidence shows that the Middle Pleistocene sites in the Darb el Arba'in desert were associated with a period of deposition at least two orders of magnitude wetter than the present (Wendorf et al. 1993), while the combined extents of Lake Megachad and Lake Megafezzan, joined by extensive wadis, not only created a continuous waterway between the Mediterranean and Central Africa, but submerged 6% of the Sahara (Drake et al., in press). The presence of permanent (at the scale of hominin generations) bodies of water in the Sahara during certain periods would have allowed the colonization of North Africa by sub-Saharan hominins. These Middle Pleistocene expansion and dispersal movements across the Sahara were not trans-Saharan faunal exchanges, as the few Eurasian elements that appear in the southern Mediterranean faunas (such as the presence of *Ursus* cf. *arctos* and possibly *Mammuthus meridionalis* at Tighenif, see Jaeger 1975; Geraads 1982) do not seem to have dispersed southwards (as also shown by the recent work on the molecular phylogenies of North African elephant shrews by Douady et al. 2003). Robert Foley and myself have argued that this 'Out of Africa' directionality is consistent with several dispersals of African Middle Pleistocene Acheulean (and later MSA) stone-tool making hominin populations into North Africa during wet episodes (Foley and Lahr 1997; Lahr and Foley 1998); although not all of these dispersals would have been of great extent, some would reach the Levant and lead to the recurrent immigration of Middle and Late Pleistocene hominins (and humans) into Eurasia.

'Out of Africa I': Different Routes and Directions

Given that there is undisputed evidence of hominins in Eurasia at and after 1.8 Ma, if Plio-Pleistocene hominins did not disperse out of Africa across northern Africa, they would have to have done so through the Bab el Mandab strait, across from the Horn of Africa to the Arabian Peninsula. The Bab el Mandab strait and the Red Sea were flooded in the Early Pliocene, and appear to have remained so until the present (Tchernov 1992b). However, the local geology is extremely complex due to the extent of tectonic activity, and it has been suggested that a land bridge was exposed at certain times (Haq et al. 1987) and that faunal exchanges took place (Tchernov 1992b; Turner 1999).

Several Early Paleolithic sites have been identified in the Arabian Peninsula along the margins of ancient river drainage systems and lakes (see Petraglia 2003 for a comprehensive review). Mode 1 sites are few, but significantly more numerous than those of the vastly greater area of North Africa, including a complex of six Oldowan sites in the proximity of the Bab el Mandab (Whalen and Schatte 1997). Acheulean artifacts and/or assemblages have been found in sites along both East and West southern margins of the Red Sea (Faure and Roubet 1968; Zarin et al. 1981; Whalen et al. 1988; Whalen and Pease 1992; Walter et al. 2000), but have so far been only identified along the eastern (Arabian) margin further North (for example, at the site of Saffaqah; see Whalen et al. 1984). Together with the East African character of the African fauna dispersing into Eurasia at ~1.8 Ma (Tchernov 1992b), the occupation of Arabia by hominins manufacturing Early Paleolithic artefacts makes a dispersal across the Bab el Mandab during the Plio-Pleistocene likely. In this scenario, the eastern coast of the Red Sea could have acted as a low carrying capacity corridor (i.e., promoting a rapid linear movement) leading to the Levant, while the coast of Yemen and Oman, or alternatively across the middle of Arabia (North of the Rub' al-Khali desert) could have taken other hominin populations to the Strait of Hormuz and thus the coastal corridor (also of low carrying capacity) towards South Asia.

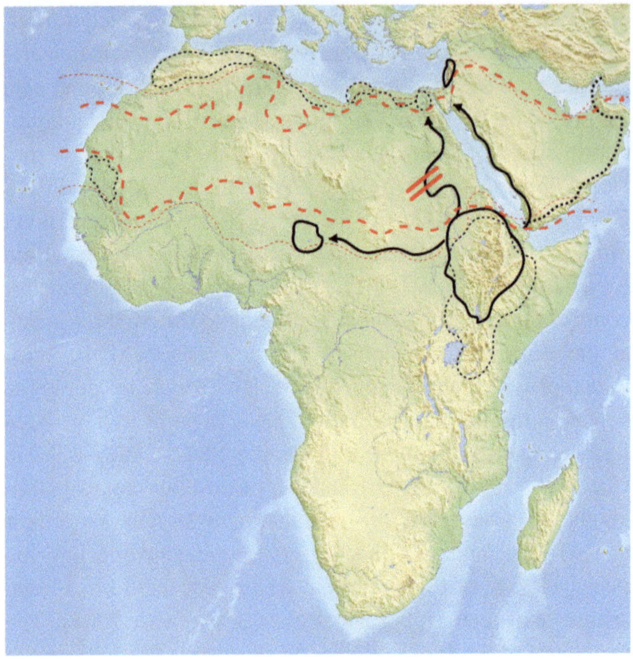

Fig. 3.6 Out of East Africa model, stage I. See Fig. 3.1a–c for climatic patterns

Integrating Model and Data

The above discussion can be used to assess and refine the corridor/refugium model presented in Fig. 3.1.

1. In this model, five immediate potential corridors from East Africa were identified (ignoring a southward dimension) – (a) towards Central Africa along a Sahelian highland edge; (b) towards the Mediterranean along the Nile; (c) towards the Sinai Peninsula along the western coast of the Red Sea (or through the Red Sea Hills); and (d) towards the Sinai Peninsula along the eastern coast of the Red Sea across the Bab el Mandab; and (e) towards the Strait of Hormuz, along the southern coast of Arabia. The latter case will not be considered further here, as the relevant data were not discussed. The first four corridors would have taken hominins into two high carrying capacity environments – the Lake Chad basin, and the Delta-Sinai-Levant area (Fig. 3.6).

   The available record provides only circumstantial evidence that the first corridor (towards the Lake Chad Basin) was used. This evidence consists of the earlier, mid-Pliocene use of this dispersal route by an australopithecine; and the presence (but undated) of both Oldowan and Acheulean sites in the area to the North and South of the Lake Chad Basin (Beli, Jos Plateau, Yayo). Of the three northward corridors, the Nile did not exist as such until the Middle Pleistocene, after which time the evidence suggests that it was indeed used; the western coast of the Red Sea is insufficiently known to be assessed (although the possible Mode 1/Oldowan site at Thébes, if confirmed, could change this picture); finally, the use of the eastern coast of the Red Sea and immediate highlands is supported by the substantial number of Early Paleolithic sites and the possibly greater affinities of Levantine faunas towards East rather than North African communities, as well as the presence of Levantine forms in East but not North Africa. However, the crossing of the Bab el Mandab remains speculative at this point, and if found to be unrealistic it would strongly point towards the western coast of the Red Sea as the main Early Pleistocene dispersal route from East Africa to Eurasia.
2. If hominin populations reached and survived in the area of the Lake Chad Basin, they would again disperse when precipitation increased and or if conditions at the refugium deteriorated. Dispersals from the Lake Chad Basin could take any of three corridors – (a) the Sahelian highland edge corridor towards East Africa; (b) a corridor towards the Central Sahara following the megalake paleodrainage system into the Central Sahara highlands and from there along wadis towards the southern foothills of the Atlas Mountains and the Mediterranean; and (c) a corridor along the northern edge of the Niger River. Only the second and third of these corridors are discussed (Fig. 3.7).

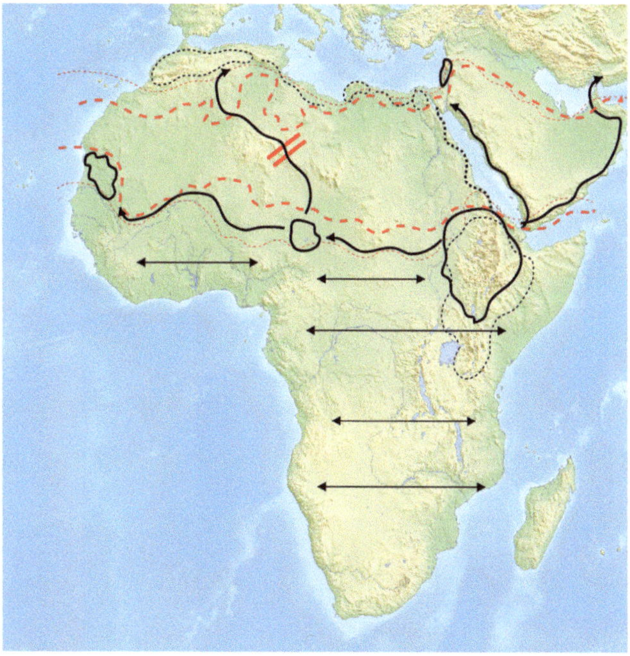

Fig. 3.7 Out of East Africa model, stage II. See Fig. 3.1a–c for climatic patterns

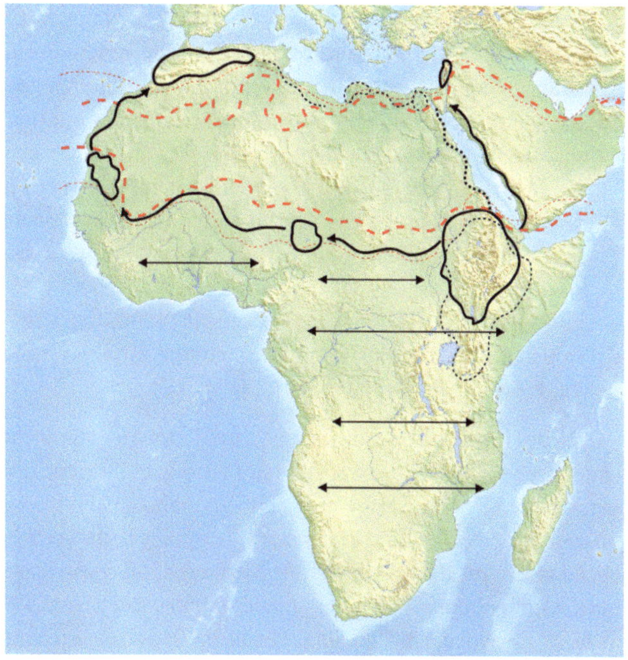

Fig. 3.8 Out of East Africa model, Stage III. See Fig. 3.1a–c for climatic patterns

The record of the Central Sahara supports only a very ephemeral or rare use of this route during the Early Pleistocene. Although a few Mode 1 and early Mode 2 sites have been identified in Fazzan in 2007–2008, these are rare in comparison with more recent localities, and remain so far undated. Middle Pleistocene sites are more frequent, and the few that have paleontological remains, such as Tihodaïne, these are associated with a riverine/lacustrine fauna of African affinities. It is likely that only during the Middle Pleistocene the magnitude of climatic change from glacial to interglacial conditions allowed for the formation of the Saharan megalakes and associated paleorivers, and thus a more substantial hominin occupation of the area. The westward corridor is supported by circumstantial evidence in terms of the number of Early Paleolithic sites in the Senegal River drainage system and Mauritanian highlands. This corridor would have taken hominins to a high carrying capacity environment, and was a likely dispersal route of early hominins.

3. If hominin populations survived in the western Saharan refugium, the model predicts that under conditions similar as those described above, hominins would disperse further. Dispersals from the Western Saharan refugium could take place along two corridors – (a) towards the Lake Chad Basin from where they had originated; and (b) towards the Maghrebian Atlantic plain along waterways paralleling the Atlantic coast. The second of these is discussed below (Fig. 3.8).

   The presence of two Early Pleistocene sites in the Maghreb (Aïn Hanech, Thomas Quarry 1, Unit L) dated to ~1.8 and ~1.0 Ma suggests that this route was used, albeit very rarely. It is likely that this route, like all trans-latitudinal corridors, would have been open for very short periods of time, since renewed aridification would have affected the mid-Saharan latitudes first, thus cutting-off the parental and daughter populations. This corridor would have taken hominins towards a high carrying capacity refugium, and yet the Early Pleistocene occupation of which does not appear to have become permanent. It is possible that the distance of this area from other, more populous hominin localities would have restricted the gene pool of the dispersing groups to the point of compromising their survival. If this hypothesis is correct, it has implications for the interpretation of the dispersals into Eurasia. The Early Pleistocene colonization of Eurasia, although probably not permanent in the very long-term, shows that hominin populations did survive for a significant period of time in certain areas. This implies that they would have had to either sustain demographically a period of continuous contact with their parental population until the daughter group reached a sufficient size, or have dispersed as one of many groups following the same route and destination. In the case of the first occupations of northwestern Africa, neither condition seems to have been fulfilled.

4. Finally, hominin populations in the Maghreb could have dispersed yet further, perhaps in relation to low population density resource stress. In this case, it is unlikely that hominins could have used again the western Saharan corridor. Indeed, the endemism of North African faunas, the absence

in sub-Saharan Africa of the few Eurasian forms that reach the southern Mediterranean coast, the deep phylogenetic history of some North African species studied, and the fact that the onset of the effects of climatic amelioration (and thus faunal and hominin demographic growth) occurs earlier in the tropics, support the view that most dispersals between sub-Saharan and North Africa had a northward direction. Therefore, if Maghrebian hominins were to disperse further, they could have used three corridors – (a) across the Strait of Gibraltar into Iberia; (b) across the Sicily-Tunisian Strait; and (c) along the southern Mediterranean coast towards northeastern Africa (Fig. 3.9).

As discussed before, dispersals across the Mediterranean straits are very unlikely, except perhaps for a Late Pliocene dispersal between Northwest Africa and Iberia, although if this took place it pre-dated the first evidence for hominins in the area. A corridor towards the East, along the Mediterranean coast, could have taken hominins into other smaller refugia, and potentially as far as the Nile Delta and Sinai Peninsula. This has to remain an open possibility at present, since this corridor was clearly used in an opposite direction by the small number of Holarctic species that are found in Maghrebian sites. Nevertheless, it should be stressed that the Libyan desert (between Cyrenaica and the Nile oases) is today the most arid part of the Sahara, and it would have constrained hominins to using an extremely narrow band of coastal plains as a dispersal corridor.

The summary of the discussion presented above is in Fig. 3.10. The main aspect of this particular model is the geographic direction and consequence of the Early versus Middle Pleistocene hominin dispersals out of East Africa. The former, would have, on the one hand, taken more than one population of hominins to North Africa, but without resulting in the permanent colonization of the region and leading to the eventual extinction of these lineages. The short duration of these early North African occupations probably implies that the dispersing groups did not adapt successfully to their new environments, and would thus appear morphologically (and culturally?) relatively undifferentiated. On the other hand, Early Pleistocene dispersals would have also taken hominin groups into Eurasia for the first time, resulting in the establishment of a number of populations distributed across a vast area, from Southeast Asia to eventually the Iberian Peninsula. These various Eurasian early hominins would have also eventually become extinct, but at least in some areas, after the long-term adaptation to new conditions. Therefore, this Eurasian dispersal would probably be best described as an Early Pleistocene adaptive radiation, leading to the evolution of a number of differentiated hominin lineages. Especially in Europe and Southeast Asia, where significant subsequent dispersals probably did not take place until the Middle and Late Pleistocene respectively, local populations would show the greatest degree of differentiation among these early groups (such as *H. antecessor*, Javanese *H. erectus*, and *H. floresiensis*).

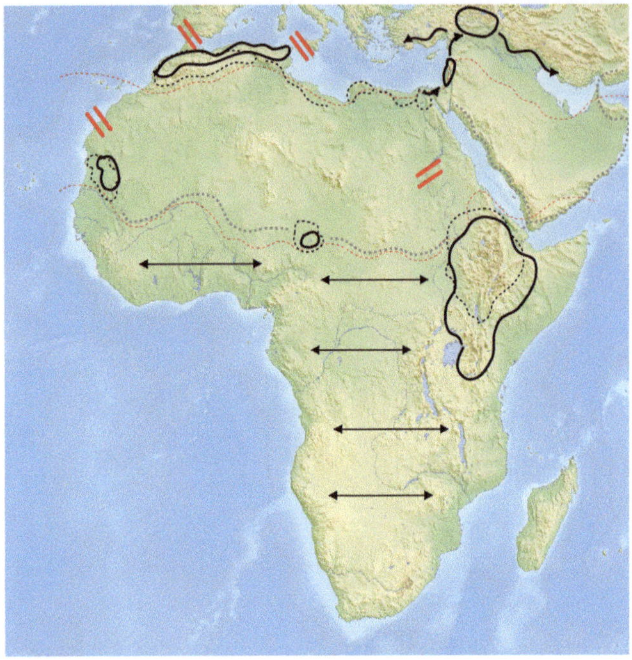

Fig. 3.9 Out of East Africa model, Stage IV. See Fig. 1a–c for climatic patterns

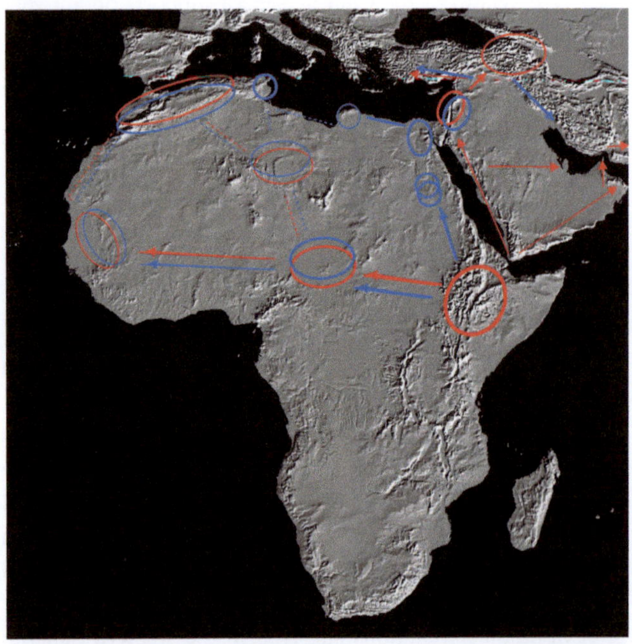

Fig. 3.10 Model for Early and Middle Pleistocene dispersals of hominins out of East Africa. Red line indicates Early Pleistocene hominin dispersals out of East Africa associated with Mode 1 and Mode 2 Industries. Blue line indicates Middle Pleistocene hominin dispersals out of East Africa associated with derived Mode 2 industries

By contrast, Middle Pleistocene hominin movements were more frequent and of lesser longitudinal and greater latitudinal geographical extent. Their frequency was probably driven by the cyclicity of climatic change after 0.8 Ma. The limits of these early Middle Pleistocene dispersals are probably responsible for the archeological distribution of the Acheulean in Eurasia or the Movius Line.

Causes, Conditions, Constraints and Consequences

The aim of this paper was to develop a model that addressed the problem of the early dispersals of hominins out of Africa at a scale closer to that at which the process took place, i.e., the scale of populations and ecological units, rather than species and continental areas. In order to do so, the areas surrounding East Africa to the North and East were described in terms of corridors, barriers and refugia, which when integrated with the record for hominin occupation of North Africa and African-Eurasian faunal exchanges, led to the proposal of a new hypothesis regarding the spatial and temporal pattern of these early hominin movements. In order to finish this discussion, I would like to use Rob Foley's "4 Cs" (Foley 1990, 1995) to explore the evolutionary context of the population movements under discussion (Fig. 3.11).

Conditions

The first dispersals into North Africa and Eurasia took place against a background of major climatic change. The model places much emphasis on the role of climatic change in altering the competitive environments of hominin (and other) populations which would have led to demographic processes of microevolutionary consequence. Climatic change clearly shaped the conditions in which these dispersals could take place, but was not the driving force. The most significant climatic shift of the Plio-Pleistocene period occurred 2.5 Ma, but it was the period between 2.5 and 1.8 Ma that set the ecological competitive conditions behind hominin transcontinental dispersals. These dispersals involved hominins that could successfully exploit grasslands, and indeed is in such open environments that many of the first hominin sites in Eurasia are found (Dennell 2004). The reconstruction of East African Pliocene hominin sites suggests woodland environments (Cerling 1992; Kappelman et al. 1997; Reed 1997), and that the evolution of extensive grasslands was one of the consequences of the onset of global glaciation 2.5 Ma. However, the paleoenvironmental reconstructions of East African hominin sites would suggest that hominins (or a population among them) did not evolve a successful grassland adaptation until 1.8 Ma (as argued by Dennell), and thus lagging behind the actual climatic driven environmental change by a period of time (Foley 1993, 1994). The evolution of the Late Pliocene East African grasslands would have changed the competitive circumstances of hominins; increased interspecific competition leads to ecological differentiation as a means of conflict avoidance (MacArthur and Levins 1967; MacArthur 1968, 1970; Pianka 1978). This process eventually led to the evolution of a dietary strategy among one group of hominins more strongly based on animal food (Foley 1987, 2001). Therefore, one of the ancestral adaptations of the hominins that dispersed out of Africa was a carnivory-dependent strategy (Stiner 2002). Stone tools were probably part of this adaptation, allowing certain hominins to break into the carnivoran guild (Brantingham 1998). As argued by Rob Foley, Mary Stiner and others, it was probably this carnivoran, stone-tool making adaptation that established the right conditions for dispersals out of East Africa.

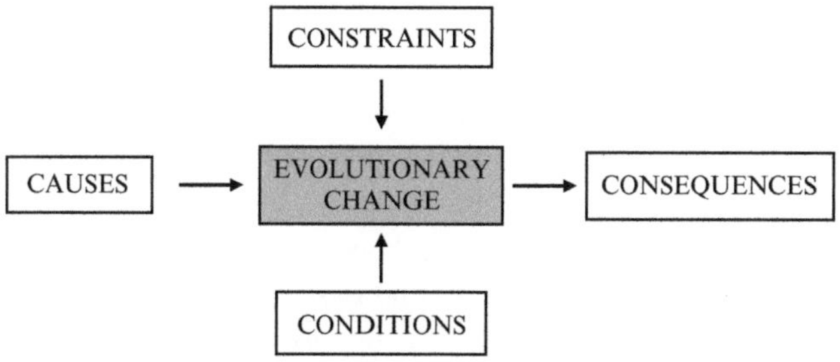

Fig. 3.11 Rob Foley's '4 Cs' evolutionary ecology model

The evolution of carnivory had major biological consequences with feedback effects and shaped a striking novel adaptive package (Fig. 3.12). This involved changes in life-history, body size and proportions, brain and gut size, and most likely cognition. Although these other traits, particularly when found in combination, came to define some of the most successful species in the planet, I would argue that they were not a necessary condition for the early dispersal of hominins out of Africa. Indeed, the fossil remains from Dmanasi strongly suggest that hominins could disperse out of Africa without having the entire adaptive complex.

Causes

What caused certain populations of hominins to disperse has been asked many times. The causes of evolutionary change are established though proximate mechanisms. As argued before, dispersals into North Africa and Eurasia could only take place during a period of increased precipitation, so that the Saharan barrier could be crossed. Therefore, the immediate mechanism leading hominins to disperse towards the North could be the reduction and fragmentation of their range as equatorial forests expanded (see Adams 2005 for a discussion on this semi-controversial topic), leading to increased inter- and intraspecific competition in East Africa among non-woodland adapted groups, and the consequent ecological differentiation through dispersal into neighbouring

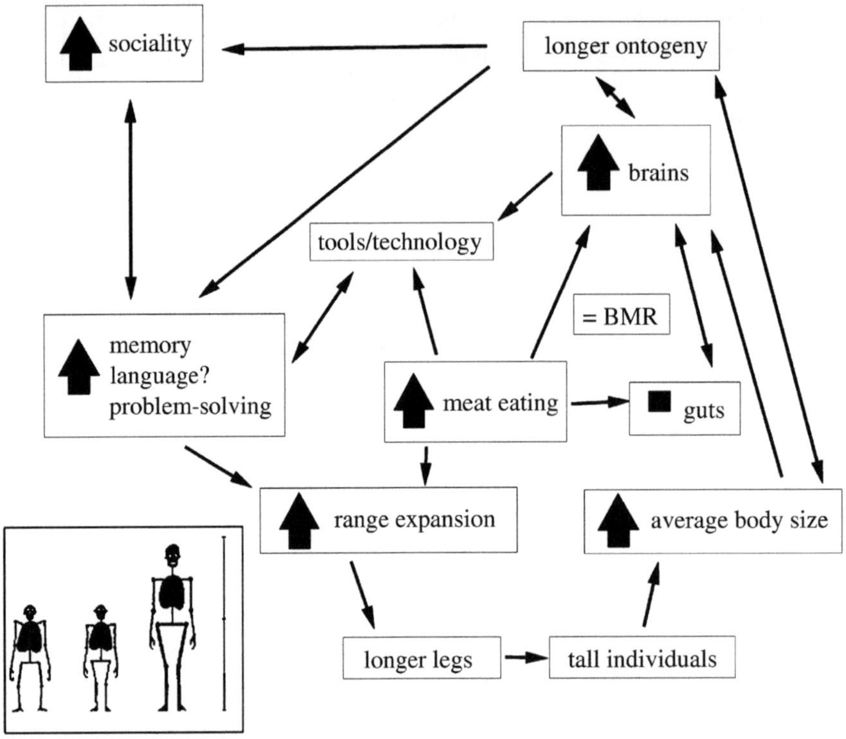

Fig. 3.12 *Homo* adaptive package. A consequence of carnivory?

zones. Similar processes did occur before, even the use of one of the immediate corridors out of East Africa (*A. bahrelghazali*). The difference in the case of Plio-Pleistocene *Homo* was probably on the success with which the ecological shift met, leading to demographic growth that sustained an extensive dispersal event.

Constraints

The pattern of dispersals suggests a strong dependence on water on the part of the hominins (maybe as one of the consequences of carnivory dependence?). This water dependence means that, in areas without significant lake basins, the primary areas of colonization were often away from plains and in highlands. This preference for higher grounds in which springs and streams could be found would also fit other important selective pressures, pressure for availability of raw material for stone tools, shelter and possibly defense.

Consequences

Hominin dispersals had many consequences, from evolutionary to ecological. However, the main immediate consequence of the Pleistocene dispersals out of Africa was the structuring of the species into a number of populations, many of which developed independent evolutionary trajectories. Population structure increases significantly the genetic variance of a species, and thus the rate of evolutionary change. In the case of Pleistocene *Homo*, that change involved both biological and cultural dimensions.

This interpretation of the evolutionary ecology of the early dispersals of hominins out of East Africa is summarized in Fig. 3.13.

Concluding Thoughts

This paper tries to make two main points. The first of these is that the evolutionary geography of populations is the correct scale at which to analyze evolutionary problems of dispersals and contractions. As a theoretical framework, it brings together ecology, geography, phylogeny and biology, making it a powerful theoretical and analytical tool. The second key point is that the answers to the problem of how hominins were capable of colonizing Eurasia need to be sought in African ecology, not Eurasian. The evolutionary geography model developed in this paper attempts to trace the pattern of early hominin dispersals into North Africa; its comparison to the first occupation of Eurasia shows that the pattern of hominin colonization of areas beyond East Africa differed markedly from one area to another. In particular, it shows that, contrary to what happened in Eurasia, hominins were possibly not able to colonize North Africa successfully until the

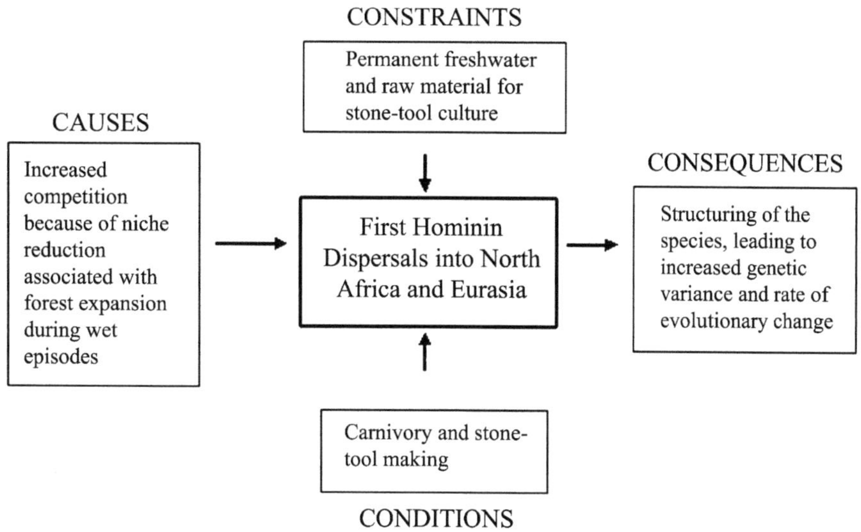

Fig. 3.13 Rob Foley's '4 Cs' evolutionary ecology model applied to the early dispersals of hominins out of East Africa

Middle Pleistocene. It is probably the case that until the Middle Pleistocene there were major constraints on the ability of hominins to establish populations in Africa beyond the East African Rift Valley and the South African Transvaal, possibly as a result of the joint effects of forests and deserts. Ironically, Eurasian migrant hominin populations were probably demographically more successful and stable that their African ancestors, but then again, it was that African instability that set the conditions for the evolution of novelties which dispersed out of Africa again and again.

References

Adams J. M. (1997). Global land environments since the last interglacial. Oak Ridge National Laboratory, TN, USA. http://www.esd.ornl.gov/ern/qen/nerc.html

Agnesi, V., Bonfiglio, L., Ciurcina, C., Conoscenti, C., Di Maggio, C., Di Patti, C., et al. (2004). *Quaternary eustatic fluctuations and biochronology of vertebrate-bearing deposits correlated with marine terraces in Sicily*. 32nd International geological congress, Florence, Italy, 20–28 August, 2004.

Alemseged, Z., & Geraads, D. (1998). *Theropithecus atlanticus* (Cercopithecidae, Mammalia) from the late Pliocene of Ahl al Oughlam, Casablanca, Morocco. *Journal of Human Evolution, 34*, 609–621.

Alexander, D. (1988). A review of the physical geography of Malta and its significance for tectonic geomorphology. *Quaternary Science Review, 7*, 41–43.

Alimen, H. (1975). Les "Isthmes" hispano-marocain et Sicilo-Tunisien aux temps Acheuléens. *L'Anthropologie, 79*, 399–436.

Arribas, A., & Palmqvist, P. (1999). On the ecological connection between sabre-tooths and hominids: Faunal dispersal events in the Lower Pleistocene and a review of the evidence for the first human arrival in Europe. *Journal of Archeological Science, 26*, 571–585.

Aumassip, G. (2004) Préhistoire du Sahara et de ses abords – Tome 1, Au temps des chasseurs: Le Paléolithique. Paris: Maisonneuve & Larose.

Azzaroli, A. (1995). The "Elephant-*Equus*" and the "End-Villafranchian" events in Eurasia. In E. S. Vrba, G. H. Denton, T. C. Partridge, & L. H. Burckle (Eds.), *Paleoclimate and evolution, with emphasis on human origins* (pp. 311–318). New Haven: Yale University Press.

Belmaker, M. (2002). First evidence of the presence of *Theropithecus* sp. in the Southern Levant. *Israel Journal of Zoology, 48*, 165.

Bianchini, G. (1973). Gli "hacheraux" nei giacimenti paleolitici della Sicilia sud-occidentale. Atti della XV Riunione Scientifica dell'Istituto Italiano di Preistoria e Protostoria, 11–25 Ottobre 1972.

Bonfiglio, L., & Berdar, A. (1970). Gli elefanti delle ghiaie pleistoceniche di Messina. *Quaternaria, 21*, 139–177.

Bonfiglio, L., & Piperno, M. (1996). Early faunal and human populations. In R. Leighton (Ed.), *Early societies in Sicily* (pp. 21–30). Accordia: University of London.

Bonfiglio, L., Insacco, G., Marra, A. C., & Masini, F. (1997). Large and small mammals, amphibians and reptiles from a new late Pleistocene fissure filling deposit of the Hyblean Plateau (South-eastern Sicily). *Bolletino della Società Paleontologica Italiana, 36*, 97–122.

Bovier-Lapierre, R. P. (1925). Le paéolithique stratifié des environs du Caire. *L'Anthropologie, 35*, 37–46.

Brantingham, J. P. (1998). Hominid-carnivore coevolution and invasion of the predatory guild. *Journal of Anthropological Archeology, 17*, 327–353.

Brunet, M., Beauvilain, A., Coppens, Y., Heintz, E., Moutaye, A. H. E., & Pilbeam, D. (1995). The First Australopithecine 2,500 kilometers west of the Rift- valley (Chad). *Nature, 378*, 273–275.

Brunner, C. A., & Maniscalco, R. (1998). Late Pliocene and Quaternary paleoceanography of the Canary Island Region Inferred from Planktonic Foraminifer Assembalges of Site 953. In P. P. E. Weaver, H.-U. Schmincke, J. V. Firth, & W. Duffield (Eds.), *Proceedings of the Ocean Drilling Program – Scientific Results* (vol. 157, pp. 73–82). College Station, TX: Ocean Drilling Program.

Cerling, T. E. (1992). Development of grasslands and savannahs in East Africa during the Neogene. *Palaeogeography, Palaeoclimatology, Palaeoecology, 97*, 241–247.

Chmielewski, W. (1968). Early and Middle Paleolithic sites near Arkin, Sudan. In F. Wendorf (Ed.), *The Prehistory of Nubia* (pp. 110–147). Dallas: Fort Burgwin Research Center and Southern Methodist University Press.

Clark, J. D. (1975). A comparison of the Late Acheulian industries of Africa and the Middle East. In K. W. Butzer & G. L. Isaac (Eds.), *After the Australopithecines: Stratigraphy, ecology, and culture change in the Middle Pleistocene* (pp. 605–659). The Hague: Mouton.

Coppens, Y., Gouze, R., Le Floch, R., & Paquet, M. (1972). Découverte d'un gisement de vertébrés fossiles avec industrie acheuléenne près de Zouerate. 6e Congrès Panafricain de Préhistoire, 457–461.

De Menocal, P. B. (1995). Plio-Pleistocene African climate. *Science, 270*, 53–59.

Delson, E. (1993). *Theropithecus* fossils from Africa and India and the taxonomy of the genus. In N. G. Jablonski (Ed.), *Theropithecus: The rise and fall of a primate genus* (pp. 157–189). Cambridge: Cambridge University Press.

Dennell, R. W. (2004). Hominid dispersals and Asian biogeography durig the Lower and Early Middle Pleistocene, ca. 2.0–0.5 Mya. *Asian Perspectives, 43*, 205–226.

Douady, C. J., Catzeflis, F., Raman, J., Springer, M. S., & Stanhope, M. J. (2003). The Sahara as a vicariant agent, and the role of Miocene climatic events, in the diversification of the mammalian order Macroscelidea (elephant shrews). *Proceedings of the National Academy of Sciences of the United States of America, 100*, 8325–8330.

Drake, N. A., El-Hawat, A. S., Turner, P., Armitage, S. J., Salem, M. J., White, K. H. & McLaren, S. (2008) Palaeohydrology of the Fazzan Basin and surrounding regions: the last 7 million years. *Palaeogeography, Paaleoclimatology, Palaeoecology 263*, 131–145.

Faure, H., & Roubet, C. (1968). Découvert d'un biface Acheuléen dans les calcaires marins du golfe Pleistocene d'Afar (Mer Rouge, Ethiopie). *Comptes Rendus de l'Académie des Sciences de Paris, Série II, 267*, 18–21.

Foley, R. A. (1987). *Another unique species: Patterns in human evolutionary ecology*. Harlow: Longman.

Foley, R. A. (1990). The causes of brain enlargement in human evolution. *Behavioural and Brain Sciences, 13*, 354–356.

Foley, R. A. (1993). African terrestrial primates: The comparative biology of *Theropithecus* and the Hominidae. In N. G. Jablonski (Ed.), *Theropithecus: The rise and fall of a primate genus* (pp. 245–270). Cambridge: Cambridge University Press.

Foley, R. A. (1994). Speciation, extinction and climatic change in hominid evolution. *Journal of Human Evolution, 26*, 275–289.

Foley, R. A. (1995). The causes and consequences of human evolution. *Journal of the Royal Anthropological Society, 1*, 67–86.

Foley, R. A. (1999). Evolutionary geography of Pliocene African hominids. In T. Bromage & F. Schrenk (Eds.), *African biogeography, climatic change, and hominid evolution* (pp. 328–348). Oxford: Oxford University Press.

Foley, R. A. (2001). The evolutionary consequences of increased carnivory in hominids. In C. Stanford & H. Bunn (Eds.), *The early human diet: The role of meat* (pp. 305–331). Oxford: Oxford University Press.

Foley, R. A., & Lahr, M. M. (1997). Mode 3 technologies and the evolution of modern humans. *Cambridge Archeological Journal, 7*, 3–36.

Foley, R. A., & Lahr, M. M. (2007). In P. Mellars, C. Stringer, O. Bar-Yosef, & K. Boyle (Eds.), *Rethinking the human revolution*. Cambridge: McDonald Institute.

Freeman, L. G. (1975). Acheulean sites and stratigraphy in Iberia and the Maghreb. In K. W. Butzer & G. L. Isaac (Eds.), *After the Australopithecines: Stratigraphy, ecology, and culture change in the Middle Pleistocene* (pp. 661–743). The Hague: Mouton.

Garcea, E. A. A. (1997). Prehistoric surveys in the Libyan Sahara. *Complutum, 8*, 33–38.

Geraads, D. (1982). Paléogiogéographie de l'Afrique du Nord depuis le Miocène terminal, d'après les grands mammifères. *Geobios Mémoire Spécial, 6*, 473–481.

Geraads, D. (1997). Carnivores du Pliocéne terminal de Ahl al Oughlam (Casablanca, Maroc). *Géobios, 30*, 127–164.

Geraads, D., Raynal, J. P., & Eisenmann, V. (2004). The earliest human occupation of North Africa: A reply to Sahnouni et al. (2002). *Journal of Human Evolution, 46*, 751–761.

Gobert, E. G. (1950). Le gisement paléolithique de Sidi Zin. *Karthago, 1*, 1–51.

Goren-Inbar, N., & Saragusti, I. (1996). An Acheulian biface assemblage from Gesher Benot Ya'aqov, Israel: Indications of African affinities. *Journal of Field Archeology, 23*, 15–30.

Guerrero-Alba, S., & Palmqvist, P. (1997). Morphometric study of the horse from Venta Micena (Orce, Granada) and its comparison with both extant and Plio-Pleistocene equids from Europe and Africa. *Paleontologia i Evolucio, 30–31*, 93–148.

Haq, B. U., Hardenbol, J., & Vail, P. R. (1987). Chronology of fluctuating sea levels since the Triassic. *Science, 235*, 1156–1167.

Haynes, C. V., Maxwell, T. A., El Hawary, A., Nicoll, K. A., & Stokes, S. (1997). An Acheulian site near Bir Kiseiba in the Darb el Arba'in Desert, Egypt. *Geoarcheology, 12*, 819–832.

Hill, C. L. (2001). Geologic contexts of the Acheulian (Middle Pleistocene) in the Eastern Sahara. *Geoarcheology, 16*, 65–94.

Hooijer, D. A. (1958). An Early Pleistocene mammalian fauna from Bethlehem. *Bulletin of the British Museum (Natural History) Geology, 3*, 267.

Jaeger, J.-J. (1975). The mammalian faunas and hominid fossils of the Middle Pleistocene of the Maghreb. In K. W. Butzer & G. L. Isaac (Eds.), *After the Australopithecines: Stratigraphy, ecology, and culture change in the Middle Pleistocene* (pp. 399–418). The Hague: Mouton.

Kappelman, J., Plummer, T., Bishop, L., Duncan, A., & Appleton, S. (1997). Bovids as indicators of Plio-Pleistocene paleoenvironments in East Africa. *Journal of Human Evolution, 32*, 229–256.

Kurtén, B., & Crusafont, M. (1977). Villafranchian Carnivores (Mammalia) from La Puebla de Valverde (Teruel, Spain). *Commentationes Biologicae, 85*, 1–39.

Lahr, M. M. (1996). *The Evolution of Human Diversity*. Cambridge: Cambridge University Press.

Lahr, M. M., & Foley, R. A. (1994). Multiple dispersals and modern human origins. *Evolutionary Anthropology, 3*, 48–60.

Lahr, M. M., & Foley, R. A. (1998). Towards a theory of modern human origins: Geography, demography, and diversity in recent human evolution. *Yearbook of Physical Anthropology, 41*, 137–176.

Lahr, M. M., Foley, R., Armitage, S., Barton, H., Crivellaro, F., Drake, N., Hounslow, M., Maher, L., Mattingly, D., Salem, M., Stock, J. & White, K. (2008) DMP III: Pleistocene and Holocene palaeoenvironments and prehistoric occupation of Fazzan, Libyan Sahara. *Libyan Studies, 39*, 263–294.

Lahr, M. M., Foley, R., Crivellaro, F., Okumura, M., Maher, L., Davies, T., Veldhuis, D., Wilshaw, A. & Mattingly, D. (2009) DMPVI: Preliminary results from 2009 fieldwork on the human prehistory of the Libyan Sahara. *Libyan Studies, 40*, 143–162.

Larrasoana, J. C., Roberts, A. O., Rohling, E. J., Winklhofer, M., & Wehausen, R. (2003). Three million years of monsoon variability over the northern Sahara. *Climate Dynamics, 21*, 689–698.

Le Houerou, H. N. (1992). Outline of the biological history of the Sahara. *Journal of Arid Environments, 22*, 3–30.

Leakey, M. D. (1975). Cultural patterns in the Olduvai sequence. In K. W. Butzer & G. L. Isaac (Eds.), *After the Australopithecines: Stratigraphy, ecology, and culture change in the Middle Pleistocene* (pp. 35–52). The Hague: Mouton.

MacArthur, R. H. (1968). The theory of the niche. In R. C. Lewontin (Ed.), *Population biology and evolution* (pp. 159–176). Syracuse: Syracuse University Press.

MacArthur, R. H. (1970). Species packing and competitive equilibrium for many species. *Theoretical Population Biology, 1*, 1–11.

MacArthur, R. H., & Levins, R. (1967). The limiting similarity, convergence, and divergence of coexisting species. *The American Naturalist, 101*, 377–385.

Martinet, G., & Searight, S. (1994). Le Maghreb préhistorique et la navigation. *Bulletin de la Société d'Etudes et de Recherches Préhistoriques des Eyzies, 43*, 85–111.

Martínez-Navarro, B., & Palmqvist, P. (1995). Presence of the African machairodont *Megantereon whitei* (Broom, 1937) (Felidae, Carnivora, Mammalia) in the Lower Pleistocene site of Venta Micena (Orce, Granada, Spain), with some considerations on the origin, evolution and dispersal of the genus. *Journal of Archeological Science, 22*, 569–582.

Martínez-Navarro, B., & Palmqvist, P. (1996). Presence of the African saber-toothed felid *Megantereon whitei* (Broom, 1937) (Mammalia, Carnivora, Machairodontidae) in Apollonia-1 (Mygdonia basin, Macedonia, Greece). *Journal of Archeological Science, 23*, 869–872.

McBurney, C. B. M. (1960). *The stone age of northern Africa*. London: William Clowes & Sons.

McCauley, J. F., Breed, C., & Schaber, G. G. (1997). *The Sahara paleodrainages: SIR-C/X-SAR Flights*. Final Report to NASA.

Monod, T. (1976). Trois gisements de galets aménagés dans l'Adrar de Mauritanie. *Provence Historique, 99*, 87–97.

Morel, A. (1992). The disappearance of hot, humid climates south of the Sahara at the Pliocene-Quaternary boundary. *Zeitschrift fur Geomorphologie, 36*, 191–205.

Napoleone, G., Albianelli, A., Azzaroli, A., Bertini, A., Magi, M., & Mazzini, M. (2003). Calibration of the Upper Valdarno Basin to the Plio-Pleistocene for correlating the Apennine Continental Sequences. *Il Quaternario, Italian Journal of Quaternary Sciences, 16*, 131–166.

O'Regan, H. J., Turner, A., & Wilkinson, D. M. (2002). European Quaternary refugia: A factor in large carnivore extinction? *Journal of Quaternary Science, 17*, 789–795.

Palmqvist, P., & Arribas, A. (2001). Taphonomic decoding of the paleobiological information locked in a lower Pleistocene assemblage of large mammals. *Paleobiology, 27*, 512–530.

Petraglia, M. D. (2003). The Lower Paleolithic of the Arabian Peninsula: Occupations, adaptations, and dispersals. *Journal of World Prehistory, 17*, 141–179.

Pianka, E. R. (1978). *Evolutionary ecology*. New York: Harper & Row.

Potts, R. (1989). Olorgesailie: New excavations and findings in Early and Middle Pleistocene contexts, southern Kenya Rift Valley. *Journal of Human Evolution, 18*, 477–484.

Raynal, J. P., Sbihi Alaoui, F. Z., Geraads, D., Magoga, L., & Mohi, A. (2001). The earliest occupation of North-Africa: The Moroccan perspective. *Quaternary International, 75*, 65–75.

Raynal, J. P., Sbihi Alaoui, F. Z., Magoga, L., Mohib, A., & Zouak, M. (2003). The Lower Palaeolithic sequence of Atlantic Morocco revisited after recent excavations at Casablanca. *BAM, 20*, 1–18.

Rea, D. K. (1994). The paleoclimatic record provided by eolian deposition in the deep sea: The geologic history of wind. *Reviews of Geophysics, 32*, 159–196.

Reed, K. E. (1997). Early hominid evolution and ecological change through the African Plio-Pleistocene. *Journal of Human Evolution, 32*, 289–322.

Rhodes, E., Raynal, J. P., Geraads, D., & Sbihi Alaoui, F. Z. (1994). Premières dates RPE pour l'Acheuléen du Maroc atlantique (Grotte des Rhinocéros, Casablanca). *Comptes Rendus de l'Académie des Sciences de Paris, Série II, 319*, 1109–1115.

Rightmire, G. P. (1996). The human cranium from Bodo, Ethiopia: Evidence for speciation in the Middle Pleistocene. *Journal of Human Evolution, 31*, 21–39.

Rook, L., Martìnez-navarro, B., & Howell, F. C. (2004). Occurrence of *Theropithecus* sp. in the Late Villafranchian of Southern Italy and implication for Early Pleistocene "out of Africa" dispersals. *Journal of Human Evolution, 47*, 267–277.

Roth, V. L. (1990). Insular dwarf elephants: A case study in body mass estimation and ecological inference. In J. Damuth & B. J. MacFadden (Eds.), *Body size in mammalian paleobiology* (pp. 151–179). Cambridge: Cambridge University Press.

Rzóska, J. (Ed.). (1976). The Nile, biology of an ancient river. *Monographiae Biologicae* (vol. 29, pp. 2–7). The Hague: Junk.

Sahnouni, M., de Heinzelin, J., Brown, F., & Saoudi, Y. (1996). Rècentes recherches dans le gisement oldwayen d'Ain Hanech, Algérie. *Comptes Rendus de l'Académie des Sciences de Paris, Série II A, 323*, 639–644.

Sahnouni, M., Hadjouis, D., van der Made, J., Derradji, A., Canals, A., Medig, M., et al. (2002). Further research at the Oldowan site of Aïn Hanech, North-eastern Algeria. *Journal of Human Evolution, 43*, 925–937.

Said, R. (1981). *The geological evolution of the river Nile*. New York: Springer.

Sarnthein, M., Thiede, J., Pflaumann, U., Erlenkeuser, H., Futterer, D., Koopman, B., et al. (1982). Atmospheric and oceanic circulation patterns off Northwest Africa during the past 25 million years. In U. von Rad, K. Hinz, M. Sarntheim, & E. Seibold (Eds.), *Geology of the Northwest African continental margin* (pp. 545–604). Berlin: Springer.

Shackleton, N. J., Berger, A., & Peltier, W. A. (1990). An alternative astronomical calibration of the lower Pleistocene timescale based on ODP Site 677. *Transactions of the Royal Society of Edinburgh, Earth Sciences, 81*, 251–261.

Simpson, G. G. (1961). *Principles of animal taxonomy*. New York: Columbia University Press.

Smith, J. R., Giegengack, R., Schwarcz, H. P., McDonald, M. M. A., Kleindienst, M. R., Hawkins, A. L., et al. (2004). A reconstruction of Quaternary Pluvial environments and human occupations using stratigraphy and geochronology of fossil-spring tufas, Kharga Oasis, Egypt. *Geoarcheology, 19*, 407–439.

Stein, R. (1985). Late Neogene changes of paleoclimate and paleoproductivity off Northwest Africa (DSDP Site 397). *Palaeogeography, Palaeoclimatology, Palaeoecology, 49*, 47–59.

Stewart, K. M. (2001). The freshwater fish of Neogene Africa (Miocene–Pleistocene): Systematics and biogeography. *Fish and Fisheries, 2*, 177–230.

Stiner, M. C. (2002). Carnivory, coevolution, and the geographic spread of the genus *Homo*. *Journal of Archeological Research, 10*, 1–63.

Strauss, L. G. (2001). Africa and Iberia in the Pleistocene. *Quaternary International, 75*, 91–102.

Tchernov, E. (1988). The paleobiogeographical history of the Southern Levant. In Y. Yom-Tov & E. Tchernov (Eds.), *The zoogeography of Israel: The distribution and abundance at a zoogeographical crossroad* (pp. 159–250). Dordrecht, The Netherlands: Junk.

Tchernov, E. (1992a). Dispersal: A suggestion for a common usage of this term. *Courier Forschungsinstitut Senckenberg, 153*, 21–25.

Tchernov, E. (1992b). Eurasian-African biotic exchanges through the Levantine corridor during the Neogene and Quaternary. *Courier Forschungsinstitut Senckenberg, 153*, 103–123.

Thaler, L. (1972). Les rongeurs (Rodentia et Lagomorpha) du Monte Pellegrino et la question des anciens isthmes de la Sicile. *Comptes Rendus de l'Académie des Sciences de Paris, Série D, 274*, 188–190.

Tiedemann, R., Sarntheim, M., & Stein, R. (1989). Climatic changes in the western Sahara: Aeolo-marine sediment record of the last 8 million years (sites 657–661). In W. Ruddiman & M. Sarntheim (Eds.), *Proceedings of the ocean drilling program, scientific results* (pp. 241–277). College Station, TX: ODP.

Turner, A. (1999). Assessing earliest human settlement of Eurasia: Late Pliocene dispersions from Africa. *Antiquity, 73*, 63–570.

Vaufrey, R. (1955). *Préhistoire de l'Afrique. Tome I. Maghreb*. Tunis: Publications de l'Institut des Hautes Études de Tunis.

Veermersch, P. M. (2001). 'Out of Africa' from an Egyptian point of view. *Quaternary International, 75*, 103–112.

Vernet, R., & Naffé, B. o. M. (2003). Dictionnaire Archéologique de la Mauritanie. Université de Nouakchott, Nouakchott.

Verosub, K. L., Goren-Inbar, N., Feibel, C., & Saragusti, I. (1998). Location of the Matuyama/Brunhes boundary in the Gesher Benot Ya'aqov archeological site, Israel. *Journal of Human Evolution, 34*, A22.

Villa, P. (2001). Early Italy and the colonization of Western Europe. *Quaternary International, 75*, 113–130.
Vrba, E. (1995). The fossil record of African antelopes (Mammalia, Bovidae) in relation to human evolution and paleoclimate. In E. Vrba, G. M. Denton, T. C. Partridge, & L. M. Burkle (Eds.), *Paleoclimate and evolution, with emphasis on human origins* (pp. 385–424). New Haven: Yale University Press.
Walter, R. C., Buffler, R. T., Bruggemann, J. H., Guillaume, M. M. M., Berhe, S. M., Negassi, B., Libsekal, Y., Cheng, H., Edwards, R. L., von Cosel, R., Néraudeau, D. & Gagnon, M. (2000) Early human occupation of the Red sea coast of Eritrea during the last interglacial. *Nature, 405*, 65–69.
Wendorf, F., Schild, R., & Close, A. E. (1993). *Egypt during the last interglacial: The Middle Paleolithic of Bir Tarfawi and Bir Sahara East*. New York: Plenum.
Wendorf, F., Schild, R., Close, A. E., Schwarcz, H. P., Miller, G. H., Grun, R., et al. (1994). A chronology for the Middle and Late Pleistocene wet episodes in the Eastern Sahara. In O. Bar-Yosef & R. S. Kra (Eds.), *Late Quaternary chronology and paleoclimates of the Eastern Mediterranean* (pp. 147–168). Tucson: University of Arizona Press.
Werdelin, L. (1999). *Pachycrocuta* (hyaenids) from the Pliocene of east Africa. *Paläontologische Zeitschrift, 73*, 157–165.
Werner, N. Y., & Mokady, O. (2004). Late Pliocene aquatic dispersal out of Africa. *Biological Journal of the Linnean Society, 82*, 103–109.
Whalen, N. M., & Pease, D. W. (1992). Archeological survey in Southwest Yemen, 1990. *Paléorient, 17*, 129–133.
Whalen, N. M., & Schatte, K. E. (1997). Pleistocene sites in southern Yemen. *Arabian Archeology and Epigraphy, 8*, 1–10.
Whalen, N. M., Siraj-Ali, J., & Davis, W. (1984). 1- Excavation of Acheulean sites near Saffaqah, Saudi Arabia, 1403 AH 1983. *Atlal, 8*, 9–24.
Whalen, N. M., Siraj-Ali, J., Sindi, H. O., Pease, D. W., & Badein, M. A. (1988). A complex of sites in the Jeddah-Wadi Fatimah area. *Atlal, 11*, 77–85.
Zarin, J., al-Jawad Murad, A., & Al-Yish, K. S. (1981). The comprehensive archeological survey program. The second preliminary report on the Southwestern Province. *Atlal, 5*, 9–42.

Chapter 4
Stone Age Visiting Cards Revisited: A Strategic Perspective on the Lithic Technology of Early Hominin Dispersal

John J. Shea

Abstract This paper examines the stone tool technology of dispersing Plio-Pleistocene hominins. The traditional division of Early Paleolithic assemblages into Oldowan, Developed Oldowan, Early Acheulean, and related industries obscures a more fundamental axis of strategic variation between pebble-core and large cutting tool technology. In Eurasia, as in Africa before it, first appearances of fossils of the genus *Homo* occur together with "Oldowan" pebble-core technology. Acheulean assemblages featuring large symmetrical cores/cutting tools, if they appear at all, do so after a considerable period of time. The patterning of the Early Paleolithic industrial variability may reflect a strategic shift between an initial "frontier" phase and a subsequent "settling in" phase of hominin dispersal.

Keywords Stone tools • Strategic modeling • Hominin dispersal • Oldowan • Acheulean • Early Paleolithic

Introduction

Glynn Isaac once likened Early Paleolithic stone tools to "Stone Age visiting cards" (Isaac 1981). In earlier times, ladies and gentlemen would leave cards inscribed with their names when they visited someone who was not at home. Upon returning, the homeowner could easily reconstruct who had visited them, and the purpose of their visit, from the cards that had accumulated in their absence. In this one turn of phrase, Isaac captured both the promise and the problem of using stone tools to research hominin evolution. The promise is that stone tools, the most durable and ubiquitous residues of hominin behavior, provide information about the activities and identities of their authors. The problem is that archeologists have a longstanding habit of thinking about Early Paleolithic stone tools as evidence first and foremost about the identities of their makers and only to a lesser extent about how tool-using activities structure the lithic record. Evaluating the role that variation in tool-using activities may have played in Early Paleolithic industrial variability requires us to think about stone tool use strategically, less in terms of the tools' mechanical function (cutting things) and more in terms of strategic variables, such as time, energy, and risk (Tooby and DeVore 1987; Nelson 1988). Such "strategic modeling" has transformed debate about Plio-Pleistocene zooarcheological evidence from a simple dichotomy of hunting vs. scavenging to a far more nuanced appreciation of hominin subsistence variability (Domínguez-Rodrigo and Pickering 2003; Pobiner 2007). A similar strategic perspective on Early Paleolithic stone tool technology may have a positive effect, focusing debate on early hominin behavioral variability away from simple questions about stone tool production and use and towards more anthropologically-interesting ones about hominin land use and technological variability (Potts 1991, 1998; Potts and Teague 2010).

Efforts to understand the forces that shaped Early Paleolithic stone tool variation in African Plio-Pleistocene contexts are limited by small numbers of excavated assemblages and complicated by the presence of more than one plausible tool-making hominin species. The addition of archeological evidence from a broader range of chronological and geographic contexts during early hominin dispersal into Eurasia between 2.0 and 1.0 Ma clarifies some of the strategic sources of variability in Early Paleolithic stone tool technology.

Between 2.5 and 2.0 Ma, stone tools and hominins are conclusively documented solely in African contexts. These pebble-core-based "Oldowan" stone tool assemblages are associated both locally and regionally with *Australopithecus garhi*, various species of *Paranthropus*, and *Homo habilis* (Susman 1991; Plummer 2004; Schick and Toth 2006). Between 1.6 and 1.3 Ma Early Acheulean stone tool assemblages featuring large, usually bifacially-flaked, cores/cutting tools appear together with early forms of *Homo ergaster/erectus*. Australopithecines and paranthropines became extinct around this time without a measurable effect on the nature of the Early Paleolithic archeological record.

By 1.0–0.9 Ma, hominin populations were discarding Acheulean tools from the shores of the Mediterranean to the

J.J. Shea (✉)
Department of Anthropology & Turkana Basin Institute, Stony Brook University, Stony Brook, NY 11794-4364, USA
e-mail: John.Shea@sunysb.edu

J.G. Fleagle et al. (eds.), *Out of Africa I: The First Hominin Colonization of Eurasia*, Vertebrate Paleobiology and Paleoanthropology,
DOI 10.1007/978-90-481-9036-2_4,

Cape of Good Hope (Klein 1999). Early Acheulean stone tools are sometimes portrayed as a part of a "breakthrough" technology that fueled *Homo erectus*' adaptive radiation (Schick and Toth 1993). Yet, Early Acheulean stone tools are not among the earliest generally accepted paleoanthropological sites in Europe, western Asia, India/Pakistan, Southeast Asia, or China (Roebroeks 2001; Straus and Bar-Yosef 2001; Langbroek 2004). Eurasian Early Pleistocene habitats were not intrinsically inimical to Acheulean stone tool production. Between 0.5 and 0.2 Ma, Europe and much of western Asia came to be as thoroughly littered with Acheulean stone tools as any part of Africa. Rather, it seems as though the circumstances surrounding *H. erectus* dispersal to Eurasia encouraged the production of evolutionarily "primitive" Oldowan-like assemblages even after more "derived" Acheulean technology was in broad use throughout much of Africa.

The "Oldowan" stone tools associated with *H. erectus* at early Eurasian sites differ little from stone tools produced nearly a million years previously in Africa. Such differences as these largely reflect raw material variation – more quartz and coarse-grained volcanic rocks in Africa, more use of cryptocrystalline silicates, such as flint and chert in Eurasia. One might reasonably expect stone tools made in the Plio-Pleistocene Transvaal to differ from those discarded in Middle Pleistocene southern China, either because they were made by different hominins or because of differences in the ways stone tools were used. Yet one would be hard pressed to reach either conclusion from the lithic evidence alone.

This paper explores archeological explanations for variation among the stone tool assemblages associated with early hominin dispersal out of Africa. Past explanations have focused on adaptive-functional differences and differences in the cultural or biological identities of hominins associated with particular named Early Paleolithic industries. This paper argues that it is more likely that the poor fit between hominin morphological variation, Early Paleolithic industrial variability, and the chronology of early hominin dispersal reflects an intrinsically high level of variability among early hominins' technological strategies (Potts 1998).

Background

The first thing one has to realize in trying to investigate early hominin behavior with lithic evidence is that many of the major typological distinctions archeologists make among stone tools and lithic assemblages (or "industries") were not intended to be useful in this task. Most named tool types and Paleolithic industries were defined decades ago, largely on an intuitive basis by archeologists with only a superficial familiarity with making and using stone tools, or, with systematic observations of stone-tool-using people (François Bordes and Louis and Mary Leakey being obvious exceptions). It would be nice if these intuitively-derived distinctions coincidentally aligned themselves with our current research goals, but we ought not expect them to do so. Indeed, the seemingly unpatterned relationships between stone tool industries and either hominin morphological variation or contextual differences in space and time suggest traditional ways of conceptualizing variation among Early Paleolithic stone tool industries may not monitor evolutionarily significant dimensions of early hominin behavior. It is possible that stone tools are relevant to major questions about hominin behavioral variability, but we have to prove this, rather than assuming it to be true because we want it to be. Investigating behavioral variability requires us to re-examine lithic variation in terms of variables relevant to the behavioral questions we are asking.

Early Paleolithic Stone Tools

Flakes and flake fragments comprise the overwhelming majority of artifacts in all Early Paleolithic stone tool assemblages. Variation in the size, cortex coverage, and morphology of these "detached pieces" provide clues about raw material economy and technological strategies, but they are not generally major factors in higher-order groupings of Early Paleolithic assemblages. Most assemblages contain a wide range of retouched flakes (heavy- and light-duty scrapers, notches, denticulates, awls, burins, truncations, and cores-on-flakes). Variation in the occurrence and relative frequencies of particular retouched tool types is a major factor in the systematics of Early Paleolithic assemblages (Bordes 1961; Leakey 1971; Debénath and Dibble 1994). Archeologists are increasingly skeptical about the significance of such distinctions. Mechanical damage (trampling, fluvial transport, sediment compaction) can create edge damage on flakes that is very similar to retouch (McBrearty et al. 1998). Although much effort has been devoted to the problem of telling retouch from various other modes of edge-damage, there is no objective measurement-based method for doing so. In addition, excavations of broad horizontal surfaces at Early and Middle Paleolithic sites have shown that retouched flakes are not randomly distributed through archeological deposits (Binford 1982; Henry et al. 2004). It follows that relative frequencies of retouched flakes can vary depending on the size and spatial pattern of archeological excavation.

The distinctions among Early Paleolithic stone tool assemblages that are most likely to shed light on strategic variation in early hominin adaptive strategies involve differences among cores and core-reduction strategies. Most Early Paleolithic cores can usefully be arrayed along a sliding scale

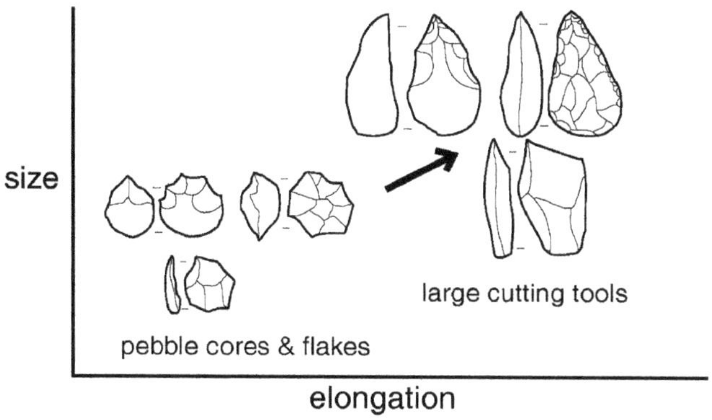

Fig. 4.1 Early Paleolithic Pebble cores and large cutting tools arrayed along a scale of increasing size and elongation

of size and elongation ranging from pebble-cores to large cutting tools (Fig. 4.1). The definitions of these artifacts listed below are adapted from Leakey (1971) and Debénath and Dibble (1994).

Pebble-Cores

Pebble-cores are subspherical, hemi-spherical, or roughly cuboid artifacts that have been shaped by varying amounts of hard-hammer percussion (Fig. 4.2). Striking platform and flake-release surfaces are interchangeable on such cores. As the term "pebble-core" suggests, many such artifacts were originally rounded clasts (pebbles or cobbles). This can sometimes be seen in smoothly-abraded cortical surfaces preserved on the core surface. As their name implies, most pebble-cores are made from rounded clasts, but they can include cores made from angular rock fragments or even flake fragments. The principal named groups of pebble-cores include choppers, core-scrapers, discoids, polyhedrons, and "tested" pieces (cobbles with one or two flaked detachment scars).

A *chopper* is a pebble-core that has had flakes removed from at least a quarter and not more than half of its circumference. Flakes can have been removed from one face of an edge (a unifacial chopper) or both faces of the same edge (a bifacial chopper). A significant portion of the remaining core circumference is unmodified. Bifacial choppers whose edges are asymmetrical (one flat, the other convex) are sometimes designated as "core scrapers" or "heavy-duty scrapers". Though the term "chopper" implies use as a cutting tool, it remains unclear if such tools were anything other than sources of flakes.

A *discoid* is a pebble-core that has had flakes removed from its entire circumference, or very nearly so. Most discoids are more-or-less symmetrical and either biconvex or plano-convex in cross section. If there are remnant cortical surfaces on a discoid, they are likely to be at the center of either the dorsal or ventral surface, framed by the distal ends of fractures originating from the core periphery.

A *polyhedron* is a blocky core with more than one discrete worked edge, or, it may be a core whose worked edge bifurcates at some point along the core periphery. To some degree, "polyhedron" is something of a catch-all for heavily-worked pebble-cores.

A *subspheroid* is a core whose surface preserves numerous fracture scars whose ridges and other prominent points

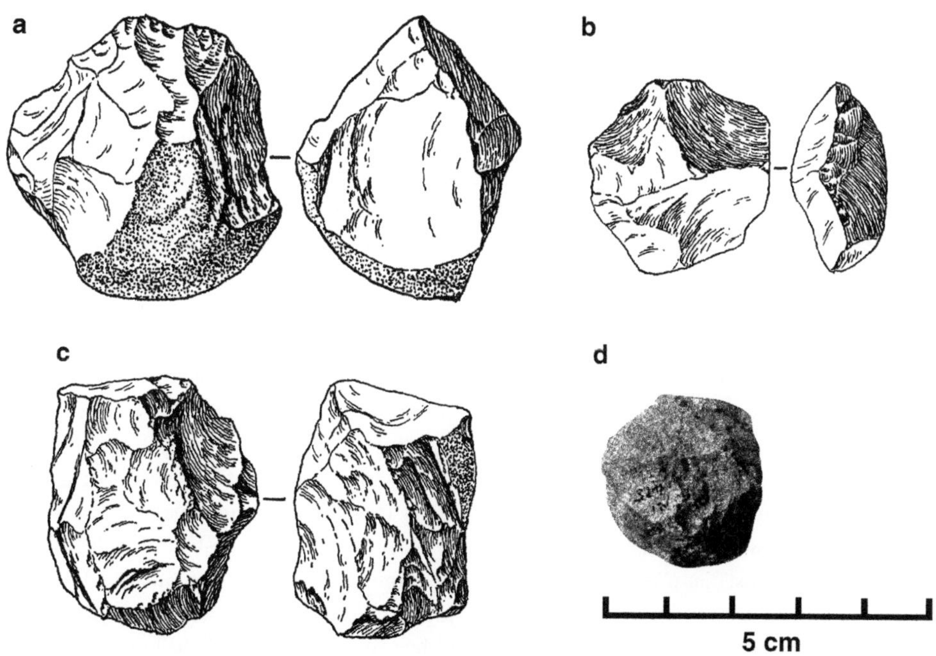

Fig. 4.2 Pebble-cores from Early Paleolithic contexts at Olduvai Gorge: (**a**) chopper, (**b**) discoid, (**c**) polyhedron, (**d**) subspheroid (After Leakey 1971, Figures 10.1, 12.6, 13.3, and Plate 16)

have been flattened by percussion and abrasion. Subspheroids are typically made of materials with poor conchoidal fracture properties. Many of these artifacts are probably best viewed as either "failed cores" started by hominins using inappropriate raw materials that were abandoned when they became too rounded to yield flakes (Sahnouni 1997) and cores or other angular rock fragments that were used as hammerstones (Schick and Toth 1994).

Some typologies of African and Eurasian Early Paleolithic tools divide the broad categories listed here into a wide range of sub-types, based on their size, shape, and alignment of working edge and morphological long axis (Leakey 1971). It has yet to be shown that these more-contingent classifications capture any significant dimension of hominin behavior. Choppers, discoids, polyhedrons, and subspheroids seem to comprise a series of increasingly-modified forms (Potts 1988), but at least one experimental study suggests that the initial shape of pebbles and cobbles exerts a powerful influence on the shape of the resulting pebble-cores (Toth 1985).

There does not seem to be much support for the hypothesis that pebble-cores were designed for some specific mode of use. Apart from generalized edge-damage, some of which may be geological in origin (De Beaune 1993), few function-specific wear patterns have been identified on Early Paleolithic pebble-cores. This might lead one to view pebble-cores primarily, or even solely, as cores – simply sources of flakes that were discarded after variable periods of use. This would probably be a mistake. Pebble-cores can be useful in many cutting, scraping, or chopping tasks, but (at least in this author's experience) they are not particularly more efficient in such tasks than a sharp-edged rock of equivalent mass and raw material. From a functional standpoint, pebble cores seem designed for no specific purpose.

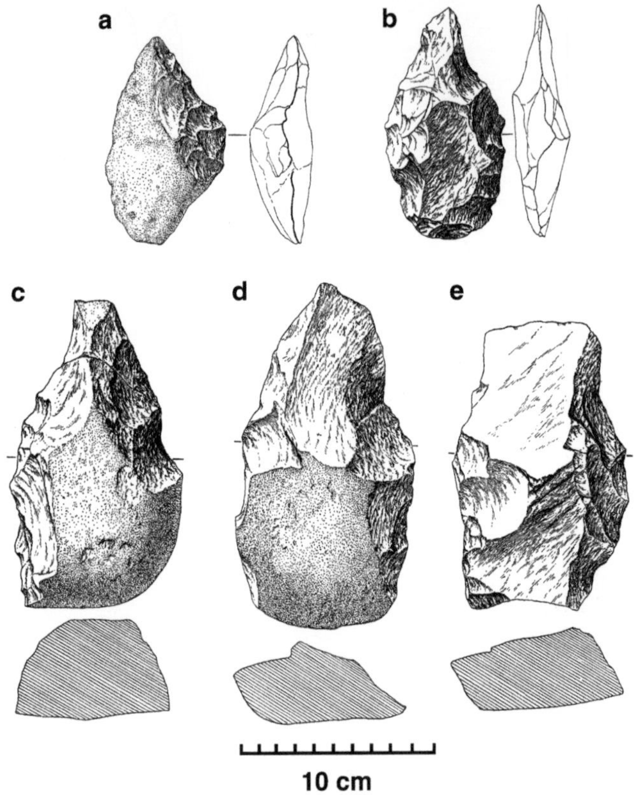

Fig. 4.3 Large cutting tools from Early Paleolithic contexts at Olduvai Gorge: (**a**) protobiface, (**b**) ovate, (**c**) pick, (**d**) biface/handaxe, (**e**) cleaver (After Leakey 1971, Figures 51.1, 97.2, 68.1, 66.2, and 66.1)

Large Cutting Tools

Large Cutting Tools (hereafter LCTs, *pace* Ambrose 2001) (Fig. 4.3) are symmetrical artifacts (usually >10 cm long) whose carefully retouched edges suggest they were not only cores, but also cutting tools in their own right. Such tools are usually flaked on both dorsal and ventral faces, but slab-like forms with unifacial retouch are also subsumed by this tool category. LCTs share with pebble-cores the interchangeability of striking platform and flake-release surfaces. Typologies for Early Paleolithic LCTs usually subdivide them into many subtypes on the basis of their relative width and degree of elongation, as well as other criteria. The most common and significant forms of LCTs in Early Paleolithic assemblages include protobifaces, picks, handaxes/bifaces, ovates, and cleavers.

A *protobiface* is a core that is in a sense intermediate between pebble-cores and LCTs. It is usually relatively small (<10 cm) with edges that converge to a point along the tool's morphological long axis. The edges are often jagged in profile, suggesting that immediately prior to being discarded, they were exploited as sources of flakes. Often, the proximal end of a protobiface is blunt and has remnant cortex. Isaac et al. (1997) argued that many protobifaces are merely elongated discoids, or elongated cobbles flaked in the same manner as discoidal pebble-cores. In support of this argument, they show that frequencies of discoids and protobifaces co-vary closely with one another among Early Paleolithic assemblages from East Turkana (Kenya). Jones (1994) has proposed that some protobifaces may be the worn-out remnants of other LCTs discarded after prolonged curation. He supports this hypothesis with evidence from Olduvai that protobifaces are more common at sites located far from raw material sources. These hypotheses are not mutually exclusive. Each probably captures some significant dimension of protobiface variation in Early Paleolithic assemblages.

A *pick* is a large (>10 cm long) elongated LCT whose distinguishing feature is a thick distal tip formed at the conjunction of two slightly concave lateral edges. This tip can be either rectangular or plano-convex in cross section, and is usually plano-convex in profile. Some *double-pointed picks* feature two such points at opposite ends of the tool.

A *handaxe* (also known as a "biface") is a LCT whose relatively straight lateral edges converge to a sharp symmetrical distal point. Handaxes are typically more symmetrical in cross-section at points closer to this tip. The proximal end is either rounded or angular, and there is often a remnant cortical surface in this area. The planform shape of handaxes varies widely, and they are frequently subdivided into myriad sub-types along lines that differ among regional traditions of prehistoric research.

An *ovate* is a large symmetrical LCT with curved convex lateral edges. In all other respects, ovates are similar to handaxes. As with protobifaces, one could view some ovates as merely large and somewhat elongated discoidal cores, or as the initial form of tools later resharpened into more sharply-tipped handaxes or picks (McPherron 1999).

A *cleaver* is a LCT with an acute, broad edge running more or less transversely to the long axis at the distal end. This distal edge is usually unretouched. There is wide variation in the extent to which cleavers are retouched. Some artifacts designated as cleavers are little more than large flakes with a broad and unretouched lateral edge.

Among the archeologists who have attempted to use replicated LCTs for various tasks, there is a consensus that they work particularly well as butchery tools, especially for the heavy-duty cutting involved in skinning and disarticulating large mammal carcasses (Shea 2007b). The edges of some Acheulean LCTs preserve wear patterns consistent with butchery, but most of these wear patterns are on tools from European Later Middle Pleistocene contexts. We cannot be certain that this is the principal activity in which LCTs from other contexts were used. Indeed, Dominguez-Rodrigo and colleagues (2001) report finding phytoliths suggesting woodworking of LCTs from Peninj (Tanzania) dating to 1.5 Ma.

The hypothesis that LCTs were designed as projectile weapons thrown sidelong (like a discus), has some proponents (O'Brien 1981; Calvin 2002), but this hypothesis has been challenged by experimental evidence (Whittaker and McCall 2001) and enjoys little support among the majority of Paleolithic archeologists. As with pebble-cores/"choppers", the additional design features that define LCTs as a group do not seem likely to measurably improve the effectiveness of a projectile weapon comprised of an unmodified rock of equivalent shape and mass.

Kohn and Mithen (1999) have proposed that LCTs were instruments of sexual selection by which male hominins engaged in "costly signaling" to attract potential mates and allies. LCTs are the most extensively modified of Early Paleolithic stone tools, and there is probably a signaling dimension to their variability. Most of the examples of "costly signaling" cataloged by the Zahavi and Zahavi (1997) are physiologically "hard-wired" and thus stable across the length and breadth of a species' distribution. LCTs, in contrast vary widely in size, shape, and frequency across time and space. This suggests there are other factors driving LCT variation in addition to their conjectural role in sexual selection.

Perhaps the least controversial hypothesis about LCTs is that they were, like pebble-cores, sources of flakes. Understanding why some hominins chose to make and discard pebble cores while others chose to make and discard LCTs may hinge on differences in the ways these core technologies deliver flakes.

Early Paleolithic Industries

African and Eurasian Early Paleolithic assemblages are usually assigned to either Oldowan, Developed Oldowan, or Early Acheulean industries (or "industrial complexes"). These assignments are based largely on greater or lesser frequencies of pebble-cores and LCTs in particular assemblages (see Table 4.1).

Table 4.1 Differences between East African Oldowan, Developed Oldowan, and Early Acheulean industries (Leakey 1971: 3; Klein 1999: 158)

Industry	Core technology	Exemplary contexts
Oldowan (>1.6 Ma)	Choppers, discoids, scrapers, occasional subspheroids and burins, together with hammerstones, utilized cobbles and light-duty utilized flakes	Olduvai Gorge Bed I and Lower Bed II East Turkana, Koobi Fora Formation (KBS Member) Kada Gona West Turkana, Nachukui Formation (Lokalalei 1, 2C sites) Omo Shungura Member F
Developed Oldowan (<1.6 Ma)	Type A: Same as Oldowan but with marked increase in spheroids and subspheroids and in the number and variety of light-duty tools Type B: Same as Developed Oldowan A, but with a few rare bifaces	Olduvai Gorge Middle and Upper Bed II, Beds III and IV East Turkana, Koobi Fora Formation (Okote Member) Karari Industry Sites
Early Acheulean (1.6> × > 0.7 Ma)	Bifaces amount to 40% or more of the tools, and include irregular ovates, oblong picks and cleavers	Olduvai Gorge Upper Middle and Upper Bed II, Beds III and IV Olorgesailie Peninj Kilombe Kariandusi

The Oldowan Industry

The Oldowan Industry was defined by the Leakeys based on their excavations in Beds I and II of Olduvai Gorge, Tanzania (Leakey 1966). It has since become common practice to extend the term to similar assemblages from African contexts dating between 2.6 and 1.5 Ma (Plummer 2004; Schick and Toth 2006). The key feature of Oldowan assemblages is that they are comprised primarily of pebble-cores and flake byproducts of pebble-core reduction. Retouched tools are rare. Most Oldowan assemblages are found close to water sources, either streams or lake margins. The term "Oldowan" is also applied to Eurasian assemblages from much younger contexts than those in Africa.

The Developed Oldowan Industry

Developed Oldowan assemblages first appear around 1.6–1.7 Ma. These assemblages are similar to Oldowan ones in that they are dominated by evidence of pebble-core reduction. Their significant derived features include increased numbers of retouched flake tools, increased spheroids and subspheroids (Developed Oldowan A), and the appearance of protobifaces (Developed Oldowan B). Other Acheulean LCTs are rare or absent. At Olduvai, Developed Oldowan A assemblages precede and are partly contemporary with Early Acheulean ones. In Olduvai Beds III and IV and elsewhere in Africa, Developed Oldowan and Early Acheulean assemblages overlap completely in time and space (Isaac 1984; Leakey and Roe 1994). Though Leakey originally argued that Oldowan, Developed Oldowan, and Early Acheulean assemblages were made by correspondingly different hominin populations (Leakey 1971), most prehistorians regard the Developed Oldowan and Early Acheulean as different modalities within a broader Acheulean Industrial Complex (Toth and Schick 2000).

The Early Acheulean Industry

The term "Acheulean" comes from the site of St. Acheul, France. There, in the early nineteenth century, gravel deposits were discovered that preserved large bifacial cutting tools together with fossils of extinct large mammals (proboscidians, hippopotamus, and rhinoceros). Acheulean LCTs were among the first artifacts recognized as having a Pleistocene antiquity (Grayson 1983). Similar finds of bifaces, cleavers and other LCTs with large mammal remains at African sites led to the widespread adoption of the term "Acheulean" in African prehistory (Leakey 1936). The term Acheulean has come to be applied to any African or Eurasian Early Paleolithic assemblage that contains "significant" numbers of bifaces or other LCTs. (What quantity constitutes "significant" varies widely among regional research traditions.) The oldest appearances of Acheulean bifaces date to 1.6–1.4 Ma in East Africa. Early Acheulean assemblages are found in riverine, lake-edge, and coastal contexts over much of Africa and southern Eurasia from 1.5 Ma to around 0.2–0.1 Ma. Acheulean assemblages appear somewhat later in Europe (<0.5 Ma)(Piperno 1999) where they grade into a variety of local/regional Middle Paleolithic industries that retain large biface core technology (Gamble 1999). LCTs are rare in China and East Asia, appearing in only a few rare contexts after 0.8 Ma (Hou et al. 2000; Norton et al. 2006). LCTs are unknown in mainland Southeast Asia and the Indo-Malaysian Archipelago (Movius 1948; Schick 1994).

Early Paleolithic Industrial Variability as Strategic Variabiliity

What do these Early Paleolithic industries represent? Early Twentieth Century prehistorians envisioned the Acheulean and its predecessors as universal stages of human behavioral evolution. As geochronological improvements around mid-century pointed to regional variation in the timing of transitions among these stages, named Early Paleolithic industries were increasingly seen as quasi-cultural entities, the "material culture" of *Homo habilis*, *Homo erectus*, and other hominins (Leakey 1960). With improvements in the fossil record, putative correlations between named Paleolithic industries and hominin morphological taxa began to break down. Such industry-hominin correlations as remain "in play" in Paleolithic archeology are those structured in very general terms and/or in contexts where there was only one hominin species with which to correlate a particular industry (but see Foley 1987; Foley and Lahr 1997). These correlations have become weaker still in recent years with the growth of an ever more species-rich hominin fossil record (Lieberman and Bar-Yosef 2005). There is a growing consensus that differences among Early Paleolithic industries do not reflect differences in the biological identities of the hominins who created them. Rather, differences among industries are increasingly thought to reflect differences in hominin behavior. Early Pleistocene hominin species almost certainly differed from one another behaviorally, but whether and how those behavioral differences affected strategies for stone tool production remains an open question.

Core Reduction Strategies

First and foremost, the differences between Oldowan, Developed Oldowan, and Early Acheulean industries involve differences in core-reduction strategies. Whatever other factors influenced their variability, it is beyond dispute that both pebble-cores and LCTs were sources of flakes that were themselves used for various cutting tasks. Any effort to explain variability in Early Paleolithic stone tool technology has to start considering the costs and benefits that influence strategic variation in core reduction. Oldowan assemblages are dominated by the byproducts of pebble-core reduction. Early Acheulean assemblages differ from Oldowan ones in their emphasis on the production of LCTs. Developed Oldowan assemblages occupy an intermediate position, with some use of both core-reduction strategies. These differences are not arrayed along a single axis of variabilty. Many Early Acheulean assemblages also feature byproducts of pebble-core technology. Pebble-core technology is a kind of shared technological substrate to which the products of other, more derived, strategies of stone tool production are added in successive prehistoric periods.

As strategies for provisioning hominins with useful cutting tools, pebble-cores and LCTs differ from one another in terms of their identifiable costs and benefits.

Pebble Core Reduction Strategies: Costs and Benefits

As a source of cutting edge, pebble-core technology has many readily-identifiable benefits. Using it, one can provision oneself with sharp flakes anywhere there are rocks amenable to conchoidal fracture. It can even be applied to rocks with only an approximation of conchoidal fracture (e.g., gneiss, quartz). It requires minimal technological expertise beyond the grip, arm power and hand-eye coordination necessary to initiate conchoidal fracture in brittle rock. In teaching flintknapping to college students for more than 15 years, the author has never had a student fail to produce passable replicas of Oldowan cores within the first few minutes of instruction. Other flintknapping instructors report similar results (N. Toth, personal communication, 2005).

Clearly, the "skill costs" of pebble-core technology are not particularly great. Requiring few skills, deployable wherever knappable stone is to be found, pebble-core technology is the quintessential "instant technology". Pebble-core technology is plausibly the most versatile method ever devised for provisioning hominins with cutting edge. Predictably, pebble-core technology is known from every period of Stone Age prehistory, into historic times (Rosen 1997), and in ethnographic contexts as well (Gould et al. 1971).

There are some significant constraints on the benefits of pebble-core technology. Reducing a pebble-core by hard hammer percussion yields only a small number of usable flakes (Kimura 1999; Kimura 2002; Braun et al. 2005). These flakes tend to be relatively short and thick. Consequently, pebble-cores have a relatively large irreducible minimum mass and individually short potential use lives. Furthermore, their worked edges are generally steep and jagged in profile, making them ill-suited for tasks involving prolonged cutting. There is little to be gained by transporting pebble-cores over great distances, except in areas where knappable stone is absent or rare. A hominin seeking efficiently-transportable byproducts among pebble-core technology would be better off transporting flakes, which have individually smaller irreducible minima, rather than pebble cores (Kuhn 1994).

The circumstances that seem most likely to elicit pebble-core technology among knappers are those in which the demand for stone cutting tools was rare, unpredictable in time and space, and variable in terms of the kinematics of the tasks involved.

LCT Reduction Strategies: Costs and Benefits

One of the benefits of LCT core technology is that it can yield enormous quantities of useful flakes. A flintknapper replicating even the least complex of Early Paleolithic LCTs will create dozens of broad, thin flakes whose circumferences preserve tens, if not hundreds of centimeters of useful cutting edge. This contrasts markedly with the relatively modest output of thick, short flakes typically produced by pebble-core reduction (Kimura 1999; Braun et al. 2005). LCTs' potential to yield large-quantities of potentially-useful flakes while maintaining their own potential as cutting tools in their own right may account for the popularity of LCTs and similar bifacial tool forms among mobile hominin and human populations (Kelly 1988). In essence, whether one chose to carry an LCT as part of "personal gear", to cache it at habitation/logistical sites, or to merely abandon it on the landscape near strategic resources or along a pathway, doing so would be more advantageous from the standpoint of recovering potential tool utility in the future than carrying, caching, or proximity-jettisoning of pebble-cores.

The most obvious costs associated with LCT technology probably involved differences in mass, raw material quality, and (albeit arguably) production skill. LCTs are usually larger than pebble cores. Transporting them was, in principle, more energetically expensive than transporting smaller pebble cores, but this cost was probably offset by LCTs larger potential yield of flakes and double-utility as tools in their own right. LCT's larger size probably required hominin toolmakers to select higher quality raw materials for the production of LCTs. A pebble core with internal flaws could have been abandoned at little cost, because the core would have yielded relatively few flakes. Abandoning a LCT that turned out to have a significant internal flaw would have involved a greater loss of potential tool utility.

Wynn (1995) has argued, using Piagetian analysis of stone tool design, that LCTs require more skill to produce than pebble cores. The experience of most knapper-teachers dovetails well with this argument. Students tasked with replicating even relatively simple LCTs require much more instruction and practice than they do in replicating pebble-cores (author's personal experience, N. Toth, pers.com).

The circumstances most likely to encourage the production of LCTs probably included increases in the frequency and predictability of tool use.

Explaining Early Paleolithic Industrial Variability

Viewing the differences among Early Paleolithic industries in terms of variation in core-reduction strategies can help us make sense of the puzzling geographic and chronological patchwork of archeological occurrences associated with early hominin dispersal out of Africa. Table 4.2 presents a list of the earliest generally accepted paleoanthropological occurrences (fossil and/or archeological sites) in Europe, western Asia, the Indian Subcontinent, Indonesia, and China. Each regional sample is presented in descending geochronological order. The presence or absence of stratigraphically-associated hominin fossils is also indicated. The stone tool technology associated with each individual occurrence is classified as pebble-core technology or LCT technology. This presentation, admittedly, oversimplifies a complex paleoanthropological record. Nevertheless, it shows a clear and consistent pattern.

In each region the oldest sites preserve evidence of pebble-core technology. LCTs make their appearance somewhat later, albeit at times that differ among these regions. These data suggest that the first wave of hominin dispersal out of Africa was accomplished by hominins equipped with pebble-core technology like that seen in African Oldowan assemblages. Acheulean LCT technology appears later on in each region, more-or-less in order of a particular region's proximity to the nearest land route to Africa. The earliest LCTs in non-African contexts occur in the Levant, at the sites of 'Ubeidiya, Revadim, and Evron (Bar-Yosef 1999). The next-oldest occurrences date to 1.0–0.5 Ma in Europe and Southern Asia, long after the oldest dated hominin fossils and stone tools in these regions (Langbroek 2004). Though they occur sporadically, LCTs remain rare in northern Europe, northern Asia, China, and Southeast Asia (Schick and Zhuan 1993; Bosinski 1995; Hoffecker 2002).

Does the strategic perspective on Early Paleolithic core technology outlined above make sense of this pattern? A strong case can be made, but it requires us to break the habit of thinking about early hominin dispersal in terms of human colonization. Most of the colonization events of the last five centuries have involved industrial states, the use of complex seafaring technology, and movements of new populations into areas already occupied by an indigenous human population (e.g., the Americas, Australia, South Africa). None of these qualities are relevant to models of early hominin dispersal. Dispersing populations of early hominins were probably small in number, moved primarily overland from contiguous geographic regions, and involved regions with no prior history of hominin occupation.

One of the first problems early hominins would have faced in any dispersal would have been the problem of unfamiliar terrain and, to a lesser extent, of food sources that differed from their area of geographic origin. In most recent colonial dispersals, the colonists were able to become quickly familiar with what their new habitats by consulting guides among the local population. Early hominin "pioneers" would have had no such guides. They would have had to conduct their own reconnaissance and survey. Under such circumstances of wide resource variability and uncertainty, the toolkits with which these "pioneer" hominins equipped themselves likely emphasized versatility. Pebble-core technology, as outlined above, is probably the most versatile technology possible. Therefore, it is hardly surprising that pebble-core technology is the earliest durable trace of hominin dispersal out of Africa. (The only places to which humans have dispersed without recourse to pebble-core technology are Antarctica and the Moon!).

Why would hominins using a versatile technology as pebble-core technology switch to making LCTs, which involve increased costs in terms of raw material procurement, transport costs, and (arguably) skill? The answer to this question requires one to imagine what would have happened during the "settling in" phase of hominin occupation. Assuming that a pioneer population was successful in reproducing itself, over time hominin populations would have increased. Earlier, small hominin populations probably focused their subsistence strategies on food sources with relatively low handling costs, and minimal needs for stone tools (either to process the food source directly or to create a secondary technology [wood tools] to aid in collecting and processing it). As populations grew, pressure on low-cost food sources would have increased, leading hominins to more regularly exploit food sources that had higher search and handling costs. As greater proportions of hominin diets came to be derived from such higher-cost food sources, the need for stone tools would have become more predictable. LCTs probably emerged in such circumstances, not through some sudden flash of insight or neurological mutation, but rather through the simple phenomenon of allometry, of merely "scaling up" the size of pebble-cores and flakes by elongation (see Fig. 4.1). Lengthening these tools increases their potential yield of cutting edge in relation to the "irreducible minimum" of each tool (the fraction of an original core volume below which a knapper can no longer detach useful flakes or find useful cutting edge).

Table 4.2 Early Paleolithic sites from Africa and Eurasia arranged by descending age within various regions

Region	Site	Age (Mya)	Basis for date	Lithics	Hominins
South Africa	Sterkfontein, Member 5	2.0–1.5	BS	PBC	P, H
	Kromdraai A & B	2.0–1.0	BS	PBC	P
	Swartrans Member 1	1.8–1.5	BS	PBC	P, H
	Swartrans Member 2	1.5	BS	PBC	P, H
	Swartrans Member 3	1.0	BS	LCT?	P, H
	Sterkfontein, Member 5 West	1.7–1.4	BS	LCT	H
	Gladysvale, South Africa	>0.8	BS, ESR	LCT	A, H
	Elandsfontein	1.0–0.6	BS	LCT	H
East Africa	Gona Sites, Ethiopia	2.7–2.6	K-AR, BS	PBC	A
	Hadar, Ethiopia	2.3	K-AR, BS	PBC	H
	Middle Awash, Ethiopia	2.3	K-AR, BS	PBC	A
	Lower Omo Valley Members E-F	2.4–2.3	K-AR, BS	PBC	P, H
	Kanjera, Kenya	2.0	K-AR, BS	PBC	
	East Turkana KBS Member	1.9–1.7	K-AR, BS	PBC	P, H
	Olduvai Bed I	1.8–1.6	K-Ar, BS	PBC	P, H
	East Turkana Okote Member, Kenya	1.7–1.3	K-Ar, BS	LCT	P, H
	Peninj, Tanzania	1.7–1.3	K-Ar	LCT	P, H
	Olduvai Lower-Middle Bed 2	1.6–1.4	K-Ar, BS	LCT	P, H
	Olduvai Upper Bed 2	1.4–1.2	K-Ar, BS	LCT	H
	Konso Gardula, Ethiopia	1.4–1.3	K-Ar	LCT	H
	Gadeb, Ethiopia	1.4	K-Ar	LCT	
	West Turkana, Lower Natoo, Kenya	1.5	K-Ar, BS	LCT	H
	Olorgesailie, Kenya	1.0–0.9	K-Ar, Ar/Ar	LCT	H
Northwest Africa	El Kherba, Algeria	1.8–1.4	PM, BS	PBC	
	Ain Hanech A & B, Algeria	1.8–1.4	PM, BS	PBC	
	Casablanca 1, Morocco	1.0–1.1	PM, BS, SC	PBC	
	Thomas Quarry1 Level L, Morocco	0.9	BS, PM	LCT	H
	Tighenif, Ternefine, Algeria	0.7	BS, SC	PBC	H
Southwestern Asia	Yiron, Israel	2.4	PM, K-Ar	PBC	
	Erq el-Ahmar, Israel	1.8–2.2	PM, BS	PBC?	
	'Ubeidiya (Li Member), Israel	1.6–1.4	PM, BS	LCT	H
	'Ubeidiya (Fi Member), Israel	1.6–1.4	PM, BS	LCT	H
	Evron Quarry, Israel	1.0–1.2	PM	LCT	
	Nahal Zehor, Israel	>0.8	PM	PBC	
	Bizat Ruhama (2–4), Israel	>0.8	PM	PBC	
	Gesher Benot Ya'aqov, Israel	0.7–0.9	PM	LCT	H
	Latamne, Syria	>0.8	PM, BS	LCT	
Southern Europe	Pirro Nord, Italy	1.6–1.3	BS	PBC	
	Venta Micena, Spain	1.5–1.6	PM, BS, SC	PBC	H?
	Fuente Nueva, Spain	1.3–1.2	PM, BS	PBC	
	Barranco León, Spain	1.3–1.2	PM, BS	PBC	

(continued)

Table 4.2 (continued)

Region	Site	Age (Mya)	Basis for date	Lithics	Hominins
	Atapuerca, Sima de l'Elefante, Spain	1.3–1.2	PM, BS	PBC?	H
	Cueva Victoria, Spain	1.0?	PM, BS	PBC?	
	Monte Poggiolo, Italy	1.0	PM	PBC	
	Huescar 1, Spain	1.0–0.9	PM, BS	PBC?	
	Estrecho del Quipar	>0.9	PM, BS	LCT	
	Le Vallonet, France	0.92–1.12	PM, BS	PBC?	
	Ceprano, Italy	0.8–0.6	PM, SC		H
	Atapuerca, Gran Dolina TD4, TD6, Spain	0.8	PM, BS, K-Ar	PBC	H
	Solana del Zamborino	0.8	PM, BS	LCT	
	Notachirichio, Italy	0.5–0.6	BS	LCT	
	Cullar Baza 1, Spain	0.6	PM, BS	PBC?	
	Isernia La Pineta, Italy	0.6	PM, BS, ESR	PBC	
	Fontana Ranuccio, Italy	0.5	K-Ar	LCT	
South Asia	Riwat, Pakistan	2.0–1.9	PM	PBC?	
	Pabbi Hills, Pakistan	2.0–1.0	PM	PBC?	
	Jammu, Pakistan	>1.8	FT	PBC?	
	Isampur, India	1.2	ESR	LCT	
	Bori, India	0.5–0.7	K-Ar, Ar/Ar	LCT	
	Dina, India	<0.5–0.7	PM	LCT	
	Jalalpur, India	<0.5–0.7	PM	LCT	
	Didwana, India	>0.4	US	LCT	
	Umrethi, India	0.2	US	LCT	
	Nevasa, India	>0.3	US	LCT	
	Yedurwadi, India	>0.35	US	LCT	
	Teggihalli, India	>0.35	US	LCT	
	Sadab, India	0.3	US	LCT	
	Hathnora (Narmada), India	>0.040	PM, US, C-14	LCT	H
Indonesia	Perning/Modjokerto	1.8	Ar/Ar, PM		H
	Sangiran Bapang Fmn.	1.66–1.22	Ar/Ar, PM, MBS	PBC	H
	Sangiran Fm above lake fmn	1.5	Ar/Ar, PM, MBS		H
	Perning/Modjokerto	1.4	Ar/Ar, PM		H
	Sambungmachan lower	L-M. Pleist.	US, ESR	PBC?	H
	Rancah	L-M. Pleist.	BS		H
	Trinil	M. Pliest.	BS		H
	Matamenge, flores	0.8	PM	PBC	
	Sambungmachan upper	0.03	US, ESR	PBC?	H
	Ngandong	<0.05	BS	PBC?	H
South China	Renzidong	>2	PM, BS	PBC?	
	Longguopo IV	2.0?	BS, PM, ESR	PBC?	H?
	Yuanmou	1.70–1.71	PM, SC	PBC	H?
	Lantian–Gongwangling	1.15	PM, SC	PBC	H

	Longguopo II	1.0??	BS, PM, ESR	PBC?	
	Mohui (Bose)	0.803 ± 3K	Ar/Ar, BS	LCT	H
	Yunxian	0.7–1.0	PM, BS	PBC?	H
	Zhoukoudian	0.7–0.25	PM, BS	PBC	
	Lantian – Chenjiawo	0.6	SC		H
Northeast China & Korea	Majuanguo I, China	<1.54	PM, SR	PBC	
	Majuanguo II, China	1.54	PM, SR	PBC	
	Majuanguo III, China	1.6	PM, SR	PBC	
	Majuanguo, Banchan, China	1.36	PM, SR	PBC	
	Xiaochangliang, China	1.34	PM, SC	PBC	
	Donggutuo, China	1.1	PM	PBC	
	Chongokni, S. Korea	0.3–0.4	K-Ar, SR	LCT	
Western Asia	Dmanisi, Georgia	1.75	MBS, MP, Ar/Ar	PBC	H
	Dursunlu, Turkey	1.0	PM	PBC	
	Ahkalkalaki, Georgia	0.8–0.9	PM, BS	LCT	H
	Kuldara, Tajikistan	0.8–0.9	PM	PBC	
	Treugolnaya, Georgia	0.583	TL	LCT	
	Kudaro I, Georgia	>0.5	PM, BS	LCT	
	Kudaro III, Georgia	>0.5	PM, BS	LCT	H
	Azykh, Azerbaijan	>0.5	PM, BS	LCT	H
	Yarimburgaz, Turkey	<0.5	BS, ESR	PBC	
	Karatau & Lakhuti, Tajikstan	0.4–0.6	PM, SC	PBC	
	Sel'-Ungur Cave, Kyrgystan	0.13	US	LCT?	H
Northern Europe	Pakenham, UK	0.65	BS, SC	PBC	
	Cagny, France	0.6	SC	LCT	
	Mauer, Germany	0.5?	BS, SC	PBC	H
	Miesenheim, Germany	0.5	BS, SC	PBC	
	Kent's Cavern, UK	0.5	BS, SC	PBC	H
	High Lodge, UK	0.5	BS, SC	PBC	
	Boxgrove, UK	0.4–0.5	BS	LCT	H
	Bilzingsleben, Germany	0.3–0.4	US	LCT	H

Dating Methods: BS = Biostratigraphy, PM = Paleomagnetism, RS = Regional stratigraphic correlations, SC = local stratigraphic correlation, SR = sedimentation rate, K-Ar = Potassium-argon, Ar-Ar = Single-crystal argon dating, US = Uranium Series, ESR = Electron spin resonance, TL = Thermoluminescence, C-14 = Radiocarbon

Lithics: PBC = Pebble-core technology, LCT = Large cutting tool technology

Hominin Fossils: A= Australopithecus, P = Paranthropus, H = Homo

Sources: South Africa: Kuman (1998), Klein (1999); East Africa: Klein (1999), Schick and Toth (2006), Leakey and Werdelin (2010); Northwest Africa: Sahnouni (2006), Lahr (2010); Southwestern Asia: Bar-Yosef (1994), Ronen (2006), Europe: Gamble (1999) Roebroeks (2001), (Santoja and Villa 2006) Martinez-Navarro (2010); South Asia: Dennell and Roebroeks (2005) Dennell (2010); Petraglia (2010), Chauhan (2010), Schick and Toth (1993); Indonesia: Larick et al. (2001), Ciochon (2010) Zaim (2010); China & Korea: Schick and Zhuan (1993), Norton (2000), Hou et al. (2000), Hou and Zhao (2010), Zhu et al. (2003), Weiwen and Pu (2007); Western & Central Asia: Hoffecker (2004)

Increased population size should have intensified intra-specific competition for food, habitation sites, social allies and mates. The use of LCTs as instruments of "costly signaling" and sexual selection envisioned by Kohn and Mithen (1999) might have emerged in the context of such intensified competition.

Why do both pebble-cores and LCTs continue to be made in the same regions? Early Paleolithic assemblages of any significant size and younger than 1.6 Ma do not consist solely of LCTs. Pebble-core technology is usually indicated as well by the presence of choppers, core-scrapers, discoids, polyhedrons, and/or subspheroids. Frequencies of LCTs can vary widely, and seemingly in a patterned way thorough regional archeological successions (Jelinek 1982; see papers in Soressi and Dibble 2003). If LCT production is linked to increased population size and niche breadth, as suggested above, then variation in LCT frequencies may be monitoring short-term changes in local or regional hominin niche breadth, and indirectly, hominin population size. Wide shifts in the productivity of terrestrial ecosystems during Middle Pleistocene glacial cycles probably caused frequent reductions in hominin population size. Hominins surviving these demographic downturns in ecogeographic refugia are unlikely to have needed toolkits that emphasized transport efficiency in the way that LCTs do. Instead, they probably reverted to pebble-core production, resuming the production of LCTs when population sizes increased and niches broadened again.

Why are LCTs persistently rare or absent from some regions and not others? Those parts of Eurasia in which LCTs remain rare, northern Europe, northern Asia, China, and Southeast Asia, are those furthest from Africa, both in terms of actual distance as well as in terms of ecogeographic similarity. It seems reasonable to infer that African-derived hominin populations living in these regions remained relatively small and their ecological foothold relatively precarious. If the adoption of LCT technology was a technological shift substantially driven by demographic factors, that is to say if it was a density-dependent phenomenon, the absence of quintessentially Acheulean technology in these areas may reflect ephemeral hominin occupation and persistently low population densities (see also Lycett and von Cramon-Taubadel 2008). Conversely, it is possible that lithic raw materials in some of these regions may have been too small or otherwise inadequate to encourage the production of LCTs. This latter issue needs to be clarified for many parts of East and Southeast Asia that would seem logical places for high hominin population density but that currently do not preserve evidence for LCT production.

In a perfect world, one could test this hypothesis with an independent measure of hominin population density. It could be falsified by showing that hominin population densities did not differ between a region that persistently lacked Acheulean LCT technology (e.g., southern China or Indonesia) and one in which LCTs were common (e.g., the East African Rift Valley). Unfortunately, accurately estimating prehistoric human and hominin population density remains a formidable methodological challenge for archeologists. It is difficult enough to formulate such estimates for recent prehistoric periods using ethnographic hunter-gatherer demographic data (Binford 2001; Shea 2007a). It is almost inconceivable that we will ever be able to express Early and Middle Pleistocene hominin population densities in terms of ratio or interval-scale measurements without attaching so many untestable assumptions and qualifications to them as to render them analytically worthless. The most plausible clues about relative population densities can be found in settlement patterns (occupation of ecologically marginal zones) and in the increased use of lower-ranked prey species (Boserup 1990).

Such evidence has been forthcoming in the archeological record of Late Pleistocene and Holocene contexts, and it is not unreasonable to expect similar evidence from Early and Middle Pleistocene contexts as well. One datum does not a pattern make, but early evidence for systematic LCT production at Gesher Benot Ya'aqov (0.9–0.7 Ma in Israel) is associated with evidence for fire and systematic processing of hard-seeded plant foods with pounding tools (Goren-Inbar et al. 2000). The use of fire to detoxify plants and to render meat more edible and the use of pounding tools to pre-masticate plant foods, all involve increases in handling costs for prey species. Such increases are consistent with economic intensification of the sort one might associate with local population growth. The quality of preservation at Gesher Benot Ya'aqov is remarkable and exceptional for Eurasian Early Paleolithic sites. Whether other Eurasian Early Acheulean contexts preserve parallel evidence for economic intensification that might signal population growth remains to be revealed by future research.

This apparent "Pioneer/Settler" model of Early Paleolithic core technology is not free of problems. The most serious one (in this author's opinion) is that natural processes can imitate the byproducts of pebble-core technology. Fluvial or colluvial transport of rocks with conchoidal fracture properties can create objects that, selectively gathered and presented, can be mistaken for artifacts (Grayson 1986). Archeologists have developed many criteria for segregating artifacts from naturally-fractured rock (Barnes 1939; Patterson 1983; Peacock 1991; Gillespie et al. 2004) (see Table 4.3). The data necessary to rule out a plausible geological origin is not uniformly known for stone tools from early Eurasian sites, or indeed for many early African ones, either. A systematic assessment of the oldest claimed Early Paleolithic assemblages from various parts of Africa and Eurasia would improve our knowledge about the temporal gap between the appearance of pebble-core and LCT technologies by winnowing away assemblages that were not clearly and convincingly products of hominin activity.

Table 4.3 Criteria for identifying human agency in the production of stone artifact assemblages (Adapted from Barnes 1939; Patterson 1983; Peacock 1991; Gillespie et al. 2004) with additions from various members of the "Out of Africa I: The First Hominin Colonization of Eurasia" conference

Rank and inference	Cores/flaked pieces	Flakes/detached pieces	Other considerations for assessing human origin of artifact assemblage
Human origin probable, natural origin improbable	Large sample size (*n* = >30) Extensive and symmetrical scarring showing imposition of symmetry and asymmetry on different axes of core (e.g, picks, handaxes) Noncortical cores predominate More than one negative flake scar on majority of cores	Large sample size (*n* > 100) Ventral radial lines common Bulbar scars (eraillures common) Surfaces mostly unweathered Predominantly noncortical flakes More than one dorsal flake scar >1cm long on most flakes Majority of flakes have dorsal flake scars aligned parallel one another Negative dorsal bulb scars on 59% or more flakes	Cut marks assessed by zooarcheologist using microscopy Flaked stone artifacts burnt by fire Refitting sets of artifacts in close proximity Low energy depositional context Majority of sample recovered from controlled excavation
Equivocal, no way to objectively decide between human vs. natural origin	Small sample size (*n* = <5) Noncortical cores account for less than 50% of sample More than one negative flake scar on minority of cores	Moderate sample size (*n* >10, <100) Even proportions of flakes with bulbar scars (eraillures) Even mix of weathered, unweathered surfaces Even proportions of cortical/noncortical flakes More than one dorsal flake scar >1 cm long on less than 50% of flakes Fewer than 30% of flakes have dorsal flake scars aligned parallel one another Negative dorsal bulb scar on 30–50% of flakes	Claimed cut marks/linear scratches on bone Majority of sample selected judgmentally from surface
Natural origin not refutable	One or fewer artifacts Noncortical cores predominate Few/no cores with negative flake scars	Small sample size (*n* = <10) Ventral radial lines rare Bulbar scars (eraillures) rare Surfaces heavily weathered Predominantly cortical flakes More than one dorsal flake scar >1 cm long on less than 20% of flakes Negative dorsal bulb scar on <10% of flakes	Collection unavailable for study (for any reason) Majority of sample selected judgmentally from surface Stratigraphic provenience unclear/unconfirmable

Alternatives to the Strategic Variation Model

The differences among Oldowan, Developed Oldowan, and Early Acheulean industries reflect contrasting emphasis on pebble-core and LCT technology. Geographic and chronological variation among the occurrences of these industries during the course of early hominin dispersal are explicable in terms of strategic variation alone. This hypothesis does not preclude the possibility that Early Paleolithic industrial variation is meaningfully correlated with biological differences among hominin toolmakers, with social-cultural variation among hominin populations, or with specific activities performed with particular tools, but the evidence for such correlations is weak and problematical.

Biological Variation?

There are no one-to-one correspondences between named major groupings of named Early Paleolithic stone tool industries and hominin morphotypes. In Africa, prior to 1.8 Ma, Oldowan pebble-core technology is associated with a range of hominin taxa, including *Australopithecus garhi*, *Paranthropus robustus/boisei*, and *Homo habilis*. Outside of Africa and after 1.8 Ma, "Oldowan" pebble-core technology is associated with *Homo erectus* (*sensu lato*). Pebble-core technology persists into the ethnographic present in Africa and elsewhere, and thus it is also associated with all known species of the Genus *Homo*. The hypothesis that the distribution of the pebble-core technology in the Early Pleistocene indicates the range of one hominin taxon can be rejected. It is possible that there are subtle strategic differences among pebble-core strategies that may be particular to one or more hominin taxa, but this remains a subject for future research. At the very least, it will require detailed information about the operational sequences (*chaînes opératoires*) at far more Early Paleolithic site than are currently available.

Most of the hominin fossils associated (at a regional level anyway) with African and Eurasian Early Acheulean assemblages are *Homo erectus*. However, *Homo erectus* populations in East and Southeast Asia are not associated with Acheulean LCT technology. Bifacially-shaped LCTs like those found in Early Acheulean assemblages also persist into more-recent time periods. They are associated with *Homo heidelbergensis* in Africa and Europe (Schick and Zhuan 1993; Gamble 1999), *H. neanderthalensis* (Soressi and Hays 2003), the oldest-dated *Homo sapiens* (Shea et al. 2007) and both recent prehistoric and ethnographic humans in the Americas and Australia (Holmes 1897; Tindale 1941). LCT technology is no more correlated with taxonomic differences among hominins than pebble-core technology is.

Cultural Variation?

If one is knapping pebbles or cobbles by hard-hammer percussion, it is impossible to avoid replicating Oldowan-like pebble-cores. The same cannot necessarily be said of Acheulean LCTs. These artifacts are more complex forms that have plainly been imposed on raw material (Roche 2005). Variation among Acheulean LCTs has frequently been plumbed for regionally or chronologically distinct patterns of variation. Few other Paleolithic artifacts have been so assiduously measured in hopes of turning up plausibly "cultural" patterns of design variation. In a just world, this search would long ago have borne fruit. It has not. Except for variation arising from different raw materials and raw material availability, and a few regional trends towards smaller tools (plausibly part of a larger trend in stone tool diminution through the course of the Paleolithic), the range of Acheulean LCT forms remain remarkably similar over the course of hundreds of thousands of years and among assemblages separated by thousands of miles (Isaac 1977; Otte 2003; Roe 2003; Sharon 2009).

The morphological stability of Acheulean LCTs so differs from even the most conservative dimensions of recent human material culture that it is impossible to accept the hypothesis that LCT formal variability was governed by anything even remotely approximating ethnographic human "culture". Human culture is flexible and variable. It creates an archeological record that is complex and highly-contingent across time and space.

One could argue that LCT morphological stability reflects a "paleo-culture", perhaps one operating among hominins with a physiologically-limited capacity to promote, sustain and communicate technological innovation (Jelinek 1977; Mithen 1996). Invoking such an implausibly-constrained "paleo-culture" raises more questions than it answers. For example, what kind of physiological constraint could resist runaway selection for even minor improvements in capacities to innovate and communicate innovations with positive fitness benefits among kin? Conjectural "paleo-culture" ignores uniformitarian theory and it adds nothing of substance to our understanding of Early Paleolithic industrial variability. It is an explanation designed to fit the evidence, not one derived from first principles.

Functional Variation?

There is a longstanding assumption linking the origins and variability of Early Paleolithic stone tools to change and variability in hominin carnivory (for a critical review of this assumption, see Shea 2007b). The numerous sites at which Early Paleolithic stone tools are juxtaposed with broken and cut-marked fossils of large- to medium-sized terrestrial vertebrates certainly shows that stone tools were used for butchery. However, the perceived strength of this link is substantially influenced by preservation bias. Stone tools are virtually indestructible, and vertebrate fossils only slightly less so. Wood, soft plant matter, animal hides, and other substances demonstrably processed with stone tools in recent ethnographic and archeological contexts do not preserve as well. Consequently, we are almost certainly receiving a false negative signal about the more versatile role that stone tools played in Early Pleistocene hominin technological strategies. Finds of plant phytoliths attached to the edges of Acheulean LCTs (Dominguez-Rodrigo et al. 2001) and carved wooden artifacts from rare Early and Middle Pleistocene waterlogged contexts (Theime 1997; Goren-Inbar et al. 2002) clearly point to the use of Early Paleolithic stone tools in tasks other than butchery. Lithic microwear analysis of Early Paleolithic stone tools adds further support to this hypothesis (Keeley 1977; Keeley and Toth 1981).

Acheulean LCTs, in particular, are often portrayed as specialized butchery tools. Experiments by modern archeologists using replicas of these tools uniformly attest to their utility as aids to butchery (Jones 1994; Toth 1997). It is conceivable that the emergence of LCT technology reflects an increasingly predictable role for butchery, and perhaps greater carnivory, in Early Pleistocene hominin adaptations. On the other hand, if there was a strong link between Acheulean LCT production and butchery/carnivory, one would expect these artifacts to exhibit regional ecogeographically-patterned variation in much the same way as recent human subsistence aids. Yet, LCTs are just as conspicuously rare in the colder steppic habitats of northern Eurasia, where hominins would have had to have depended heavily on fats and protein from animal sources, as they are in the jungles of Southeast Asia, where plant foods probably played a more significant dietary role. While there might have been some link between LCTs and butchery (or some other task), these tools do not display the kind of ecogeographically-patterned variation we see among functionally-specialized components of recent human material culture. Again, it is possible that Early Paleolithic technological organization plays by some different set of "rules" from recent human material culture, but unless we are prepared to abandon uniformitarian principles of scientific inference altogether, we have to base our models of past technological variability on the rules by which contemporary material culture appears to be organized.

Conclusion

This paper has explored a strategic perspective on changes in stone tool technology during the period of early hominin dispersal into Eurasia. The broad pattern of Early Paleolithic industrial variability in Africa and Eurasia between 2.0–1.0 Ma can be explained in terms of differences between toolmaking strategies in the "frontier" and "settling in" phases of Early Pleistocene hominin adaptive radiation (Fig. 4.4). "Oldowan" pebble-core technology appears earliest in the Eurasian regions colonized by hominins after 2.0 Ma because its versatility conferred the kind of adaptive flexibility dispersing hominins would have needed in order to cope with unfamiliar landscapes, food sources, and other challenges to their survival. The appearance of Acheulean LCT technology in Africa and in the more temperate parts of Eurasia after 1.5 Ma reflects the consequences of hominins "settling in" to these regions. The key specific variables involved in "settling in" involve increasing predictability of circumstances requiring stone tool use, regional population growth, and probably some degree of intensified intra-specific competition.

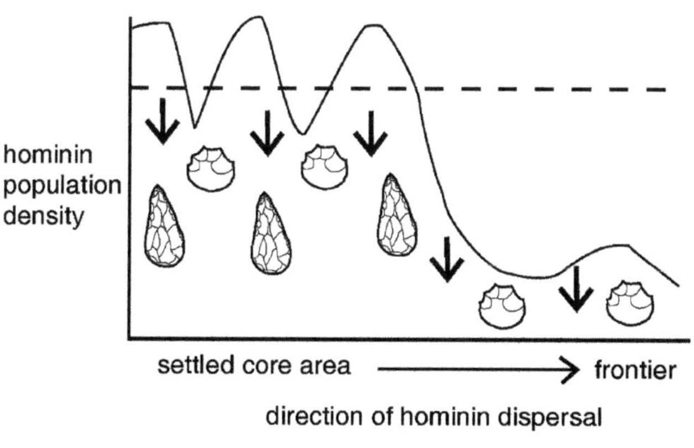

Fig. 4.4 Schematic representation of the proposed relationship between hominin dispersal, demography, and core-reduction strategies. The *x-axis* indicates the direction of hominin dispersal. The *y-axis* is a relative scale of local hominin population density. At the *frontier* of hominin dispersal and *behind* it in regions with low population density the lithic "fallout" is dominated by pebble-core reduction products: cutting tools. The *broken line* indicates a conjectural population density "threshold" above which intraspecific competition, subsistence intensification, and other factors encouraging the production of large cutting tools

Much work remains to be done in refining these strategic models. Choices among strategic alternatives respond to many kinds of energetic and reproductive costs and benefits. Though most archeologists are rightly skeptical that we will ever glean much about hominin reproductive strategies from stone tool variation, it is possible to refine our models of Early Paleolithic behavioral variability by improving the quality of evidence we have that touches on energetic costs and benefits. The area of the catchment over which raw materials were procured, variation in mechanical properties of the rocks being knapped, measurements of cores and flakes that inform us about actual and potential tool utility – all of these need to be more systematically documented and published before we can gauge whether the strategic model outlined here moves us closer to an understanding of Early Paleolithic assemblage formation processes and their relationship to larger patterns of hominin evolution.

Curtailing the use of the terms "Oldowan", "Developed Oldowan", and "Acheulean", to industries widely scattered over time and space would be an important step towards an improved understanding of Early Paleolithic industrial variability. This practice reflects an outdated and unrealistic view that these named industries are substantially analogous to named archeological and ethnographic "cultures". They are not. The strategic differences in core-reduction underlying Oldowan/Developed Oldowan/Acheulean variation have little or no predictive value whatsoever for inferring either their geological antiquity, the identities of the hominin fossils associated with them, or the activities in evidence at sites where they are found. With what descriptive terms to replace these named industries is a complex question. Some form of binomial nomenclature would probably work well, perhaps one combining Grahame Clark's (1977) technological "Modes 1–5" with a local/regional names and more nuanced registers of variability. Deriving these names systematically using cladistic principles would probably also improve the clarity of archeological lithic terminology. The precise nature of such a new taxonomy lies beyond the scope of this paper, but something clearly needs to be done to make our descriptive terms for Early Paleolithic stone tool industries more transparent and anthropologically useful.

Strategic models of Early Paleolithic variability do not rule out the possibility that there are dimensions of variability plausibly referable to cognitive differences or to such social factors as biased transmission of learned patterns of toolmaking behavior. Such "cultural" variation has been amply documented among panin (chimpanzee and bonobo) technology (McGrew 2004). It would be highly counterintuitive for it to have played no role in variation among the byproducts of hominin tool use. Because of the reductive nature of knapped stone technology and the relatively superficial degree to which Early Paleolithic stone tools are modified, identifying dimensions of variability plausibly referable to cognitive or cultural differences is not an easy task. On the other hand, hypotheses invoking variation in cognition and culture as factors in Early Paleolithic industrial variability will be a lot more credible if one can rule out strategic factors of the sort outlined above as plausible alternative explanations.

Acknowledgements I thank the late Glynn Isaac for exciting my interest in Early Paleolithic technology and for making it possible for me to study it with him. I thank Ofer Bar-Yosef and our late colleague, Eitan Tchernov, for countless conversations about the issues discussed here. The data presented in Tables 4.2 and 4.3 reflect the results of discussions among all of the participants in the 2005 Stony Brook Human Evolution Conference. John Fleagle, Rick Potts, and Mike Petraglia made helpful comments on an earlier draft of this paper. The opinions expressed in this paper and any errors remaining are my own responsibility.

References

Ambrose, S. H. (2001). Paleolithic technology and human evolution. *Science, 291*, 1748–1753.

Barnes, A. S. (1939). The difference between natural and human flaking in prehistoric flint implements. *American Anthropologist, 41*, 99–112.

Bar-Yosef, O. (1994). The Lower Paleolithic of the near east. *Journal of World Prehistory, 8*, 211–265.

Bar-Yosef, O. (1999). Lower Paleolithic sites in South-Western Asia-evidence for "Out of Africa" movements. *Anthropologie, 37*, 51–69.

Binford, R. L. (1982). The archaeology of place. *Journal of Anthropological Archaeology, 1*, 5–31.

Binford, L. R. (2001). *Constructing frames of reference: An analytical method for archaeological theory building using hunter-gatherer and environmental data sets*. Berkeley: University of California Press.

Bordes, F. (1961). *Typologie du Paléolithique ancien et moyen*. Bordeaux: Delmas.

Boserup, E. (1990). Population, growth, innovation and resource exploitation. In E. Boserup (Ed.), *Fertility and Resources* (pp. 118–125). Cambridge: Cambridge University Press.

Bosinski, G. (1995). The earliest occupation of Europe: Western Central Europe. In W. Roebroeks & T. Van Kolfschoten (Eds.), *The Earliest Occupation of Europe* (pp. 103–128). Leiden: University of Leiden.

Braun, D. R., Tactikos, J. C., Ferraro, J. V., & Harris, J. W. K. (2005). Flake recovery rates and inferences of Oldowan hominin behavior: a response to Kimura 1999, 2002. *Journal of Human Evolution, 48*, 525–531.

Calvin, W. H. (2002). *A brain for all seasons: human evolution and abrupt climate change*. Chicago: University of Chicago Press.

Chauhan, P. R. (2010). The Indian Subcontinent and 'Out of Africa I'. In J.G. Fleagle, J.J. Shea, F.E. Grine, A.L. Baden, & R.E. Leakey (Eds.), *Out of Africa I: The First Hominin Colonization of Eurasia* (pp. 145–164). Dordrecht: Springer.

Ciochon, R.L. (2010). Divorcing hominins from the *Stegodon-Ailuropoda* Fauna: New views on the antiquity of hominins in Asia. In J.G. Fleagle, J.J. Shea, F.E. Grine, A.L. Baden, & R.E. Leakey (Eds.), *Out of Africa I: The First Hominin Colonization of Eurasia* (pp. 111–126). Dordrecht: Springer.

Clark, G. (1977). *World prehistory in new perspective (third edition)*. Cambridge: Cambridge University Press.

De Beaune, S. A. (1993). Nonflint stone tools of the early Upper Paleolithic. In H. Knecht, A. Pike-Tay, & R. White (Eds.), *Before Lascaux: The complex record of the early Upper Paleolithic* (pp. 163–192). Boca Raton, FL: CRC Press.

Debénath, A., & Dibble, H. L. (1994). *Handbook of Paleolithic Typology* (Vol. 1: Lower and Middle Paleolithic of Europe). Philadelphia: University of Pennsylvania Press.

Dennell, R. (2010). "Out of Africa I": Current problems and future prospects. In J.G. Fleagle, J.J. Shea, F.E. Grine, A.L. Baden, & R.E. Leakey (Eds.), *Out of Africa I: the first hominin colonization of Eurasia* (pp. 247–273). Dordrecht: Springer.

Dennell, R., & Roebroeks, W. (2005). An Asian perspective on early human dispersal from Africa. *Nature, 438*, 1099–1104.

Dominguez-Rodrigo, M., & Pickering, T. R. (2003). Early hominid hunting and scavenging: a zooarchaeological review. *Evolutionary Anthropology, 12*, 275–282.

Dominguez-Rodrigo, M., Serralonga, J., Juan-Tresserras, J., Alcala, L., & Luque, L. (2001). Woodworking activities by early humans: a plant residue analysis on Acheulian stone tools from Peninj (Tanzania). *Journal of Human Evolution, 40*, 289–299.

Foley, R. A. (1987). Hominid species and stone-tool assemblages: how are they related? *Antiquity, 61*, 380–392.

Foley, R., & Lahr, M. M. (1997). Mode 3 technologies and the evolution of modern humans. *Cambridge Archaeological Journal, 7*, 3–36.

Gamble, C. (1999). *The palaeolithic societies of europe*. New York: Cambridge University Press.

Gillespie, J. D., Tupakka, S., & Cluney, C. (2004). Distinguishing between naturally and culturally flaked cobbles: a test case from Alberta, Canada. *Geoarchaeology, 19*, 615–633.

Goren-Inbar, N., Feibel, C. S., Verosub, K. L., Melamed, Y., Kislev, M. E., Tchernov, E., et al. (2000). Pleistocene milestones on the out-of-Africa corridor at Gesher Benot Ya'aqov, Israel. *Science, 289*, 944–947.

Goren-Inbar, N., Werker, E., & Feibel, C. S. (Eds.). (2002). *The Acheulian site of Gesher Benot Ya'acov, Israel: the wood assemblage*. Oxford: Oxbow Books.

Gould, R. A., Koster, D. A., & Sontz, A. (1971). The Lithic assemblage of the western desert aborigines of Australia. *American Antiquity, 36*, 149–168.

Grayson, D. K. (1983). *The establishment of human antiquity*. New York: Academic.

Grayson, D. K. (1986). Eoliths, archaeological ambiguity, and the generation of "Middle-Range" research. In D. J. Meltzer, D. D. Fowler, & J. A. Sabloff (Eds.), *American Archaeology Past and Future* (pp. 77–133). Washington, DC: Smithsonian Institution Press.

Henry, D. O., Hietala, H. J., Rosen, A. M., Demidenko, Y., Usik, V., & Armagan, T. L. (2004). Human behavioral organization in the Middle Paleolithic: were Neanderthals different? *American Anthropologist, 106*, 17–31.

Hoffecker, J. F. (2002). *Desolate landscapes: ice-age settlement in eastern Europe*. New Brunswick, NJ: Rutgers University Press.

Hoffecker, J. F. (2004). *A prehistory of the north: human settlement of the higher latitudes*. Piscataway, NJ: Rutgers University Press.

Holmes, W. H. (1897). Stone implements of the Potomac-Chesapeake tidewater province. *Bureau of American Ethnology Annual Report, 15*, 13–152.

Hou, Y., Zhao, L.X. (2010). New archeological evidence for the earliest hominin presence in China. In J.G. Fleagle, J.J. Shea, F.E. Grine, A.L. Baden, & R.E. Leakey (Eds.) *Out of Africa I: the first hominin colonization of Eurasia* (pp. 87–99). Dordrecht: Springer.

Hou, Y., Potts, R., Baoyin, Y., Zhengtang, G., Deino, A., Wei, W., et al. (2000). Mid-pleistocene Acheulean-like stone tTechnology of the bose basin, South China. *Science, 287*, 1622–1626.
Isaac, G. L. (1977). *Olorgesailie: archaeological studies of a middle pleistocene Lake Basin in Kenya*. Chicago: The University of Chicago Press.
Isaac, G. L. (1981). Stone Age visiting cards: approaches to the study of early land-use patterns. In I. Hodder, G. Isaac, & N. Hammond (Eds.), *Past in perspective* (pp. 131–155). Cambridge: Cambridge University Press.
Isaac, G. L. (1984). The earliest archaeological traces. In J. D. Clark (Ed.), *The Cambridge history of Africa* (Vol. 1, pp. 157–247). Cambridge: Cambridge University Press.
Isaac, G., Harris, J. W. K., & Kroll, E. M. (1997). The stone artefact assemblages: a comparative study. In G. Isaac & A. B. B. Isaac (Eds.), *Koobi Fora research project series* (Plio-Pleistocene Archaeology, Vol. 5, pp. 262–362). Oxford: Clarendon.
Jelinek, A. J. (1977). The Lower Paleolithic: current evidence and interpretations. *Annual Review of Anthropology, 6*, 11–32.
Jelinek, A.J. (1982). The Middle Paleolithic in the Southern Levant with comments on the appearance of modern Homo sapiens. In A. Ronen (Ed.), *The transition from Lower to Middle Paleolithic and the origins of modern man* (pp. 57–104). Oxford: British Archaeological Reports International Series 151.
Jones, P. R. (1994). Results of experimental work in relation to the stone industries of Olduvai Gorge. In M. D. Leakey & D. A. Roe (Eds.), *Olduvai Gorge* (Excavations in Beds III, IV and the Masek Beds, 1968–1971, Vol. 5, pp. 254–298). Cambridge: Cambridge University Press.
Keeley, L. H. (1977). The functions of Paleolithic flint tools. *Scientific American, 237*, 108–126.
Keeley, L. H., & Toth, N. P. (1981). Microwear polishes on early stone tools from Koobi Fora, Kenya. *Nature, 293*, 464–465.
Kelly, R. L. (1988). The three sides of a biface. *American Antiquity, 53*, 717–734.
Kimura, Y. (1999). Tool-using strategies by early hominids at Bed II, Olduvai Gorge, Tanzania. *Journal of Human Evolution, 37*, 807–831.
Kimura, Y. (2002). Examining time trends in the Oldowan technology at Beds I and II, Olduvai Gorge. *Journal of Human Evolution, 43*, 291–321.
Klein, R. G. (1999). *The human career (2nd ed.)*. Chicago: University of Chicago Press.
Kohn, M., & Mithen, S. (1999). Handaxes: products of sexual selection? *Antiquity, 73*, 518–526.
Kuhn, S. L. (1994). A formal approach to the design and assembly of mobile toolkits. *American Antiquity, 59*, 426–442.
Kuman, K. (1998). The earliest South African industries. In M. Petraglia & R. Korisettar (Eds.), *Early human behaviour in global context* (pp. 151–186). New York: Routledge.
Lahr, M. M. (2010). Saharan corridors and their role in the evolutionary geography of 'Out of Africa I'. In J.G. Fleagle, J.J. Shea, F.E. Grine, A.L. Baden, & R.E. Leakey (Eds.), *Out of Africa I: the first hominin colonization of Eurasia* (pp. 27–46). Dordrecht: Springer.
Langbroek, M. (2004). *Out of Africa: an investigation into the earliest occupation of the old world*. Oxford, UK: Archaeopress (British Archaeological Reports International Series 1244).
Larick, R., Ciochon, R., Zaim, Y., Sudijono, S., Rizal, Y., Aziz, F., et al. (2001). Early Pleistocene 40 Ar/39 Ar ages for Bapang formation hominins, Central Jawa, Indonesia. *Proceedings of the National Academy of Sciences of the United States of America, 98*, 4866–4871.
Leakey, L. S. B. (1936). *Stone Age Africa*. London: Oxford University Press.
Leakey, L. S. B. (1960). *Adam's ancestors (fourth edition)*. New York: Harper & Row.
Leakey, M. D. (1966). A review of the Oldowan culture from Olduvai Gorge, Tanzania. *Nature, 210*, 462–466.
Leakey, M. D. (1971). *Olduvai Gorge: excavations in Beds I and II, 1960–1963*. Cambridge: Cambridge University Press.
Leakey, M. D., & Roe, D. A. (Eds.). (1994). *Olduvai Gorge, Volume 5: Excavations in Beds III, IV and the Masek Beds, 1968–1971*. Cambridge: Cambridge University Press.
Leakey, M., & Werdelin, L. (2010). Early Pleistocene mammals of Africa: background to dispersal. In J.G. Fleagle, J.J. Shea, F.E. Grine, A.L. Baden, & R.E. Leakey (Eds.), *Out of Africa I: the first hominin colonization of Eurasia* (pp. 3–11). Dordrecht: Springer.
Lieberman, D. E., & Bar-Yosef, O. (2005). Apples and oranges: morphological versus behavioral transitions in the Pleistocene. In D. E. Lieberman, R. J. Smith, & J. Kelly (Eds.), *Interpreting the past: Essays on human, primate, and mammal evolution in honor of David Pilbeam* (pp. 275–296). Boston: Brill.
Lycett, S. J., & von Cramon-Taubadel, N. (2008). Acheulean variability and hominin dispersals: a model-bound approach. *Journal of Archaeological Science, 35*, 553–562.
Martínez-Navarro, B. (2010). Early Pleistocene faunas of Eurasia and hominin dispersals. In J.G. Fleagle, J.J. Shea, F.E. Grine, A.L. Baden, & R.E. Leakey (Eds.), *Out of Africa I: the first hominin colonization of Eurasia* (pp. 207–224). Dordrecht: Springer.
McBrearty, S., Bishop, L., Plummer, T., Dewar, R., & Conard, N. (1998). Tools underfoot: Human trampling as an agent of lithic artifact edge modification. *American Antiquity, 63*, 108–130.
McGrew, W. C. (2004). Primatology: advanced ape technology. *Current Biology, 14*, R1046–1047.
McPherron, S. P. (1999). Ovate and pointed handaxe assemblages: two points make a line. *Préhistoire Européenne, 14*, 9–32.
Mithen, S. (1996). *The prehistory of the mind*. New York: Thames & Hudson.
Movius, H.L., Jr. (1948). *The Lower Paleolithic cultures of Southern and Eastern Asia*. Transactions of the American Philosophical Society (n.s.) 38 Philadelphia, PA.
Nelson, M. C. (1988). The study of technological organization. In M. B. Schiffer (Ed.), *Archaeological method and theory* (pp. 57–100). Tucson: University of Arizona Press.
Norton, C. J. (2000). The current state of Korean paleoanthropology. *Journal of Human Evolution, 38*, 803–825.
Norton, C. J., Bae, K., Harris, J. W. K., & Lee, H. (2006). Middle Pleistocene handaxes from the Korean Peninsula. *Journal of Human Evolution, 51*, 527–536.
O'Brien, E. M. (1981). The projectile capabilities of an Acheulian Handaxe from Olorgesailie. *Current Anthropology, 22*, 76–79.
Otte, M. (2003). The pitfalls of using bifaces as cultural markers. In M. Soressi & H. L. Dibble (Eds.), *Multiple approaches to the study of bifacial technologies* (pp. 183–192). Philadelphia, PA: University of Pennsylvania Museum.
Patterson, L. W. (1983). Criteria for determining the attributes of man-made lithics. *Journal of Field Archaeology, 10*, 296–307.
Peacock, E. (1991). Distinguishing between artifacts and geofacts: a test case from eastern England. *Journal of Field Archaeology, 18*, 345–361.
Petraglia, M. D. (2010). The Early Paleolithic of the Indian subcontinent: Hominin colonization, dispersals, and occupational history. In J.G. Fleagle, J.J. Shea, F.E. Grine, A.L. Baden, & R.E. Leakey (Eds.), *Out of Africa I: the first hominin colonization of Eurasia* (pp. 165–179). Dordrecht: Springer.
Piperno, M. (Ed.). (1999). *Notarchirico: Un sito del Pleistocene medio iniziale nel bacino di Venosa*. Venosa, Italy: Edizioni Osanna.
Plummer, T. (2004). Flaked stone and old bones: biological and cultural evolution at the dawn of technology. *Yearbook of Physical Anthropology, 47*, 118–164.
Pobiner, B.L. (2007). *Hominin-carnivore interactions: evidence from modern carnivore bone modification and Early Pleistocene archaeofaunas*. Doctoral Dissertation. New Brunswick, NJ: Rutgers University.

Potts, R. (1988). *Early hominid activities at Olduvai*. New York: Aldine de Gruyter.

Potts, R. (1991). Why the Oldowan? Plio-pleistocene toolmaking and the transport of resources. *Journal of Anthropological Research, 47*, 153–176.

Potts, R. (1998). Variability selection and hominid evolution. *Evolutionary Anthropology, 7*, 81–96.

Roche, H. (2005). From simple flaking to shaping: Stone knapping evolution among early hominins. In B. Brill & V. Roux (Eds.), *Stone knapping: the necessary conditions for a uniquely hominid behaviour* (pp. 35–52). McDonald Institute Monographs, Cambridge: Cambridge University.

Roe, D. (2003). An overview, with some thoughts on the study of bifaces. In M. Soressi & H. L. Dibble (Eds.), *Multiple approaches to the study of bifacial technologies* (pp. 273–285). Philadelphia, PA: University of Pennsylvania Museum.

Roebroeks, W. (2001). Hominid behaviour and the earliest occupation of Europe: An exploration. *Journal of Human Evolution, 41*, 437–461.

Ronen, A. (2006). The oldest human groups in the Levant. Comptes Rendus Palevol (Special Issue: CLIMATS-CULTURES-SOCIETES aux temps prehistoriques, de l'apparition des Hominides jusqu'au Neolithique) 5, 343–351.

Rosen, S. A. (1997). *Lithics after the Stone Age: a Handbook of stone tools from the levant*. Walnut Creek, CA: AltaMira Press.

Sahnouni, M. (1997). An Experimental Investigation into the Nature of Faceted Limestone "Spheroids" in the Early Paleolithic. *Journal of Archaeological Science 24*(8), 701–713.

Sahnouni, M. (2006). The North African Early Stone Age and the sites at Ain Hanech, Algeria. In N. P. Toth & K. Schick (Eds.), *The Oldowan: case studies into the earliest Stone Age* (pp. 77–112). Bloomington, IN: Stone Age Institute Press.

Santoja, M., & Villa, P. (2006). The Acheulian in Western Europe. In N. Goren-Inbar & G. Sharon (Eds.), *Axe age: Acheulian toolmaking from quarry to discard* (pp. 429–779). London: Equinox.

Schick, K. D. (1994). The Movius Line reconsidered: Perspectives on the Early Paleolithic of Eastern Asia. In R. S. Corruccini & R. L. Ciochon (Eds.), *Integrative paths to the past: Paleoanthropological advances in honor of F* (pp. 569–595). Englewood Cliffs, NJ: Clark Howell. Prentice-Hall.

Schick, K. D., & Toth, N. P. (1993). *Making silent stones speak: human evolution and the dawn of technology*. New York: Simon & Schuster.

Schick, K. D., & Toth, N. (1994). Early Stone Age technology in Africa: A review and case study into the nature and function of spheroids and subspheroids. In R. S. Corruccini & R. L. Ciochon (Eds.), *Integrative paths to the past: Paleoanthropological advances in honor of F Clark Howell* (pp. 429–449). Englewood Cliffs, NJ: Prentice-Hall.

Schick, K., & Toth, N. P. (2006). An overview of the Oldowan industrial complex: The sites and the nature of the evidence. In N. P. Toth & K. Schick (Eds.), *The Oldowan: case studies into the Earliest Stone Age* (pp. 3–42). Bloomington, In: Stone Age Institute Press.

Schick, K., & Zhuan, D. (1993). Early Paleolithic of China and Eastern Asia. *Evolutionary Anthropology, 2*, 22–35.

Scott, G. R., & Gilbert, L. (2009). The Oldest Handaxes in Europe. *Nature, 461*, 82–85.

Sharon, G. (2009). Acheulian giant core technology: a worldwide perspective. *Current Anthropology, 50*, 335–367.

Shea, J. J. (2007a). The boulevard of broken dreams: Evolutionary discontinuity in the Late Pleistocene Levant. In P. Mellars, K. Boyle, O. Bar-Yosef, & C. Stringer (Eds.), *Rethinking the Human Revolution* (pp. 219–232). Cambridge: McDonald Institute for Archaeological Research Monographs.

Shea, J. J. (2007b). Lithic archaeology, or, what stone tools can (and can't) tell us about early hominin diets. In P. Ungar (Ed.), *Evolution of the human diet: the known, the unknown and the unknowable* (pp. 321–351). Oxford: Oxford University Press.

Shea, J. J., Fleagle, J. G., & Assefa, Z. (2007). Context and chronology of early *Homo sapiens* fossils from the Omo Kibish Formation, Ethiopia. In P. Mellars, K. Boyle, O. Bar-Yosef, & C. Stringer (Eds.), *Rethinking the human revolution* (pp. 153–162). Cambridge: McDonald Institute for Archaeological Research Monographs.

Soressi, M., & Dibble, H. L. (Eds.). (2003). *Multiple approaches to the study of bifacial technologies*. Philadelphia, PA: University of Pennsylvania Museum.

Soressi, M., & Hays, M. A. (2003). Manufacture, transport, and use of Mousterian bifaces: A case study from the Perigord (France). In M. Soressi & H. L. Dibble (Eds.), *Multiple approaches to the study of bifacial technologies* (pp. 125–145). Philadelphia, PA: University of Pennsylvania Museum.

Straus, L. G., & Bar-Yosef, O. (2001). Out of Africa in the Pleistocene: an introduction. *Quaternary International, 75*, 1–5.

Susman, R. L. (1991). Who made the Oldowan tools? Fossil evidence for tool behavior in Plio-Pleistocene hominids. *Journal of Anthropological Research, 47*, 129–151.

Theime, H. (1997). Lower Paleolithic hunting spears from Germany. *Nature, 385*, 807–810.

Tindale, N. B. (1941). The hand axe used in the Western Desert of Australia. *Mankind, 3*, 37–41.

Tooby, J. M., & DeVore, I. (1987). The reconstruction of hominid behavioral evolution through strategic modeling. In W. G. Kinzey (Ed.), *The evolution of hominid behavior: primate models* (pp. 183–237). Albany, NY: State University of New York Press.

Toth, N. P. (1985). The Oldowan reassessed: a close look at early stone artifacts. *Journal of Archaeological Science, 12*, 101–120.

Toth, N. P. (1997). The artifact assemblages in the light of experimental studies. In G. Isaac, & A.B.B. Isaac (Eds.), *Koobi Fora research project series, Volume 5: Plio-Pleistocene archaeology* (pp. 363–402). Oxford: Clarendon.

Toth, N., & Schick, K. (2000). Early Paleolithic. In E. Delson, I. Tattersall, J. A. Van Couvering, & A. Brooks (Eds.), *Encyclopedia of human evolution and prehistory* (pp. 225–229). New York: Garland.

Weiwen, H., & Pu, Z. (2007). Les plus anciennes occupations humaines en Chine. *L'Anthropologie, 111*, 166–181.

Whittaker, J. C., & McCall, G. (2001). Handaxe-hurling hominids: an unlikely story. *Current Anthropology, 42*, 566–572.

Wynn, T. (1995). Handaxe enigmas. *World Archaeology, 27*, 10–24.

Zahavi, A., & Zahavi, A. (1997). *The handicap principle: a missing piece of Darwin's puzzle*. Oxford: Oxford University Press.

Zaim, Y. (2010). Geological evidence for the earliest appearance of hominins in Indonesia. In J.G. Fleagle, J.J. Shea, F.E. Grine, A.L. Baden, & R.E. Leakey (eds.), *Out of Africa I: the first hominin colonization of Eurasia* (pp. 97–110). Dordrecht: Springer.

Zhu, R., An, Z., Potts, R., & Hoffman, K. A. (2003). Magnetostratigraphic dating of early humans in China. *Earth-Science Reviews, 61*, 341–359.

Part II
Eastern Asia

Introduction

In "Behavioral and environmental background to 'Out of Africa I' and the arrival of *Homo erectus* in East Asia", Richard Potts and Robin Teague provide a transition between the papers in the previous part and those documenting early hominin presence in Eastern Asia. In an effort to assess the mobility of early hominins at the beginning of the Pleistocene, they review what is known of limb proportions, distances of archeological sites from available source rocks, and distribution of sites through geological evidence of environmental change. All suggest that early *Homo* was capable of greater geographic and ecological mobility or adaptability than earlier hominins, thus showing greater potential for intercontinental dispersal. Turning to the earliest evidence of hominins in East Asia, they find that the earliest stone tools from East Asia are similar to Oldowan artifacts from Africa, although their abundance varies considerably among sites. The identification of early hominins at many sites in East Asia is currently being reassessed (see Ciochon, this volume), but incisors from Yuanmou are similar to the same teeth in *Homo erectus*. The authors then review alternative geographical routes for hominin dispersal from Africa into Eurasia as well as different models of the ecological and geographical relationships between dispersing hominins and other fauna.

In "New archeological evidence for the earliest hominin presence in China", Ya-Mei Hou and Ling-Xia Zhao describe the results of recent and ongoing excavations at three Early Pleistocene sites in China with possible evidence of early hominin presence. Renzidong consists of two fissure-filling deposits in Anhui Province estimated to be of Early Pleistocene age. They have yielded a large mammalian fauna and 59 stone artifacts identified as crude tools, mostly scrapers and flakes. In addition, several artifacts have been identified as bone tools. Longgudong is a well-known cave site in southwest Hubei Province that has yielded a large mammalian fauna, including *Gigantopithecus*. Recent excavations have yielded over 600 stone artifacts identified as the result of hominin activity, including cores, flakes, and small tools. Longgupo is another well-known cave site from southern China that has been the focus of intermittent excavations for over 20 years. Dated to the Early Pleistocene, the site has yielded a large mammalian fauna of 92 species including a mandible and several teeth attributed to *Homo erectus*. Twenty-six stone artifacts, mostly large tools, have been identified from four different layers. There is much ongoing debate over the identification of many of the stone tools and hominid fossils from these sites. Nevertheless, it is clear that the Early Pleistocene dispersals out of Africa reached China and much more research is needed to clarify the role of this vast area in early human evolution.

The early hominins of Indonesia were discovered well before we had any evidence of a hominin fossil record in Africa, and ongoing discoveries continue to provide new insights into human evolution from that country. In "Geological evidence for the earliest appearance of hominins in Indonesia", Yahdi Zaim reviews the Quaternary geology of Java. He discusses not only famous hominin sites such as Trinil and Ngandong, but also many other paleontological sites from the Pleistocene of Java, and places the geology of Java in a regional context for both geology and biogeography of the mammalian faunas.

In "Divorcing hominins from the *Stegodon-Ailuropoda* fauna: new views on the antiquity of hominins in Asia", Russell Ciochon provides a personal overview and revision of the record of early hominin fossils from the Early and Middle Pleistocene of China as well as a review of the early hominins from Java. The recent discovery of a large sample of fossil ape teeth from Mohui Cave in Guangxi Province now indicates that in the Early Pleistocene of southern China, *Gigantopithecus* is found in association with another fossil ape, similar in dental features to the late Miocene and Pliocene *Lufengpithecus*. On the basis of this new material, Ciochon argues that the purported early hominin teeth from Longgudong and Longgupo are most likely to be from this fossil ape rather than from the genus *Homo*. In addition, he suggests that the hominin specimens previously identified from similar aged-deposits in Vietnam need to be reviewed in this context. While this interpretation brings into question evidence of Early Pleistocene hominins in southern China, excavations in the Nihewan Basin support an early hominin presence in northern China. Turning to the well-documented early hominins of Java, Ciochon argues for a coastal dispersal of hominins from Africa to Southeast Asia and offers a new reconstruction of the habitat occupied by *Homo erectus* in the Solo Basin.

The Editors

Chapter 5
Behavioral and Environmental Background to 'Out-of-Africa I' and the Arrival of *Homo erectus* in East Asia

Richard Potts and Robin Teague

Abstract Current evidence of hominin fossils and artifacts in China and Indonesia points to the arrival and persistence of the genus *Homo* in East Asia by 1.7 million years ago (Ma). By at least 1.66 Ma, East Asian hominins had spread across a wide range of biotic and climatic zones, spanning 7°S–40°N on the basis of well-constrained age data from the Nihewan and Yuanmou basins, China, and Sangiran, Indonesia. Archeological assemblages and fragmentary hominin morphology show strong similarities with African Oldowan hominin toolmakers and early *Homo erectus*, although the taxonomic status of the oldest known Asian hominins is not yet securely established. Despite this apparent derivation of East Asian from African hominins, an initial comparison of large mammal faunas offers little evidence of 'fellow travelers', i.e., a set of African mammalian species that co-dispersed with *Homo* to East Asia. We offer three hypotheses to account for the existing data: (1) unique hominin dispersal, in which no other African mammals were involved; (2) African fellow travelers, in which *Homo* and a small number of other mammals reached western Eurasia, yet hominins dispersed independently to more distant regions; and (3) relay dispersal, in which *Homo* always dispersed as part of an ecological community but in association with a different set of mammalian species from one region to another.

Keywords Adaptability • East Asia • Geographic dispersal • *Homo erectus* • Nihewan • Paleoenvironment • Plio-Pleistocene fauna • Yuanmou

R. Potts (✉)
Human Origins Program, National Museum of Natural History, Smithsonian Institution, Washington, DC, 20013-7012, USA
e-mail: pottsr@si.edu

R. Teague
Human Origins Program, National Museum of Natural History, Smithsonian Institution, Washington, DC, 20013-7012, USA
e-mail: teaguer@si.edu

Introduction

This paper explores several datasets and their implications relevant to the spread of earliest Pleistocene hominins to eastern Asia. We first examine the African setting of stone toolmakers between ~2.6 and 1.7 million years ago (Ma), which is the era before and immediately subsequent to the first appearance datum (FAD) of hominins outside of Africa. We have assembled a range of information bearing on:

- The mobility of early human toolmakers (e.g., archeological stone source-to-site distances).
- The paleogeographic diversity of African archeological sites, which relates to constraints on the mobility of Oldowan toolmakers and when these began to be lifted.
- The adaptability of those toolmakers to habitat variability, which we consider an important factor in the eventual dispersal of African hominins to novel environments in eastern Asia.

Initial comparison of mammalian fossil data and archeological occurrences from East Africa, China, and intervening regions leads us to identify several hypotheses regarding the original timing, faunal context, and routes of the earliest dispersal out of Africa. Age constraints on the earliest Pleistocene archeological sites in China provide the oldest definitive evidence of stone toolmaking and animal processing in eastern Asia. These new findings from the Yuanmou and Nihewan basins also enable us to characterize the environments encountered by the first hominins to arrive in this region of the world, by ~1.7 Ma. We end by discussing several ramifications of early *Homo erectus*' long-distance expansion, mainly concerning the range of morphological variation in this species and its adaptability to novel habitats.

The African Context for 'Out-of-Africa I'

The mobility of Late Pliocene hominins can be assessed using at least two lines of evidence – first, skeletal morphology and, second, archeological data on rock transport distances

J.G. Fleagle et al. (eds.), *Out of Africa I: The First Hominin Colonization of Eurasia*, Vertebrate Paleobiology and Paleoanthropology,
DOI 10.1007/978-90-481-9036-2_5,

and the spread of stone tools to new paleogeographic zones. Archeological data presented in this section help to assess the adaptability of African Oldowan hominins, and ultimately relates to their ability to disperse out of Africa and into novel habitats of East Asia.

Morphological Indicators of Terrestrial Mobility

A change in limb proportions evident in Plio-Pleistocene hominins is pertinent to the question of long distance dispersal. Analysis of limb bones ~2.5 Ma from the Middle Awash, Ethiopia, suggests that elongation of the femur occurred prior to shortening of the forearm in at least one lineage of late Pliocene hominins (Asfaw et al. 1999). If confirmed by further discoveries, this finding implies that an important shift in limb proportions (at least compared to *A. afarensis*) was initiated by ~2.5 Ma. *H. habilis* may also have had a human-like, elongated femur coupled with an ape-like, long forearm. According to Haeusler and McHenry's (2004) analysis of the fragmentary partial skeletons OH 62 and KNM-ER 3735 (~1.8–1.9 Ma), *H. habilis* exhibited a modern human pattern of limb shaft proportions, with an elongated hindlimb relative to *A. afarensis* and *A. africanus*, yet with similar brachial proportions to these taxa. If this finding is also borne out by future studies, the elongated hindlimb of *H. habilis* suggests similarities to *H. erectus* that may imply an anatomical commitment to terrestrial bipedality over longer distances than in early *Australopithecus*.

An even greater commitment to long distance terrestrial locomotion is suggested in the ruggedly-built innominate KNM ER-3228, thought to date to ~1.9 Ma, and is further evidenced by femora assigned to early African *H. erectus* at ~1.7 Ma (Ruff and Walker 1993) and by the overall skeletal proportions of this species (Walker and Leakey 1993).

Archeological Measures of Mobility: Lifting of Landscape Constraints

Tracing the sources of rocks found in archeological sites provides a more direct line of evidence regarding mobility. Although rock source-to-site distances offer a very limited estimate of actual home range, the development of longer transport distances may signal greater mobility associated with the dispersal of Oldowan toolmakers. The diversity of paleogeographic settings in which stone artifact sites occur is another important measure. It indicates the degree to which the ranging distances of the toolmakers were tethered to, or decoupled from, particular lithic, water, or other resources that occurred in specific depositional environments. Key observations in this regard are as follows:

- The oldest known archeological sites in the Middle Awash, ~2.6 Ma at Gona, Ethiopia, and in the Turkana Basin (West Turkana) ~2.3 Ma, were situated only tens to a few hundreds of meters away from paleo-conglomerates where the source rocks for making tools were located (Semaw et al. 2003; Kibunjia 1994; Delagnes and Roche 2005). At Turkana, archeological sites in the Late Pliocene are quite rare, and the sites are distinctly limited to places where marginal drainages intersected the axial drainage (Proto-Omo) (Rogers et al. 1994).
- Stone tools excavated from Kanjera South, ~2.1 Ma, were made of rocks obtainable from conglomerate sources up to 12 km away from the archeological sites, although most rock sources occurred within shorter distances (Plummer 2004; Braun et al. 2005; Braun 2006).
- In Bed I and lower Bed II Olduvai, ~1.85–1.77 Ma, all published sites occur within about 10 km of the most distant rock source (Kelogi gneiss). Yet all such sites and rock sources were confined to the lake margin lithofacies, which suggests a strong tethering of stone-tool activities to the lake margin zone until about 1.77 Ma (Hay 1976).
- Data for the interval 1.77–1.50 Ma, available chiefly from Bed II Olduvai and the Okote Member, East Turkana, show substantial paleogeographic diversification in the location of archeological sites for the first time in both basins. At Olduvai, hominin tool activities were situated in both the lake-margin and fluvial facies; these sites also record the first use of raw materials (e.g., Engelosin phonolite) obtainable beyond the lake margin zone (Hay 1976). At Turkana, stone tool sites occurred in the channels and on proximal and distal floodplains of the marginal drainages away from the axial river system (Rogers et al. 1994). However, the maximum transport distance of stone remains the same, with all lithic sources available within 12–15 km of the archeological sites.
- The next available East Africa dataset in the temporal sequence comes from Member 1 Olorgesailie, Kenya, ~0.99 Ma. Our detailed study of the Olorgesailie region shows that 98% of the stone tools were made from rocks obtainable within about 5 km; however, 2% were made from sources up to about 46 km away (Isaac 1977; Potts et al. 1999; Noll 2000). This fourfold increase in maximum transport distance represents a notable change in the mobility of hominin groups and the decoupling of their activities from highly localized landscape features.

Adaptability to Novel Environments

The spread of early humans to eastern Asia meant that the dispersing hominins faced new climatic and biogeographic zones. This encounter with novel landscapes and adaptive settings almost certainly placed the adaptability of those populations at a premium. The degree of adaptability of pre-existing populations of African Plio-Pleistocene hominins is thus interesting to measure, and one way to do this is by examining the persistence of archeological and fossil material across stratigraphic boundaries that mark substantial changes in environment. The lengthy stratigraphic and paleoanthropological records of the Koobi Fora and Olduvai regions are particularly helpful in this regard.

Figure 5.1 synthesizes the stratigraphic, environmental, and archeological record of the Koobi Fora region, an exercise that has benefited greatly from Craig Feibel's input (see also Brown and Feibel 1991). Our main observations are as follows:

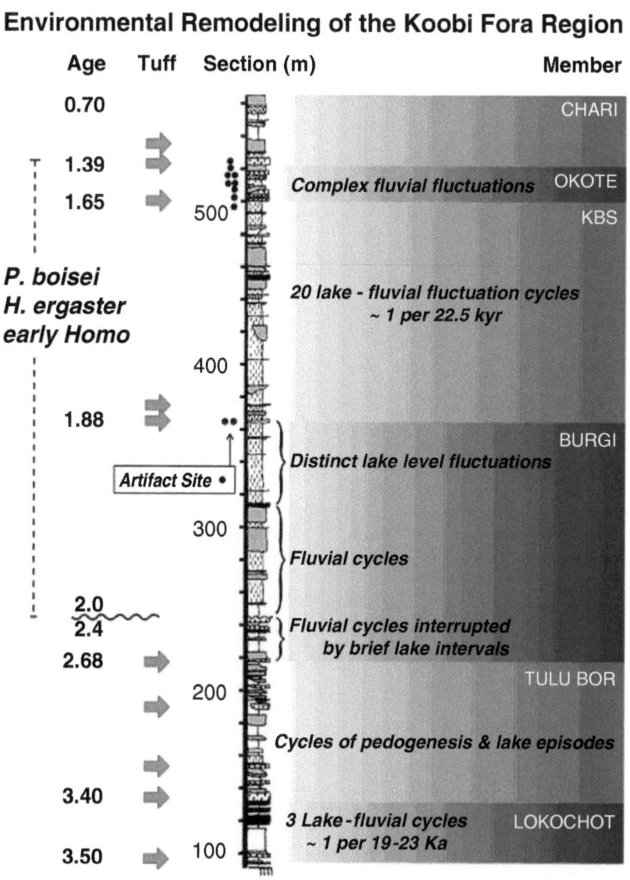

Fig. 5.1 Composite stratigraphic section of the Koobi Fora Formation, Turkana Basin, northern Kenya (based on Brown and Feibel 1991). *Arrows* indicate important tephra layers. Artifact sites are indicated by the *dots* along the *left* margin of the section, and show the stratigraphic concentration of archeological sites after ~1.7 Ma. Early hominin fossils occur throughout the section from 2.0 to 1.4 Ma. Each portion of the section is characterized by a combination of climatic and tectonically-induced environmental variability

- The Turkana basin was subject to frequent landscape remodeling due to (1) alternation between an expanded lake and widespread fluvial conditions on a precessional cycle timescale (19–23 kyr) and (2) variability in river conditions due to the effects of volcanic ash input. According to Feibel's reconstructions, a large meandering river became clogged with volcanic ash, causing the axial river system to become divided into a complex braided stream system, which eventually recovered its original meandering state (Rogers et al. 1994).
- The sporadic occurrence of archeological sites over the time period of 2.3–1.7 Ma prevents evaluation of the persistence of toolmakers across paleoenvironmental shifts.
- By the oldest part of the Okote Member, ~1.7 Ma, when environmental fluctuations were accentuated by frequent volcanic input (Feibel 1997), there is evidence that the stone toolmakers were generally able to persist through episodic disruptions of the landscape. As shown in Fig. 5.1, archeological sites greatly increase in number at this time, and the stratigraphic persistence of stone tools across intervals of major remodeling of the landscape is considerably greater than in earlier periods.

The record from Olduvai Gorge shows the persistence of hominin toolmakers through a variety of landscape, vegetation and climatic changes (e.g., Hay 1976; Bonnefille and Riollet 1980; Cerling and Hay 1986). The stratigraphic interval from 1.9 to 1.77 Ma (Fig. 5.2) shows three significant environmental transitions: Initially wet, marshy and wooded conditions gave way to arid, open vegetation, followed by the return of moist, wooded habitats, and then followed by very dry conditions with sparse vegetation in the Lemuta Member. One major environmental shift took place every 17–37 kyr, with the Lemuta Member aridity lasting approximately 20–50 kyr (Hay 1976). Although stone tools are present throughout the sequence, the sites and lithic source rocks were evidently confined to the lake margin zone.

After this time, the broad, perennial lake intermittently broke up into a series of ponds and marshes in an arid landscape (Hay 1976). Hominin sites outside the lake margin zone occurred for the first time, indicating that toolmakers were no longer tethered to this particular environmental setting. It appears that a limited repertoire of tool activities confined to specific landscape features (evident prior to 1.77 Ma) was replaced by a more diverse array of activities able to accommodate to a less stable landscape (Potts et al. 1999).

Overall, the geographic and ecological constraints that shaped and tethered the activities of Oldowan toolmakers in East Africa apparently began to be relaxed by about the upper Olduvai-Matuyama transition at 1.78–1.77 Ma, or slightly thereafter. Evidence of the first definite appearance of hominins outside of Africa, and their spread to East Asia, now appears to be consistent with this date.

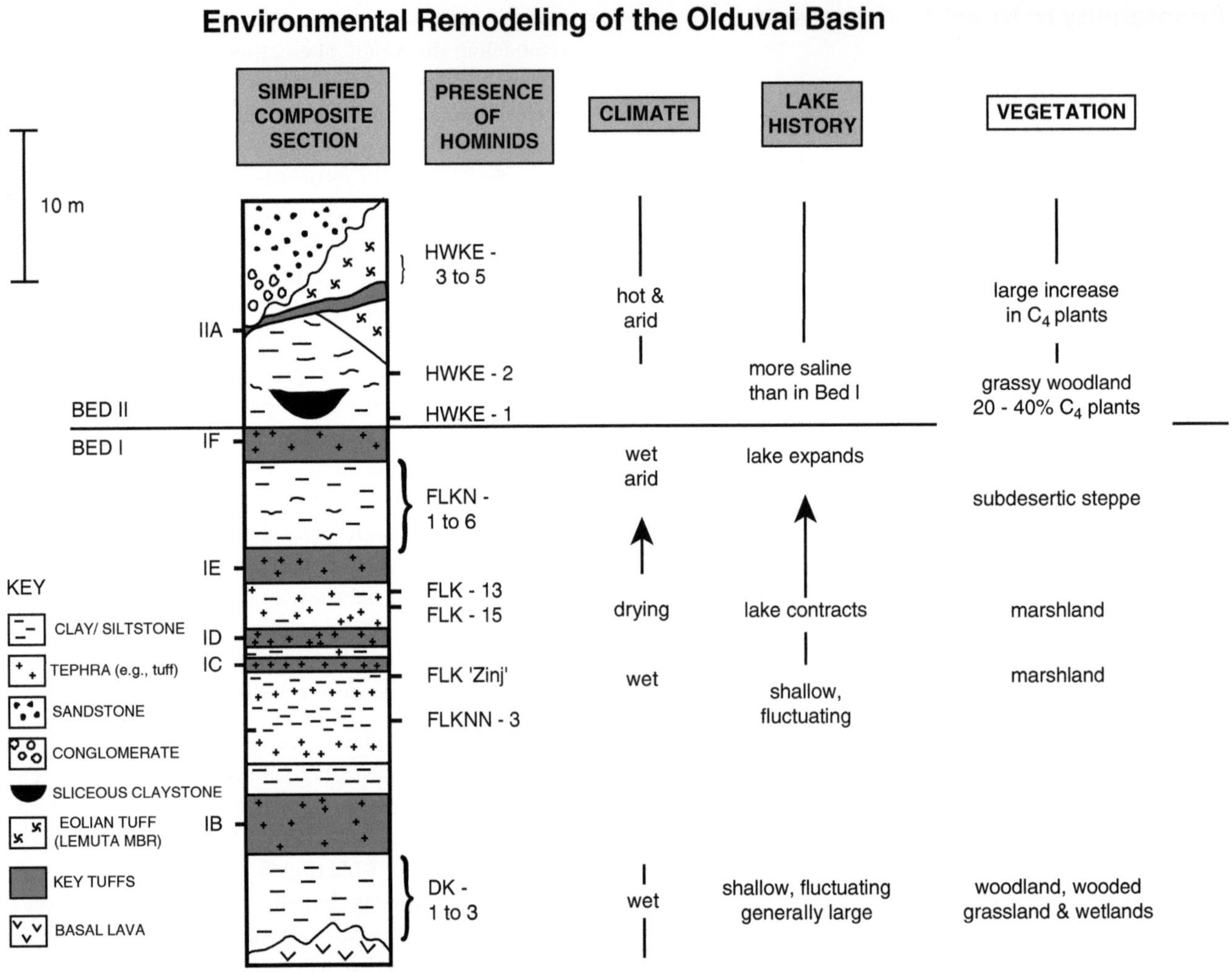

Fig. 5.2 Simplified stratigraphic section of Bed I and lower Bed II Olduvai, northern Tanzania. Also shown are the names of archeological sites, which illustrate the continuous presence of hominin toolmakers throughout the sequence, and three lines of evidence regarding environmental change

Arrival in East Asia: Evidence from Yuanmou and Nihewan, China

A team lead by Zhu Rixiang (Institute of Geology and Geophysics, Beijing) and R. Potts has undertaken a detailed re-investigation of the magnetostratigraphic ages of potentially the oldest artifacts and fossil hominins in China. This work has yielded age estimates of ~1.34 to ~1.71 Ma for the oldest archeological sites and hominin fossils in North and South China (Zhu et al. 2001, 2003, 2004, 2008).

As summarized in Fig. 5.3, the sites of Xihoudu, Gongwangling, and Donggutuo are dated between ~1.3 and 1.1 Ma, and reflect solid evidence of hominin toolmakers in North China during that interval. Dates from the older sediments of the Nihewan basin extend the known time range of hominin stone tools and animal-processing activities in North China back to ~1.36 Ma at the Xiaochangliang site, and ~1.34–1.66 Ma for four archeological levels at the site of Majuangou. Recent work at Yuanmou, South China, has resulted in the oldest estimated ages for definite stone artifacts and hominin fossils (two upper central incisors) in China, at ~1.71–1.70 Ma. The claim of older hominin evidence from Longgupo, China, is based on published fossils (Huang et al. 1995) that are increasingly viewed as an ape possibly related to *Lufengpithecus*, which is known from the Late Miocene of South China (e.g., Schwartz and Tattersall 1996; Wu 2000; Ciochon 2010). Stone artifacts have also been recovered from Longgupo (see Hou and Zhao 2010).

Yuanmou Basin Hominin Fossils and Stone Tools

The upper central incisors of what is probably a single individual of *H. erectus* and four quartzite artifacts were

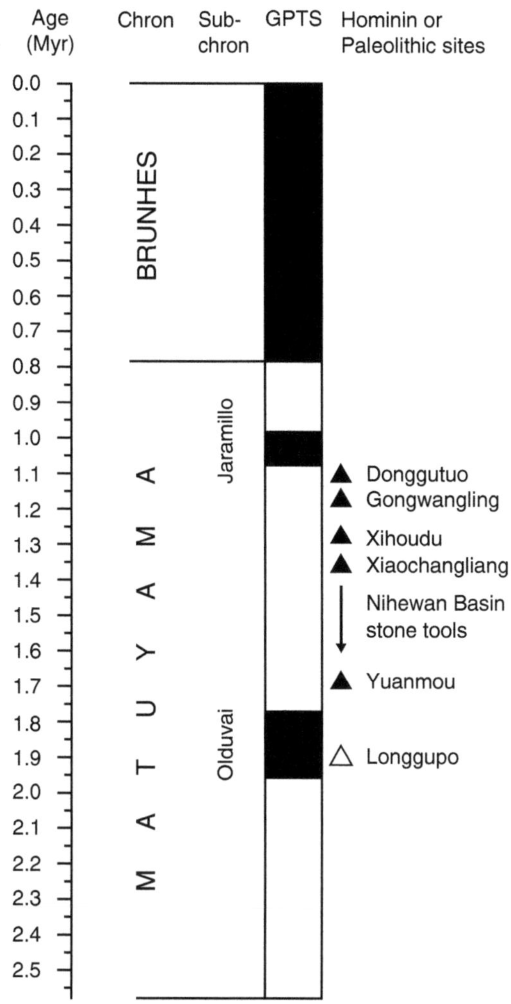

Fig. 5.3 Age estimates for the oldest Chinese paleoanthropological sites, adapted from Zhu et al. (2003). *Filled triangles* indicate sites with tight age constraints; *open triangle* (Longgupo) indicates an uncertain age constraint or the uncertain hominin status of the remains. Further excavations at Xihoudu are needed to re-examine whether its purported stone tools do indeed reflect repetitive, purposeful conchoidal fracture consistent with Oldowan capabilities

recovered in situ in silty-clay sediments within a 4-m-high hill of Member 4 sediments of the Yuanmou Formation (Hu 1973; Wen 1978; Yuan et al. 1984). The assignment of the teeth to hominins, specifically *H. erectus*, is based on observations of multiple investigators (e.g., Hu 1973; Wu and Poirier 1995; Zhu et al. 2008) that the Yuanmou incisors are essentially identical in metrical and qualitative characteristics to later *H. erectus* I^1s from Zhoukoudian. The Yuanmou I^1s also strongly resemble those of KNM-WT 15000 (1.6 million-year-old *H. ergaster*, West Turkana, Kenya) in overall and detailed morphology, including the degree of lingual cervical crown swelling, development of the central pillar (both Yuanmou I^1 crowns, only on the left I^1 of WT 15000), and the degree of enamel wrinkling. The Yuanmou incisors, furthermore, are dissimilar to the I^1 crowns of all extant great apes, especially in the lack of a cervical cingulum, the degree of expansion of the lingual cervical surface, and the overall splaying (or rounded arch) typical in the great apes. In the only definitive fossil ape comparisons possible at present, the mesiodistal and labiolingual dimensions (Yuanmou left I^1: 11.4 and 8.1 mm; right I^1: 11.5 and 8.6 mm) of the Yuanmou incisors fall outside the 95% confidence intervals estimated from the large sample of I^1 crowns of *Lufengpithecus*, a late Miocene ape known from the Yuanmou basin (Liu et al. 2000; Wood and Xu 1991; Zhu et al. 2008).

Based on more recent interpretation of the hominoid gnathic remains from Longgupo (see above), it is valid to wonder whether one or more lineages of poorly documented Plio-Pleistocene, East Asian apes converged in their dental crown morphology with that of early *Homo* in Africa (Wang et al. 2007; Ciochon 2010). The strong match of the Yuanmou incisors with those of even later *H. erectus* – coupled with the presence of artifacts in fine-grained, *in situ* context and indicative of precise stone-on-stone percussion, competent use of striking platform angles, and the production of multiple, overlapping flake scars typical of African Oldowan and slightly later Chinese archeological assemblages (Majuangou, Nihewan basin [Zhu et al. 2004]) – together present reasonably distinct evidence of hominin presence.

A previous publication claimed that the Yuanmou finds had occurred in a magnetostratigraphically normal interval that was likely the Brunhes, implying an age of <790–780 ka for the hominin material (Hyodo et al. 2002). Although it is impossible to determine from the information provided in his publication, the detailed field records show that Hyodo's magnetostratigraphic study sampled the upper 88 m of Yuanmou Member 4 in only six sampling levels, and these samples were from a section that ranged from several kilometers to 600 m away from the discovery site of the hominin fossils and artifacts (Zhu et al. 2008).

The Yuanmou work led by Zhu and Potts from 2003 to 2005 has yielded a far more detailed magnetostratigraphic record, which encompasses parallel sections throughout the entire Yuanmou sequence, including three sections surrounding the original fossiliferous hillside (which was destroyed in the process of excavation). The stratigraphic interval of the hominin finds were 12.7–14.2 m above the upper boundary of the Olduvai subchron, and the entire section of Member 4 represents a continuously aggrading floodplain. Using the range of sedimentation rates for these deposits of 29.1 cm kyr^{-1} to 21.3 cm kyr^{-1} (a relatively consistent rate estimated from various portions of the record between geopolarity boundaries), we estimated that the age of the hominin-bearing stratigraphic interval is ~1.71–1.70 Ma, with ~1.7 Ma as the best approximation of the hominin remains (Zhu et al. 2008). We have made a considerable effort to determine exactly which of the diverse mammalian fossil material is from Yuanmou Member 4, a rich assemblage derived from the same hill as the hominin material. The fauna is entirely consistent with a late Pliocene to early Pleistocene age, and the Yuanmou fauna is constrained to an age of roughly 1.8–1.6 Ma.

On the basis of the combined fossil, archeological, and paleomagnetic evidence, therefore, the teeth and artifacts found together in Member 4 of the Yuanmou Formation comprise, for now, the oldest documented record of associated *Homo* and hominin stone tools in East Asia. Although further survey is needed, it appears that stone artifacts at Yuanmou are rare. In contrast to the situation in the Nihewan basin (see below), the dearth of artifacts at Yuanmou suggests that the toolmakers there depended very little on stone tools or deposited them while merely "passing through" an ephemeral stopping point, rather than "settling in" a region for a substantial period of time.

The age estimate for the Yuanmou teeth and artifacts is only ~50–70 kyr younger than the Dmanisi material, assuming that the latter currently represents, at a maximum age of 1.75–1.77 Ma (Rightmire et al. 2006), the oldest undeniable evidence of hominins outside the African continent. Taking these dates at face value as the first appearance datums (FADs) in western Eurasia and East Asia, and given the direct distance between Dmanisi and Yuanmou of about 6,000 km, a minimum (and highly underestimated) average rate of spread across Asia would have been roughly 1.7–2.4 km per generation (20 years). By comparison, a recent paper considers a "diffusion wave" of "something more than 3 km per generation" as a model for the rapid spread of modern humans during the late Pleistocene (Eswaran et al. 2005).

Two problems exist with any such a calculation for the initial spread of hominins to East Asia. The first is that the hominin FAD error margins for western Eurasia and East Asia must be very large given the very few sites that help constrain the first appearances in these regions. In the case of western Eurasia, Dmanisi is the only relevant site so far, while in eastern Asia, the sites are few even though Yuanmou, Nihewan (Majuangou), and Sangiran all converge on a date of roughly 1.7 Ma. Dennell and Roebroeks (2005; Dennell 2010) caution that there is little firm evidence to show that the hominin dispersal out of Africa occurred at 1.9–1.7 Ma and that an earlier dispersal, possibly involving a species of hominin other than *Homo erectus* or *Homo ergaster*, is possible. The real problem, in fact, is that criteria for signaling a significant geographic expansion have not yet been adequately defined. Such criteria have never been addressed in a statistical sense by Dennell and Roebroeks, or by those who claim that a single, isolated discovery site represents a true first appearance by hominins in a new region of the world. Given the serendipitous nature of discovery and taphonomic factors such as sedimentary burial, fossilization, and erosion, the study of hominin biogeography may be better served by focusing not on isolated discovery sites and dates, but rather on evidence of successful colonization. The oldest appearance of stone tools or hominin fossils in a prolonged stratigraphic sequence of hominin sites, as is evident in Nihewan and Sangiran beginning by 1.66 Ma, may be considered to reflect success in infiltrating a new region. Such evidence signals a far more significant biogeographic and ecological event than is indicated by an isolated discovery in one stratum.

The second problem is that the notion of a direct spread from west to east, as if hominin populations knew where they were going, is flawed. Although dispersal can be defined as the unidirectional movement of an individual away from its birth place (Bullock et al. 2002), long-distance dispersal as part of the evolutionary history of a lineage is generally thought to involve numerous stochastic movements (Hubbell 2001; Nathan 2006). The resulting process of movement, thus, makes the calculation of a lineage's dispersal rate in a single direction based on two endpoints rather misleading.

Nihewan Basin Archeological Sites

The sparse hominin record from Yuanmou contrasts considerably from that in the Nihewan basin. In the Nihewan, the oldest archeological occurrence, Majuangou III, is stratigraphically located 10 m above the upper boundary of the Olduvai subchron, and is one of four stone tool levels uncovered at Majuangou between the Olduvai and the Jaramillo subchrons. Based on a highly consistent average rate of deposition in two parallel sections of predominantly lacustrine sediments, the interpolated ages of the four levels are 1.32, 1.55, 1.64, and 1.66 Ma (Zhu et al. 2004).

The significance of the Majuangou and Xiaochangliang archeological sites is as follows:

- The stone tool assemblages are consistent with African Oldowan technology, both in terms of tool types and approaches to stone flaking. Each of the Nihewan assemblages is readily divided into flakes, flaked pieces, and hammerstones; the flaked pieces can be placed into artifact categories such as chopper, scraper, and polyhedron that are distinctive of African Plio-Pleistocene stone tool assemblages. The main differences between the Nihewan artifacts and the Oldowan of East Africa can be explained by differences in the initial form of the raw material. The Nihewan cores were chipped from angular fragments of Triassic basement rock, and thus differ from typical choppers and other core forms of the East African Oldowan, which are generally made from rounded cobbles.
- Fractured animal bones damaged by stone tools are precisely associated with the stone tools at Majuangou III (MJG-III). Although some faunal remains occur in each of

the archeological levels at Majuangou and Xiaochangliang, mammalian bones ranging in size from deer to elephant are abundant at MJG-III and correspond spatially and stratigraphically exactly to the distribution of the stone artifacts. The animal remains accumulated at the site include horse *Equus sanmeniensis*, deer *Cervus* sp., elephant *Elephas* sp., bovid *Gazella* sp., and rhinoceros *Coelodonta antiquitatis*. Both the artifacts and faunal remains occur in fine-grained silty clay associated with preserved aquatic plants, which together indicate a low-energy lake margin or marsh environment. Most importantly, the animal bones show evidence of tool percussion marks indicative of marrow extraction. Surface marks are also very abundant; while some of these are very likely butchery marks, the overwhelming proportion appear to have been caused by animal trampling, which hinders the definitive recognition of cut marks. Nonetheless, evidence of marrow processing, the exact association with stone tools, and the body size and taxonomic diversity of mammals transported to the site, all represent strong similarities to the early archeological record of East Africa. The Nihewan toolmakers gained access to animal tissues in ways that suggest continuities with the Late Pliocene and Early Pleistocene Oldowan hominins of Africa.

- The stratigraphic recurrence of stone artifacts suggests frequent or continuous occupation of the Nihewan basin. In contrast to the one layer of hominin remains currently known so far in the Yuanmou Basin, the stone tools present in multiple stratigraphic horizons at Majuangou – corresponding to intervals of lake-margin sediments in a mostly lacustrine sequence – suggest that the Nihewan was an area of consistent hominin occupation for a long period. The density of archeological remains within sites and their stratigraphic recurrence are in accord with an archeological signal of groups "settling in" rather than merely "passing through" an area.
- The Nihewan basin is situated at ~40°N latitude, approximately equivalent to Dmanisi, but is nearly adjacent to a permanent area of aridity, the desert sources that supply wind-blown loess to the Loess Plateau. Furthermore, the latest age estimates for archeological sites in the Nihewan and elsewhere in North China indicate a persistence of hominins in this region from ~1.66 to 1.1 Ma, a span of ~500 kyr. This begs the question of how, beginning at ~1.66 Ma, hominins with an Oldowan technology could withstand the seasonal and longer-term oscillations in climate that must have operated in this location. The presence of a permanent lake in the basin may have had a critical ameliorating influence, serving to buffer seasonal and larger-scale fluctuations in plant and animal resources. A study of the fossil pollen sequence in the Nihewan lake sediments is planned to assess vegetational change over time.

Biogeographic Context of Out-of-Africa Dispersal

On the basis of faunal data, there were two periods of potentially marked dispersal of large mammals from Africa during the Late Pliocene and Early Pleistocene. These dispersal events are constrained to ~2.6–2.5 Ma and shortly after 2.0 Ma. Both correspond to periods of faunal turnover (Martínez-Navarro 2010; Vrba 1995). Faunal turnover in Africa and Eurasia around 2.6 Ma was associated with the onset of colder temperatures and loess deposition, and the extinction of warm-adapted taxa in the Eurasian temperate zone (Kukla and An 1989; Liu 1985; Azzaroli 1995). In this section we briefly examine the possible first appearance of African taxa in other regions immediately east of the African continent during the Plio-Pleistocene and the implications regarding potential dispersal routes.

Levant

After 1.77 Ma (upper boundary of the Olduvai subchron), African taxa are found at Dmanisi and 'Ubeidiya. It is commonly assumed that dispersal out of Africa into Eurasia took place across the Sinai Peninsula into the Levant or across the straits of Bab-el-Mandeb into the Arabian Peninsula, although the latter idea has received criticism (Derricourt 2005). East African species are found at 'Ubeidiya, along with species derived from several other biogeographic regions (Tchernov 1987, 1992). Belmaker (2006, 2010), concludes on the basis of multivariate comparison that the Levant fauna had less of an African character than previously assumed and was more similar to Mediterranean woodland faunas. Belmaker further notes that the African taxa present at 'Ubeidiya are represented by few specimens. A further reservation about the role of the Levant as a departure zone for other points in Eurasia concerns the barriers surrounding the region, which is bordered by the Taurus-Zagros mountains, and by deserts of the Arabian peninsula. Despite the mixture of biogeographic faunal elements, it appears that few faunal elements from the Ethiopian region penetrated these mountains. African species such as *Equus* cf. *tabeti*, *Pelorovis oldowayensis*, and *Oryx* cf. *gazella* are found in the Levant, but have not been found elsewhere in northern Eurasia (Martínez-Navarro 2010).

Bab-el-Mandeb

Some faunal exchange between Africa and southwestern Arabia is also proposed to have taken place at the straits of Bab-el-Mandeb (Tchernov 1992; Turner 1999). Undated

Acheulean archeological sites with low artifact densities have been identified near the straits of Bab-el-Mandeb (Petraglia 2003). Kinematic reconstruction of the Afar region suggests, however, that connectivity between the Afar triangle and the Arabian plate was lost after 6.2 Ma, although the land between remained subaerial until approximately 4.5 – 3.2 Ma (Redfield et al. 2003). Geochemical analysis of deuterium concentrations, indicative of a connection between the Red Sea and the Gulf of Aden during the latest Miocene to Early Pliocene (Friedman and Hardcastle 1974), supports Redfield et al.'s (2003) conclusion that the land bridge disappeared in the Pliocene or had become only an intermittent feature. Although glacial conditions caused the lowering of sea level several times during the last 500,000 years (Rohling et al. 1998), the low stands of earlier times in that region are not yet known. Evaporite formation, which would occur when salt became concentrated in the Red Sea due to the formation of a land barrier, is not known to have occurred since the Miocene (Fernandes et al. 2006; Orszag-Sperber et al. 2001). Derricourt (2005) argues that a dispersal involving sea crossing, as required of a route involving the Bab-el-Mandeb straits, was unlikely prior to the technological capabilities of Late Pleistocene humans.

Arabian Peninsula

Oldowan and Acheulean artifact assemblages are found in the Arabian Peninsula (summarized by Petraglia 2003). Faunal remains are not reported to occur with these archeological assemblages, which are undated. A small faunal assemblage from the An Nafud Desert in Saudi Arabia contains a few African taxa, including *Crocuta crocuta*, *Pelorovis oldowayensis* and *Oryx* sp (Thomas et al. 1998). Significantly, these are species identified as African dispersants in the Levant also.

Many Arabian artifact sites are found in montane areas and near ancient lake beds and wadis, as well as along the Red Sea and Arabian Sea (Petraglia 2003). However, Petraglia cautions that the distribution of presently known sites is related to visibility and geomorphological conditions as much as to potential dispersal corridors. Artifacts in Oman have been interpreted as part of a route to the Zagros Mountains (Whalen et al. 2002). Petraglia (2003) hypothesizes an eastward expansion on the margins of the Zagros Mountains and over the Iranian plateau. If the artifact distribution does contain a signal about the dispersal route, it is possible that hominins expanded mainly into areas that had local raw material sources for stone toolmaking.

Indian Subcontinent and Central Asia

Information from this area is limited. From a faunal perspective, Africa and India share several species of small mammals, including gerbillids, which dispersed from Africa to India in the Late Pliocene (Patnaik and Sahni 2000). From the evidence of shared taxa, they conclude that during the Pliocene, there was a means of dispersal between Africa and India suitable for small mammals. Martínez-Navarro (2010) also notes evidence that African antelopes dispersed to the Siwaliks in the Pliocene. According to Badam (1984), by the middle and late Pleistocene, many African mammalian species occurred in India, and a faunal connection existed between India and Europe.

The ability of hominins to reach sites at the latitude of Dmanisi and the Nihewan basin (40°N) raises the possibility that hominins could have spread further east into Eurasia along a Palearctic mammalian dispersal corridor running between mountain systems, as suggested by Rolland (1997). This route would also have kept hominins close to raw material sources. A second route of dispersal to East Asia might have occurred along the southern margin of Asia. A third route, which requires strong consideration, is suggested by the likely co-occurrence of excellent source rocks for making tools, perennial fresh-water springs and water courses, and the potential to harbor a high mammalian biomass in the Purana basins of north-central India (Korisettar 2007). This region would seem to have high potential for paleoanthropological investigations related to the spread of early hominins both eastward and westward across Asia (Fig. 5.4).

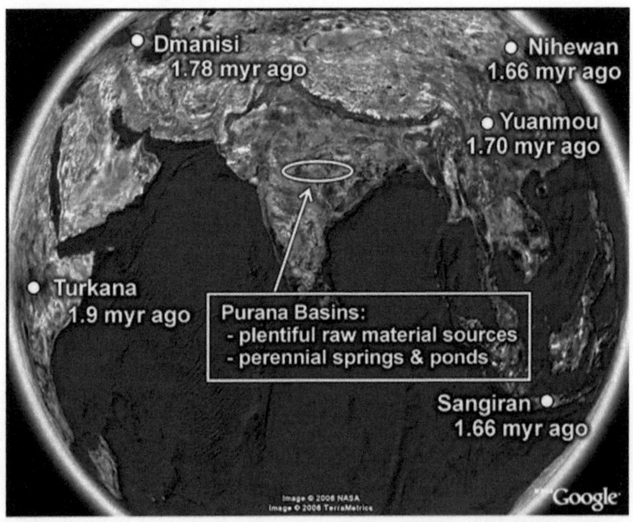

Fig. 5.4 The Purana basins of north-central India, situated between East African and East Asian sites of earliest *Homo erectus sensu lato*, are posited by Korisettar (2007) to have been areas attractive to early and middle Pleistocene hominin populations. Due to abundant food, water, and lithic sources, these basins may have also offered a viable route of population movement between the western and eastern parts of Eurasia. Google Earth image

Fellow Travelers and Three Hypotheses of Hominin Dispersal

Fellow travelers – i.e., mammalian populations that dispersed from Africa and through Eurasia at the same time as hominins – are thought to be important for understanding the environments into which hominins dispersed and the ecological circumstances of the dispersal (Turner 1984; Martínez-Navarro 2010). Table 1 presents those taxa of African origin that are found in the Levant and the known first occurrences of these taxa in other regions. We summarize here several hypotheses to account for patterns of co-occurrence of hominins and possible fellow travelers.

1. Unique dispersal hypothesis: Hominin dispersal out of Africa was a unique event, and did not involve the co-dispersal of any other African species.
2. African fellow travelers hypothesis: The out-of-Africa dispersal represented a small ecological opportunity, allowing hominins and other African taxa to reach a surrounding region. After that regional dispersal, hominin groups dispersed independently to distant regions.
3. Relay hypothesis: Hominins have always dispersed with other species; however, the movement from one region to another was associated with different animal species. For example, some African species may have participated in the initial out-of-Africa dispersal; west Asian taxa may have accompanied hominins into central or southern Asia; and species endemic to these latter regions dispersed with hominins into eastern Asia. The implication is that early human dispersers were always part of a local/regional ecological community, and their dispersal was associated with, and perhaps facilitated by, a relay of fellow travelers from one region to another.

The first hypothesis proposes a case in which only hominins could take advantage of the conditions offering a very limited dispersal opportunity. This hypothesis rules out the spread of vegetational types, such as savannas, which other African species could have followed. The implication is that hominins may have had unique capabilities compared with other animals. Given the presence of some African species at 'Ubeidiya (~1.6 to 1.2 Ma), it may be argued that early Pleistocene dispersals beyond Africa were not unique to hominins, and that other African taxa dispersed into regions of western Eurasia at roughly the same time.

The second hypothesis involves an initial regional dispersal made with fellow travelers. For example, an ecological opportunity may have allowed hominins to travel with several African species to the neighboring regions of the Levant, the Arabian Peninsula, and even as far as India. After dispersing with these African species, the hominins had adapted sufficiently to Eurasian environments that they were then able to disperse to areas such as East Asia without the benefit of fellow travelers. In regions close to Africa, hominins would appear with African species, while hominins in more remote regions would not be found with African taxa. Both *Megantereon* and *Crocuta* have been reported from many sites out of Africa, including East Asia; however, the timing of the initial dispersal of the species ancestral to the Eurasian forms is uncertain (for further discussion, see Lewis and Werdelin 2010). A great deal of further research would also be required to determine whether the timing of the dispersal of *Theropithecus* to India matches that of the arrival of hominins in that region.

The third hypothesis involves a change in the taxonomic identity and biogeographic affinities of the species dispersing with hominins. When dispersing to distant regions, such as East Asia, hominins may have dispersed with non-African species. If this were the case, the fossil record would show hominins appearing with African animals in some regions, but with exotic, non-African species in other, more distant regions. To determine whether this is a possibility in relation to East Asia, the faunas of potential intermediate regions such as the Arabian Peninsula and India must be dated and analyzed thoroughly to see whether faunal exchange occurred with distant regions such as East Asia. This evidence could also be used to understand routes of dispersal.

Preliminary evidence (Table 5.1) shows that although some African species found in the Levant are also found in some other regions, there is no clear pattern indicating a group movement of all these species to other regions, such as India or the Caucasus. Individual species did, however, succeed in colonizing other regions. The exceptions to this are the carnivores *Crocuta* and *Megantereon*, which are found in many areas. Carnivores are good dispersers, and hominins may have followed their biogeographic pattern rather than that of herbivores that did not disperse as far from Africa. However, as noted earlier, the phylogenetic relationships and timing of dispersal events of *Crocuta* and *Megantereon* in East Asia and East Africa are not certain (Lewis and Werdelin 2010).

Another possibility is that not all of the potential African fellow travelers have been identified. An overall taxonomic comparison of the faunas of East Asia and Africa (Table 5.2) shows those species that are shared. Common to both regions were representatives of the Carnivora, Proboscidea, and Equidae – i.e., those mammalian groups known to have species with excellent dispersal abilities and wide ranges. Some micromammals are also found in both regions. Micromammalian connections also occur between Africa and India, with a proposed dispersal of gerbillids from India to Africa in the Late Pliocene when fewer ecological barriers existed (Patnaik and Sahni 2000). *Equus* entered Africa from Eurasia around 2.5 Ma, and thus the genus-level similarity between East African and East Asian equids is probably

Table 5.1 These African species have been found in the Levantine site of 'Ubeidiya (Tchernov 1987; Martínez-Navarro 2004; Belmaker 2010), and are identified as potential fellow travelers for the initial hominin dispersal. Following the interregional approach discussed in the text, the other regions in which these taxa have been identified are indicated. This evidence suggests that the dispersals of *Giraffa* and *Oryx* predate current evidence for the earliest hominin dispersal to Eurasia.

	Equus cf. *tabeti*	*Pelorovis oldowayensis*	*Oryx* cf. *gazella*	*Kolpochoerus*	*Giraffa sp.*	*Hippopotamus*	*Theropithecus*	*Crocuta crocuta*	*Megantereon*	References
Arabia		*Pelorovis* sp. (Early Pleistocene)	*Oryx* sp. (Early Pleistocene)					*Crocuta crocuta* (Early Pleistocene)		Thomas et al. 1998
India			*Oryx sivalensis* (2.6 Ma)				*T. delsoni* (Mirzapur, 1.0 Ma)	*Crocuta* sp. (Pinjor Formation?)	*Megantereon falconeri* (Siwaliks, upper Pliocene)	Martínez-Navarro 2004, 2010; Gupta and Sahni 1981; Delson 1993; Pickford 1993; Lewis and Werdelin 2010
Europe						*H. antiquus* (Venta Micena, 1.5 Ma)	(Pirro Nord, 1.5 Ma)	(Gran Dolina, Atapuerca, 0.8 Ma)	*M. whitei* (Venta Micena, 1.6 Ma)	Martínez-Navarro 2004, 2010; Rook et al. 2004; Alberdi and Ruiz-Bustos 1985; Martínez-Navarro et al. 2004; García 2003
Caucasus					(Kvabebi 2.5Ma)	(Akhalkalaki, 0.98–0.78 Ma)		*Crocuta sinensis* (Akhalkalaki, 0.98–0.78 Ma)	*Megantereon* sp.(Dmanisi)	Vekua 1972, in Martínez-Navarro 2010; Tappen et al. 2002; Gabunia et al. 2000
North Africa	(Ain Hanech)	*Pelorovis* sp. (Ain Hanech, 1.8 Ma)	*Oryx* sp. (Ain Hanech, 1.8 Ma)	*K. phacochoeroides* (2.5 Ma)	*G. pomeli* (Ain Hanech, 1.8 Ma)	*H. amphibious* (Ain Hanech, 1.8 Ma)	*T. atlanticus* (2.5 Ma)	*C. dbaa* (2.5 Ma)		Arambourg 1970; Martínez-Navarro 2004; Geraads 1993; Alemseged and Geraads 1998; Geraads 1997
Far East Asian mainland	*Equus* sp; relationship to *Equus* cf. *tabeti* unknown							*Crocuta honanensis* (Longdan, Late Pliocene)	*Megantereon* sp. (Haiyan Fm., Yushe Basin, 2.5–1.9 Ma); *Megantereon nihowanensis* (Longdan, Late Pliocene)	Flynn 2002; Flynn et al. 1991; Qiu et al. 2004

Table 5.2 Chinese early Pleistocene hominin and archeological sites are listed, along with evidence of dating, and genera that are found in both Africa and East Asia. Asterisks indicate taxa that are only found in North Africa. The Pliocene site of Ahl al Oughlam is provided for comparison as a richly fossiliferous site that does not have evidence of hominins.

Site	Hominin Evidence	Date	Method	Number mammal species	Genera also found in Africa	References
Xiaochangliang, Nihewan Basin, North China	Artifacts	1.36 Ma	Paleomagnetic stratigraphy	11	*Equus, Gazella, Hipparion, Hyaena*	Tang et al. 1995; Zhu et al. 2003
Majuanguo, Nihewan Basin, North China	Artifacts; tool-modified bone	Sequence dated 1.34 to 1.66 Ma	Paleomagnetic stratigraphy	6	*Elephas, Equus, Gazella, Pachycrocuta*	Zhu et al. 2004
Gongwangling, Lantian North China	*Homo erectus* cranium	1.15 Ma	Paleomagnetic stratigraphy	43	*Canis, Equus, Felis, Gerbillus, Hyaena, Hystrix, Megantereon, Micromys, Panthera, Stegodon, Ursus*?	An and Ho 1989; An et al. 1990; Zhu et al. 2003
Yuanmou South China	Hominin incisors; stone artifacts (n=4)	1.7 Ma	Paleomagnetic stratigraphy; biochronology	35	*Canis, Equus, Gazella, Megantereon, Pachycrocuta, Panthera, Stegodon, Vulpes*	Zhu et al., in review; Han and Xu 1 985
Jianshi (Longgudong Cave), South China	Hominin teeth; artifacts	Early Pleistocene	Biochronology	87	*Crocidura, Equus, Homotherium, Pachycrocuta, Micromys, Rhinolophus, Ursus*?	Zheng 2004
Xihoudu, North China	Stone artifacts	1.27 Ma	Paleomagnetic stratigraphy	22	*Elephas, Equus, Gazella, Hipparion, Hyaena, Stegodon*	Zhu et al. 2003
Gigantopithecus Cave, Liucheng, South China (Juyuandong)	No clear evidence of hominin fossils or artifacts	Early Pleistocene	Biochronology	40	*Equus, Felis, Herpestes delibis, Hyaena (Pachycrocuta?), Hystrix, Megantereon, Micromys, Panthera pardus, Viverra*	Han and Xu 1985; Zhang 1985
Renzidong, Fanchang, Anhui (Renzi Cave) eastern China	Purported hominin artifacts	Late Pliocene	Biochronology	67	*Equus, Felis, Hipposideros, Homotherium, Hystrix, Megantereon, Micromys, Nyctereutes, Pachycrocuta, Rhinolophus, Ursus*?	Jin et al. 2000
Longgupo Cave (Wushan), South China	Purported hominin teeth Stone artifacts	Fossils/artifacts in a normal magnetozone, possibly 1.95–1.77 Ma	Biochronology; paleomagnetic stratigraphy; ESR	68 (genera)	*Equus, Felis, Hipposideros, Homotherium, Hystrix, Nyctereutes, Pachycrocuta, Rhinolophus, Ursus*?	Huang et al. 1995
Zhoukoudian 1, North China	Hominin remains Stone artifacts	500–200 kyr	Geological, faunal, pollen correlation with glacial chronology	89	*Canis, Crocidura, Crocuta crocuta, Equus, Felis, Gazella, Gerbillus, Homotherium, Hystrix, Lepus, Lutra, Macaca, Megantereon, Micromys, Miniopterus schreibersi, Mus, Nyctereutes, Pachycrocuta, Panthera pardus, Rhinolophus, *Spiroceros, Ursus?, Vulpes*	Dong et al. 2000; Aigner 1986; Wu 1985
Mohui cave, Bubing Basin, South China	Hominoid teeth (hominin status uncertain)	Plio-Pleistocene	Biochronology	30	*Felis, Hystrix, Macaca, Stegodon, Ursus?*	Wang et al. 2007
Dmanisi, Caucasus	Hominin remains, artifacts	1.8 Ma	K-Ar; $^{40}Ar/^{39}Ar$	36	Shared with Africa *Canis, Equus, Gazella, Gerbillus, Homotherium, Megantereon, Mungos, Pachycrocuta, Panthera, Ursus?* Shared with China *Apodemus, Archidiskodon, Canis, Cervus, Dama, Dicerorhinus, Equus, Eucladoceros, Gazella, Gerbillus, Homotherium, Hypolagus, Kowalskia, Lynx, Marmota, Martes, Megantereon, Mimomys, Ochotona, Pachycrocuta, Panthera, Sorex, Sus, Ursus*	Gabunia et al. 2000, 2001

(continued)

Table 5.2 (continued)

Site	Hominin Evidence	Date	Method	Number mammal species	Genera also found in Africa	References
Ahl al Oughlam, North Africa	None	2.5 Ma	Biochronology	52	Shared with Sub-Saharan Africa *Acinonyx, Anancus, Beatragus, Camelus, Canis, Ceratotherium, Chasmaporthetes, Crocuta, Dinofelis, Elephas, Felis, Gazella, Genetta, Gerbillus, Herpestes, Hipparion, Homotherium, Hyaenictitherium? (only in very old sites), Kobus, Kolpochoerus, Lepus, Lutra, Mellivora, Mus, Nyctereutes, Panthera, Parmularius, Praomys, Serengetilagus, Sivatherium, Suncus, Theropithecus, Tragelaphus, Ursus?, Viverra* Shared with China *Camelus, Canis, Chasmaporthetes, Crocuta, Elephas, Erinaceus, Felis, Gazella, Gerbillus, Herpestes, Hipparion, Homotherium, Hystrix, Lepus, Lutra, Lynx, Macaca, Mus, Nyctereutes, Panthera, Ursus, Viverra*	Geraads, 1993, 1994, 1995, 1997; Geraads and Amani 1998; Geraads and Metz-Muller 1999; Alemseged and Geraads 1998

Artiodactyla, Bovidae: *Beatragus*, *Gazella*; *Kobus*, *Parmularius*, *Spiroceros*, *Tragelaphus*; Camelidae: *Camelus*; Cervidae: *Cervus*, *Dama*, *Eucladoceros*; Giraffidae: *Sivatherium*
Suidae: *Kolpochoerus*, *Sus*
Carnivora, Canidae: *Canis*; *Nyctereutes*, *Vulpes*; Felidae: *Acinonyx*, *Dinofelis*, *Felis*, *Homotherium*, *Lynx*, *Megantereon*, *Panthera*
Hyaenidae: *Chasmaporthetes*, *Crocuta*, *Hyaena*, *Hyaenictitherium*, *Pachycrocuta*; Mustelidae: *Lutra*, *Martes*, *Mellivora*
Ursidae: *Ursus*; Viverridae: *Genetta*, *Herpestes*, *Mungos*, *Viverra*
Chiroptera, Hipposideridae: *Hipposideros*; Rhinolophidae: *Rhinolophus*; Vespertilionidae: *Miniopterus*
Insectivora, Soricidae: *Crocidura*, *Sorex*, *Suncus*; Erinaceidae: *Erinaceus*
Lagomorpha, Leporidae: *Hypolagus*, *Lepus*, *Serengetilagus*
Perissodactyla, Equidae: *Equus*, *Hipparion*; Rhinocerotidae: *Ceratotherium*, *Dicerorhinus*
Primates, Cercopithecidae: *Macaca*, *Theropithecus*
Proboscidea, Elephantidae: *Archidiskodon*, *Elephas*; Gomphotheriidae: *Anancus*; Stegodontidae: *Stegodon*
Rodentia, Arvicolidae: *Mimomys*; Hystricidae: *Hystrix*; Muridae: *Apodemus*, *Gerbillus*, *Kowalskia*, *Micromys*, *Mus*; *Praomys*; Sciuridae: *Marmota*

ancient. *Equus* cf. *tabeti* found at 'Ubeidiya has also been found at Koobi Fora, and is related to the African equid *E. oldowayensis* (Eisenmann 1983). The shared genera *Hipparion, Camelus*, and *Gazella* also represent ancient similarities rather than a dispersal event associated with *H. erectus*. Movements, taxonomic similarities, and interconnectivity among the regions need further investigation. At present, none of the three hypotheses summarized here receives overwhelming support.

Environmental Contexts of the Oldest Hominins in East Asia

The oldest hominin evidence in China converges on the oldest age estimates for hominin fossils from Java in Southeast Asia (Swisher et al. 1994; Larick et al. 2001; Antón and Swisher 2004; Zaim 2010). By at least ~1.66 Ma, East Asian hominins had dispersed over a latitudinal range from 7°S to 40°N, and across habitats ranging from tropical woodland/forest to seasonally cool grassland. In this section we summarize the diversity of East Asian fauna, vegetation, and climate dynamics that formed part of the context of the oldest known Asian dispersal of hominins.

Distinctive Faunas and Regional Diversity

Early Pleistocene hominins arriving in East Asia encountered a novel array of mammalian species relative to East Africa. East Asian faunas were taxonomically diverse, combining species from the Palearctic and Oriental realms and from North America. One potentially important difference between the two faunal communities was the presence of a diverse bovid community in East Africa while East Asia was characterized by woodland and forest cervids that were unknown in Pleistocene sub-Saharan Africa.

After a period of high faunal similarity between China and Europe (at the genus and species levels) prior to 2.5 Ma (Tedford 1995), North China became separated from other Eurasian temperate faunas due to uplift, climatic changes, and the expansion of a desert-steppe belt across the southern region of Inner Mongolia (Zhang 1988). During the Pleistocene, the faunas of North and South China also became increasingly divergent, apparently due to climatic differences between the regions and uplift of the Tibetan Plateau and Qinling Mountains (Xu 1988; Zhang 1988; Cao 1994).

North and South China are divided by the Yangtze River and the Qinling Mountains, which pose a formidable barrier to faunal interchange. Early Pleistocene sites of southern China are generally assigned to tropical or subtropical forest habitats, whereas northern Chinese sites supported a steppe and grassland fauna that included forest elements during the Early Pleistocene (Han and Xu 1985). The northern forests included floral species such as *Quercus, Ulmus* and *Carpinus* (Cao 1994). Sites in the Nihewan basin, North China, include taxa that reflect a variety of habitats, including woodland, grassland and arid environments. Examples of open country and steppe taxa include gazelles, *Proboscidipparion sinense, Equus, Myospalax tingi* and camelids, while cervids may indicate the presence of woodland (Zhang 1988; Qiu et al. 2004; Cao 1994; Geist 1998).

A transitional zone between North and South China existed between the middle and lower Yangtze River, the Qinling Mountains and the Huai River. The early Pleistocene site of Gongwangling, which is in the transitional zone, contains several southern taxa in its faunal assemblage, possibly reflecting the ability of these animals to expand their ranges northward during favorable climatic conditions. Gongwangling also contains montane and forest indicators, such as *Capricornis sumatraensis, Myospalax fontanieri* and *Elaphodus cephalophus*, which are found at higher elevations, and *Scaptochirus moschatus, Ailuropoda* and *Stegodon*, which are found in forests (Zhang 1988; Nowak 1999).

Early Pleistocene sites of South China are characterized by forest indicators, such as *Stegodon* and *Ailuropoda*. The fauna and paleobotanical remains associated with the Yuanmou hominin fossils suggest that this basin hosted a diverse mosaic of habitats. Twenty-one of 35 mammalian taxa (for example, *Equus yunnanensis*) belong to families, genera, or species typical of an open grassland environment. Bushland habitat is also suggested by the presence of *Rhizomys* sp., *Sus* sp., *Nestoritherium* sp. and *Viverricula malaccensis fossilis*, whereas fossils of *Megantereon nihowanensis, Stegodon elephantoides* and *Stegodon* sp. are considered indicative of forest. Numerous mollusks indicate a low-energy lakeshore or marsh setting, and the occurrence of *Cyriuus caspio* and *Testudo* sp. is characteristic of a perennial open aquatic environment (Zhu et al. 2008). Fossil pollen was also recovered in the sedimentary layer bearing the *H. erectus* fossils and artifacts at Yuanmou. The pollen samples are dominated by *Pinus* (33.3%), *Alnus* (13%) and herbaceous vegetation (40%) (Pu and Qian 1977; Qian and Zhou 1991), indicative of locally extensive herbaceous cover and patches of forest surrounding the ancient lake or swamp. The faunal assemblage and pollen from the hominin site thus imply that the diverse habitats encountered by the Yuanmou hominins included open grassland, bushland, forest, marsh and fresh water – i.e., vegetation not at all typical of that assumed to characterize the Early Pleistocene of southern China (Ciochon 2010; see Wang et al. 2007).

Climate Dynamics

Environmental proxies derived from coring of the South China Sea (e.g., $\delta^{18}O$, K/Si and Ba/Al ratios) and the analysis of Loess Plateau sediments (e.g., Fe isotope ratio, magnetic susceptibility) provide a lot of information about East Asian climate history. A substantial shift in elemental ratios in sediments younger than 3.0 Ma resulted from stronger seasonality in East Asia, which appears to have been caused by a change in atmospheric circulation due to uplift of the Himalayan-Tibetan complex. Tectonic changes after 3.2 Ma are likely to have led to an intensified Siberian high-pressure system that resulted in colder, drier, and less stable climate conditions after 2.65 Ma (Wehausen and Brumsack 2002; Müller et al. 2001). Onset of the loess-paleosol sequence of the Loess Plateau at ~2.6 Ma is related to the intensification of the winter monsoon.

One commonality in the climate histories of East Africa and East Asia, therefore, is insolation-driven monsoon variability, with a major low-latitude influence. Low-latitude insolation variability appears to have asserted major control on the pace and intensity of climate oscillation. Beyond this, however, the details of climate change in East Asia, the zonation of habitats, and the expansion/contraction of habitats in response to climate variability were the result of tectonic features, pressure systems, and monsoon wind intensities unique to that region of the world.

Magnetic susceptibility data from the Loess Plateau (Guo Zhengtang, personal communication) and planktonic foraminifera $\delta^{18}O$ from the South China Sea (Steven Clemens, personal communication) show that relatively low-amplitude climate oscillation characterized the period from at least 2.0 to ~1.2 Ma, which was followed by substantial increases in oscillation amplitude at ~1.2 Ma and 550 ka. In other words, the first appearance of hominin populations in East Asia occurred during a period of relatively low climate variability.

In sum, upon arrival in East Asia, early hominin populations encountered climatic and biotic regimes that differed substantially from those of East Africa. In East Asia, hominins inhabited temperate woodlands and grasslands and tropical/subtropical wooded landscapes that were under the influence of Asian monsoonal conditions and that harbored plant and animal communities unlike those of Plio-Pleistocene East Africa. Evidently, early *H. erectus* possessed the ability to adjust to these differences and also to the wide variety of climatic and biotic settings throughout its range in eastern Asia. The ecological, behavioral, and physiological means by which *H. erectus* accommodated to this breadth of environments is not yet well understood, even though such means of adaptability have been characteristic of the genus *Homo* ever since.

Discussion

The implications we draw based on this wide variety of data sets can be summarized as follows:

1. The earliest hominins in East Asia show morphological and behavioral connections with East African hominins.
   - Stone flaking: The oldest definite stone artifacts of East Asia can be described as Oldowan, and show details of percussion flaking and core form that match those of the African Oldowan.
   - Morphology: The morphology of the oldest distinctively hominin dental remains in China (from Yuanmou) – possibly the oldest known hominin fossils in all of eastern Asia – show strong similarities to the dental remains of early African *H. erectus*.
   - Acquiring large mammal carcasses: Archeological evidence from the Nihewan basin closely matches that from East African Plio-Pleistocene sites. The similarities include: access by the toolmakers to a wide variety of mammalian taxa (e.g., equids, bovids, cervids, rhinos, elephants) across a broad range of body sizes; accumulation of animal bones and stone tools in distinct concentrations; and the butchery and extraction of bone marrow from these animals.
2. Animal foods may have been a critical resource to dispersing hominin groups because plants produce potentially toxic secondary compounds, whereas animal tissues have relatively similar nutritional and digestive properties across diverse biotic zones.
   - Access to animal tissues was almost certainly instrumental in enabling hominins to enter and pass through a wide variety of habitats, harboring many unfamiliar plant species, from Africa to East Asia. Despite the taxonomic differences of the prey, techniques of animal food foraging were probably transferable across faunal communities.
   - The earliest hominin toolmakers in North China evidently persisted long-term in a seasonally cool and dry habitat; consuming animal tissues would have been an important aspect of the survival strategy in temperate latitudes.
3. The combination of current dating and mammalian faunal evidence from Dmanisi and East Asia suggests that *Homo erectus* took advantage of a general mammalian dispersal opportunity to reach western Eurasia, and was then able to overcome geographic barriers and adapt to new habitats beyond the capabilities of most other mammalian taxa. The dispersal of hominins through eastern Asia especially required an ability to adjust to a wide array of novel environments.

4. Potentially rapid, long-distance dispersal across Asia probably increased the geographic isolation and morphological variability of *H. erectus* populations.
   - Beginning ~1.7 Ma, the heightened degree of intergroup isolation may have been unprecedented to that point in human evolutionary history. East Asian *H. erectus* was isolated not only from African populations, but also from other Eurasian and East Asian populations. During the early Pleistocene, temperate woodland populations of East Asia were periodically cut off from those in Europe. There is also evidence that North and South China diverged over time in their faunal composition; the degree of isolation of hominin groups across these regions is unknown, but could have been substantial.
   - Wide geographic dispersion and small population sizes may have increased the degree of morphological variation due to genetic drift. Like species of carnivores, *Homo erectus* occurred over a wide range and may have had a low speciation rate.
5. The Asian archeological record between 1.77 and 1.66 Ma is very sparse and/or hard to detect – essentially nonexistent across vast areas of Asia – which may reflect the fast pace of dispersal of relatively small populations.
   - In contrast to the Asian record, the African archeological record between ~1.8 and 1.5 Ma is highly visible because Oldowan toolmakers recurrently visited or occupied specific regions and discrete places on the landscape. This is known as "the archeology of settling in" – prolonged habitation of a region by stone toolmakers leaves an easily-detected signal in the stratigraphic record.
   - The archeological record of Yuanmou, where only four artifacts were found in an excavation of a hillside that preserved many mammalian fossils, including two *H. erectus* teeth, is consistent with what is called "the archeology of passing through" – i.e., toolmakers moving through a region leave an archeological record that is highly dispersed in space, possibly confined to a single stratigraphic layer. (No other stone tools have been found in the Yuanmou basin.) Such a record is nearly invisible archeologically in comparison to the rich stratigraphic series of early Pleistocene Oldowan sites in certain East African basins. "The archeology of passing through" implies the dispersal of small, widely spaced groups that were not tethered to specific places on the landscape for multiple generations.
   - The oldest archeological evidence in Asia for "settling in" occurs in the Nihewan basin, beginning at least ~1.66 Ma. There we see the familiar African Oldowan evidence of multiple archeological strata reflecting a lengthy period of re-occupation. Stone-tool evidence from the Nihewan and Yuanmou basins implies that the dispersing populations reaching East Asia had maintained the ability to make stone tools from local source rocks. This idea implies that the Oldowan toolmakers of Asia, as in Africa, could adapt their stone flaking capabilities to a large variety of rock types.
6. There are several possibilities for determining possible dispersal corridors from East Africa to East Asia.
   - Oldowan hominins are likely to have dispersed into regions and moved through areas where stone raw materials were locally available, enabling them to maintain a basic toolmaking tradition. Likely dispersal routes may thus have been constrained by this factor, along with continuous access to fresh water and animal meat/marrow resources.
   - Overall faunal biogeographic patterns can be broken down regionally, and suggest which areas may have been connected by dispersal corridors. Taxa shared between widely separated locations, such as *Equus* and genera of proboscideans and carnivores, are good dispersers. Most African species found in the Levant during the early Pleistocene have so far not been found in Eurasia outside this region. Exceptions to this are *Hippopotamus* and *Theropithecus*. Early Pleistocene African taxa found in the Levant, therefore, may have spread not in response to the expansion of a specific type of habitat across Asia so much as to a more specific geographic opportunity (i.e., a favorable dispersal corridor) to spread beyond Africa.
7. The dispersal from Africa to East Asia, along with evidence regarding the climate dynamics, habitat diversity, and distinctive faunas of East Asia, all suggest that earliest Pleistocene *H. erectus* adjusted to highly-varied adaptive conditions across disparate types of environment – and had thus evolved a high degree of adaptability, with important implications for its evolutionary history.
   - Earliest Asian *H. erectus* evidently occurred from northern China to Java, occupying open and forested settings, and accommodating to a wide variety of mammalian communities and changing environmental conditions.
   - The implication is that *H. erectus* responded to disparate adaptive conditions (in time and space) by a combination of population movement and local adaptability. Given the paleoclimate evidence of alternating periods of increased-decreased moisture, warmer-cooler temperatures, and higher-lower climate variability, we may envision that *H. erectus* experienced alternating periods of population isolation and contact/mixing (as well as local extinction). That is, considerable opportunity would have occurred for phenotypic experimentation,

without any one period necessarily lasting long enough to promote the development of permanent reproductive barriers.

- *H. erectus* spread to new habitats within African basins, then took advantage of an opportunity in its initial move to western Eurasia, and did so during a time of faunal turnover. Whatever characteristics enabled this initial dispersal probably also aided *H. erectus*' survival during its spread across Asia and expansion into the diverse and dynamic environments of East Asia.

Conclusion

The African background: Between 2.5 and 1.8 Ma, shifts in East African hominin limb proportions and locomotor morphology began to occur that were related to long-distance bipedal mobility. While these changes may have presented an opportunity for hominins to engage in a greater range of mobility, they appear to precede archeological behavioral evidence signaling (a) longer distance (>12 km) transport of stone raw material, and (b) an expansion of tool activities into a wider diversity of environmental zones.

Fellow travelers: Preliminary analysis shows that the African taxa found immediately outside Africa in the Levant are not found with *H. erectus* in East Asia, with the exception of the carnivores *Megantereon* and *Crocuta*. However, these may not be the same species as those found in western Asia and in Africa. Comparison of the faunas found in the intervening regions may show corridors of faunal dispersal, and might identify non-African fellow travelers with whom hominins may have spread simultaneously into East Asia (the Relay Hypothesis).

The timing of dispersal: The faunal evidence for a broad dispersal of fellow travelers is unclear on this point. East African archeological evidence shows, however, that the expansion of hominin toolmakers into new landscapes and depositional zones began ~1.77–1.70 Ma, a period that corresponds with the oldest definite evidence of hominins outside of Africa and the spread of *H. erectus*, in particular, to East Asia.

Speed of dispersal: Based on new age estimates of the oldest stone tools and hominin fossils known so far in China, the dispersal across Asia may have been rapid, and led hominin toolmakers into biogeographic zones of different taxonomic makeup from those in Africa.

Adaptability: The oldest recorded spread of hominins across Asia apparently culminated in an ability to adapt to novel habitats and climatic regimes in East Asia. This degree of adaptability likely enabled the persistence of *H. erectus* in East Asia for a very long time prior to the arrival of *H. sapiens*.

Acknowledgements Recent research in Africa and China has been supported by National Science Foundation HOMINID Program grant BCS-0218511, the Smithsonian's Human Origins Program, the National Natural Science Foundation of China, and the Chinese Academy of Sciences. RP thanks Zhu Rixiang, Wang Wei, and other colleagues based in China for research permission and collaboration in our Nihewan, Yuanmou, and Bose projects. We also thank Jennifer Clark for help in preparing the figures.

References

Aigner, J. S. (1986). Age of Zhoukoudian locality 1: The newly proposed O^{18} correspondences. *Anthropos, 23*, 157–173.

Alberdi, M. T., & Ruiz-Bustos, A. (1985). Descripción y significado bioestratigráfico del *Equus* e Hipopótamo en el yacimiento de Venta Micena (Granada). *Estudios Geológicos, 41*, 251–261.

Alemseged, Z., & Geraads, D. (1998). *Theropithecus atlanticus* (Cercopithecidae, Mammalia), from the late Pliocene of Ahl al Oughlam, Casablanca, Morocco. *Journal of Human Evolution, 345*, 609–621.

An, Z., & Ho, C. K. (1989). New magnetostratigraphic dates of Lantian *Homo erectus*. *Quarternary Research, 32*, 213–221.

An, Z., Gao, W., Zhu, Y., Kan, X., & Wang, J. (1990). Magnetostratigraphic dates of Lantian *Homo erectus*. *Acta Anthropologica Sinica, 9*, 1–7.

Antón, S. C., & Swisher, C. C., III. (2004). Early dispersals of *Homo* from Africa. *Annual Review of Anthropology, 33*, 271–296.

Arambourg, C. (1970). Les vertébrés du Pleistocène de l'Afrique du Nord. *Archives du Museum National d'Histoire Naturelle, serie 7, 10*: 1–127.

Asfaw, B., White, T., Lovejoy, O., Latimer, B., Simpson, S., & Suwa, G. (1999). *Australopithecus garhi*: A new species of early hominid from Ethiopia. *Science, 284*, 629–635.

Azzaroli, A. (1995). The "Elephant-*Equus*" event and the "End-Villafranchian" events in Eurasia. In E. Vrba, G. H. Denton, T. C. Partridge, & L. H. Burkle (Eds.), *Paleoclimate and evolution with emphasis on human origins* (pp. 311–318). New Haven: Yale University Press.

Badam, G. L. (1984). Pleistocene faunal succession of India. In R. O. Whyte (Ed.), *The evolution of the East Asian environment, Vol. II. Palaeobotany, palaeogeology and palaeoanthropology* (pp. 746–775). Hong Kong: University of Hong Kong.

Belmaker, M. (2006). *Community structure through time: 'Ubeidiya, a Lower Pleistocene Site as a case study*. PhD dissertation, The Hebrew University of Jerusalem, Jerusalem.

Belmaker, M. (2010). Early Pleistocene faunal connections between Africa and Eurasia: An ecological perspective. In J. G. Fleagle, J. J. Shea, F. E. Grine, A. L. Baden, & R. E. Leakey (Eds.), *Out of Africa I: The first hominin colonization of Eurasia* (pp. 183–205). Dordrecht, The Netherlands: Springer.

Bonnefille, R., & Riollet, G. (1980). Palynologie, vegetation et climats de Bed I et Bed II a Olduvai, Tanzania. *Proceedings of the Eighth PanAfrican Congress of Prehistoric and Quaternary Studies 1977*. Nairobi, pp. 123–127.

Braun, D. R. (2006). The ecology of Oldowan technology: Perspectives from Koobi Fora and Kanjera South. Ph.D. dissertation, Rutgers University, New Jersey.

Braun, D. R., Plummer, T., Ferraro, J. V., Bishop, L. C., Ditchfield, P., Potts, R., et al. (2005). Oldowan technology at Kanjera South, Kenya: The context of technological diversity. Abstracts of the PaleoAnthropology Society 2005 Meetings. *PaleoAnthropology 2005*: A23.

Brown, F. H., & Feibel, C. S. (1991). Stratigraphy, depositional environments, and palaeogeography of the Koobi Fora Formation. In J. M. Harris (Ed.), *Koobi Fora Research Project* (Vol. 3, pp. 1–30). Oxford: Clarendon.

Bullock, J. M., Kenward, R. E., & Hails, R. (Eds.). (2002). *Dispersal ecology*. Malden, MA: Blackwell.

Cao, B. (1994). Quaternary. In H. Yin (Ed.), *The palaeobiogeography of China* (pp. 293–304). Oxford: Clarendon.

Cerling, T. E., & Hay, R. L. (1986). An isotopic study of paleosol carbonates from Olduvai Gorge. *Quaternary Research, 25*, 63–78.

Ciochon, R. (2010). Divorcing hominins from the *Stegodon-Ailuropoda* fauna: New views on the antiquity of hominins in Asia. In J. G. Fleagle, J. J. Shea, F. E. Grine, A. L. Baden, & R. E. Leakey (Eds.), *Out of Africa I: The first hominin colonization of Eurasia* (pp. 111–126). Dordrecht, The Netherlands: Springer.

Delagnes, A., & Roche, H. (2005). Late Pliocene hominid knapping skills: The case of Lokalalei 2C, West Turkana, Kenya. *Journal of Human Evolution, 48*, 435–472.

Delson, E. (1993). *Theropithecus* fossils from Africa and India and the taxonomy of the genus. In N. G. Jablonski (Ed.), *Theropithecus: The rise and fall of a primate genus* (pp. 157–189). Cambridge: Cambridge University Press.

Dennell, R., & Roebroeks, W. (2005). An Asian perspective on early human dispersal from Africa. *Nature, 438*, 1099–1104.

Derricourt, R. (2005). Getting "out of Africa": Sea crossings, land crossings and culture in the hominin migrations. *Journal of World Prehistory, 19*, 119–132.

Dong, W., Jin, C.-Z., Xu, Q-q, Liu, J-y, Tong, H-w, & Zheng, L-t. (2000). A comparative analysis on the mammalian faunas associated with *Homo erectus* in China. *Acta Anthropologica Sinica, 19* (Suppl), 246–256.

Eisenmann, V. (1983). Family Equidae. In J. M. Harris (Ed.), *Koobi Fora Research Project Vol 2: The Fossil Ungulates: Proboscidae, Perissodactyla, and Suidae* (pp. 156–214). Oxford: Clarendon.

Eswaran, V., Harpending, H., & Rogers, A. R. (2005). Genomics refutes an exclusively African origin of humans. *Journal of Human Evolution, 49*, 1–18.

Feibel, C. S. (1997). Debating the environmental factors in hominid evolution. *GSA Today, 7*, 1–7.

Fernandes, C. A., Rohling, E. J., & Siddall, M. (2006). Absence of post-Miocene Red Sea land bridges: Biogeographic implications. *Journal of Biogeography, 33*, 961–966.

Flynn, L. J. (2002). Can the wolf event be recognized in north China? *Annales Geologiques des Pays Helleniques, 39*(A), 317–326.

Flynn, L. J., Tedford, R. H., & Qiu, Z. (1991). Enrichment and stability in the Pliocene mammalian fauna of North China. *Paleobiology, 17*, 246–265.

Friedman, I., & Hardcastle, K. (1974). Deuterium in interstitial waters from Red Sea cores. ODP Preliminary Reports Leg 23, Init. Resp. DSDP XXIII, pp. 969–974.

Gabunia, L., Vekua, A., Lordkipanidze, D., Swisher, C., Ferring, R., Justus, A., et al. (2000). Earliest Pleistocene hominid cranial remains from Dmanisi, Republic of Georgia: Taxonomy, geological setting, and age. *Science, 288*, 1019–1025.

Gabunia, L., Antón, S. C., Lordkipanidze, D., Vekua, A., Justus, A., & Swisher, C. (2001). Dmanisi and dispersal. *Evolutionary Anthropology, 10*, 158–170.

García, N. (2003). *Osos y otros carnívoros de la Sierra de Atapuerca*. Asturias, Spain: Fundación Oso de Asturias.

Geist, V. (1998). *Deer of the world*. Mechanicsburg, PA: Stackpole books.

Geraads, D. (1993). *Kolpochoerus phacochoeroides* (THOMAS, 1884) (Suidae, Mammalia) du Pliocène supérieur de Ahl al Oughlam (Casablanca, Maroc). *Geobios, 26*, 731–743.

Geraads, D. (1994). Rongeurs et Lagomorphes du Pléistocène moyen de la "Grotte des Rhinocéros", carrière Oulad Hamida 1, à Casablanca, Maroc. *Neues Jahrbuch. für Geologie und Paläontologie Abhandlungen, Tübingen, 191*, 147–172.

Geraads, D. (1995). Rongeurs et Insectivores du Pliocène final de Ahl al Oughlam, Casablanca, Maroc. *Geobios, 28*, 99–115.

Geraads, D. (1997). Carnivores du Pliocène terminal de Ahl al Oughlam (Casablanca, Maroc). *Geobios, 30*, 127–164.

Geraads, D., & Amani, F. (1998). Bovidae (Mammalia) du Pliocène final d'Ahl al Oughlam, Casablanca, Maroc. *Paläontologische Zeitschrift, Stuttgart, 72*, 191–205.

Geraads, D., & Metz-Muller, F. (1999). Proboscidea (Mammalia) du Pliocène final d'Ahl al Oughlam (Casablanca, Maroc). *Neues Jahrbuch. für Geologie und Paläontologie Monatshefte, 1*, 52–64.

Gupta, V. J., & Sahni, A. (1981). *Theropithecus delsoni*, a new cercopithecine species from the Upper Siwaliks of India. *Bulletin of the Indian Geological Association, 14*, 69–71.

Haeusler, M., & McHenry, H. M. (2004). Body proportions of *Homo habilis* reviewed. *Journal of Human Evolution, 46*, 433–465.

Han, D., & Xu, C. (1985). Pleistocene mammalian faunas of China. In R. Wu & J. W. Olsen (Eds.), *Palaeoanthropology and Palaeolithic Archaeology in the People's Republic of China* (pp. 267–289). New York: Academic.

Hay, R. L. (1976). *Geology of the Olduvai Gorge*. Berkeley, CA: University of California Press.

Hou, Y. M., & Zhao, L. X. (2010). New archeological evidence for the earliest hominid presence in China. In J. G. Fleagle, J. J. Shea, F. E. Grine, A. L. Baden, & R. E. Leakey (Eds.), *Out of Africa I: The first hominin colonization of Eurasia* (pp. 87–95). Dordrecht, The Netherlands: Springer.

Hu, C. Z. (1973). Ape-man teeth from Yuanmou, Yunnan. *Acta Geologica Sinica, 1*, 65–72.

Huang, W., Ciochon, R., Gu, Y., Larick, R., Rang, Q., Schwarcz, H., et al. (1995). Early *Homo* and associated artifacts from Asia. *Nature, 378*, 275–278.

Hubbell, S. P. (2001). *The unified neutral theory of biodiversity and biogeography*. Princeton, NJ: Princeton University Press.

Hyodo, M., Nakaya, H., Urabe, A., Saegusa, H., Xue, S., Yin, J., et al. (2002). Paleomagnetic dates of hominid remains from Yuanmou, China, and other Asian sites. *Journal of Human Evolution, 43*, 27–41.

Isaac, G. L. (1977). *Olorgesailie*. Chicago: University of Chicago Press.

Jin, C.-Z., Dong, W., Xing, G., Wu, L., Yi, L.-J., Ting, Z.-L., et al. (2000). A preliminary study on the Early Pleistocene deposits and the mammalian fauna from the Renzi Cave, Fanchang, Anhui, China. *Acta Anthropologica Sinica, 19* (Suppl), 235–245.

Kibunjia, M. (1994). Pliocene archaeological occurrences in the Lake Turkana basin. *Journal of Human Evolution, 27*, 159–171.

Korisettar, R. (2007). Toward developing a basin model for Paleolithic settlement of the Indian subcontinent: geodynamics, monsoon dynamics, habitat diversity and dispersal routes. In M. D. Petraglia, & B. Allchin (Eds.), *The evolution and history of human populations in South Asia* (pp. 69–96). New York: Springer.

Kukla, G., & An, Z. S. (1989). Loess stratigraphy in central China. *Palaeogeography, Palaeoclimatology, Palaeoecology, 72*, 203–225.

Larick, R., Ciochon, R. L., Zaim, Y., Sudijono, S., Rizal, Y., Aziz, F., et al. (2001). Early Pleistocene $^{40}Ar/^{39}Ar$ ages for Bapang Formation hominins, Central Java, Indonesia. *Proceedings of the National Academy of Sciences of the United States of America, 98*, 4866–4871.

Lewis, M., & Werdelin, L. (2010). Carnivoran dispersal out of Africa during the early Pleistocene: Relevance for hominins? In J. G. Fleagle, J. J. Shea, F. E. Grine, A. L. Baden, & R. E. Leakey (Eds.), *Out of Africa I: The first hominin colonization of Eurasia* (pp. 13–26). Dordrecht, The Netherlands: Springer.

Liu, T. (1985). *Loess and the environment*. Beijing: China Ocean Press.

Liu, W., Zheng, L., & Jiang, C. (2000). Statistical analyses of metric data of hominoid teeth found in Yuanmou of China. *Chinese Science Bulletin, 45*, 936–942.

Martínez-Navarro, B. (2004). Hippos, pigs, bovids, saber-toothed tigers, monkeys, and hominids: Dispersals through the Levantine Corridor during the Late Pliocene and Early Pleistocene times. In N. Goren-Inbar & J. D. Speth (Eds.), *Human paleoecology in the Levantine Corridor* (pp. 37–51). Oxford: Oxbow Books.

Martínez-Navarro, B. (2010) Early Pleistocene Faunas of Eurasia and Hominin Dispersals. In J. G. Fleagle, J. J. Shea, F. E. Grine, A. L. Baden, & R. E. Leakey (Eds.), *Out of Africa I: The first hominin colonization of Eurasia* (pp. 207–224). Dordrecht, The Netherlands: Springer.

Martínez-Navarro, B., Rook, L., Segid, A., Yosief, D., Ferretti, M. P., Shoshani, J., et al. (2004). The large fossil mammals from Buia (Eritrea): Systematics, biochronology and paleoenvironments. *Rivista Italiana di Paleontologia i Stratigrafia, 110*, 61–88.

Müller, J., Oberhänsli, H., Melles, M., Schwab, M., Rachold, V., & Hubberten, H.-W. (2001). Late Pliocene sedimentation in Lake Baikal: Implications for climate, tectonic change in SE Siberia. *Palaeogeography, Palaeoclimatology, Palaeoecology, 174*, 305–326.

Nathan, R. (2006). Long-distance dispersal of plants. *Science, 313*, 786–788.

Noll, M. P. (2000). *Components of Acheulean lithic assemblage variability at Olorgesailie, Kenya*. PhD dissertation, University of Illinois, Urbana-Champaign, IL.

Nowak, R. M. (1999). *Walker's mammals of the world*. Baltimore: Johns Hopkins University Press.

Orszag-Sperber, F., Plaziat, J.-C., Baltzer, F., & Purser, B. H. (2001). Gypsum salina-coral reef relationships during the Last Interglacial (Marine Isotopic Stage 5e) on the Egyptian Red Sea coast: A Quaternary analogue for Neogene marginal evaporites? *Sedimentary Geology, 140*, 61–85.

Patnaik, R., & Sahni, A. (2000). Palaeobiogeography of Pliocene Siwalik murid and Gerbillid rodents. *Geological Survey of India Miscellaneous Publications, 64*, 43–47.

Petraglia, M. D. (2003). The Lower Paleolithic of the Arabian peninsula: Occupations, adaptations and dispersals. *Journal of World Prehistory, 17*, 141–179.

Pickford, M. (1993). Climatic change, biogeography, and *Theropithecus*. In N. G. Jablonski (Ed.), *Theropithecus: The rise and fall of a primate genus* (pp. 227–243). Cambridge: Cambridge University Press.

Plummer, T. (2004). Flaked stones and old bones: Biological and cultural evolution at the dawn of technology. *Yearbook of Physical Anthropology, 47*, 118–164.

Potts, R., Behrensmeyer, A. K., & Ditchfield, P. (1999). Paleolandscape variation and Early Pleistocene hominid activities: Members 1 and 7, Olorgesailie Formation, Kenya. *Journal of Human Evolution, 37*, 747–788.

Pu, Q. Y., & Qian, F. (1977). Study on the fossil human strata – the Yuanmo Formation. *Acta Geologica Sinica, 1*, 89–100.

Qian, F., & Zhou, G. X. (1991). *Quaternary geology and paleoanthropology of Yuanmou, Yunnan, China*. Beijing: Science Press.

Qiu, Z., Deng, T., & Wang, B. (2004). Early Pleistocene mammalian fauna from Longdan, Dongxiang, Gansu, China. *Palaeontologia Sinica, New Series C, 191*(27), 1–198. (English summary).

Redfield, T. F., Wheeler, W. H., & Often, M. (2003). A kinematic model for the development of the Afar depression and its paleogeographic implications. *Earth and Planetary Science Letters, 216*, 383–398.

Rightmire, G. P., Lordkipanidze, D., & Vekua, A. (2006). Anatomical descriptions, comparative studies and evolutionary significance of the hominin skulls from Dmanisi, Republic of Georgia. *Journal of Human Evolution, 50*, 115–141.

Rogers, M. J., Feibel, C. S., & Harris, J. W. K. (1994). Changing patterns of land use by Plio-Pleistocene hominids in the Lake Turkana Basin. *Journal of Human Evolution, 27*, 139–158.

Rohling, E. J., Fenton, M., Jorissen, F. J., Bertrand, P., Ganssen, G., & Caulet, J. P. (1998). Magnitude of sea-level lowstands of the past 500,000 years. *Nature, 394*, 162–165.

Rolland, N. (1997). Early hominid expansion into Eurasia: Biogeographical and ecological issues. *Anthropologie, 35*, 101–107.

Rook, L., Martínez-Navarro, B., & Howell, F. C. (2004). Occurrence of *Theropithecus* sp. in the Late Villafranchian of Southern Italy and implication for Early Pleistocene "out of Africa" dispersals. *Journal of Human Evolution, 47*, 267–277.

Ruff, C. B., & Walker, A. (1993). Body size and body shape. In A. C. Walker, & R. E. Leakey (Eds.), *The Nariokotome Homo erectus skeleton* (pp. 234–265). Cambridge, MA: Harvard University Press.

Schwartz, J. H., & Tattersall, I. (1996). Whose teeth? *Nature, 381*, 201–202.

Semaw, S., Rogers, M. J., Quade, J., Renne, P. R., Butler, R. F., Dominguez-Rodrigo, M., et al. (2003). 2.6-Million-year-old stone tools and associated bones from OGS-6 and OGS-7, Gona, Afar, Ethiopia. *Journal of Human Evolution, 45*, 169–177.

Swisher, C. C., Curtis, G. H., Jacob, T., Getty, A. G., Suprijo, A., & Widiasmoro. (1994). Age of the earliest known hominids in Java, Indonesia. *Science, 263*, 1118–1121.

Tang, Y. J., Li, Y., & Chen, W. Y. (1995). Mammalian fossil and the age of Xiaochangliang paleolithic site of Yangyuan, Hebei. *Vertebrata Palasiatica, 33*, 74–83.

Tappen, M., Adler, D. S., Ferring, R., Gabunia, M., Vekua, A., & Swisher, C. (2002). Akhalkalaki: The taphonomy of an early Pleistocene locality in the Republic of Georgia. *Journal of Archaeological Science, 29*, 1367–1391.

Tchernov, E. (1987). The Age of the 'Ubeidiya formation, an Early Pleistocene hominid site in the Jordan valley, Israel. *Israel Journal of Earth Science, 36*, 3–30.

Tchernov, E. (1992). Eurasian-African biotic exchanges through the Levantine Corridor during the Neogene and Quaternary. *Courier Forsch. Inst. Senckenberg, 153*, 103–123.

Tedford, R. H. (1995). Neogene mammalian biostratigraphy in China: Past, present and future. *Vertebrata Palasiatica, 33*, 277–289.

Thomas, H., Geraads, D., Janjou, D., Vaslet, D., Memesh, A., Billou, D., et al. (1998). First Pleistocene faunas from the Arabian peninsula: An Nafud desert, Saudi Arabia. *Comptes Rendus de l'Academie des Sciences Paris, 326*, 145–152.

Turner, A. (1984). Hominids and fellow travelers: Human migration into high latitudes as part of a large mammal community. In R. Foley (Ed.), *Hominid evolution and community ecology* (pp. 193–217). London: Academic.

Turner, A. (1999). Assessing earliest human settlement of Eurasia: Late Pliocene dispersions from Africa. *Antiquity, 73*, 563–570.

Vekua, A. (1972). Kvabebskaja fauna akcagylskich pozvonocnych. Moskow. (in Russian).

Vrba, E. S. (1995). The fossil record of African Antelopes (Mammalia, Bovidae) in relation to human evolution and paleoclimate. In E. Vrba, G. H. Denton, T. C. Partridge, & L. H. Burckle (Eds.), *Paleoclimate and evolution with emphasis on human origins* (pp. 385–424). New Haven, CT: Yale University Press.

Walker, A. C., & Leakey, R. E. (Eds.). (1993). *The Nariokotome Homo erectus skeleton*. Cambridge, MA: Harvard University Press.

Wang, W., Potts, R., Yuan, B., Huang, W., Hai, C., Edwards, R. L., et al. (2007). Sequence of mammalian fossils, including hominoid teeth, from the Bubing Basin caves, South China. *Journal of Human Evolution, 52*, 370–379.

Wehausen, R., & Brumsack, H.-J. (2002). Astronomical forcing of the East Asian monsoon mirrored by the composition of Pliocene South China Sea sediments. *Earth and Planetary Sciences Letters, 201*, 621–636.

Wen, B. H. (1978). *Contributions to paleoanthropology* (pp. 126–135). Beijing: Science Press.

Whalen, N. M., Zoboroski, M., & Schubert, K. (2002). The Lower Palaeolithic in southwestern Oman. *Adumatu, 5*, 27–34.

Wood, B. A., & Xu, Q. (1991). Variation in the Lufeng dental remains. *Journal of Human Evolution, 20*, 291–311.

Wu, R. (1985). New Chinese *H. erectus* and recent work at Zhoukoudian. In E. Delson (Ed.), *Ancestors: The hard evidence* (pp. 245–248). New York: Alan R. Liss.

Wu, X. (2000). Longgupo mandible belongs to ape. *Acta Anthropologica Sinica, 19*, 1–10.

Wu, X. & Poirier, F. E. (1995). *Human evolution in China: A metric description of the fossils and a review of the sites*. Oxford: Oxford University Press.

Xu, Qinqi. (1988). Southward migration events of mammals in East Asia during the Pleistocene, and their relations with climatic changes. In P. Whyte, J. S. Aigner, N. G. Jablonski, G. Taylor, D. Walker, & P. X. Wang (Eds.), *The palaeoenvironment of East Asia from the mid-tertiary* (pp. 873–882). Hong Kong: Centre of East Asian Studies.

Yuan, Z. X., Lin, Y. P., Zhou, G. X., Zhang, X. Y., Wen, B. H., Jian, C., et al. (1984). Field report of an excavation at Yuanmou Man's site. In G. X. Zhou & X. Y. Zhang (Eds.), *Yuanmou Man* (pp. 12–22). Kunming, China: Yunnan People's Press.

Zaim, Y. (2010). Geological evidence for the earliest appearance of hominins in Indonesia. In J. G. Fleagle, J. J. Shea, F. E. Grine, A. L. Baden, R. E. Leakey, (Eds.), *Out of Africa I: the first hominin colonization of Eurasia* (pp. 97–110). Dordrecht: Springer.

Zhang, Y. (1985). *Gigantopithecus* and "*Australopithecus*" in China. In R. Wu & J. W. Olsen (Eds.), *Palaeoanthropology and palaeolithic archaeology in the People's Republic of China* (pp. 69–78). Orlando: Academic.

Zhang, Y. (1988). Preliminary analysis of the Quaternary zoogeography of China based on distributional phenomena among land vertebrates. In P. Whyte, J. S. Aigner, N. G. Jablonski, G. Taylor, D. Walker, & P. X. Wang (Eds.), *The palaeoenvironment of East Asia from the Mid-Tertiary* (pp. 883–896). Hong Kong: Centre of East Asian Studies.

Zheng, S. (Ed.). (2004). *Jianshi hominid site*. Beijing: Science Press.

Zhu, R., An, Z., Potts, R., & Hoffman, K. A. (2003). Magnetostratigraphic dating of early humans in China. *Earth Science Reviews, 61*, 341–359.

Zhu, R., Potts, R., Xie, F., Hoffman, K. A., Deng, C. L., Shi, C. D., et al. (2004). New evidence regarding the earliest human presence at high northern latitudes in northeast Asia. *Nature, 431*, 559–562.

Zhu, R. X., Hoffman, K. A., Potts, R., Deng, C. L., Pan, Y. X., Guo, B., et al. (2001). Earliest presence of humans in northeast Asia. *Nature, 413*, 413–417.

Zhu, R. X., Potts, R., Pan, Y. X., Yao, H. T., Lü, L. Q., Zhao, X., et al. (2008). Early evidence of the genus *Homo* in East Asia. *Journal of Human Evolution, 55*, 1075–1085.

Chapter 6
New Archeological Evidence for the Earliest Hominin Presence in China

Ya-Mei Hou and Ling-Xia Zhao

Abstract Although China is literally at the end of Eurasia, there is increasing evidence that some of the earliest records of hominins outside of Africa are from China. Recent work at three archeological sites in southern China provide artifactual evidence of hominins in this region near the beginning of the Pleistocene. Renzidong in Anhui Province is a cave site dated to approximately 2.6 Ma. It contains a large mammalian fauna and 59 specimens that have been identified as stone artifacts. Longgudong, in Hubei Province, is dated to the Matuyama Epoch by paleomagnetic studies. It has yielded a large mammalian fauna, including *Gigantopithecus* as well as 592 stone artifacts in stratigraphic context and some bone artifacts. Longgupo, in Wushan County is a well-known site attributed to the Early Pleistocene. It has yielded 26 stone artifacts and a large mammalian fauna including *Gigantopithecus* and a second hominoid that has been considered by some authorities to belong to a species of *Homo* and by others to be a small ape related to *Lufengpithecus*. The identity of many Early Pleistocene hominoid fossils from China and their phylogenetic relationship with other hominoids, including *Homo* is a topic of considerable discussion and debate. The resolution of these issues and a better understanding of the place of China in early human evolution will come from continued field work and new discoveries.

Keywords Renzidong • Longgudong • Longgupo • Nihewan - Stone artifacts • Early human evolution • Asia

Y.M. Hou (✉)
Institute of Vertebrate Paleontology and Paleoanthropology & Joint Laboratory of Human Evolution and Archaeometry,
Chinese Academy of Sciences, No. 142, Xizhimenwai Dajie,
Beijing 100044, PR China
e-mail: houyamei@ivpp.ac.cn

L.X. Zhao
Institute of Vertebrate Paleontology and Paleoanthropology & Joint Laboratory of Human Evolution and Archaeometry,
Chinese Academy of Sciences, No. 142, Xizhimenwai Dajie,
Beijing 100044, PR China
e-mail: zhaolingxia@ivpp.ac.cn

Introduction

In addressing the issues surrounding the initial expansion of hominins out of Africa, China is a critical part of the world that must be discussed. Although China is literally at the end of Eurasia, there is increasing evidence, much of it very recent, documenting that the earliest dispersal of hominins out of Africa clearly involved China.

Following the important discoveries at Zhoukoudian in the early part of the twentieth century, China has continued to produce evidence of human evolution in the Pleistocene. Many new early sites have been discovered throughout the country, from the south to the north and from the east to the west. They include sites yielding fossil humans, sites yielding Paleolithic archeology, and sites yielding both. New techniques have also been applied for new or revised dating work at many sites. All this new information is making evidence of human evolution in China more detailed and more reliable than previously understood, though it can still be complicated and confused in some respects. Nevertheless, we believe that evidence from China is essential if we are to have an understanding of human evolution and early hominin biogeography. In this paper we focus on several archeological sites with new evidence bearing on the earliest presence of hominins in Asia, in particular, the sites of Renzidong, Longgudong, and Longgupo (Fig. 6.1).

The Natural Landscape and Chronological Scale of China

Along the western margin of the Pacific Ocean, China has a very diverse geology, much of it dating to the Quaternary. The topography of China is characterized by elevated highlands in the west and vast lowlands in the east. The west and southwest are comprised mostly of plateaus and basins, while the east is characterized mostly by rolling hills and several dissected plains. The extensive East Asian Monsoon dominates the climate in China. It ranges from a subtropical

J.G. Fleagle et al. (eds.), *Out of Africa I: The First Hominin Colonization of Eurasia*, Vertebrate Paleobiology and Paleoanthropology,
DOI 10.1007/978-90-481-9036-2_6,

Fig. 6.1 A general map of the geography of China and positions of some sites discussed in the article

zone in the south and temperate zone in the north to areas of high humidity in the east, gradually becoming colder and drier to the west. Pleistocene deposits of continental facies are well developed, preserved and exposed in many areas.

At the latitude of ca. 34°N, the Qinling Range in Central China is conventionally taken as a physiographic boundary between the north and south. Quaternary deposits are generally different on these two sides. In northern China, the principal deposits are composed of widely distributed, thick loess deposits and well-developed fluvio-lacustrine basin formations. On the other hand, the south and southwest are characterized by laterite fluviatile sediments and cave deposits in karstic limestones. Excepting the laterites, many of the above-mentioned strata often contain abundant human and mammalian fossils as well as cultural remains, and these geological circumstances provide prime conditions for preserving evidence regarding the behavior of early humans.

The Paleolithic chronological sequence in China has been constructed on the basis of biostratigraphy and lithostratigraphy often in conjunction with isotopic dating and paleomagnetic analysis. Generally speaking, the more than 2 million year time span referred to by geochronologists as the Early and Middle Pleistocene in China, is equivalent to the Marine Isotope Stages 103 to 6, ca. 2.6 Ma to 128 ka. During this period, the sedimentological record is characterized by the formation of the Wucheng and Lishi loess deposits, and is further identified by the presence of unique faunal units in northern China, including the Nihewan fauna of the Early Pleistocene, the Gonwangling fauna of the later Early Pleistocene, and Zhoukoudian fauna of the Middle Pleistocene.

In southern China, in addition to earlier Pleistocene fluviatile and fluvio-lacustrine formations and cave deposits, lateritic sediments characterize the period. Representative mammal units include the Yuanmou and *Gigantopithecus* faunas of the Early Pleistocene and the Yanjinggou *Ailuropoda-Stegodon* fauna of the Middle Pleistocene. However, lateritic fluviatile sediments are strongly acidic and not conducive to the preservation of mammalian fossils to provide evidence of biostratigraphy for determinate age of many sites. Therefore, chronology is a critical puzzle for archeologists in many sites and greatly restricts our understanding of the Chinese pebble-tool sites which are distributed in this geological belt. However, this difficult circumstance has recently changed greatly because of the advances in isotopic dating and other methods.

Three Newly Discovered Early Archeological Sites in Southern China

Renzidong Site in Fanchang County of Anhui Province

These Late Cenozoic, two cave-fissure filling deposits were discovered in 1998 in Fanchang County of southeast Yangtze River. The site is located at 118°5′46″E and 31°5′23″N. It is developed in a folium of Triassic limestone. In two seasonal excavations, stone artifacts, bone artifacts, and a great variety of vertebrate fossils, including *Procynocephalus*, were found *in situ*.

The vertebrate fossils from Renzidong include turtles, birds and 67 mammal species showing many primitive characters and consists mainly of extinct species. The mammalian fauna is very rich and rather old showing many transitional forms linking southern and northern China. The Tertiary residual species include *Homotherium, Sinomastodom,* and *Nestoritherium*, among others. Micromammals, including *Beremendia, Hypolagus, Mimomus, Kowaskia,* and *Brachyrhizomys shansius* are also numerous. The large mammals *Pachycrocuta licenti, Equus sanmenensis, Muntiacus nanus, Metacervulus capreolinus, Cervavitus ultimus*, and *Cervus* cf. *phyilisi*, among others are common members of the Late Tertiary and Early Quaternary mammalian faunas in northern China. Among them *Cervus* cf.*phyilisi* can be compared with those of European Villafranchian faunas. Fifteen species are commonly shared by the faunas of Renzidong and Longgupo (see below). Compared with the other Early Pleistocene faunas from northern and southern China, the percentage of extinct genera and species is relatively higher. On the other hand, the appearance of *Equus* in the fauna shows that the fauna should be younger than 2.6 Ma. If 2.6 Ma can be considered as the lowest boundary of the Pleistocene recommended by many Chinese stratigraphers (Mascarelli 2009),

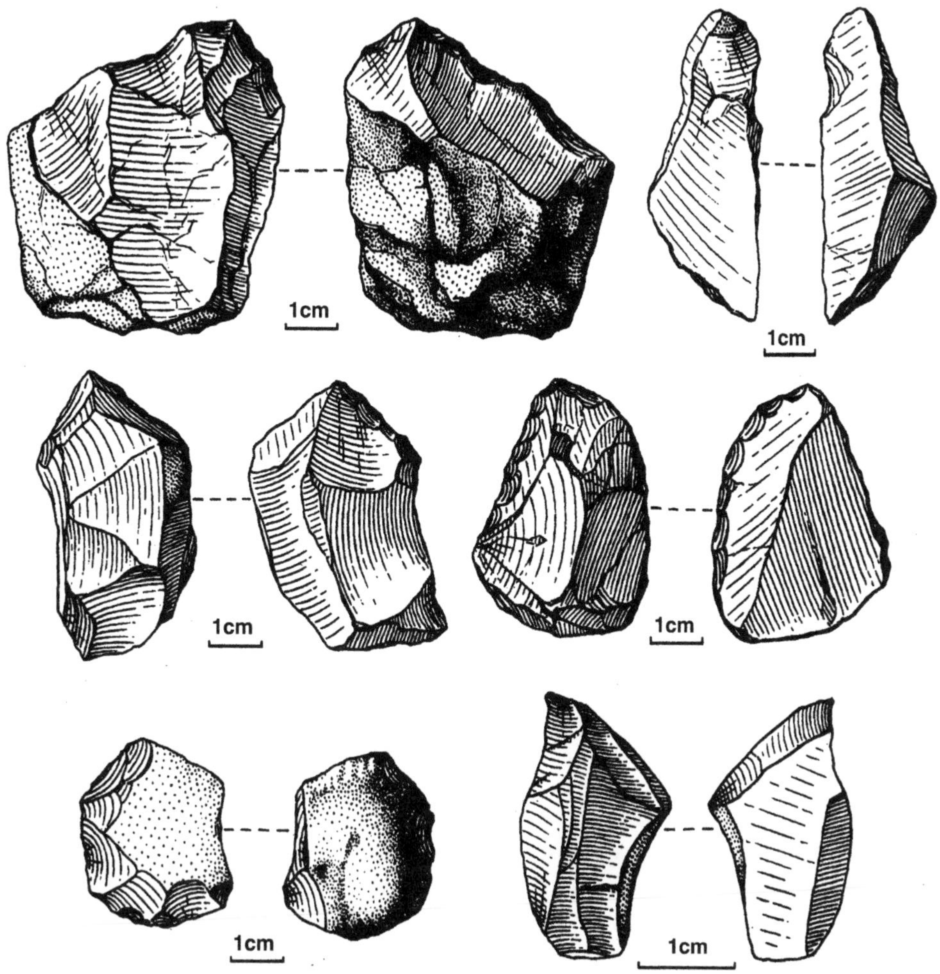

Fig. 6.2 Some stone artifacts from the Renzidong site (After Zhang et al. 2000, *upper row*: core and flake; *middle row*: scrapers; *lower row*: scraper and burin)

the geological age of the Renzidong fauna is very likely to be the early stage of the Early Pleistocene. The much closer character of the Renzidong fauna to those of northern China implies a cold period in China in the early stage of the Early Pleistocene and suggests that the changed climate caused the northern faunas to migrate southward. This geographic area of the southern Yangtze became a transitional zone between the Palaearctic and Oriental Regions (Jin et al. 1999). The new discovery of the Renzidong site is very helpful to understanding the zoogeography and paleoclimate of China in the Quaternary and also provides a very important clue about early human evolution in China.

In 1998, excavation in the same layer containing this fauna group produced 59 specimens of stone artifacts that were recognized from 575 unearthed nonlimestone pieces (Zhang et al. 2000). Iron ore is the dominant raw material, and it was used for making 52.5% of the tools. The other four raw materials are siliceous mudstone, quartz-sandstone, siliceous limestone and gneiss. The artifacts, including cores (17%), flakes (30.5%), scrapers (45.7%), burins (5.1%) and undetermined (1.7%), are mostly small. The cores are single-platformed and double-platformed for producing irregular flakes by simple hammer percussion (Fig. 6.2). Most tools (64.4%) were made of chunks and cores while 35.5% were flakes. These tools were crudely modified and the majority of them were retouched on the dorsal face. Among scrapers, single-edged pieces are more frequent than double-edged ones and include straight, convex and concave shapes. The original researchers think that the general character of these artifacts is close to Paleolithic cultures of northern China in many ways. However, experimental work would help determine if these are indeed human-produced artifacts (Zhang et al. 2000). Several bone tools are also identified including one made on a piece of long bone and another made on a mandible of *Rhinoceros* sp. Biostratigraphic data and geological comparison indicate that Renzidong likely dates to the Early Pleistocene.

Longgudong Site in Jianshi County of Hubei Province

In Southwest Hubei province, a small hill (30°39′14.9″E, 110°04′29.1″N) in a village named Jintang in the Gaoping town of Enshi city is well known for its large mammalian fauna and *Gigantopithecus* fossils. It has long been known as Longgudong (Dragon Bone Cave) or Juyuandong (Giant Ape Cave). The cave resides at an elevation of 794 m, and in front of the cave the Longdong River is 85 m lower than the cave itself, flowing underground and 1.5 km out to the Yesanhe River of Qingjiang River, a southern tributary of the Yangtze River. At a direct distance of 63 km to the north it reaches another famous paleoanthropological site, Longgupo in Wushan, Chongqing city (Gu and Fang 1991).

As is typical of other older cave sites in China, Longgudong Cave contains many mammalian fossils, including 87 species, 5 of which were only recently described (Zheng 2004). Half of these are micromammals and appear to have biochronological significance, including animals such as *Anourosorex quadratidens, Sericolagus, Sciurotamias teilhardi, Belomys parapearsoni, Allocricetus ehiki, Eothenomys hubeiensis, Allphairmys terrae-rubrae, Hyperacrius jianshiensis, Typhlomys intermedius, Rhizomys brachyrhizomysoides, Hystrix magna, Vernaya prefulva, Apodemus asianicus, Wushanomys ultimus,* and *Niviventer preconfucianus.* Based on comparisons with the small mammal assemblage from Loc.18, Huangkan, Sunjiashan, Zibo in North China, the Longgudong microfauna is considered to be Early Pleistocene. However, some species, like the derived *Wushanomys* species, suggest that the Longgudong Cave fauna is slightly later in time.

The 35 species of large mammals show strong Early Pleistocene characteristics with the coexistence of *Gigantopithecus blacki, Cuon dubius, Ailuropoda wulingshanensis, Pachycrocuta licenti, Sivapanthera pleistocaenicus, Equus, Tapirus sanyuanensis, Hesperotherium* sp., *Sus xiaozhu, Sus peii, Sinomastodon,* and *Stegodon preorientalis.* Comparison with seven localities containing the *Gigantopithecus blacki-Ailuropoda* fauna l in South China shows that the common genera and species can reach 63.33% and 42.86%, respectively with the Wushan fauna group, and 70% and 42.88% with the Liucheng *Gigantopithecus* fauna group.

Taking the biostratigraphy correlations above with the results from paleomagnetic work indicates that the age of Longgudong Cave is correlated with the Matuyama Epoch. The upper normal polarity is correlated to the Olduvai Event, and the lower normal polarity corresponds to the Reunion Event (Zheng 2004).

In the reported 632 pieces of stone artifacts from the 1999–2000 excavation period, 592 pieces were from two parts of the strata (L2–L11) in the cave and 40 were collected from non-strata. The artifacts have been described as manuports (natural stones moved inside artificially), cores, flakes, tools, chunks and chips. Cores, flakes and tools make up more than half of the collection. Most of them retain part of the cortex. Direct hammer percussion and bipolar techniques were used and adopted for core reduction and tool making. Most tools are small-sized. Tools can be classified into scrapers (94.2%), points (1.93%), burins (0.49%) and hammer stones (3.38%) (Li 2004). Bone artifacts are also identified as bone flakes, points and shovels. On some of them, cut and chopper marks can be observed.

Longgupo Site in Wushan County of Chongqing City

Located in Longping village of Miaoyu town in Wushan county, the Longgupo site (109°4′50″E, 30°21′25″N) is a cave formation developed in Triassic Jialingjiang limestone. It became a sloping deposit after its collapse (Huang and Fang 1991). The karstic topography provides good conditions for having Plio-Pleistocene deposits preserved in the valley of Miaoyu. The deposits have been dated from 1 to 2.6 Ma by paleomagnetic and other different methods including ESR, Th, U, and Amino Acid analyses. The cave has yielded lithic artifacts and the remains of *Gigantopithecus* and numerous other mammalian species confirming that these deposits were deposited during the Pliocene and Early Pleistocene. But these deposits have also furnished a mandible fragment and several teeth that have been successively attributed to *Homo erectus wushanensis* (Huang et al. 1995), to *Homo habilis* or to *Homo ergaster* (Wood and Turner 1995), in levels estimated to date to 1.9 Ma (Huang et al. 1995). In addition, the Wushan fauna is composed of 92 species. Some species are very specific to the Early Pleistocene such as *Gigantopithecus blacki, Ailuropoda microta, Pachycrocuta licenti*, and *Nestoritherium* sp., confirming the antiquity of the site.

Three stages of excavation were in 1985–1986, 1997–1998, and 2003–2006. The earlier two stages were excavated using the horizontal method by the interval of one meter and divided into 20 levels. A total of 26 artifacts from four levels (5–8 levels) have been reported (Huang et al. 1995; Hou et al. 1999; 2006). They were made from different limestones from the local area, including metamorphic and siliceous limestones. The hard hammer percussion technique was applied during the production of these stone tools. The tool blanks were procured by splitting the pebble in its long axis through the bipolar technique. Most choppers were made from natural stones of pebbles or blocks except some retouched ones on large flakes. The artifacts include hammers, picks, choppers, proto-adze, proto-cleavers, knives and flakes (large or small sizes). Unifacial and alternative retouches were often used for making a tool edge into traversed, pointed, notched or convex shapes. Knives, chisels and a habit of preparing a tool butt are particularly characteristic of this industry. The tool sizes are mostly large (around 10–20 cm). Some flakes are smaller than 10 cm (Fig. 6.3). The artifacts from this site show both the diversity and the particularity of tool typology among the early human artifacts in southern China from 2 Ma. These stone tools indicate the level of development in material culture that early man in China had achieved by the Early Pleistocene.

The most recent stage of work at Longgupo is being conducted by a Sino-Franco team co-lead by the first author, Eric Boëda and Huang Wanbo. In this new period (2003–2006), the archeological excavations are carefully based on the geological sequence. Clearer archeological layers are for the first time distinguished. Much new information regarding the oldest human behavior has been recorded in more detail than any previous excavation at the site. The renewed excavations and further research work at Longgupo offer great potential for providing more significant information for understanding the oldest human behavior in China. These results further confirm the paleontological and archeological interest of this site. New publications about this work are in preparation.

Hominoid Coexistence and Other Relevant Information

With many old archeological sites from the Late Pliocene-Early Pleistocene in China, we need to consider what hominins are associated with this material. However, this is an area of ongoing research with many unresolved issues (Ciochon 2010). Up to now, no hominin fossils have been

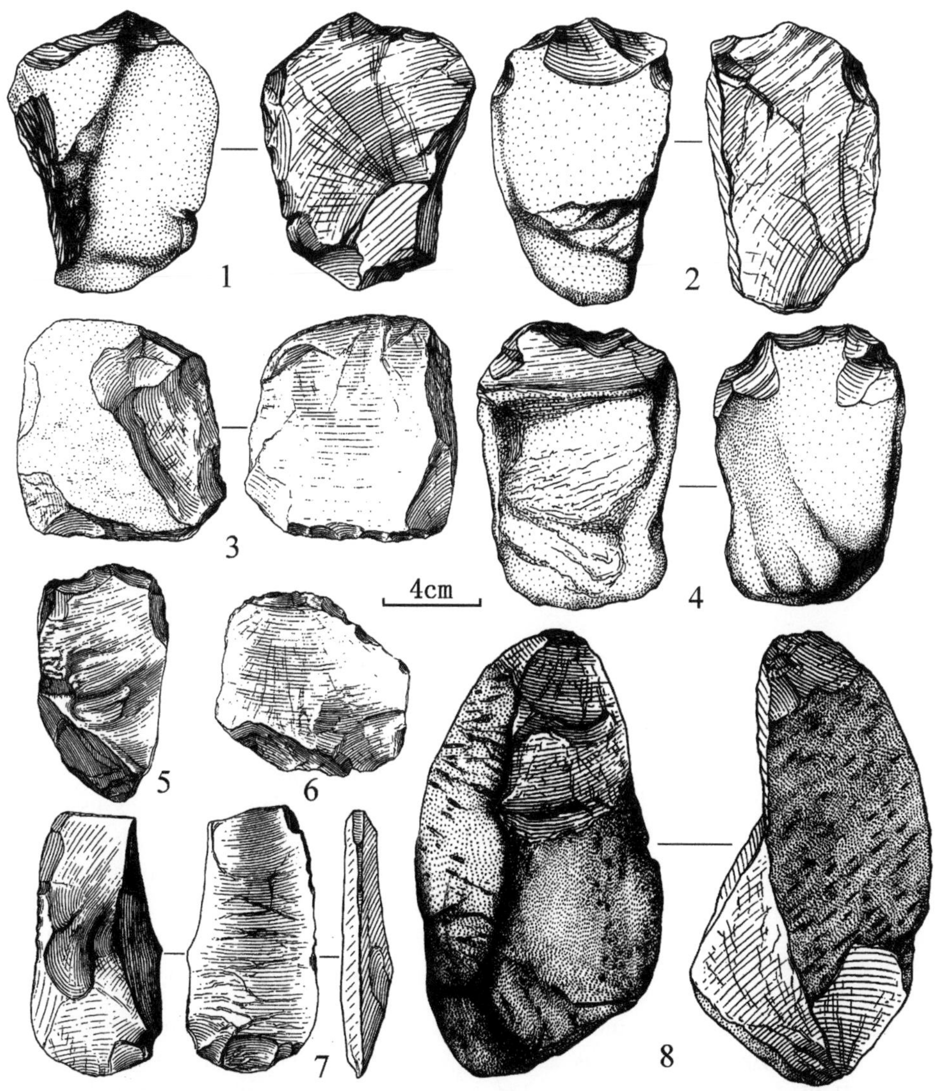

Fig. 6.3 Stone artifacts from the Longgupo site (1–5. Choppers; 6. Flake; 7. Knife of a Kombewa flake; 8. Pick)

found in the Early Pleistocene sites of the Nihewan Basin, even though there are many stone artifacts and reliable paleomagnetic dating in several sites such as Donggutuo, Xiaolangliang and Majuangou (Wang et al. 2005; Zhu et al. 2001, 2004).

However, in central and southern China there are a number of hominoid fossils that have often been linked to the archeological materials there, including a mandibular fragment from Longgupo in Wushan, six isolated teeth from Longgudong in Jianshi, and two incisors from Yuanmou. Longgupo is one of the most exciting and heavily debated early sites because of its archeological evidence and questionable hominoid fossils. At least two hominoid species have been discovered at this site *in situ*. The first is *Gigantopithecus blacki*, known from more than 14 isolated teeth. A second taxon is known from a mandible. There are various opinions regarding the proper allocation of the mandible from Longgupo. Huang et al. (1995) thought that the mandible belonged to *Homo*, whereas Wu (2000) insisted that it belongs to an ape related to *Lufengpithecus* (see also Ciochon 2010). The second author agrees with this assessment and thinks that the morphology of the mandible is very similar to that of *Lufengpithecus hudienensis* from Yuanmou (Fig. 6.1). As discussed later, the phylogeny of *Lufengpithecus* is still uncertain.

Longgudong in Jianshi is another important Early Pleistocene site. More than 100 isolated teeth of *Gigantopithecus blacki* were discovered in the cave deposits along with six teeth assigned to other taxa. Three teeth discovered in 1970s were suggested to belong to *Australopithecus* by Gao (1975) or *Homo erectus* by Zhang (1984), and three other teeth discovered in the 1990s were assigned to *Meganthropus paleojavanicus* by Zhang et al. (2004). Whatever species these teeth belong to, it is believed by many Chinese scholars and others that they are hominins, not pongines. Up to now, these isolated teeth have not attracted much attention because of the lack of skulls and long bones which are more reliable for determining phylogenetic position.

Gigantopithecus blacki is an interesting taxon that needs to be discussed here because of its widespread occurrence in Longgupo and Longgudong as well as several other Early Pleistocene cave deposit sites. Weidenrich (1946) proposed *Gigantopithecus blacki* as the forerunner of *Homo*, at a time when only several isolated teeth had been collected from drugstores. Based on three mandibles and nearly a thousand teeth of *Gigantopithecus blacki* from Liucheng in Guangxi, Woo (1962) suggested that *G. blacki* was an early extinct branch of hominin, not the direct forerunner of later hominins. The morphology of *Gigantopithecus* is between apes and hominins. However, most authorities, including the authors of this paper regard *Gigantopithecus* as a fossil pongine (Zhao 2006). Until recently, many authors have suggested that *Homo* (or some other hominin) and *Gigantopithecus blacki* coexisted in southern China in the Early and Middle Pleistocene, at sites such as Longgupo in Wushan and Longgudong in Jianshi. However, the identity of this other hominoid, known from fragmentary remains has recently been called into question, and several authorities now argue that it is not a hominin, but rather some sort of ape, a younger relative of *Lufengpithecus*.

Lufengpithecus is a fossil ape from Yunnan Province of China. Since the mid-1970s, a large quantity of Late Miocene hominoid fossils have been found in Yunnan Province, including skulls, mandibles, maxillary, teeth and postcranial bones. Different opinions exist regarding the exact phylogenetic placement of *Lufengpithecus*. While some believe that *Lufengpithecus* is on the evolutionary line toward Homininae (Gao 1998; Gao et al. 2004, 2006), others think that *Lufengpithecus* is a branch of the Ponginae related to the orangutan or a conservative hominin stock that preceded the split between Ponginae and Homininae. Based on enamel microstructure and dental development, the second author of this paper (Zhao et al. 2000, 2002, 2003, 2008; Zhao and He 2005), has suggested a potential relationship between *Lufengpithecus* and early hominins.

Conclusions

So far, the limited evidence in China does not provide any definite new conclusions regarding human evolution in this region. However, the evidence urges us to keep our minds open in thinking of wider possibilities regarding early human evolution in this region. Asia is a large region whose paleontology and archeology remain poorly known and understood. Many currently popular views must be tentative and the weight of evidence is not enough to fully accept or reject certain hypotheses. Early human behavior may be well known in Africa, but it is not well-known yet in Asia. The early culture in Asia and the early living environment in Asia are not widely known. The longer people keep working primarily in Africa, the longer we will have to wait for clearer answers in Asia. The current "Out of Africa" scenario may turn out to be just wishful thinking. As noted by other scholars (Dennell 2003; Dennell and Roebroeks 2005), the necessity of considering Asia in discussions of human origins needs to be taken more seriously and not be put aside. Only additional research on associated faunas, stone artifacts and hominin or hominoid remains can help us to obtain a better understanding of the important events that took place in Asia. Some final conclusions can be drawn:

1. Though the evidence of the time span of the earliest humans from Asia has not been comparable with that from East Africa, there is evidence of hominins in Asia around 2 Ma in southern China. This relatively limited information

is certainly not enough to change any traditional theories or hypotheses about human origins, but they do offer suggestions of early human activity in the East.

2. Outside of China, there is additional old evidence from areas of Asia showing connections with Africa, including Georgia in the north and Indonesia in the south. It is not yet clear how the evidence from these areas relates to that in China.
3. The unique geographical position of China causes it to have a very diverse environmental history relevant to early human evolution. For the past 30 million years the Qinghai-Tibetan Plateau has a long history of uplift and subsidence. Between 8 and 4 million years ago, the Qinghai-Tibetan Plateau and the area east of it was stable with a dry climate (Li et al. 1979). Between 2.4–0.5 Ma the intensive uplift of Qingghai-Tibetan Plateau led to many mountains in western China and formed "The Three Steps" geomorphology in eastern China (Huang et al. 1980). This completely changed the climate in the entire area, including the new setting of the monsoon system. The loess plateau in northern China and the river system from western to eastern China had all been formed. Such an enormous transformation influenced the local environment dramatically and led to the appearance of many new habitats and new species.
4. There is cultural evidence of early humans in China, not only from the south, but also from the north, such as in Nihewan. The emergence of new sites raises new questions in addition to those discussed above. The recently reported Goudi-part of Majuangou site (Gao et al. 2005) contains a stone artifact layer similar to or even earlier than the previously reported site of Majuangou (Zhu et al. 2004). The artifacts at both sites show clear "debitage" industry. However, in the south, for example at Longgupo, the industry has the strong character of "façonage" which is a consistent feature with later Chinese southern Paleolithic culture over a wide region. Such differences between the south and north in a comparably early period raise the question of where they are from and whether they have the same origin. According to distributed areas of Levallois and Acheulean industries, Boëda and Hou (2004) proposed that a later developed Acheulean culture in China was a local innovation. He believes that the early human culture was developed convergently in the East and West (Mulot 2006). The study of the newly excavated materials from Longgupo site (Fig. 6.4) may reveal some new behavior of early humans here. Some scholars who have doubted the evidence from the Longgupo site (Gao et al. 2005) have become more positive after new observations from recently excavated stone artifacts (Wei Qi, personal communication). The artifacts from more than dozens of archeological layers of Longgupo show much diversity and a somewhat mature organization of technology. We doubt if it could be the earliest culture in the East. On the contrary, the technology seems to show that it had already been developed for quite a long time.
5. No matter how popular one theory has become, any study of human origins must be tested by new discoveries. It needs to stand up to new evidence; otherwise, it needs

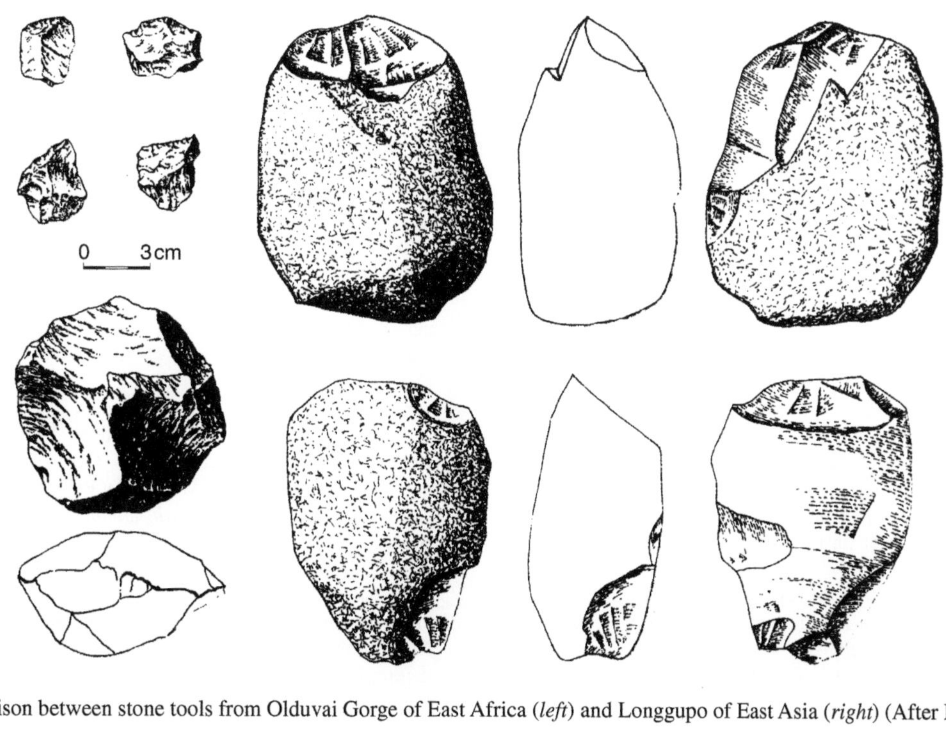

Fig. 6.4 Comparison between stone tools from Olduvai Gorge of East Africa (*left*) and Longgupo of East Asia (*right*) (After Mulot 2006)

to be replaced by a new and more convincing theory. Having previously argued that human evolution is the result of two instances of "Out of Africa" migration (Templeton 2002); Templeton (2005) has subsequently argued for three expansions of human evolution, the first of which occurred at 1.9 Ma. This date is close to the estimated age of the earliest evidence of hominins in China. This implies that cosmopolitan gene flow has happened many times. The present record is only partly understood. Dennell and Roebroeks (2005) go even further based on some very recent evidence of human evolution from different parts of Asia. They suggest the need for a new "Asian perspective" to fully understand early human evolution. The way of looking for the earliest tool maker in China is to keep exploring.

Acknowledgements This research is supported by the project of NSFC (No. 40872023), the Major Basic Research Projects (2006CB806400) of MST of China, Open project program of State Key Laboratory of Palaeobiology and Stratigraphy (Nanjing Institute of Geology and Paleontology, Chinese Academy of Sciences) and the foundation of the French Embassy in Beijing. The first author thanks Stony Brook conference organizer for the very kind invitation and the opportunity to participate in the conference. We thank Dr. John G. Fleagle for his kind help in correcting the English in this manuscript and Miss Guangcai Liu for her management of some figures of this paper.

References

Boëda, E., & Hou, X. (2004). Relationship between east and west Asia in Paleolithic age. *Quaternary Sciences, 3*, 255–264.

Ciochon, R. L. (2010). Divorcing Hominins from the *Stegodon-Ailuropoda* Fauna: New views on the antiquity of hominins in Asia. In J. G. Fleagle, J. J. Shea, F. E. Grine, A. L. Baden, & R. E. Leakey (Eds.), *Out of Africa I: the first hominin colonization of Eurasia* (pp. 111–126). Dordrecht: Springer.

Dennell, R. (2003). Dispersal and colonization, long and short chronologies: how continuous is the Early Pleistocene record for hominids outside East Africa? *Journal of Human Evolution, 45*, 421–440.

Dennell, R., & Roebroeks, W. (2005). An Asian perspective on early human dispersal from Africa. *Nature, 438*, 1099–1104.

Gao, J. (1975). Australopithecine teeth associated with *Gigantopithecus*. *Vertebrata PalAsiatica, 13*, 81–88.

Gao, F. (1998). Phylogeny for the large-bodied hominoid of Yunnan (China) its significance in human origin. In J. Chen, Z. Jiang, & X. Ji (Eds.), *Collected works for "The 30th anniversary of Yuanmou man discovery and the international conference on Palaeoanthropological studies"* (pp. 50–65). Kunming: Yunnan Science and Technology Press.

Gao, F., Kelley, J., Zheng, L., & Liu,W. (2004). An infant skull of *Lufengpithecus* from Yuanmou, Yunnan Province, China. *American Journal of Physical Anthropology 123*(S38), 98.

Gao, X., Wei, Q., Shen, C., & Keates, S. (2005). New light on the earliest hominid occupation in East Asia. *Current Anthropology (Suppl.), 46*, 115–120.

Gao, F., Kelley, J., Zheng, L., & Liu, W. (2006). An infant skull of *Lufengpithecus* from Yuanmou, Yunnan province, China. In D. Yang & S. Yang (Eds.), *Collected works for "The 40th anniversary of Yuanmou man discovery and international conference on paleoanthropological studies"* (pp. 40–63). Kunming: Yunnan Science and Technology Press.

Gu, Y., & Fang, Q. (1991). Hominid Fossils. In W. & Q. Fang (Eds.), *Wushan Hominid Site* (pp. 16–19). Beijing: Ocean Press.

Hou, Y., Xu, Z., & Huang, W. (1999). Some new stone artifacts discovered in 1997 at Longgupo, Southern China. *Longgupo Prehistoric Culture 1*, 69–80.

Hou, Y., Xu, Z., & Huang, W. (2002). Study on stone artefacts of level 8 from 1998 excavation at Longgupo, Southern China. *Longgupo Prehistoric Culture, 4*, 1–10.

Hou, Y., Huang, W., Boëda, E., Zhao, L., Xu, Z., Daussy, A., et al. (2006). Le site de Longgupo et la naissance du genre humain en Asie. In X. Chen, & M. Bussotti (Eds.), *Sinologie Française* (pp. 11, 7–22). Beijing: Zhonghua Shuju.

Huang, W. and Fang Q. (Eds.) (1991). *Wushan Hominid site*. Beijing: Ocean Press.

Huang, J., Ren, J., Jiang, C., Zhang, Z., & Qin, D. (1980). *The earth structure and evolvement of China*. Beijing: Science Press.

Huang, W., Ciochon, R., Gu, Y., Larick, R., Fang, Q., Schwartz, H., et al. (1995). Early *Homo* and associated artifacts from Asia. *Nature, 378*, 275–278.

Jin, C., Han, L., & Wei, G. (1999). The discovered Paleolithic site of Renzidong in Fanchang County of Anhui Province. *Acta Anthropologica Sinica, 18*, 70–71.

Li, C. (2004). Stone artifacts. In S. Zheng (Ed.), *Jianshi Hominid site* (pp. 37–66). Beijing: Science Press.

Li, J., Wen, S., Zhang, Q., Wang, F., Zheng, B., & Li, B. (1979). The discussion about the age, range and form of the uplift of Qinghai-Tibet Plateau. *Science in China, 6*, 608–616.

Li, Y. (1991). Stone Artifacts. In W. Huang & Q. Fang (Eds.), *Wushan Hominid Site* (pp. 20–23). Beijing: Ocean Press.

Mascarelli, A. L. (2009). Quarternary geologists win timescale vote. *Nature, 459*, 624.

Mulot, R. (2006). *Contre-enquête sure le berceau chinois* (pp. 56–61). Avril: Science et Avenir.

Templeton, A. R. (2002). Out of Africa Again and Again. *Nature, 416*, 45–51.

Templeton, A. R. (2005). Haplotype trees and modern human origins. *Yearbook of Physical Anthropology, 48*, 33–59.

Wang, H., Deng, C., Zhu, R., Wei, Q., Hou, Y., & Boëda, E. (2005). Magnetostratigraphic dating of the Donggutuo and Maliang Paleolithic sites in the Nihewan Basin, North China. *Quaternary Research, 63*, 1–11.

Weidenrich, F. (1946). *Apes, Giants and Man*. Chicago: The University of Chicago Press.

Woo, J. (1962). The mandibles and dentition of *Gigantopithecus*. *Palaeontologia Sinica, New Series D, 11*, 1–94.

Wood, B., & Turner, A. (1995). Out of Africa and into Asia. *Nature, 37*, 239–240.

Wu, X. (2000). Longgupo hominoid mandible belongs to ape. *Acta Anthropologica Sinica, 19*, 1–10.

Zhang, Y. (1984). The "*Australopithecus*" of west Hubei and some early Pleistocene hominids of Indonesia. *Acta Anthropologica Sinica, 3*, 85–92.

Zhang, S., Han, L., Jin, C., Wei, G., Zheng, L., & Xu, Q. (2000). On the artifacts unearthed from the Renzidong Paleolithic site in 1998. *Acta Anthropologica Sinica 19*, 169–183.

Zhang, Y., Zhang, Z., & Liu, W. (2004). Paleoanthropology of Jianshi Site. In: S. Zheng (Ed.), *Jianshi hominid site* (pp. 26–37). Beijing: Science Press.

Zhao, L. (2006). Comprehensive dental study on *Gigantopithecus*. Ph.D. dissertation, Graduate School of Chinese Academy of Sciences

Zhao, L., & He, Z. (2005). Dental development and ontogeny of Late Miocene large-bodied hominoids from Yunnan, China. *Anthropological Science, 113*, 79–83.

Zhao, L., Lu, Q., & Xu, C. (2000). Enamel microstructure of *Lufengpithecus lufengensis*. *Acta Anthropologica Sinica 19* (Suppl.), 77–82.
Zhao, L., Lu, Q., & He, Z. (2002). Eruption sequence of lower permanent dentition of *Lufengpithecus lufengensis*. *Acta Anthropologica Sinica, 21*, 14–18.
Zhao, L., Zheng, L., Gao, F., & Jiang, C. (2003). Preliminary study on enamel microstructure of Yuanmou Miocene hominoids of China. *Science in China (Series D), 46*, 1142–1150.
Zhao, L., Lu, Q., & Zhang, W. (2008). Age at first molar emergence in *Lufengpithecus lufengensis* and its implications for life-history evolution. *Journal of Human Evolution, 54*, 251–257.
Zheng, S. (Ed.). (2004). *Jianshi Hominid site*. Beijing: Science Press.
Zhu, R., Hoffman, K., Potts, R., Deng, C., Pan, Y., Guo, B., et al. (2001). Earliest presence of humans in northeast Asia. *Nature, 413*, 413–417.
Zhu, R., Potts, R., Xie, F., Hoffman, K., Deng, C., Shi, C., et al. (2004). New evidence on the earliest human presence at high northern latitudes in northeast Asia. *Nature, 431*, 559–562.

Chapter 7
Geological Evidence for the Earliest Appearance of Hominins in Indonesia

Yahdi Zaim

Abstract Until the end of Tertiary period (Pliocene), almost all Indonesian regions were still occupied by a marine environment. Tectonics and glacioeustatic changes took place during the Pleistocene to form the Indonesian Archipelago, including the emergent lands known as Sunda Land and the Sahul Shelf. The Sunda Land connected Asia mainland with the islands of Sumatra, Java and Kalimantan. It also acted as a land bridge and migration route for *Homo erectus* and vertebrate faunas from the Asia mainland to Java. In the Pleistocene, the sedimentations in Indonesian regions occurred mostly in non-marine environments: lakes and rivers (sometimes with marine influences) which are favorable for hominin and vertebrate occupations. The first arrival of vertebrate faunas from Asia to Indonesia through the Sunda Land at the end of the Late Pliocene and later on was followed by arrival of the early hominin – *Homo erectus paleojavanicus (Meganthropus paleojavanicus)* to Java in the Early Pleistocene (1.6–1.0 Ma). During the Middle to Late Pleistocene (1.0–0.125 Ma) there existed another hominin in Java as indicated by the presence of *Homo erectus ngandongensis/soloensis*. Recently, hominin fossils have also been discovered in eastern Indonesia, at Liang Bua, Flores island and are attributed to *Homo floresiensis* (*orang pendek* – pygmy people) dated as 12,000 BP or the end of Late Pleistocene probably coexistant with *Homo sapiens*. The very important site of hominin and vertebrate fossils in Indonesia and in Java is the Sangiran Dome. More than 152 fossils of *Homo erectus* in Indonesia are found in the Sangiran Dome, from Early Pleistocene deposits of the Sangiran Formation to Middle Pleistocene sediments of the Bapang Formation. These represent more than 77% of the total hominin specimens found in Java. The fossils represent an early human occupation since Sunda emerged from the Java Sea. Therefore, in the Quaternary, Java Island is the home for hominins in Southeast Asia. The human fossil *Homo erectus* and the vertebrate fossils found in Quaternary sediments in Java are very important for understanding human evolution in Indonesia and Southeast Asia and the initial expansion of hominins out of Africa.

Y. Zaim (✉)
Department of Geology, Institut Teknologi Bandung, Jl. Ganesa no.10, Bandung, 40132, Indonesia
e-mail: ZAIM@GC.ITB.AC.ID

Keywords Indonesia • Vertebrate Fauna • Geology • Stratigraphy

Introduction

Quaternary faunal associations (hominin and vertebrate fossils), as well as artifacts/stone tools, have been found dispersed throughout Southeast Asia, including the Indonesian Archipelago. They derived from the Asian mainland, found in sediments deposited in terrestrial, fluvial and/or lake environments. In West Indonesian regions, vertebrates and artifacts/stone tools have been found in the islands of Sumatra/Nias, Java and Kalimantan (Borneo). Crossing Wallace's Line, they have also been found in the East Indonesian islands of Sumbawa, Sumba, Flores, Timor, Sulawesi, Seram, Halmahera and Sangihe. Recently, a newly described species of hominin, *Homo floresiensis*, has been discovered in Liang Bua, Flores. These hobbit-like people (*orang pendek* – pygmy people) date to the end of the Late Pleistocene (ca. 12,000 BP). All other hominin fossils in Indonesia have been discovered in Java.

Since hominins and vertebrates are dependant on natural environments, one of them being geology, their development and appearence in Indonesian regions must be influenced by the development and geological condition of these regions. Tectonic and glacioeustatic activity during the Quaternary may also be closely related to the dispersal and evolution of hominins and vertebrates, as well as their modes of migration from Asia mainland to the Indonesian Archipelago.

Quaternary Geology of Indonesia

The development of Southeast Asia, including the Indonesian Archipelago, is related closely to plate tectonics, specifically those related to the interaction and movement of the major lithospheric plates in Eurasia in the north, India–Australia in the south and Pacific in the east. The continental crust of

J.G. Fleagle et al. (eds.), *Out of Africa I: The First Hominin Colonization of Eurasia*, Vertebrate Paleobiology and Paleoanthropology,
DOI 10.1007/978-90-481-9036-2_7,

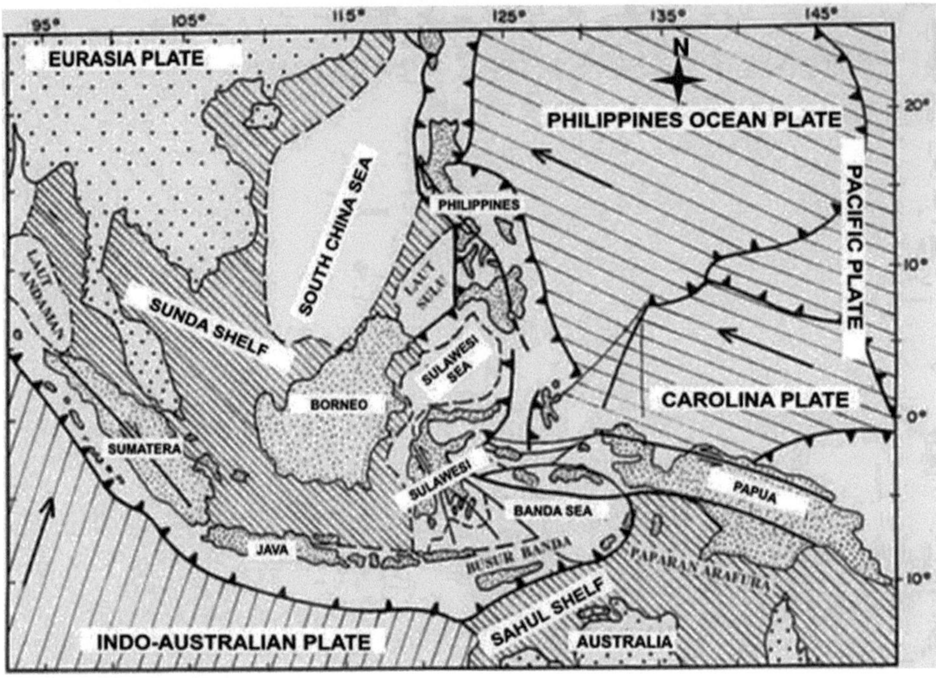

Fig. 7.1 Tectonic map of Indonesian regions (Modified from Simanjuntak and Barber 1996)

Eurasia (including Asia, Sumatra, Kalimantan and Java) moved to the southeast while the oceanic crust of India–Australia (with Australia and Papua) moved northward and the Pacific plate moved to the west (Fig. 7.1). Consequences of the interaction and movement of these plates revealed that Southeast Asia and Indonesia are very unstable regions with high intensities of volcanic activity and seismicity.

The distributions of land and sea in Southeast Asian regions during the Tertiary have been interpreted by Hall (1998). He suggests that tectonic activity at the end of the Mesozoic to Paleogene (Early Tertiary) formed Sumatra and a part of Kalimantan (Borneo) and that they were emerged as a part of Asia mainland. Asia mainland was surrounded by shallow seas and separated by the deep sea from Papua and Australia. Only a small part of West Java was emerged while other regions were still occupied by the sea. Its position was in a northwest–southeast direction, parallel to Sumatra, while Kalimantan (Borneo) was in an east–west direction.

Sedimentations in the regions mentioned above occurred in a non-marine environment. Eocene non-marine sediments which occurred in Sukabumi, West Java (Batuasih Formation) were deposited in lake and fluvial environments. Eocene lake sediments of the Sangkarewang Formation have also been found in West Sumatra. These contain the fossils of fish, snails, plants and *Protoplotus beauforti*, an Eocene avian (van Tets et al. 1989). Stromer (1931) reported the existence of the Paleogene artiodactyl, *Choeromorus* sp., from the Eocene fluvial sediments of the Melawi Formation in West Kalimantan (Borneo).

The plate's movement and subduction continued during the Late Tertiary, increasing the intensity of tectonic and volcanic activity. The tectonic activity caused subsidence of some regions in Sumatra, Java, Kalimantan (Borneo), Sulawesi and Papua. This situation was contemporaneous with periods of transgression due to a global sea level rise during the Miocene which caused all Indonesian regions to be covered by shallow seas. Only small portions of Sumatra, Java, Kalimantan (Borneo), Sulawesi and Papua remained above sea level in the form of isolated islands. Sedimentation during this period took place in the subsidence basin areas of shallow to deep marine. In Java, Early Tertiary volcanic activity produced volcanic material classified by van Bemmelen (1949) as the Old Andesite Formation. This material has been dated to the Late Oligocene to Early Miocene (Soeria-Atmadja et al. 1994). They were found along the southern part of Java. However, Ngkoimani et al. (2004) reported the existence of andesite from the Old Andesite Formation dated to the Late Cretaceous to Early Eocene.

According to Hall (1998), in the Late Miocene, about 10 Ma, the islands of Sumatra, Java and Kalimantan were rotated 30° counterclockwise into the positions they are in today. At the end of the Tertiary, ca. 5–2 Ma, Sumatra, Kalimantan (Borneo) and Asia were connected by a piece of land known as Sunda Land. The very shallow sea formed the Sunda Shelf in the west and the Sahul Shelf in eastern Indonesia. The regions between west and eastern Indonesia were still occupied by a deep marine environment (Fig. 7.2). The Pliocene was ended due to Plio-Pleistocene tectonic

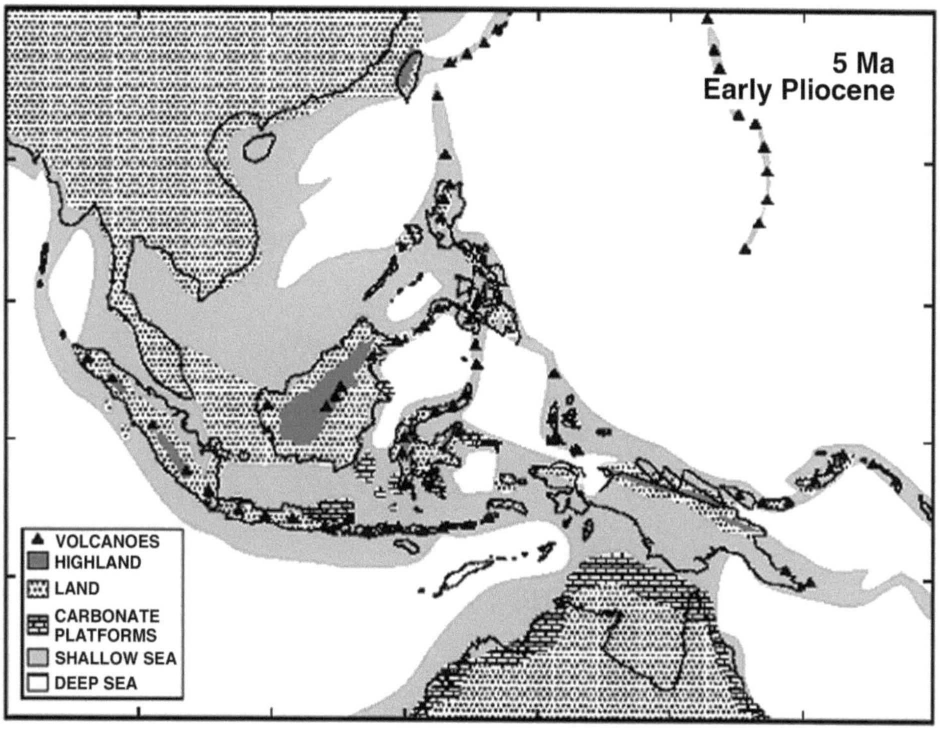

Fig. 7.2 Paleogeography of Southeast Asia during the Early Pliocene (Modified from Hall 1998)

activities which affected all Indonesian regions and caused high levels of volcanic activity.

Changes in tectonic and glacioeustatic activity took place during the Pleistocene to form the present day Indonesian Archipelago. In Indonesia, Quaternary non-marine sediments are well exposed in the islands of Java, Sumbawa, Flores, Sumba, Timor and South Sulawesi. These sediments contain hominin (only in Java) and vertebrate fossils, as well as artifacts. Very little is known about the deposits in Sumatra, Kalimantan (Borneo) and Papua. These regions are almost entirely covered by thick soils, recent volcanic products and vegetations, leaving no well exposed Quaternary outcrops. In Sulawesi, the Quaternary deposits are only exposed in the central part of South Sulawesi.

Quaternary Stratigraphy of Hominin and Vertebrate Remains in Java

In Java, Quaternary sediments are exposed west to east along the middle part of the island, forming the mountain regions. The area containing these exposures is called the Bogor Zone to the west and the North Serayu and Kendeng Zones in central and eastern Java. Until the present day, all hominin specimens (*Homo erectus*) in Java have been found at several locations in the Kendeng Zone in Central and East Java. Additional vertebrate fossils and artifacts have been found in the Early to Late Pleistocene deposits of all mountain regions (Fig. 7.3).

Quaternary Stratigraphy of West Java

During the Quaternary, sedimentations in Indonesian regions occurred mostly in non-marine environments (i.e., lakes and rivers) which were favorable for hominin and vertebrate occupations.

In West Java, all Quaternary sediments have been deposited in a non-marine environment and laid unconformably above Late Pleistocene shallow marine sediments. The marine sediments have been grouped into the Kaliwangu Formation and are characterized by bluish clays rich with mollusks. The Quaternary sediments contain vertebrate fossils and are well exposed particularly in the northern part of the Bogor Zone. In the southern part they are almost entirely covered by young volcanic products (Fig. 7.3).

In the northern part of West Java, the lower part of the Quaternary sediments consist of alternating light and dark grey clays which are sometimes carbonaceous with lignite, medium to very coarse and conglomeratic sandstones with cross cutting of trough and planar cross-bedding structures and conglomerates. These sediments, which have been

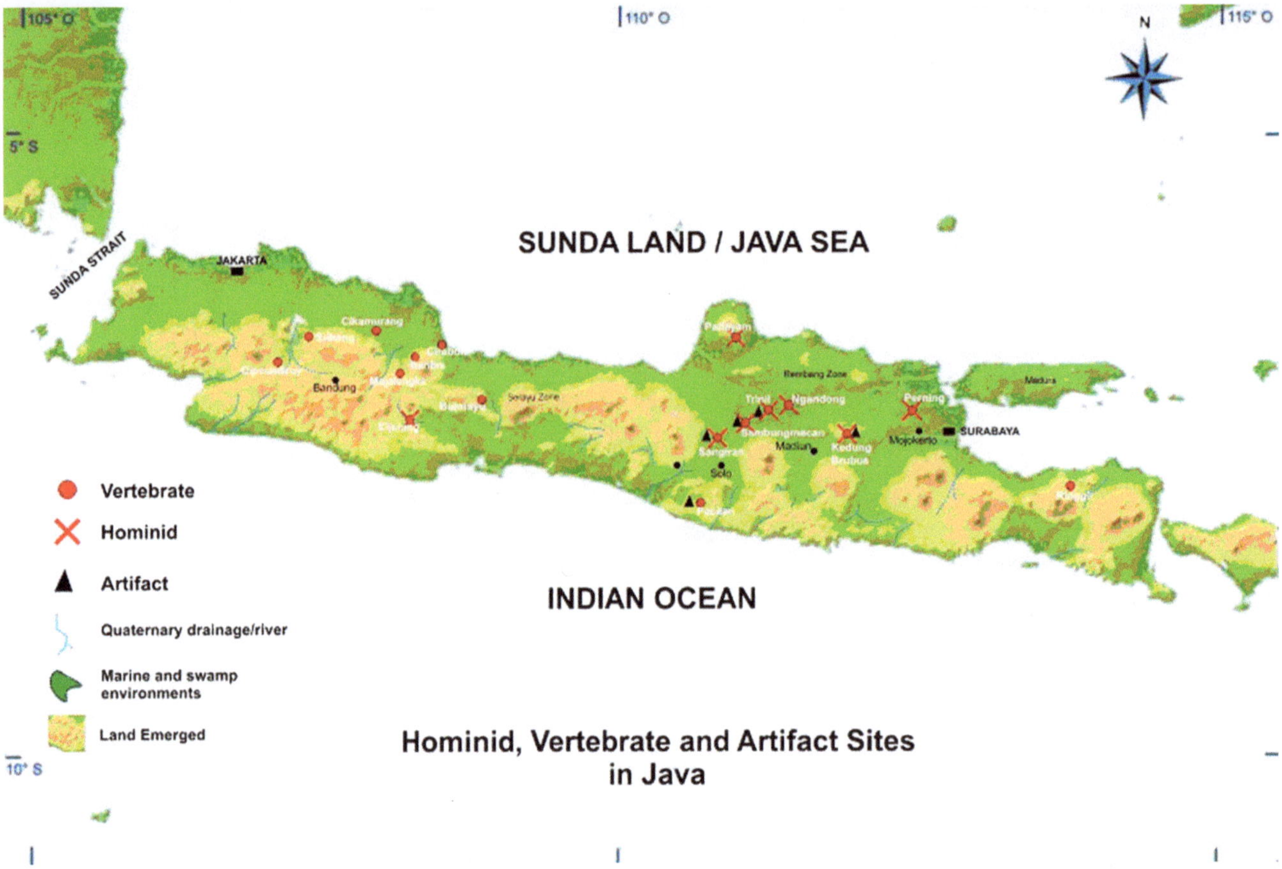

Fig. 7.3 Distribution of hominid, vertebrate and artifact sites in Java. Base map after Hertler and Rizal, 2005

deposited in meander and point bar channels, are grouped into the Citalang Formation and are Early to Middle Pleistocene in age. No hominin fossils have been found in this formation, but it does contain fossil Bovidae, Cervidae, large *Stegodon* (*Stegodon trigonocephalus*) and *Crocodylus*, which might belong to the Ci Saat or Trinil H.K. Faunas. In the upper part, the Quaternary sediments are grouped into the Tambakan Formation and date to the Middle to Late Pleistocene. This formation consists of braided stream fluvial deposits characterized by alternating black clays, high intensities of small to medium cross-cutting sandstone channel lenses and the lenses of conglomerates. The Tambakan Formation contains vertebrate fossils which are similar to those vertebrates found in the Citalang Formation. The Citalang and Tambakan Formations are well exposed in Karawang, Subang, north of Sumedang and Majalengka. Zaim and Marino (2002) reported the discovery of pygmy *Stegodon* (*Stegodon cariangensis*) in the Early Pleistocene black clay lake deposits in Majalengka, West Java. These black clays are laid unconformably above the Late Pliocene shallow marine clays of Kaliwangu Formation. In the south of Cirebon and Kuningan areas, the fluvial sediments share the same characters with the sediments of the Citalang Formation, but they are grouped into the Cijuray Formation. These Middle to Late Pleistocene deposits contain large *Stegodon*, Bovidae, *Rhinoceros*, Cervidae and *Crocodylus*.

In the southern part of West Java, Quaternary sediments are well exposed only at Cijulang, Ciamis District. Cijulang is the type locality for the oldest faunal stage in the vertebrate biostratigraphy of Java, known as the Cijulang Faunal stage (von Koenigswald 1934). The age of the Cijulang Fauna is believed to date to the Middle Pliocene based on the existence of *Merycopotamus nanus* (van Bemmelen 1949). The vertebrate faunas from this area, which consist of *Merycopotamus nanus,* small and large *Stegodon,* bovids, cervids and crocodiles, could be grouped into either the Satir or Ci Saat faunal stages (Fig. 7.4; Sondaar et al. 1996). They were recovered from bluish-grey clays rich with shallow marine mollusks, as well as in medium to coarse sandstones with medium to large cross-bedding structures. The presence of fossil *Merycopotamus nanus* suggests that these sediments date to the Middle Pliocene (van Bemmelen 1949). Additionally, the presence of shallow marine mollusks indicates that these sediments were deposited in deltaic or near shore environments. Recently, Kramer et al. (2005) reported the discovery of the first fossil hominin from West Java, a permanent, right I_2 crown, "RH1" or "Rancah Hominin 1". It was discovered during an excavation at Cisanca River in

Age		von Koenigswald (1934)	de Vos, et al. (1982)	de Vos, et al. (1985) van den Berg et al. (2004)
Holocene		Recent		Wajak (6 - 10 Ka)
		Sampung		
Pleistocene	L	Ngandong	Ngandong	Punung (60 - 125 Ka)
				Ngandong (0.4 Ma ?)
	M	Trinil	Kedung Brubus	Kedung Brubus (0.7 - 0.8 Ma)
				Trinil H.K (0.9 - 1.0 Ma)
	E	Jetis	Trinil	Cisaat (1.0 - 1.2 Ma)
Pliocene	L	Kaliglagah		Satir (1.5 - 2.0 Ma) ?
	M	Cijulang		

Fig. 7.4 Development of the vertebrate biostratigraphy of Java

Rancah village near Cijulang. Rancah Hominin 1 was recovered from blue sandstone probably Early or Middle Pleistocene in age.

Quaternary Stratigraphy of Central Java

The Quaternary sediments in Central Java are located in the North Serayu and Kendeng Zones. In the North Serayu Zone, they are well exposed in Bumiayu area, while in the Kendeng Zone they are found in Sangiran and Sambungmacan areas. These sediments are also found at Patiayam area, which is situated outside these zones (Fig. 7.3).

Bumiayu Area

In Bumiayu area, Quaternary sediments lie above Late Pliocene shallow marine blue clays of the Kaliwangu Formation. These sediments are divided into the Kaliglagah, Mengger, Gintung and Linggopodo Formations (ter Haar 1934; van Bemmelen 1949; Sumarso and Suparjono 1974; Zaim 1978).

The Kaliglagah Formation is comprised of black clays with lignite, the rare fresh water mollusk, *Sulcospira* sp., and is intercalated by coarse to very coarse grained sandstones and conglomerate. The Mengger Formation is named after the hill of Mengger laid above the Kaliglagah Formation. It consists of sediments which share the performance and sedimentological characteristics with the sediments belonging to the Kaliglagah Formation. Zaim (1978) suggests, based on their lithological and sedimentological characteristics, that these two cannot be separated into two distinct formations. Furthermore, he proposes that these formations be grouped into a single formation, the Cisaat Formation. The Mengger/Cisaat Formations are overlain by the Gintung Formation. It consists mainly of conglomerates and very coarse sandstones. The Linggopodo Formation is the youngest Quaternary formation in Bumiayu. It consists of volcanic breccias and andesite lavas derived from volcanic activity during the end of Late Pleistocene to the Early Holocene.

The Cisaat (formerly Kaliglagah and Mengger) and Gintung Formations contain vertebrate fossils such as *Sinomastodon (Mastodon) bumiajuensis*, *Stegodon trigonocephalus*, *Elephas* cf. *planifrons*, *Cervus stehlini*, *Bos sp.*, *Sus stremmi*, *Hexaprotodont (Hippopotamus) simplex*, *Geochelone* and Crocodiles. To date, the Bumiayu area has yielded no hominin fossils. Based on the vertebrate fossil association and the presence of *Sinomastodon (Mastodon) bumiajuensis*, the age of this fauna is believed to be Late Pliocene (van Bemmelen 1949). These faunas were grouped by von Koenigswald (1934) into the Kaliglagah Faunal stage based on the excavation site at the Kaliglagah River bank. However, it has been revised and regrouped into two separate faunal stages. The first and oldest stage, the Satir Fauna, dates to the Early Pleistocene and is defined by the presence of *Sinomastodon (Mastodon) bumiajuensis* and *Hexaprotodont (Hippopotamus) simplex*. The second stage, the Ci Saat Fauna, dates to the Early to Middle Pleistocene. This fauna consists of vertebrates which were recovered from the Bumiayu Area with the exception of those belonging to the Satir Fauna (Fig. 7.4; de Vos 1985; de Vos et al. 1982; de Vos and Sondaar 1994; van den Bergh et al. 2004).

Patiayam Area

Patiayam area, which is situated in the north coast of Central Java, is a small hilly region at southeast foot slope of Muria Volcano. The volcano itself was an isolated island during the Pleistocene, separated by W-E strait from the mainland of Java Island (Zaim 1989).

The first geological investigations of Patiayam area were conducted by van Es in 1931. Sartono et al. (1978) produced a geological map and established the stratigraphic framework of this area. More detailed sedimentological studies on volcanic materials have been conducted by Zaim (1989). The oldest sediments consist of bluish, shallow marine clays of the Jambe Formation and date to the Late Pliocene. Quaternary non-marine sediments in this area are laid unconformably above the Jambe Formation and are divided into four formations (Sartono et al. 1978). The Kancilan Formation, which lies above the Jambe Formation, consists of volcanic breccia and intrusion of shoshonitic composition dated to the Early Pleistocene (Zaim 1989). The Kancilan Formation is then overlain by the Slumprit Formation, which consists of sediments deposited by fluvial and deltaic braided streams. The lower part of this formation consists of tuffaceous silt and clays, as well as fine to medium sandstones. The upper part is made up of medium to coarse tuffaceous sandstones with medium to large cross-bedding structures and conglomerates. Two isolated premolars, as well as several parietal fragments of *Homo erectus* were found by Zaim in 1977 and 1978 in the tuffaceous silts of the Slumprit Formation. These have been dated to the Middle Pleistocene (Zaim and Ardan 1998). The discovery of human fossils from this region has established Patiayam area as a new hominin locality outside of the Kendeng Zone.

The Slumprit Formation is covered by the Late Pleistocene Kedungmojo and Sukobubuk Formations. The Kedungmojo Formation consists of fluvial sediments deposited in meanders and point bar channels which are made up of light to dark grey clays intercalated with medium to very coarse tuffaceous sandstones and conglomerates with cross-bedding structures. The Sukobubuk Formation consists of agglomerates containing the volcanic materials from Muria Volcano.

The vertebrate fossils from the Slumprit and Kedungmojo Formations include large *Stegodon* (*Stegodon trigonocepahalus*), *Rhinoceros*, Bovidae, Cervidae and *Sus brachygnathus*, which might belong to the Trinil H.K. or Kedung Brubus Fauna.

Sangiran Area

Sangiran, which is situated about 20 km north of Solo, has a domed structure and is commonly known as the Sangiran Dome. It is one of the most important hominin and vertebrate fossil discovery sites in Indonesia and Southeast Asia. Indriati (2004) reported the total number of hominin specimens found in Indonesia from 1889 to 2003 to be 197. Of these, 152 came from Sangiran. Thus, 77.157% of all hominin specimens in Indonesia have been found in the Sangiran Dome. They were recovered from the Early Pleistocene deposits of the Sangiran Formation, as well as the Middle Pleistocene deposits of the Bapang Formation. The fossils represent an early human occupation lasting approximately 500,000 years as Sunda emerged from the Java Sea (Larick et al. 1999).

Formerly, the lithostratigraphic units of the Sangiran Dome followed the lithostratigraphy of the Kendeng Zone as established by von Koenigswald (1940). New lithostratigraphic units of the Sangiran Dome have been proposed through joint research by the Geological Research and Development Centre (GRDC) in Indonesia and the Japan International Cooperation Agency (JICA) (Watanabe and Kadar 1985). These new units have been accepted by many authors and will be referred to in this paper.

The oldest sediment exposed in the Sangiran area belongs to the Puren (formerly Kalibeng) Formation. These sediments date to the Pliocene and consist of marine marls which were deposited in shallow marine. The formation is then overlain unconformably by the Sangiran (Pucangan) Formation, which dates to the Early Pleistocene. The Sangiran Formation, which consists of lahars at its base, known as the Lower Lahar Unit (LLU) (Watanabe and Kadar 1985), is characterized by the matrix supported by lahar deposits as the product of volcanic activity (Zaim et al. 1999). $^{40}Ar/^{39}Ar$ dating, which has been carried out recently by Bettis et al. (2004) on some samples taken from the LLU, suggests an age of 1.90 ± 0.02 Ma.

The LLU is overlain by marine sediments consisting of thin layers of diatomite. This indicates that the area was inundated by a shallow sea during a transgressive sea level rise. These marine sediments were covered by black lacustrine clays intercalated with thin layers of tuff. The presence of the tuff layers indicates the existence of volcanic reactivity during the deposition of the black clays.

The Sangiran Formation is covered by the Middle Pleistocene fluvial deposits of the Bapang (Kabuh) Formation. It consists of alternating clays, medium to very coarse and conglomeratic sandstones with cross cutting of trough and planar cross-bedding structures and conglomerates. They are deposited in meanders and point bar channels. Very hard, compacted and calcareous layers of 0.5–2.5 m thick conglomerate occur at the base of the Bapang (Kabuh) Formation forming the boundary between it and the Sangiran (Pucangan) Formation. Commonly known as the Grenzbank Layer, it is rich with hominin and vertebrate fossils dated to 1.51 ± 0.08 Ma (Larick et al. 2001).

The Pohjajar (Notopuro) Formation is the name for the Upper Lahar Unit (Watanabe and Kadar 1985) found overlying the Bapang (Kabuh) Formation. It contains large boulders of andesite within a tuffaceous matrix. Based on its stratigraphic position above the Bapang (Kabuh) Formation (Middle Pleistocene), the Pohjajar Formation is believed to have been deposited during the Late Pleistocene.

The youngest unit in the Sangiran Dome was formed during the Holocene. It consists of river terraces laid unconformably above the Upper Lahar Unit of the Pohjajar Formation.

Many hominin and vertebrate fossils have been discovered in the Sangiran Dome. The Early Pleistocene hominin specimens (S-6b, S-8, S-27, S-31) from this area, which were formerly attributed to *Meganthropus paleojavanicus* (von Koenigswald 1968; Sartono 1986; Indriati 2004), were found in the black clays of the Sangiran (Pucangan) Formation and date to older than 1.51 ± 0.08 Ma based on $^{40}Ar/^{39}Ar$ dating of pumice samples taken from the Bapang/Sangiran Formation contact (Larick et al. 2001). The vertebrate assemblage consists of *Hexaprotodon* (*Hippopotamus*) *simplex*, *Sinomastodon bumiajuensis*, Cervidae and *Geochelone*. They belong to the Satir Fauna based upon the new mammalian biostratigraphy of Java (Fig. 7.4; Sondaar 1984; Sondaar et al. 1996; Leinders et al. 1985; de Vos 1996).

Middle Pleistocene *Homo erectus*, ranging from 1.51 ± 0.08 Ma to 1.02 ± 0.06 Ma (Larick et al. 2001), has been recovered from the Bapang Formation. Additionally, the vertebrate assemblage consists of *Hexaprotodon sivalensis*, *Stegodon trigonocephalus*, *Elephas hysudrindicus*, *Axis lydekkeri*, *Duboisia santeng*, *Bos (Bibos) palaesondaicus*, *Bos (Bubalus) palaeokarabau*, *Rhinoceros sondaicus*, *Sus brachygnathus*, *Sus macrognathus*, *Semnopithecus auratus* and *Macaca fascicularis*. These fall under the Ci Saat Fauna based on faunal association with the Sangiran Formation below the Grenzbank Layer and Trinil H.K Fauna based on association with the Bapang Formation above the Grenzbank Layer (Sondaar et al. 1996). In addition to these fossils, several artifacts have been recovered from the Bapang (Kabuh) Formation.

Sambungmacan Area

The stratigraphy of Sambungmacan area follows that of the Kendeng Zone. The oldest sediments consist of platy limestones from the Klitik Formation and marls from the Kalibeng Formation. Both of these were deposited in shallow marine during the Late Pliocene. These marine sediments were overlain unconformably by terrestrial black clay deposits of the Pucangan Formation and contain vertebrate fossils such as pygmy *Stegodon*, Bovidae and Cervidae (Satir Fauna). The Kabuh Formation of the Middle Pleistocene occurs above the Pucangan Formation and consists of conglomerates and coarse to very coarse sandstones with cross-bedding structures. Vertebrate fossils recovered from the Kabuh Formation include *Stegodon trigonocephalus*, Cervidae and *Bos (Bubalus) palaeokarabau* of the Trinil H.K Fauna. Four specimens of *Homo erectus* were discovered in the Kabuh Formation at Sambungmacan area (Indriati 2004).

Quaternary Stratigraphy of the Kendeng Zone, East Java

In East Java, Quaternary sediments are well exposed in an east–west direction along the mountain regions in the Kendeng Zone. Hominin and vertebrate fossils were recovered from these sediments in the Trinil, Ngandong, Kedung Brubus, Ngawi and Perning (Mojokerto) areas. The Quaternary sediments in the Kendeng Zone overlay the Late Pliocene shallow marine sediments consisting of Klitik limestone and Kalibeng marls.

Trinil Area

The discovery of *Pithecanthropus erectus* (also referred to as *Homo erectus trinilensis*; Sartono 1986, 1987a,b), the first hominin fossil in Southeast Asia, by Dubois in 1891, as well as many additional vertebrate fossil discoveries in Trinil, has made this area well-known for its vertebrate paleontology and paleoanthropology. Trinil is also the type locality for the Middle Pleistocene faunal stage known as the Trinil Fauna (von Koenigswald 1934) as well as the Trinil H.K. Fauna of the early Middle Pleistocene (de Vos 1985; de Vos et al. 1982; de Vos and Sondaar 1994) which replaced the Early Pleistocene Djetis Fauna established by von Koenigswald (1934). The oldest sediments exposed in the Trinil area are black clays belonging to the Late Pleistocene Pucangan Formation. Above the Pucangan Formation lies the Kabuh Formation, which consists of medium to very coarse cross-bedded and conglomeratic sandstones containing remains of *Pithecanthropus erectus (Homo erectus)* and vertebrate fossils belonging to the early Middle Pleistocene Trinil H.K Fauna. The Pucangan and Kabuh Formations are covered by Old Solo terrace deposits.

Ngandong Area

In the Ngandong area, the Old Solo terrace deposits have been classified as Ngandong Terraces. These Late Pleistocene (<50 Ka) deposits are rich with hominin and vertebrate fossils. In total, seventeen hominin specimens have been recovered from the Ngandong Terraces and attributed to *Homo erectus ngandongensis* (cf. *Homo erectus soloensis*) by Sartono (1986, 1987a,b). The vertebrate fossils from the Ngandong area have been grouped into the Late Pleistocene Ngandong Fauna (von Koenigswald 1934; de Vos 1985; de Vos et al. 1982; de Vos and Sondaar 1994).

Kedung Brubus Area

In Kedung Brubus, hominin and vertebrate fossils have been found in the tuffaceous sandstone layer representing the boundary between the Pucangan and Kabuh Formations. These vertebrate fossils represent the Late Pleistocene Kedung Brubus Fauna (de Vos 1985; de Vos et al. 1982).

Perning (Mojokerto) Area

Perning is situated at the west of Surabaya, 10 km northeast of the city of Mojokerto in the eastern Kendeng Hills of East Java. A well-preserved partial skull of a juvenile early hominin was discovered at Perning in 1936 by R. Tjokrohandojo, an assistant geologist in the Geological Survey of the Netherlands. The location of the fossil site is approximately 3.5 km north of Perning village and about 500 m east of the road from Perning to Kepuhklagen. The fossil, first known as *Homo modjokertensis* (von Koenigswald 1936b–c *in*

Huffman and Zaim 2003), and sometimes referred to as "Perning 1" or "Mojokerto 1", has since been attributed to *Homo erectus* (Antón 1997, 2002).

The geology of Perning and its surrounding areas have been investigated by Duyfjes (1936). More detailed studies of the stratigraphy and geology of this area have also been carried out by several authors (i.e., Sartono et al. 1981; Zaim 1981; Kumai et al. 1985; Semah 1986; Swisher et al. 1994). The oldest sediments in this area belong to the Late Pliocene Kalibeng Formation and consist of marls and dark grey calcareous clays which were deposited in the shallow marine. This formation is then overlain by the Pucangan Formation which consists of carbonaceous black clay facies (lower part) and volcanic facies (upper part) (van Bemmelen 1949). The lower part of the volcanic facies consist of marine clays containing shallow marine mollusks and light grey tuffaceous clays rich with leaf fossils and another plant remains. The upper part contains coarser sediments and consists of alternating layers of tuff, tuffaceous clays and conglomeratic sandstones with pumice clasts.

The hominin and vertebrate fossils recovered from Pucangan Formation's volcanic facies were embedded in very coarse, cross-bedded tuffaceous sandstones and conglomerates which were deposited in a deltaic environment called the Mojokerto Delta (Huffman and Zaim 2003). The hominin and vertebrate bearing layer has been dated to the Late Pliocene (1.81 ± 0.04 Ma) through $^{40}Ar/^{39}Ar$ dating (Swisher et al. 1994). However, this age is older than those for other hominins outside Africa. Recently, Morwood et al. (2003) stated that their redating of two pumice horizons at the *Homo modjokertensis* site indicates that the age of the Mojokerto cranial vault is less than 1.49 Ma. Therefore, *Homo modjokertensis* must be referred to as *Homo erectus* (*sensu* Antón 1997, 2002).

The Kabuh Formation, which lies directly above the Pucangan Formation consists of light grey tuffaceous clays intercalated with coarse and cross-bedded tuffaceous sandstones and conglomerates with cross-cutting channel structures which were deposited by braided rivers. No hominin fossils have been discovered in this formation, but it has yielded vertebrate fossils such as *Stegodon trigonocepahlus*, *Stegodon hypsilophus*, *Elephas hysudrindicus*, *Rhinoceros sondaicus*, *Hexaprotodon sivalensis*, *Sus macrognathus*, *Duboisia santeng*, *Bubalus paleokarabau*, *Axis lydekkeri*, *Muntiacus muntjak* and several species of cervid which all are members of the Middle Pleistocene Kedung Brubus Fauna. The lateral and vertical distribution of Quaternary sediments in Java are shown in Fig. 7.5.

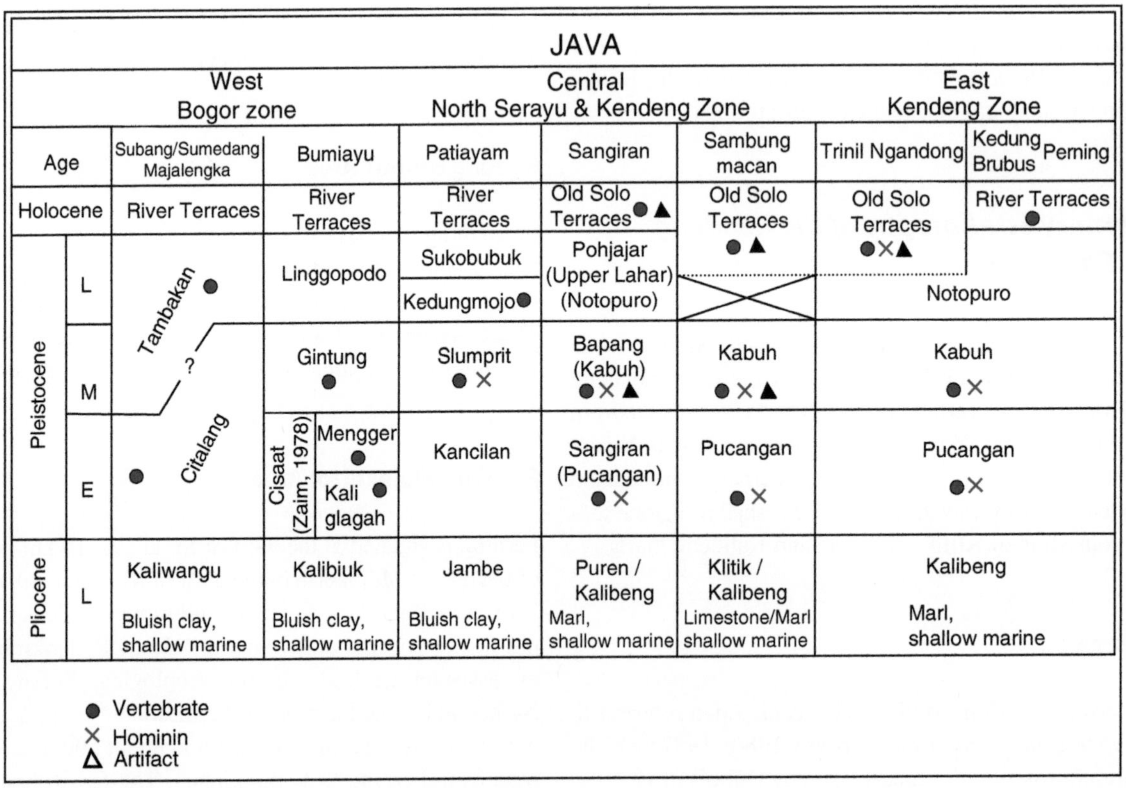

Fig. 7.5 Quaternary stratigraphy of Java (source from various authors)

Quaternary Stratigraphy of the Islands Outside Java

Quaternary deposits are not well exposed in the islands outside Java. In the Nusa Tenggara (Lesser Sunda) Islands, which comprise Bali, Lombok, Sumbawa, Sumba, Flores, Timor and several other small islands, the Quaternary deposits are well exposed only in Flores and Timor. Although very little is known about Bali, Lombok, Sumbawa and Sumba, Sartono (1979) reported the presence of *Stegodon sumbaensis*, a pygmy *Stegodon* recovered from Quaternary deposits in Sumba Island.

On Timor Island, Quaternary sediments are well exposed in the graben of Soe depression which is located in the central part of the island. The unconformable sediments lie above marine Tertiary sediments consisting of very coarse sandstones and gravel sized conglomerate deposited by fluvial systems. They contain fossils of pygmy *Stegodon timorensis* and large *Stegodon trigonocephalus* dated to the Early to Middle Pleistocene.

On Flores Island, Quaternary sediments are well exposed in Soa Plateau and consist of the Early Pleistocene volcanic deposits belonging to the Ola Kile Formation. The Ola Kile Formation is overlain unconformably by the Middle Pleistocene Ola Bula Formation. This formation consists of clay, coarse and cross-bedded sandstones and conglomerates. It was divided by van den Bergh (1999, 2004) into two members, A and B, which represent the upper and lower parts, respectively. Member A, which is dated to 0.9 Ma, contains fossils of *Varanus komodoensis*, *Crocodylus* sp., *Geochelone sp.* and pygmy *Stegodon sondaari* which have been labeled Fauna A. Member B, which is dated to 0.8–0.7 Ma, contains fossils labeled Fauna B. These comprise *Varanus komodoensis*, *Stegodon florensis*, *Hooijeromys nusatenggara* and *Crocodylus sp.* The discovery of artifacts on the island of Flores may also suggest the presence of *Homo erectus*.

Well exposed Quaternary sediments have also been found in the depression of the Walanae Valley, South Sulawesi. According to Sartono (1979), the Quaternary sediments of the Beru Formation are deposited unconformably above the late Tertiary marine sediments of the Walanae Formation. These fluvial deposits consist of light grey clays, cross-bedded sandstones and conglomerates which contain vertebrate fossils and artifacts. Van den Bergh (1999, 2004) has classified the Beru Formation (Sartono 1979) as a member in the Walanae Formation. The Beru Member was later divided into two subunits. The lower part, Subunit A, which has been dated to the Late Pliocene to Early Pleistocene (2.5 Ma), is characterized by lagoonal/estuarine and fluvio-lacustrine clays and siltstones. The upper part of the Beru Member, Subunit B, lies above Subunit A and has been dated to the Early to Middle Pleistocene. It consists of fluvial coarse and cross-bedded sandstones and conglomerates. Deposited above the Beru Member is the Middle to Late Pleistocene Tanrung Formation. This formation is characterized by fluvial conglomerates.

According to van den Bergh (1999, 2004), Subunits A and B of the Beru Member contain vertebrate fossils, such as *Geochelone atlas*, large *Stegodon*, pygmy *Stegodon sompoensis*, *Elephas celebensis*, *Celebochorus heekereni*, *Celebochorus* (shortlegged species), *Crocodile sp.* and *Tryonychidae*. These fauna have been grouped into the Walanae Faunal stage. The Tanrung Formation contains faunas belonging to the Tanrung Faunal stage. These comprise *Celebochorus* (shortlegged species), highcrowned *Elephas sp.* and *Stegodon sp.* In addition to these fossils, artifacts have also been discovered in both the Beru Member and Tanrung Formation.

Java: The Home for Hominins in Southeast Asia

The distribution of Quaternary sediments in Indonesia suggests that these regions were almost entirely occupied by a marine environment until the end of the Tertiary Period (Pliocene). All Quaternary sediments in Java, Timor and South Sulawesi are laid unconformably above Late Tertiary marine sediments.

In Indonesia, the end of the Pliocene was marked by an increase in tectonic and volcanic activity. All Tertiary and older (marine) sediments were folded and emerged due to Plio-Pleistocene tectonic activity associated with the first glacial. This resulted in a decrease in sea level and the formation of the Indonesian islands.

In Java, Tertiary marine environments emerged above sea level forming "proto-islands" (Fig. 7.6). In Sangiran, the Lower Lahar Unit (LLU) indicates that the base of the Sangiran Formation, dated to 1.90 ± 0.02 Ma (Bettis et al. 2004), was formed through terrestrial sedimentation which was influenced by volcanic activities. According to Bettis et al. (2004), the age of 1.90 Ma for the LLU of the Sangiran Formation postdates glacioeustatic sea level lowering caused by the first major continental glaciation of the Late Pliocene, but predates the more frequent glacial episodes of the Early Pleistocene. Furthermore, Bettis et al. (2004) suggests that the LLU event did not create terrestrial surfaces as it flowed into a near-shore marine or lagoonal environment in the Sangiran area. The LLU's stratigraphic and lithological characteristics suggest a deposition of a heterolithological, cohesive debris flow. Emplacement of the LLU significantly decreased the depth of shallow near-shore marine environments in the Sangiran area and set the stage for terrestrial emergence during the Early Pleistocene.

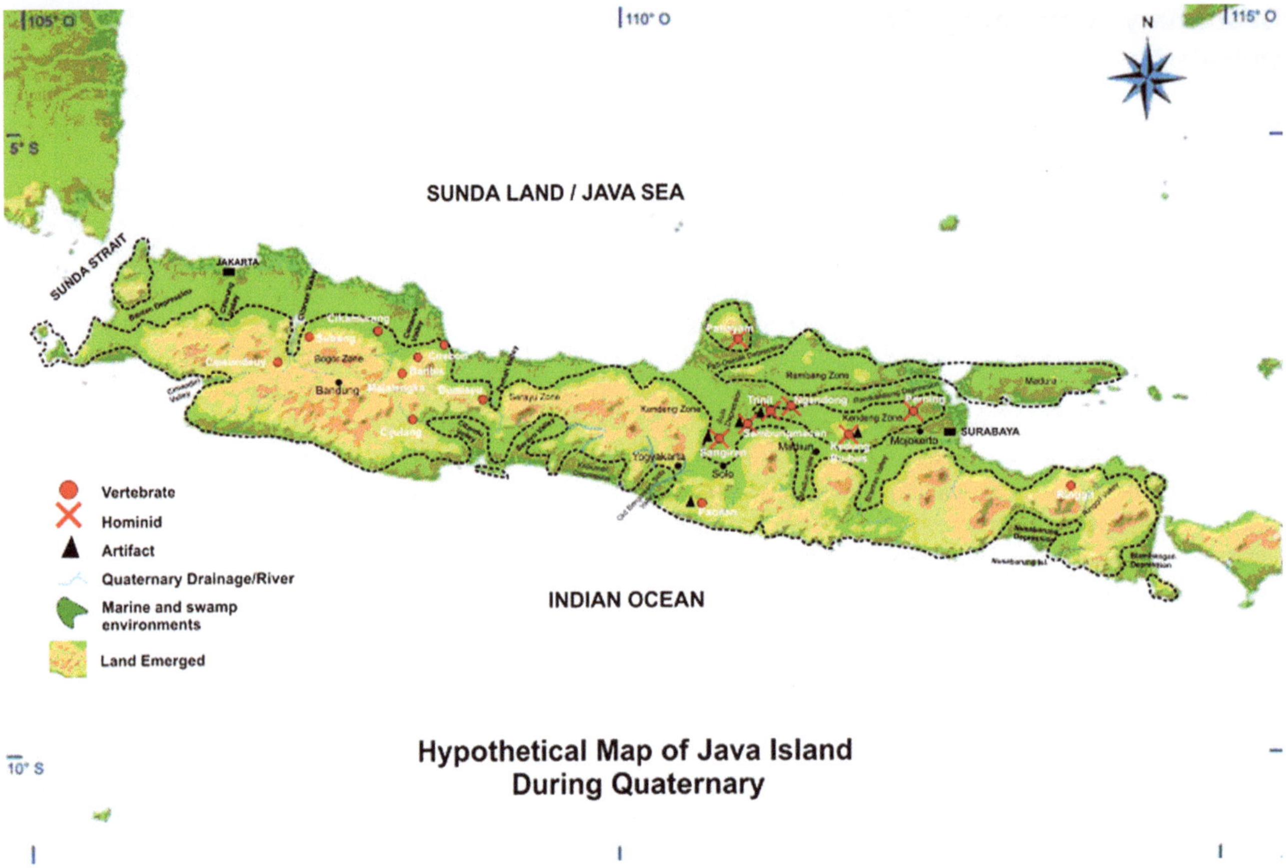

Fig. 7.6 Hypothetical map of Java Island during the Quaternary. Map based after Hertler and Rizal, 2005

The LLU, overlain by black clays which were deposited in a lacustrine environment, contains several Early to Middle Pleistocene (1.6–1.0 Ma) vertebrates belonging to the Satir Fauna as well as *Homo erectus paleojavanicus* (*Meganthropus paleojavanicus*). The presence of *Homo* is evidence for the first arrival of early hominins and vertebrates from Asia Mainland into Java through Sunda Land (Sartono 1985, 1987a, b; de Vos 1996) (Fig. 7.7).

In Sulawesi, the Late Pliocene to Early Pleistocene faunas, represented by the Walanae Faunas *Stegodon sompoensis*, *Elephas celebensis* and *Celebochorus heekereni*, migrated from Asia to Sulawesi by crossing the sea between the Sunda Shelf and Sulawesi (van den Bergh et al. 2004). However, Sartono (1985) suggests that the Early to Middle Pleistocene faunal assemblage in Sulawesi may have migrated from Asia through the Philippine Mindanao Islands as evidenced by the coexistence of pygmy *Stegodon mindanensis* from the Mindanao Islands and *Stegodon sompoensis* in West Sulawesi.

Sea level recessions due to the second global cooling, which occurred during the Early to Middle Pleistocene (1.0–0.5 Ma), created a wide corridor across the Sunda Shelf and Southeast Asia and created open woodlands in Java (van den Bergh et al. 2004). Around the same time, another *Homo erectus* (attributed to *Homo erectus trinilensis*, *Homo erectus erectus* and *Homo modjokertensis*), as well as Ci Saat and Trinil H.K. vertebrates, arrived in Java Island from Asia Mainland and dispersed using different pathways from West Java (Rancah?) to Central Java (Sangiran, Sambungmacan and Patiayam) and East Java (Trinil, Kedung Brubus and Perning – Mojokerto?). Here they continued the long trek to Flores and Timor Islands as evidenced by the artifacts found there (Fig. 7.8).

The Middle Pleistocene was ended by strong tectonic and volcanic activity. All Middle Pleistocene and older sediments were gently folded. Tectonic and volcanic activity continued and may have coincided with an additional sea level recession associated with the last glacial. During the Late Pleistocene, tectonic and volcanic activity weakened and only yielded the uplifted regions without any folding. Some of these uplifted regions are known today as river terraces. In Java, these terraces are represented by the Old Solo River deposits which can be observed in Sangiran, Sambungmacan, Trinil, Ngandong, Pitu and Ngawi Areas (Sartono 1976; Zaim 1996). They contain vertebrates of the Ngandong Fauna. Additionally, those at Ngandong (ca. <50 Ka) contain *Homo erectus* (*Homo erectus ngandongensis*, *Homo* cf. *erectus soloensis*, *Pithecanthropus soloensis*, *Homo soloensis*, *Javanthropus soloensis* after Sartono 1985).

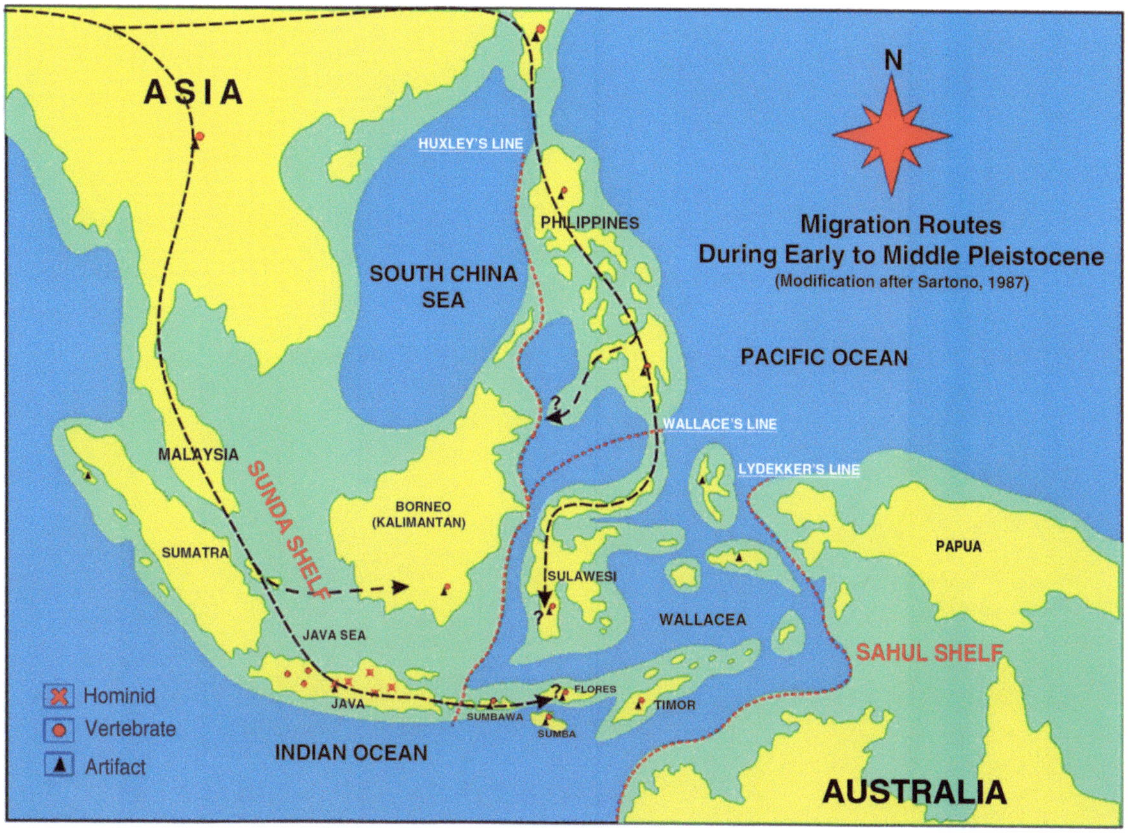

Fig. 7.7 Migration routes during the Early to Middle Pleistocene (1.8–0.5 Ma)

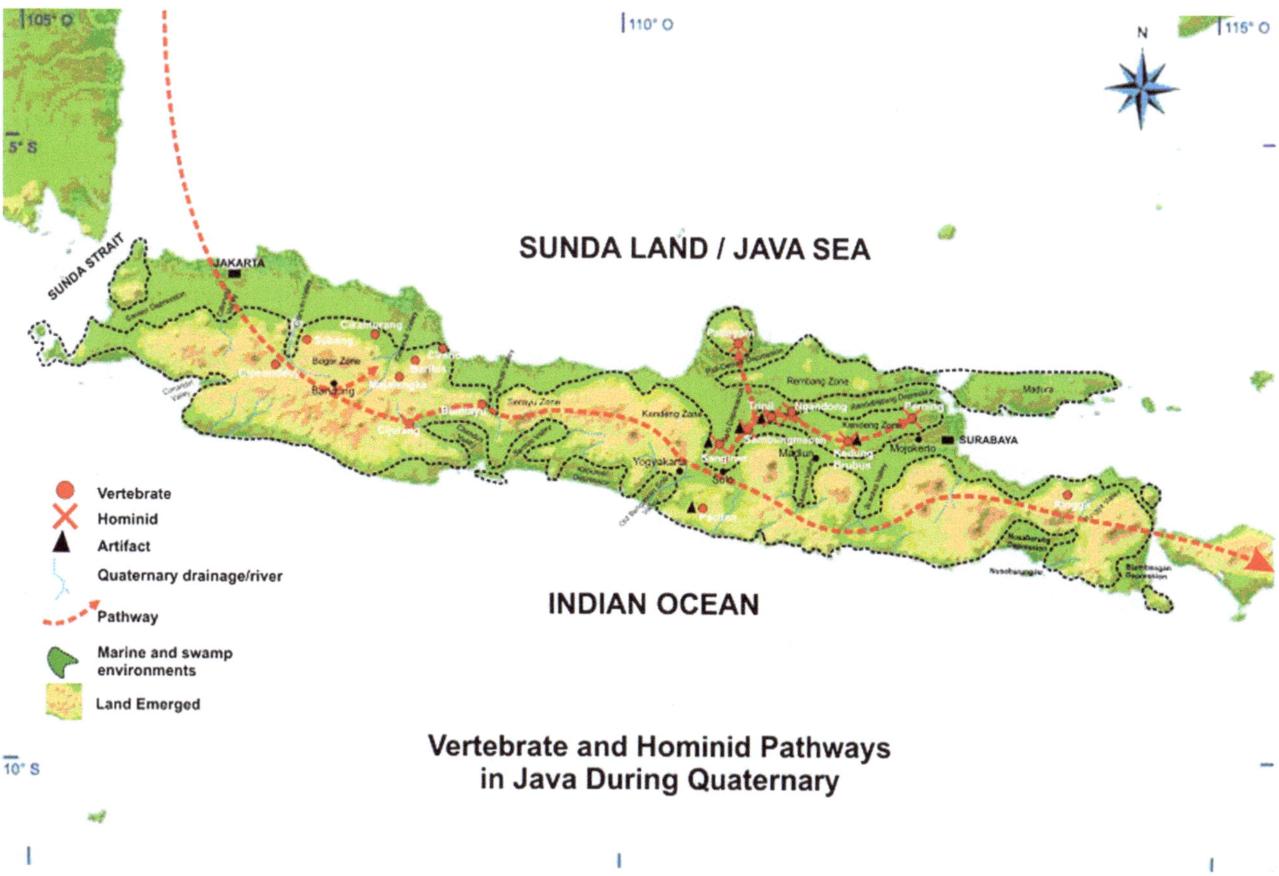

Fig. 7.8 Hominid and vertebrate pathways in Java during the quaternary. Base map after Hertler and Rizal, 2005

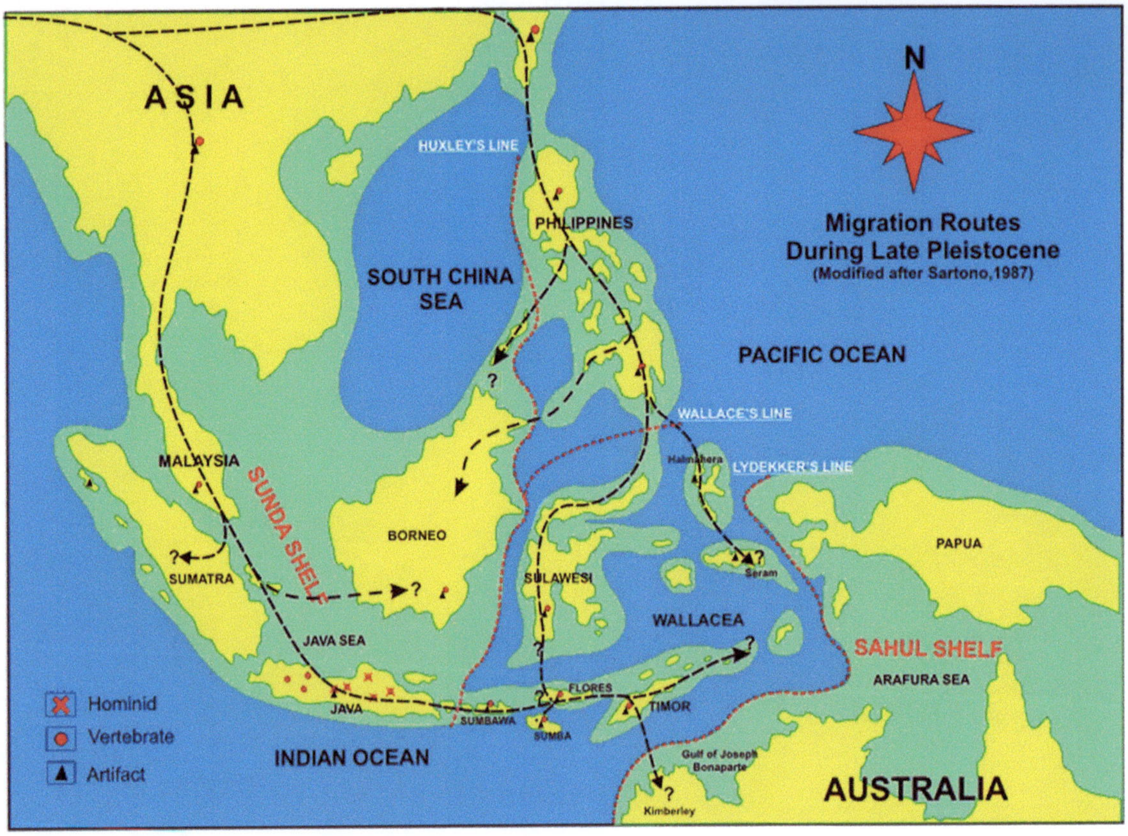

Fig. 7.9 Migration routes during the Late Pleistocene (0.5–0.125 Ma)

Late Pleistocene terrace deposits can also been found in South Sulawesi, Timor, Flores and other regions of East Indonesia. These contain vertebrate fossils and artifacts. Although there have been no hominin fossil discoveries in these regions, the existence of these artifacts suggests they may have been present.

During the Late Pleistocene, hominins and other vertebrates migrated from Asia using different routes of dispersal (Fig. 7.9). The western route went through the Sunda Shelf to Java before continuing on to the Nusa Tenggara (Lesser Sunda) Islands (i.e., Bali Lombok, Sumbawa, Sumba, Flores and Timor). The northern route went through Taiwan to the Philippines before crossing to Sulawesi and continuing on to Flores and Timor.

By the end of the Late Pleistocene to Early Holocene, *Homo sapiens* entered Java. Contemporaneous with members of the Punung and Wajak Faunas, they occupied humid forest to open woodland environments (van den Bergh et al. 2004). In addition to these faunas, these modern humans may have been contemporaneous with the newly described species, *Homo floresiensis*.

Of all the islands in Southeast Asia, Java is the most important regarding the evolution of our species and other vertebrates. Java provides new insights into human evolution and human dispersal from Africa.

Conclusions

The Late Tertiary (Pliocene) was ended by Plio-Pleistocene tectonic activity which was accompanied by increased volcanic activity. These tectonics were associated with the glacial periods and formed the Sunda Land which connected mainland Asia with the islands of Sumatra, Java and Kalimantan.

During the Quaternary, sedimentation in Indonesian regions occurred mostly in non-marine environments. These sediments are well exposed in the islands of Java, Flores, Sumbawa, Sumba, Timor and South Sulawesi. All Quaternary non-marine sediments in Java, South Sulawesi and Timor are laid unconformably above Late Pliocene marine sediments.

Quaternary sediments in Indonesia contain numerous vertebrate fossils and artifacts. In addition, those in Java contain fossil hominins (*Homo erectus*). To date, the majority of the hominin specimens found in Java have been recovered from the Kendeng Zone in several Central and East Javan locations whereas all other vertebrate fossils have been recovered in all Early to Late Pleistocene mountain regions.

Homo erectus paleojavanicus (*Meganthropus paleojavanicus*) arrived in Java from Asia in the Early Pleistocene (1.6–1.0 Ma) following the arrival of the Satir and Ci Saat Faunas to Indonesia through the Sunda Land at the end of the Late Pliocene. The arrival of *Homo erectus* (*Homo erectus trinilensis* and *Homo erectus ngandongensis/ soloensis*) occurred during the Middle to Late Pleistocene (1.0–0.125 Ma).

Additional hominin fossils have been discovered in Liang Bua, Flores. This species has been attributed to *Homo floresiensis* and dated to 12,000 BP. Evidence suggests that these hominins may have been contemporaneous with modern *Homo sapiens*.

Java Island, particularly the Sangiran Dome, is a very important hominin and vertebrate fossil site in Indonesia. The majority of *Homo erectus* fossils in Indonesia have been recovered from the Early Pleistocene deposits of the Sangiran Formation and Middle Pleistocene sediments of the Bapang Formation. These comprise more than 77% of all hominin specimens found in Java. The hominin and vertebrate fossils from Java are extremely important regarding the understanding of human evolution and dispersal out of Africa.

Acknowledgments The author is grateful to Dr. Russell L. Ciochon of Department of Anthropology, University of Iowa, USA for his helpful suggestion and valuable advice. The author also wants to thank to Dr. Bambang Priadi of Department of Geology, ITB Bandung, Indonesia for supporting some data for the paper.

References

Antón, S. C. (1997). Developmental age and taxonomic affinity of the Mojokerto child, Java, Indonesia. *American Journal of Physical Anthropology, 102*, 497–514.

Antón, S. C. (2002). Evolutionary significance of cranial variation in Asian *Homo erectus*. *American Journal of Physical Anthropology, 18*, 301–323.

Bettis III, E. A., Zaim, Y., Larick, R., Ciochon, R.L., Suminto, Rizal, Y., Reagan, M. et al. (2004). Landscape development preceeding *Homo erectus* immigration into Central Java, Indonesia: the Sangiran Formation Lower Lahar. *Journal of Palaeogeography, Palaeoclimatology, Palaeoecology 206*, 115–131.

de Vos, J. (1985). Faunal stratigraphy and correlation of the Indonesian Hominid sites. In E. Delson (Ed.), *Ancestors, the hard evidence* (pp. 215–220). New York: Liss.

de Vos, J. (1996). *Faunal turnovers in Java in relation to faunas of the continent, odontology, Vol. 1, Association for comparative odontology* (pp.32–36). Japan: Terajima Cultural Hall Foundation.

de Vos, J., & Sondaar, P. Y. (1994). Dating hominid sites in Indonesia. *Science, 266*, 1726–1727.

de Vos, J., Sartono, S., Hardjasasmita, S., & Sondaar, P. Y. (1982). The fauna from Trinil, type locality of *Homo erectus*, a reinterpretation. *Geologie en Mijnbouw, 61*, 207–211.

Duyfjes, J. (1936). Zur Geologie und Stratigraphie des Kendenggebietes zwischen Trinil und Soerabaia (Java). *De Ingineur in Nederlandsch-Indie, Mijnbouw en Geologie, De Mijningineur, 4*, 136–149.

Hall, R. (1998). The plate tectonics of Cenozoiz SE Asia and the distribution of land and sea. In R. Hall, & J. D. Hooloway (Eds.), *Biogeography and geological evolution of SE Asia* (pp. 1–33). Leiden, Netherlands: Backhuys Publishers.

Hertler, C. and Rizal, R. (2005). Excursion guide to the Pleistocene hominid sites in Central and East Java. Joint Excursion July–August 2005, JWG University, Frankfurt (Germany), and ITB, Bandung (Indonesia).

Huffman, O.F., & Zaim, Y. (2003). Mojokerto Delta, East Java: paleoenvironment of *Homo mojokertensis* – first results. Jurnal Teknologi Mineral (JTM) – ITB, vol. X, No. 1, 53–82.

Indriati, E. (2004). Indonesian fossil hominid discoveries from 1889 to 2003: catalogue and problems. *Proceedings of the 5th and 6th symposium on collection building and natural history studies in Asia and the Pacific Rim* (pp.163–177). National Science Museum Monographs 24.

Kramer, A., Djubiantono, T., Aziz, F., Bogard, J.S., Weeks, R.A., Weinand, D.C., et al. (2005). The first hominid fossil recovered from West Java, Indonesia. *Journal of Human Evolution 48*, 661–667.

Kumai, H., Itihara, M., Sudijono, Shibasaki, T., Aziz, F., Yoshikawa, S., et al. (1985). Geology and stratigraphy of the Mojokerto area. In N. Watanabe, D. Kadar (Eds.), *Quaternary geology of the hominid fossil bearing formations in Java* (pp. 55–61). Bandung-Indonesia: Geological Research and Development Centre.

Larick, R., Ciochon, R. L., & Zaim, Y. (1999). Fossil farming in Java. *Natural History, 108*, 54–57.

Larick, R., Ciochon, R. L., Zaim, Y., Sudijono, S., Rizal, Y., Aziz, F., et al. (2001). Early Pleistocene $^{40}Ar/^{39}Ar$ ages for Bapang formation hominins, Central Jawa, Indonesia. *Proceedings of the National Academy of Sciences of the United States of America, 98*, 4866–4871.

Leinders, J. J. M., Aziz, F., Sondaar, P. Y., & de Vos, J. (1985). The age of hominid-bearing deposits of Java: state of the art. *Geologie en Mijnbouw, 64*, 167–173.

Morwood, M. J., O'Sullivan, P., Susanto, E. E., & Aziz, F. (2003). Revised age for Mojokerto 1, an early *Homo erectus* cranium from East Java, Indonesia. *Australian Archaeology, 57*, 1–4.

Ngkoimani, L.O., Bijaksana, S., Utoyo, H., & Permanadewi, S. (2004). New K-Ar ages of igneous rocks from Yogyakarta Area (Abstract). *Workshop on paleogene stratigraphy of Jawa*, 28 September, Yogyakarta.

Sartono, S. (1976). Genesis of the Solo Terraces. *Modern Quaternary Research in SE Asia, 2*, 1–21.

Sartono, S. (1979). The discovery of a pygmy *Stegodon* from Sumba, East Indonesia: an announcement. *Modern Quaternary Research in SE Asia, 5*, 57–63.

Sartono, S. (1985). Pleistocene peopling of the Southeast Asian Archipelago. *Proceedings of the 12th Indo-Pacific Prehistory Association Congress* (pp.1–12). Phillipines: Penablanca – Cagayan.

Sartono, S. (1986). New light on human evolution in Southeast Asia. Proceedings of Geosea V, Vol. II. *Geological society of Malaysia bulletin 20*, 269–288.

Sartono, S. (1987a). The Long Trek to the South. *Proceedings of the Workshop on economic geology, tectonics, sedimentary processes and environment of quaternary in Souteast Asia* (pp. 193–212). Bangkok, Thailand: Department of Geology, Chulalongkorn University.

Sartono, S. (1987b). *Influence of plate tectonics on dispersals of quaternary faunas in Southeast Asia* (pp. 1–13). Technology and Arts held by Deutscher Akademischer Austauschedienst (DAAD) in ITB – Bandung: Proceedings of the Seminar on International Developments in Science.

Sartono, S., Hardjasasmita, S., Zaim, Y., Nababan, U.P., & Djubiantono, T. (1978). Sedimentasi Daerah Patiayam, Jawa Tengah. Berita Pusat Penelitian Arkeologi (in Indonesian: Sedimentation of Patiayam Area, Central Java, Centre of Archaeological Research Newsletter) 19, 1–21.

Sartono, S., Semah, F., Astadiredja, K.A.S., Sukendarmono, M. & Djubuantono, T. (1981). The age of *Homo modjokertensis*. *Modern Quaternary Research in SE Asia 6*, 91–102.

Semah, F. (1986). Le Peuplement Ancien de Java Ebouch d'un Cadre Chronologique. *L'Anthropologie, 9*, 359–400.

Simanjuntak, T. O. and Barber, A. J. (1996). Contrasting tectonic styles in Neogen orogenic belts of Indonesia, in Hall, R., and Blundell, D (Eds), Tectonic Evolution of Southeast Asia. Geol. Soc. Spec. Publ. *106*, 185–201.

Soeria-Atmadja, R., Maury, R. C., Bellon, H., Pringgoprawiro, H., & Polve, M. (1994). Tertiary magmatic belt in Java. *Journal of Southeast Asian Earth Sciences, 9*, 3–27.

Sondaar, P. Y. (1984). Faunal evolution and the mammalian biostratigraphy of Java. *Courier Forschung Institut Senckenberg, 69*, 219–235.

Sondaar, P.Y., Aziz, F., van den Bergh, G, D., & de Vos, J. (1996). Faunal change and homonid evolution during quaternary of Java. Geological Research and Development Centre, *Paleontology Series 8*, 1–10.

Stromer, E. (1931). Die ersten altterti en S getir-Reste aus den Sunda-Inseln-Wetensch. Med.v.d. Dienst v.d. Mijnbouw in Ned-Indie, no.17, pp. 11–14.

Sumarso & Suparjono, N. (1974). A Contribution to the Stratigraphy of the Bumiayu Area. Proceedings of the 3rd IAGI Convention.

Swisher III, C.C., Curtis, G.H., Jacob, T., Getty, A.G., Suprijo, A., & Widiasmoro (1994). Age of the earliest known hominids in Java, Indonesia. *Science 263*, 1118–1121.

ter Haar, C. (1934). Toelichting bij Blaad 58 (Boemiajoe). Geologische Kaart van *Java Schaal 1*, 100.000.

van Bemmelen, R. W. (1949). *The geology of Indonesia* (Vol. IA). The Hague, Netherlands: Martinus Nijhoff.

van den Bergh, G.D. (1999). The Late Neogene elephantoid-bearing faunas of Indonesia and their palaeozoogeographic implications. Scripta Geologica no.117, Nat. Natuurhistorisch, The Netherlands, pp. 419.

van den Bergh, G. D., de Vos, J., & Sondaar, P. Y. (2004). The Late Quaternary paleogeography of mammal evolution in the Indonesian Archipelago. *Paleogeography, Paleoclimatology, Paleoecology, 171*, 385–408.

van Es, L. J. C. (1931). *The age of Pithecanthropus*. Martinus Nijhoff, The Netherlands: The Hague.

van Tets, G. V., Rich, P. V., & Marino-Hadiwardoyo, R. (1989). A reappraisal of *Protoplotus beauforti* from the early tertiary of Sumatra and the basis of New Pelecaniform family. *Publishers of the Geological Research and Development Centre (GRDC), Paleontology Series, 5*, 57–75.

von Koenigswald, G. H. R. (1934). Zur Stratigraphic des Javanischen Pleistozans. *De Ingenieur in Netherland Indie, 11*, 185–201.

von Koenigswald, G. H. R. (1936b). *Ein fossiler Hominide aus dem Alpleistocän Ostjavas*. De Ingenieur in Nederlandsch-Indië, Mijnbouw & Geologie, De Mijningenieur 4 (8), 149–157.

von Koenigswald, G. H. R. (1936c). *Erste Mitteilung über einen fossilen Hominiden aus dem Altpleistocän Ostjavas*. Proceedings, Koninklijke Nederlandsche Akademie van. Wetenschappen te Amsterdam XXXIX (8), 1000–1009.

von Koenigswald, G.H.R. (1940). Neue *Pithecanthropus* – Funde 1936–1938, Eine Beitrag zur Kenntnis der Praehominiden. Diens Mijnbow Ned. Indie, We. Meded, 28 Landsdrukkerij, pp. 1–205.

von Koenigswald, G.H.R. (1968). Java: Prae-Trinil man. Proceedings of the VIIIth International Congress of Anthropological and Ethnological Sciences, Tokyo, vol. 1, Tokyo Science Council of Japan, pp.104–105.

Watanabe, N., & Kadar, D. (1985). Quaternary geology of the hominid fossil bearing formations in Java. *GRDC Special Publication, 4*, 1–378.

Zaim, Y. (1978). Paleogeografi Daerah Bumiayu, Jawa Tengah (in Indonesian: Paleogeography of the Bumiayu Area, Central Java). B. Sc. Hon Thesis, Department of Geology, ITB, Bandung, Indonesia.

Zaim, Y. (1981). Revisi Umur dan Stratigrafi Formasi Pucangan di Daerah Perning, Mojokerto, Jawa Timur (in Indonesian: Revision on Age and Stratigraphy of Pucangan Formation at Perning Area, Mojokerto, East Java). Proceedings of Xth IAGI Convention, Bandung, pp. 1–10.

Zaim, Y. (1989). Les Formations "Volcano-Sedimentaires" Quaternaires de la Region de Patiayam (Central Java, Indonesie). Milieau de Sedimentation et Mineralogie. PhD Thesis, Institut de Paleontologie Humaine, Paris, France.

Zaim, Y. (1996). The age of Pitu Terrace of Watualang Region, Ngawi (East Java) Indonesia. *Buletin Geologi, Jurusan Teknik Geologi ITB, 26*, 97–105.

Zaim, Y., & Ardan, R. (1998). A premolar of *Homo erectus* from Patiayam Region, Central Java – Indonesia. *Buletin Geologi, Jurusan Teknik Geologi ITB, 28*, 31–36.

Zaim, Y., & Marino, R. (2002). Pygmy *Stegodon* dari Desa Cariang, Kec. Tomo, Kab. Sumedang, Jawa Barat: Sebuah Pemberitahuan (in Indonesian with English summary: Pygmy Stegodon from Cariang Village, Tomo District, Sumedang Area, West Java: an Announcement). *Buletin Geologi. Jurusan Teknik Geologi, 34*, 45–52.

Zaim, Y., Larick, R., Ciochon, R.L., Suminto, Rizal, Y., & Sujatmiko (1999). Karakteristik satuan Lahar Bawah dari formasi Pucangan di Sangiran, Jawa Tengah. (in Indonesian with English summary: Characters of Lower Lahar Unit of Pucangan Formation in Sangiran, Central Java). *Buletin Geologi, Jurusan Teknik Geologi ITB* 31, 67–84.

Chapter 8
Divorcing Hominins from the *Stegodon-Ailuropoda* Fauna: New Views on the Antiquity of Hominins in Asia[1]

Russell L. Ciochon

Abstract The distinctive *Stegodon-Ailuropoda* fauna of southern China and peninsular Southeast Asia is known to include a number of ape species no longer present today. Among these apes, it is becoming increasingly clear, was a medium-bodied genus previously misattributed to the genus *Homo*. This unidentified ape is known only from dental remains, and is morphologically distinct from any Pleistocene ape or hominin in this region. For two decades, I have supported and promoted the idea that *Gigantopithecus* and *Homo erectus* co-existed in the Early and Middle Pleistocene of China and Vietnam. With the discovery of a chimpanzee-sized ape co-occurring with *Gigantopithecus* at Mohui Cave, I realized that many of the claims for early hominins in the *Stegodon-Ailuropoda* faunas of southern China and Southeast Asia were likely incorrect. This calls for a reappraisal of the remains from the so-called "human" sites of this time period, namely Mohui, Longgupo, Jianshi, Sanhe, Lang Trang and Tham Khuyen, in the context of irrefutable hominin evidence from elsewhere in Asia. Therefore, the earliest hominin record from Asia is documented in the far north of China in the Nihewan Basin at sites such as Xiaochangliang and in the far south on Java at sites within the Sangiran Dome. By studying the unquestionable *Homo erectus* sites with significant cranial remains, such as Gongwangling (Shanxi province), Hexian (Anhui province) and Tangshan (Jiangsu province), we see a clear pattern. All of these sites are found north of the *Stegodon-Ailuropoda* fauna. Early hominins may very well have inhabited parts of southern China, such as in river valleys or areas devoid of forest, but they were not part of the heavily forested, humid-climate adapted *Stegodon-Ailuropoda* mammalian fauna of the region. Additional hominin research far to the north in China, or far to the south in Java, will provide important information and valuable insights into the potential dispersal routes of early *Homo erectus* out of Africa or Georgia and the habitats these earliest Asian immigrants preferred.

Keywords *Gigantopithecus* • *Lufengpithecus* • *Homo erectus* • *"Hemanthropus"* • Mohui • Longgupo • Jianshi • Zhoukoudian • Nihewan • Sangiran

R.L. Ciochon (✉)
Department of Anthropology, University of Iowa,
Iowa City, IA, 52242–1409
e-mail: russell-ciochon@uiowa.edu

Introduction

This essay was originally written for the open discussion format of the Stony Brook Human Evolution Workshop as a personal reflection based on my 20 years of fieldwork in Asia researching Pleistocene apes and hominins. The content and context of this paper was inspired by a research trip to Nanning (Guangxi Province) that I undertook in May 2005 in order to meet with my colleague Wang Wei and to examine some unusual teeth that he had excavated from a new Early Pleistocene cave site called Mohui. We examined a total of 33 hominoid teeth, 16 of which were assigned to *Gigantopithecus*; the others were provisionally assigned to *Homo*, *Pongo* or were genus indeterminate. There was no doubt that *Gigantopithecus* was present, but the remaining group of 17 teeth was morphologically different from any known Pleistocene ape or hominin in this region. The size range of many of these 17 teeth was comparable to that of either *Australopithecus* or early *Homo,* but their morphology was decidedly ape-like. I realized I could be looking at a previously undescribed genus of Pleistocene ape, or possibly a Pleistocene descendant of *Lufengpithecus*. For two decades, I have supported and promoted the idea that *Gigantopithecus* and *Homo erectus* co-existed in the Early and Middle Pleistocene of China and Vietnam. With the discovery of a chimpanzee-sized ape co-occurring with *Gigantopithecus* at Mohui Cave, I realized that many of the claims for early hominins in the *Stegodon-Ailuropoda* faunas of East and Southeast Asia were likely incorrect. Since I was responsible for making a number of these claims (Ciochon et al. 1996; Huang et al. 1995), I decided to review and re-evaluate this evidence, in order to provide alternative interpretations and lay out plans for future research.

In the remainder of this paper, I present new and recent observations on the purported early hominins identified from

[1]A much abbreviated version of this paper entitled "The mystery ape of Pleistocene Asia" first appeared in *Nature*, volume 458:153–154 (2009).

J.G. Fleagle et al. (eds.), *Out of Africa I: The First Hominin Colonization of Eurasia*, Vertebrate Paleobiology and Paleoanthropology,
DOI 10.1007/978-90-481-9036-2_8,

cave sites in southern China and peninsular Southeast Asia, as well as the documented early hominin open-air occupations of northeast China and central Java. These observations challenge the accepted presence of early hominins in southern China and peninsular Southeast Asia. When I started my career as a paleoanthropologist, *Homo erectus* in Asia was judged to be no older than 1 Ma. Over the succeeding decades, that date has been pushed back as far as 1.8 Ma. We now have evidence of early hominin occupations in northeastern China in the Nihewan Basin (40.2°N, 114.65°E) that date to no more than 1.3 Ma (Li et al. 2008) and of the hominin presence in the Sangiran Dome of central Java (07°27.460′S, 110°50.360′E) that dates to 1.6 Ma (Larick et al. 2001; Ciochon et al. 2005; Zaim et al. 2009). Even earlier $^{40}Ar/^{39}Ar$ dates of 1.8 Ma have been reported for the *Homo erectus* child from Perning (Mojokerto) (82°33.22′N, 122°23.56′E) (Swisher et al. 1994), though the recent relocation of the 1936 skull discovery site has called this date into question (Huffman et al. 2006). As the relocated discovery bed proved to be ~20 m above the horizon that Swisher et al. (1994) dated, the skull is certainly younger than had been previously reported (Huffman et al. 2006) and now falls within the range of dates of the Sangiran Dome hominins (O.F. Huffman, personal communication, 2008).

In spite of these discoveries in the far north and far south of East Asia, the tropical forested areas of southern China and peninsular Southeast Asia may prove to be devoid of early hominins. Through the evidence presented, you will come to realize, as I have, that early hominins were generally not part of the *Stegodon-Ailuropoda* fauna of southern China and peninsular Southeast Asia. Therefore, the earliest hominin record from Asia is only documented in the far north in the Nihewan Basin at sites such as Xiaochangliang (~1.3 Ma) and in the far south on Java at sites within the Sangiran Dome (~1.6 Ma). Additional paleoanthropological fieldwork in northeast China and in central Java may provide valuable insights into the actual dispersal patterns of early *Homo erectus* out of Africa or Georgia, as well as the habitats these hominins preferred upon reaching East and Southeast Asia.

Evidence from China

Mohui Cave, Guangxi, China

Mohui Cave (107°00.13′E, 23°34.891′N) is located in southwestern Guangxi province in the Bubing Basin, which parallels the better-known Bose Basin (Fig. 8.1). The cave is located about 65 meters (m) above the valley floor and has a chamber about 50 m in length, 2–6 m in breadth and 5–6 m in height (Wang et al. 2005; Wang et al. 2007). Mohui Cave first came to the attention of the Natural History Museum of the Guangxi Zhuang Autonomous Region in March of 2001 when it yielded mammalian fossils. These discoveries led to two seasons of excavation, the first from October to December in 2002 and the second in November 2003 (Wang et al. 2005). Three areas of the cave were excavated in 2 × 2 m^2, generally in levels of 5 centimeters (cm) or 10 cm, recovering a total of 675 Early Pleistocene mammalian fossils. As is typical in karst cave excavations, most specimens are isolated teeth, though some jaw fragments and a few postcranial elements were also unearthed. The fauna contains the typical members of the *Stegodon-Ailuropoda* fauna (Matthew and Granger 1923; Colbert and Hooijer 1953; Pei 1957, 1987; Tougard et al. 1996). The *Stegodon-Ailuropoda* fauna is a paleontological assemblage characterized by the extinct elephant-like *Stegodon* and the giant panda, *Ailuropoda*. This subtropical to tropical fauna also can include such other warm, humid-climate mammals as the tapir and the orangutan. The geographical range of this fauna is throughout southern China and much of peninsular Southeast Asia (Fig. 8.1) and its temporal range is from the Early Pleistocene to the Late Pleistocene (Rink et al. 2008). Three key taxa at Mohui Cave, *Ailuropoda microta, Nestoritherium* sp., and *Sus pei,* are also found at Liucheng Cave (also known as "*Gigantopithecus* Cave") in Guangxi (Pei 1987) and at Longgupo Cave in Sichuan (Huang et al. 1995). These and other common faunal elements link the cave sites and support an Early Pleistocene age for all three.

To date, 33 hominoid teeth have been recovered from Mohui Cave, 16 of which can be assigned to *Gigantopithecus blacki*. The *G. blacki* teeth can be divided into the following categories: one RI_2, one LC_1, one LC^1 and one RC^1, three LP^3, one LP_3, one RM_{1-2} and three LM_{1-2}, one RM^{1-2} and one LM^{1-2} one LM_3 and one LM^3 (Wang et al. 2007). Many features, including the dental dimensions, very thick enamel (5–6 mm) and the distinctive accessory internal tubercles between metaconid and entoconid of the lower molars (Wang et al. 2005), match the samples of *Gigantopithecus blacki* from the Early Pleistocene sites of Liucheng, Guangxi (Woo 1962) and Longgupo, Sichuan (Huang et al. 1995). As expected, the dental dimensions of the Mohui *G. blacki* sample are smaller than those of the *G. blacki* sample recovered from the Middle Pleistocene cave of Wuming, Guangxi (see also Zhang 1982), though the sample, as a whole, exhibits substantial size variation (Wang et al. 2007). Two of the remaining 17 hominoid teeth from Mohui, a complete RM_2 (MH0001) and a fragmentary LM^{1-2} (MH0018), were provisionally attributed to *Homo erectus* (Wang et al. 2005). However, following arguments voiced by other scholars (e.g., Wu 2000), Wang et al. (2007) reassigned all of the Mohui non-*Gigantopithecus* primate teeth to Hominoidea (gen. et sp. indet.). It is important to note, however, that from the outset, the now ambiguously attributed Mohui hominoid

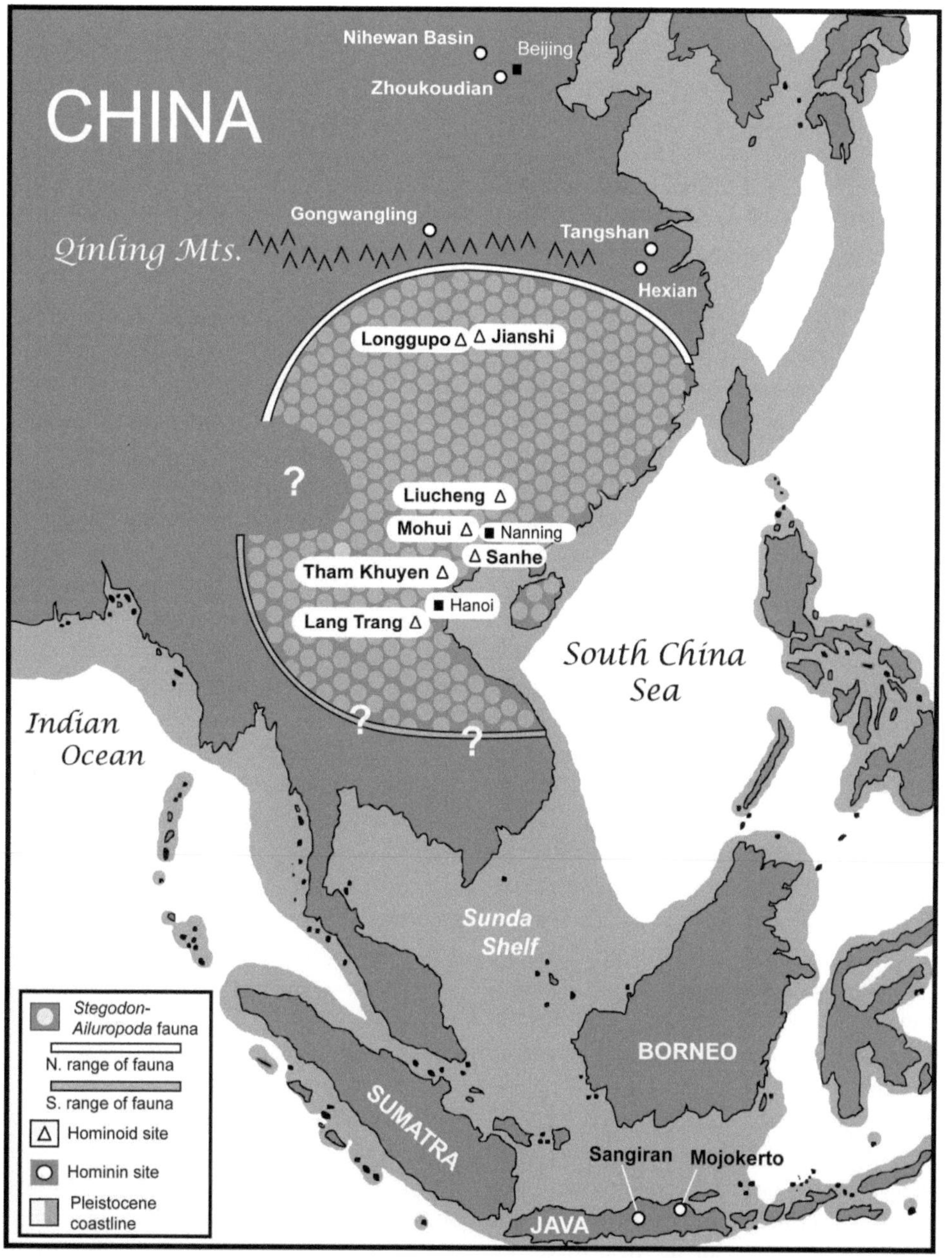

Fig. 8.1 Approximate geographic range of the *Stegodon-Ailuropoda* fauna based on various published sources (e.g., Matthew and Granger 1923; Colbert and Hooijer 1953; Pei 1957; Tougard et al. 1996; Long et al. 1996) and unpublished sources (e.g., de Vos 2008, personal communication). The hominoid and hominin sites discussed in this paper are plotted on this map from Nihewan Basin in the north to Sangiran Basin in the south using their published coordinates

specimens exceed the dental dimensions of all known Chinese *Homo erectus* and thus, compare more favorably with African *Australopithecus* or apes in this same size range.

The remaining 15 hominoid teeth from Mohui Cave, now termed Hominoidea (gen. et sp. indet.) (Wang et al. 2007), can be divided into the following groupings: one upper second incisor, three upper third premolars, one upper fourth premolar, four upper first or second molars, one upper third molar, one lower fourth premolar, one lower second molar, and three lower third molars. The preservation is good on all of these specimens, with several even retaining partial or complete roots – an unusual occurrence in karst cave environments where porcupine gnawing often reduces most teeth to nothing but enamel caps, and despite evidence of two species of *Hystrix* at the site (Wang et al. 2007). A few molar crowns exhibit slightly crenulated enamel, as do those of the late Miocene *Lufengpithecus*. The single upper second incisor is particularly distinctive in its morphology – its size and shape closely match those of *Lufengpithecus* from the Yuanmou Basin (25°40′N, 101°55′E), Yunnan (He and Jia 1997;

plate 20, no. 3, 4, 7). Additionally, all of the unworn upper molars have a prominent crista obliqua, as seen in other apes. These 15 teeth can be divided into two size morphs, perhaps representing one sexually dimorphic species or alternatively, more than one species of chimpanzee-sized fossil ape.

Lufengpithecus is a late Miocene ape known from four sites or site complexes in Yunnan, southern China. Harrison et al. (2002) have reviewed all of these sites and present approximate ages for each: Xiaolongtan (10 Ma), Yuanmou and Lufeng (8–9 Ma), Shihuiba (7–8 Ma) and Yangyi (~6 Ma). In their review, Harrison et al. (2002) recognized two species: *L. keiyuanensis*, known from Xiaolongtan and Yuanmou and *L. lufengensis*, known from Shihuiba. The Yuanmou hominoids have since been all attributed to *Lufengpithecus hudienensis* (Qi et al. 2006), though no species has yet been designated for the single mandible of *Lufengpithecus* recovered at the late Miocene site of Leilao. In 2003, Chaimanee et al. (2003) identified a new species of *Lufengpithecus* (*L. chiangmuanensis*) from a middle Miocene lignite locality in northern Thailand. This identification deemed premature by some (Kunimatsu et al. 2004, 2005), was soon revised by the authors, with *L. chiangmuanensis* being placed in the newly created genus, *Khoratpithecus* (Chaimanee et al. 2004). The holotype of *Khoratpithecus* is an early Late Miocene hominoid from a sand pit in Chalerm Prakieat District, Nakorn Ratchasima Province (Khorat) of northeastern Thailand. This new genus differs from the other proposed antecedents of the orangutan (*Lufengpithecus* and *Sivapithecus*) both in mandibular and dental morphology, as well as in its ecological setting (Kunimatsu et al. 2004). *Khoratopithecus*' tropical environment and derived features have led some to claim that it, and not *Lufengpithecus* nor *Sivapithecus*, is the closest known relative of the orangutan lineage (Chaimanee et al. 2004, 2006), though small sample sizes preclude definitive judgment (Merceron et al. 2006).

The question remains, though: Could *Lufengpithecus* or a related genus, persist in southern China into the Early Pleistocene? The answer is critical if we are to successfully determine the taxonomic placement of the new fossil ape discovered at Mohui Cave. However, even before a taxonomic identification is made, it is still possible to discuss the implications of a new chimpanzee-sized ape or apes in the Pleistocene of China.

Jianshi, Hubei, China

In 1970, three hominin teeth identified as belonging to *Australopithecus* (PA 502, PA 503, PA 504) were recovered from Jianshi (Longgudong Cave) (30°38′N, 110°04′E) in Jianshi county, Hubei province (Fig. 1), in association with five *Gigantopithecus* teeth and numerous faunal elements characteristic of the *Stegodon-Ailuropoda* fauna (Gao 1975). This site became known in the literature as the "Jianshi Hominid Site" and was often cited as evidence of Early Pleistocene hominins in China. During the 1980s and 1990s, the site was additionally mentioned as evidence for the coexistence of *Gigantopithecus* with these early humans (Wu and Poirier 1995). Since the locality was discovered during the Chinese Cultural Revolution, the researchers (Xu Chunhua, Wang Linghong and Hang Kangin) published their discovery under the pseudonym Gao Jian. They placed the Jianshi specimens, all lower first and second molars, in the genus *Australopithecus* because their dental dimensions, like those found at Mohui, greatly exceeded the range of variation observed in Chinese *Homo erectus*. However, Zhang (1984) later opted to transfer the Jianshi teeth to *Homo erectus* based on comparisons with the Javan Skk 1952.02 (Pith B) and Sangiran 9 (see Larick et al. (2000) for a discussion of new and colloquial numbering systems for the Sangiran Dome hominins).

From 1998 to 2000, new excavations were carried out at Jianshi (Zheng 2004), yielding more than 1,000 individual specimens representing 87 species of the *Stegodon-Ailuropoda* fauna. These excavations produced 28 teeth of *Gigantopithecus* and three teeth identified as hominin – a lower first molar (PA 1277), and two upper third premolars (PA 1278 and PA 1279). Again, it was observed that the dental dimensions were larger than Chinese *Homo erectus*, but similar to *Australopithecus robustus* and to some Javan hominins (Sangiran 4 and 6) (Zheng 2004). On this basis, these three teeth were assigned to Weidenreich's antiquated taxon, *"Meganthropus"* (included by most modern researchers in *Homo erectus*).

There is a distinct pattern in the identification of purported "hominins" found in the earlier period of fieldwork at Jianshi in 1970 and in the later period in 1998–2000. In both cases, the dimensions of these large teeth compare more favorably with "megadont" *Australopithecus*, than with *Homo erectus*. Though comparisons are also made with Javanese *Homo erectus*, the tooth dimensions of the Jianshi "hominins" actually exceed those samples as well (Gao 1975). At Jianshi, as well as other karst cave sites in China, there has been a tendency to allocate the larger hominoid teeth to *Gigantopithecus* and the smaller teeth to some species of hominin. *Gigantopithecus* teeth are so distinctive that there is little chance of error in their identification. However, with the discovery of the new Mohui fossil ape taxon, alternative identifications of the *smaller* hominoid component must be considered. Once full descriptions and analyses of the Mohui sample are completed, it will be necessary to re-evaluate the smaller hominoids from sites such as Jianshi.

Longgupo Cave, Sichuan, China

Longgupo Cave (30° 50′N, 109° 40′E), also known as the "Wushan Hominid Site", lies 20 km south of the Yangtze

River, near the eastern border of Sichuan province (Fig. 8.1). The site was discovered in 1984 and excavated from 1985–1988 by the Institute of Vertebrate Paleontology and Paleoanthropology (IVPP), Beijing, and the Chongqing Natural History Museum in Sichuan (Huang 1991; Huang et al. 1995). The excavations at Longgupo have yielded a diverse fauna of 116 species representing all elements of the *Stegodon-Ailuropoda* fauna, 16 teeth of *Gigantopithecus*, a fragmentary lower jaw of a small hominoid with the fourth premolar and first molar, an isolated hominoid upper lateral incisor, and two andesite stone artifacts. The site was dated by a combination of paleomagnetism and electron spin resonance (ESR) of tooth enamel, which yielded an age of 1.8 to 2.0 Ma for the main faunal-bearing levels (Huang et al. 1995).

The discoveries at Longgupo Cave received international exposure when they were published as a cover story in *Nature* (Huang et al. 1995). The authors claimed that the small hominoid jaw (CV.939.1) and isolated incisor (CV.939.2), as well as the two stone artifacts (P.6523 and P.6524) were comparable in age and morphology with early representatives of the genus *Homo* (*H. habilis* and "*H. ergaster*") and the Oldowan technology in East Africa. On this basis, they argued that Longgupo represented evidence of a pre-*Homo erectus* form of early *Homo* that must have dispersed to Asia about 2 Ma, coincident with the earliest diversification of genus *Homo* in Africa. The announcement of the Longgupo hominin site met with some skepticism. In a reply to the *Nature* article, Schwartz and Tattersall (1996) claimed that the morphology of the lower molar of the Longgupo jaw was not hominin, but more likely representative of an orang-related species. In response to Schwartz and Tattersall (1996), Huang et al. (1996) compared the dental metrics of the Longgupo first molar with 400+ orangutan molars excavated from three Pleistocene cave sites in Southeast Asia. They found that the Longgupo first molar was significantly smaller than any known fossil orangutan lower molar, falling far outside the range of variation of all specimens in their 400+ comparative sample. In a separate analysis of the Longgupo lower incisor, Wang (1996) concluded that its morphology does not resemble any known Early or Middle Pleistocene hominin, but actually represents the upper lateral incisor of a modern human from East Asia. Finally, Etler et al. (2001) re-examined the Longgupo mandibular fragment and concluded that it is a relict survivor of a late Miocene ape lineage possibly related to *Lufengpithecus*, a conclusion similar to that of Wu (2000), who also was adamant in the view that the specimen is an ape. Etler et al. (2001) base their conclusions on the double-rooted nature and ape-like cusp proportions of the Longgupo fourth premolar, and on the dental metrics, which compare favorably with late Miocene *Lufengpithecus*.

My involvement with the research at Longgupo Cave began in 1991, when Huang Wanpo extended an invitation to my research team to visit the site in an effort to provide a reliable age determination and to understand its complex site formation processes. A monograph announcing this new hominin site had just been published (Huang 1991) and it was deemed a propitious time to let other researchers view Longgupo. Therefore, in March of 1992, geochemist Charles Yonge, geoarcheologist Roy Larick, and myself as paleoanthropologist, made the journey to Longgupo accompanied by our hosts, Huang Wanpo and Gu Yumin. We were successful in obtaining samples for dating and in deciphering the site formation process, and once back in Beijing, we set about studying Longgupo's extensive faunal remains, especially the hominoid specimens, as well as the two artifacts. We compared the metrics of the Longgupo dental remains and found they did not sort with Asian *Homo erectus*. We then extended the comparison to the hominins of the Late Pliocene and Early Pleistocene of East Africa, especially *Homo habilis* and "*Homo ergaster*". Since the dates of the Longgupo site were in the range of 1.8–2.0 Ma, it seemed reasonable at the time to compare the Longgupo specimen with similarly-aged fossils in East Africa. From our extensive comparisons, we knew that the Longgupo mandibular fragment could not be assigned to any known Asian ape, since its morphology was strikingly different from the two other known Asian Pleistocene apes, *Gigantopithecus* and *Pongo*. Additionally, the enamel on the Longgupo molar and premolar was worn in such a way that it excluded any detailed comparisons. Because of these factors, we designated this specimen as coming from a hominin, although at the time we did recognize that some researchers at the IVPP had lingering doubts about the hominin status of the Longgupo mandibular fragment.

When the announcement of the Longgupo discovery in *Nature* was published, it generated a huge amount of academic and public interest. Based in part on this media attention, Huang Wanpo was able to establish a privately-funded research institute in the small village of Miao-yu adjacent to the Longgupo Cave complex and to later establish a new journal entitled *Longgupo Prehistoric Culture* (e.g., Huang and Zheng 1999). Approximately 1 year or so after the *Nature* paper appeared, the Chinese government announced a major new funding initiative for research on human origins. More recently, new excavations for artifacts began in the upper levels of Longgupo by a Chinese-French team (see Hou and Zhao 2010).

There is no doubt that Longgupo is a complex site spanning a long time range. Artifacts certainly do occur in the later levels near the top of the deposits and possibly in the earlier deposits as well. Over the years, I continued to support the hominin status of the Longgupo mandibular fragment in various publications (e.g., Larick and Ciochon 1996; Ciochon and Eaves-Johnson 2004) though I did become increasingly aware of other researchers' views that this partial mandible was an ape. My views changed abruptly in May 2005 during the visit to Nanning, when Wang Wei showed me the new hominoid teeth he had discovered in Mohui Cave. The existence of a new chimpanzee-sized fossil ape in the Early Pleistocene of southern China has swayed me towards those who call into question the hominin status of the Longgupo mandibular fragment. I believe that a new study of the Longgupo mandibular fragment in comparison with recent discoveries of the ape species at Mohui Cave and other sites in southern China could resolve this controversy once and for all.

Sanhe Cave, Guangxi, China

Recently, some of my Chinese colleagues have begun to come to similar conclusions about ape diversity within the *Stegodon-Ailuropoda* fauna. A new *Gigantopithecus* fauna discovered at Sanhe Cave in Guangxi has begun to change minds. Sanhe Cave (22°16.493′N, 107°30.663′E) is located in Chongzuo Ecological Park (Chongzuo Biodiversity Research Institute, Peking University) 16 km to the northeast of Chongzuo urban district, Jiangzhou district, Guangxi province, China (Fig. 8.1). *Gigantopithecus* and cf. Ponginae fossils have been discovered in layers 2–7 of the ~11.3 m thick lower unit. In association with *Gigantopithecus* and other elements of the *Stegodon-Ailuropoda* fauna, a right M_1 identified as cf. Homininae or cf. Ponginae was recovered (Jin et al. 2009). Though the occlusal surface is worn, this tooth falls within the size range and morphology of the Mohui non-*Gigantopithecus* hominoid teeth. The fact that it was identified, in part, as cf. Ponginae demonstrates that Chinese researchers are beginning to accept the premise that the diversity of apes in the *Stegodon-Ailuropoda* fauna is greater than previously recognized. As it stands, however, the only solid evidence of early hominin activity in Asia is known from a handful of sites, as discussed below.

Nihewan Basin, Hebei, China

Undisputed evidence of early hominin occupation in northeastern China can be found in the Nihewan Basin. The Nihewan Basin (previously Nihowan) extends for approximately 200 km^3 150 km west of Beijing, Hebei province near the village of Nihewan, for which the basin is named (Barbour et al. 1926). This region first came to the attention of archeologists in the 1920s through French missionaries who were collecting interesting artifacts from the basin. Nihewan Basin is in a transition zone between the North China Plain and the Inner Mongolian Plateau, not far from the well-known *Homo erectus* site of Zhoukoudian (Zhu et al. 2001). The archaic lake of Nihewan would have stretched over 9,000 km^2 and have been surrounded by forested mountains and arid plains. Tectonic uplift over time separated this massive lake into two, now represented by two smaller constituent basins: the Yangyuan Basin (Li et al. 1998) in the north and the Yuxian Basin in the south (Cai and Li 2004). The main archaeological localities with a confirmed Early to Middle Pleistocene age are all found on the eastern margin of the present-day basin (Schick et al. 1991) (see Fig. 8.2). These lake-bed sediments hold the scattered, ephemeral early occupation sites including Maliang, Cenjiawan, Donggutuo, Majuangou, and Xiaochangliang. Three sites within the basin have proven especially fruitful from a paleoanthropological perspective, those being Donggutuo, Xiaochangliang and Majuangou (Keates 1995). This clustering of sites around the Nihewan's archaic lacustrine deposits may reveal a preference by the

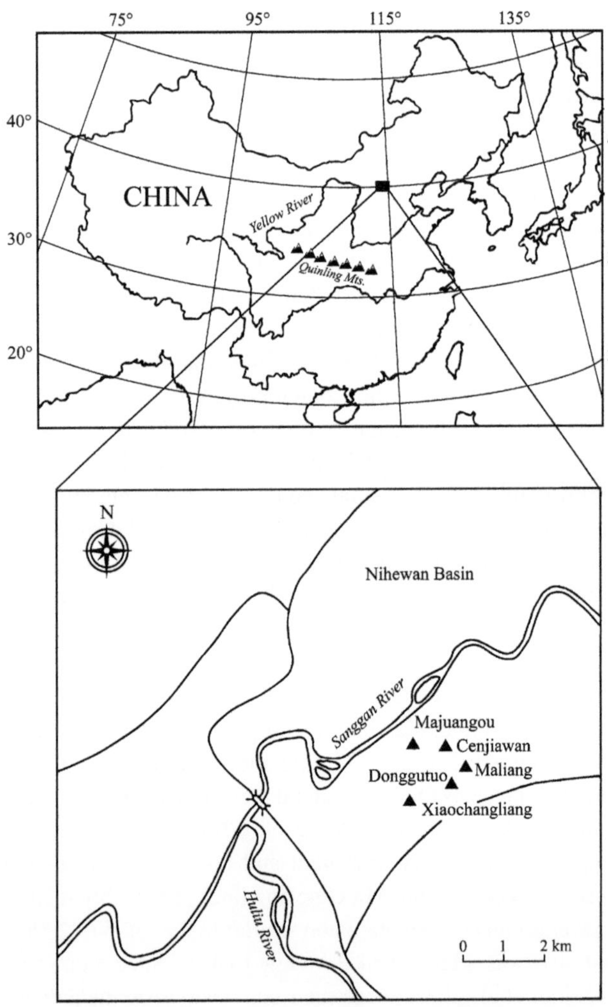

Fig. 8.2 The Nihewan Basin extends for approximately 200 km^3 150 km west of Beijing, Hebei province, in a transitional zone between the North China Plain and the Inner Mongolian Plateau (Barbour et al. 1926; Zhu et al. 2001). The main archeological localities with a confirmed Early to Middle Pleistocene age (Donggutuo, Xiaochangliang, Majuangou, Cenjiawan, and Maliang) are found on the eastern margin of the present-day basin (Schick et al. 1991), as illustrated here

hominins of northeastern China for lakeside habitation. However, as these are environments which support the preservation of fossils, we cannot rule out the possibility that these clusters are due to taphonomic factors rather than hominin preference.

The archeological site Xiaochangliang (40.2°N, 114.65°E) was discovered in the late 1970s, followed by the identification of many more Paleolithic sites in the 1980s and 1990s, though few of these have been fully explored. This site provides definite evidence of early hominin occupation and tool-making in northeastern China, though the stone tool industry uncovered at Xiaochangliang is deceptively simple and is dominated by side scrapers and notches, with only a few end scrapers, burins and disc cores (Schick et al. 1991). However, more progressive items, such as nearly cylindrical cores, blades and flat core platforms complicate the picture (Huang 1985).

The dating of Xiaochangliang has proven somewhat contentious and dates from this site have recently been revised again (Li et al. 2008). Magnetostratigraphy, which uses the Earth's known magnetic reversals preserved within local sedimentary sequences to estimate their age, has recently become one of the primary dating technique used for sites in this region (Li et al. 2008; Wang et al. 2005, 2006; Zhu et al. 2001, 2003, 2004). For Xiaochangliang, early magnetostratigraphic dating of the one artifact laden section, along with estimates of sedimentation rate, bracket the occupation between the Olduvai and Jaramillo subchrons in a layer of reverse polarity, thus, confining the age of the artifacts to roughly 1.36 Ma and making it one of the earliest sites of recognizable stone tool forms in China (Zhu et al. 2001). However, recent work has revealed problems with open-air sedimentation rates (one of the critical variables in magnetostratigraphy) (Li et al. 2008), and these early dates are being reassessed. Li et al. (2008) have revised dates on several of the Nihewan sites including Xiaochangliang which now dates to 1.26 Ma. Revised dates have also been given to Hebei sites of Maliang (0.78 Ma) (Wang et al. 2005), Cenjiawan (1.1 Ma) (Wang et al. 2006), and Donggutuo (1.2 Ma) (Li et al. 2008; see also Ciochon and Bettis 2009).

Like Xiaochangliang, Donggutuo (40° 13′ 22″N, 114° 40′ 11″E) – which lies 4 km from Nihewan Village at locality Xujiapo – is an important site in the Nihewan Basin yielding many stone tools associated with *Equus*, *Bison*, and other large mammals (Wei et al. 1985). The tools here are identified as predominantly flakes, with casual cores, denticulates and choppers also present. In total, five horizons provide artifacts, several of which are made from volcanic rocks, ignimbrites and lavas. These materials are of particular interest as potassium-argon dating may be possible, providing a bottom bracket for the human occupation of this site (Schick et al. 1991). A preliminary paleomagnetic study of the Donggutuo sequence, conducted by Li and Wang (1982), alleged that the artifact layer was located ~5 m below the Jaramillo subchron giving the site a date of roughly 1.1 Ma. This early estimate was supported by later paleomagnetic studies (Wei et al. 1985; Schick and Dong 1993), but remained controversial due to the dearth of published explicit paleomagnetic data. In Wang et al. (2005), this 1.1 Ma age was confirmed by a more comprehensive investigation combining rock-magnetic and magnetostratigraphic studies of the Donggutuo artifact layers. However, the most recent revision puts the date closer to 1.2 Ma (Li et al. 2008).

Until recently, the oldest instance of stone tool processing of fauna and the earliest definitive hominin occupation in China was believed to be found in the Majuangou (40° 13.517′N, 114° 39.844′E) section on the eastern boundary where paleomagnetic dating yielded an age of 1.66 Ma (Zhu et al. 2004). However with the revised dates for all other Nihewan sites it is probable that Majuangou is much younger as well – most likely in use during the 1.1–1.2 Ma time span currently given for the other nearby sites (F. Heller 2009, personal communication; see also Li et al. 2008). Regardless of age, this site is of great importance as it preserves four artifact layers which contain stone tools that can be identified as choppers, scrapers and polyhedrons. These tools are directly comparable to those found in the African early Pleistocene, with the exception that use of chert, sandstone, quartz and andesite was substituted for lava cobbles. Accompanying these tools are vertebrate remains attributable to *Elephas*, *Equus*, *Pachycrocuta*, *Coelodonta*, *Cervus*, *Gazella*, *Struthio* and Carnivora gen. et sp. indet., many of which show clear evidence of modification by humans.

Shanxi Province, China

To the southwest of the Nihewan basin, Shanxi province also holds one possible settlement site at Xihoudu (34.7°N, 110.7°E) dating to 1.27 Ma (Li et al. 2008) as well as important *Homo erectus* fossils found at the site of Gongwangling (34°11′N, 109°29′E) which date to 1.22 Ma (Li et al. 2008). Though there is no question about the hominin status of Gongwangling remains, many present-day Chinese archeologists regard the "artifacts" from Xihoudu as eoliths (produced by natural causes) (W. Wang 2009, personal communication). While the sites in the Nihewan Basin and Gongwangling never saw glacial ice, glacial periods brought significant cooling and drying conditions (Ciochon and Bettis 2009). Open habitats, exemplified by grasslands and mixed steppes expanded during glacial periods. Such environments favored a complex of large grazing animals with *Homo erectus* existing as one of several large predators.

In sum, the sites of northeastern China provide an important source for discerning the mode and tempo of hominin dispersal into the region, as well as for understanding local dynamics. Until recently, it was believed that the earliest hominin occupation of China was in the south at Yuanmou, Yunnan province (Li et al. 1976; but see also Hyodo et al. 2002). However, with the continuing reassessments of the age of the archeological material from the Nihewan Basin, current evidence supports an earlier hominin presence in northeastern China (Ciochon and Bettis 2009). Recently, Zhu et al. (2008) revealed new magnetostratigraphic dates for the two putative stone tools and two "hominin" incisors, reviving the debate. At 1.7 Ma, the authors claim that Yuanmou represents the earliest hominin occupation in mainland East Asia indicating a rapid southward migration of *Homo* into the region. At issue are the aforementioned contentious tools and incisors, described as having affinity to both Chinese and African *H. erectus* and attributed to *Homo* sp., more specifically, *Homo erectus* sensu lato. Based on the largely qualitative assessments of the tools and incisors presented by Zhu et al. (2008) as well as the sites' temporal incongruence with the new Nihewan dates (Li et al. 2008), it is clear that though the debate may continue, the outcome pulls sharply to a conclusion of an unknown ape species at Yuanmou, not an early human.

The main problem with the "Yuanmou hominin site" is that the remains are simply too fragmentary for accurate identification. The evidence at Yuanmou is very similar to the fragmentary remains found at Longgupo. At Yuanmou, there are two incisors and at Longgupo one finds a small jaw with two worn teeth. Each site also has just two equivocal stone artifacts. Both Longgupo and Yuanmou have each been claimed to be the earliest record of hominins in China – this is a very significant claim. Such a claim should be backed up with unequivocal evidence of hominin presence. Thus, as the evidence stands, it is my opinion that neither site should be regarded as the earliest evidence of hominins in China. Future research will likely substantiate that they are not hominin sites at all!

Historically, the dating of the Nihewan sites has proven difficult for many reasons and has complicated our ability to accurately track the movements of early hominins in this region. Chief among the dating issues is the unsuitability of the materials for accurate isotopic dating (Schick and Dong 1993), which in turn has made faunal correlation one of the most often used dating methods since work in the area began in the 1920s. Regrettably, such correlations cannot provide precise chronometric estimates, but can only confirm the Pleistocene age of the sediments. The recent employment of biostratigraphy and magnetostratigraphy in combination, however, has provided paleoanthropologists with a robust means of elucidating the chronometric dates of the Nihewan sites although, as recent date revisions prove, the application of these methods are still being refined in this region (Li et al. 2008; Ciochon and Bettis 2009). Even more promising dating techniques such as cosmogenic $^{26}Al/^{10}Be$ burial dating can be used to more accurately date the hominin sites in the Nihewan Basin, Hebei as well as Gongwangliang in Shanxi province (Ciochon and Bettis 2009; Shen et al. 2009).

One significant aspect of the new Nihewan dates is that they allow the temporal connection of several important archeological sites in the region. Previous attempts to link the sites were hindered in part by the Nihewan Basin Hebei sites' incongruent dates of 1.7–1.4 Ma (Zhu et al. 2001, 2004). These old dates made them nearly three times the previous age estimates for the important cave site of Zhoukoudian (39°41′N, 115°55′E). However, with new estimates for Zhoukoudian pushing it older (0.78 Ma) and the open-air Hebei sites younger, such an association is at last within grasp. It now seems that the early human occupation of northeast Asia began around 1.3 Ma and continued to at least 400,000 ka based on the youngest occurrence of *Homo erectus* at Zhoukoudian (Shen et al. 2009).

Evidence from Peninsular Southeast Asia

Lang Trang Caves, Vietnam

In the late 1980s, I began a series of paleoanthropological research projects in northern Vietnam (Ciochon and Olsen 1986; Olsen and Ciochon 1990). Working with University of Arizona archeologist, John Olsen, and Vietnamese researchers from the Institute of Archaeology in Hanoi, we set out to investigate cave sites and open-air sites across northern Vietnam with the express purpose of documenting the co-existence of *Gigantopithecus* and *Homo erectus* (see Ciochon et al. 1990).

The first excavation at the Middle Pleistocene site of Lang Trang Caves (20°12′N, 105°21′E), about 125 km southwest of Hanoi in Ba Thuoc district, was undertaken in January of 1989. The site consists of a complex of four caverns and several smaller openings situated on the eastern bank of the Ma River about 20 m above the river level. The first excavations of this cave complex produced 1,025 identifiable fossil specimens spanning 36 genera of the *Stegodon-Ailuropoda* fauna. Primates made up 13% of the fauna from the site, with macaque monkeys being the most common primate, followed by the orangutan, *Pongo pygmaeus*. Although no teeth of *Gigantopithecus* were among the excavated specimens, five primate teeth attributed to *Homo* sp. were also recovered. Initially, these teeth were thought to derive from the same levels as the *Stegodon-Ailuropoda* fauna (Ciochon et al. 1990), but subsequent excavation revealed that the breccia deposits on the floor of the caves were capped by a thin Hoabinhian occupation level. This Late Pleistocene to

Holocene occupation level yielded stone artifacts, a few modern faunal elements, and five teeth of *Homo sapiens*.

Due to the geologic and taphonomic palimpsest uncovered during the 1989 excavations, future excavations were delayed until 1993 in order to assemble a team of six specialists in karst geology, geochronology, geomorphology, vertebrate paleontology and geopaleontology. When research resumed at Lang Trang Caves in March of 1993, stratigraphically-controlled excavations were undertaken in all four caverns yielding more than 15,000 fossil specimens (see Long et al. 1996). The largest single faunal sample came from Cave II, breccia level 5, where 10,000 specimens were excavated from a single tightly-controlled stratigraphic interval (Long et al. 1996). ESR dates, derived by Rainer Grün (Australian National University) from all excavation levels, showed that the fauna of Lang Trang Caves were deposited in a series of different units that spanned 185 to 385 kya (R. Grün, 2005, personal communication), contrary to the published estimates of 60–80 Ka based on biostratigraphy (Long et al. 1996).

The key discoveries of the 1993 expedition were that the dominant primates at Lang Trang (*Macaca, Pongo, Presbytis*) are tropical/subtropical species and neither hominins nor *Gigantopithecus* were present. The fact that these 15,000 fossil specimens were excavated with precise stratigraphic control suggests that *Homo erectus* was not present in northern Vietnam during this time interval when the tropical *Stegodon-Ailuropoda* fauna flourished, although additional data from other sites will strengthen this claim.

Tham Khuyen Cave, Vietnam

Tham Khuyen Cave is a Middle Pleistocene karst cave site located in Lang Son province. It is situated in northern Vietnam about 125 km NNE of Hanoi and 30 km WSW of the China border (Fig. 8.1). The cave was discovered by a team of Vietnamese and German paleontologists in 1964 and was first excavated in 1965 (Bao and Kha 1966; Cuong 1971). Excavations yielded a diverse mammalian fauna typical of the *Stegodon-Ailuropoda* fauna, with the addition of *Pongo, Gigantopithecus*, *Hylobates* and *Homo erectus* (Kha and Long 1976). Most taxa were represented only by isolated teeth, with the exception of *Hylobates*, which was documented by a partial skull. During the 1980s, Vietnamese researchers often cited Tham Khuyen Cave as evidence for the early presence of humans in Southeast Asia and for the co-existence of *Homo erectus* and *Gigantopithecus* (Long and Du 1981; Cuong 1985).

Following a change in the political climate of the region, in 1993, my research team was able to travel to Lang Son province to visit Tham Khuyen and other caves in the area. Since Tham Khuyen is generally thought to contain the oldest evidence of *Homo* in Vietnam, we collected samples for U/Th series and ESR dating. The results were published in 1996, presenting a detailed geological analysis of the Tham Khuyen Cave focusing on the sedimentary dynamics and the absolute age of the faunal-bearing units (Ciochon et al. 1996). For sedimentary levels 1–3, we obtained an ESR date of 475 Ka that was constrained by a U/Th date of 117 kya on speleothems several meters above the bone-bearing levels, thus documenting a Middle Pleistocene date for the co-occurrence of *Homo erectus* and *Gigantopithecus* in Vietnam.

During the early 1990s, Vietnamese collaboration with Jeffrey Schwartz and Ian Tattersall resulted in two provocative papers on Pleistocene hominoid diversity in northern Vietnam (Schwartz et al. 1994, 1995). In the first paper, Schwartz et al. (1994) reviewed the fossil hominoids from Tham Khuyen Cave, suggesting the presence of a previously undescribed species of large-bodied hominoid. In the follow-up paper, Schwartz et al. (1995) named a new species of *Pongo*, *P. hooijeri*, and a new ape genus, *Langsonia liquidens*. They additionally recommended that some specimens allocated to *Homo erectus* at Tham Khuyen by Vietnamese researchers be placed into the new genus, *Langsonia* (Schwartz et al. 1994, 1995). These conclusions were subsequently challenged by Harrison (2000) who suggested that most (if not all) of the hominoid diversity at Tham Khuyen could be encompassed within the highly variable species of Pleistocene *Pongo*.

With the discovery and identification of a new Pleistocene ape species at Mohui Cave in Guangxi, it is now necessary to carefully re-analyze all of the hominoids from the karst cave faunas of northern Vietnam. Given our new understanding of Pleistocene hominoid diversity in southern China and northern Vietnam, it will also be crucial to re-evaluate the claims for the presence of the genus *Homo* in the *Stegodon-Ailuropoda* faunas of these Vietnamese sites. At this point, however, it now becomes fruitful to discuss those sites in Southeast Asia where *H. erectus* is clearly represented, and consider the associated biota and climate in order to better understand the early human presence within this environment.

Evidence from Java

How Did Homo erectus *Disperse from Africa to Island Southeast Asia?*

The East African Rift and extreme Southeast Asia are endpoints on a grand east-west geotectonic pathway called the Tethys corridor. *Homo erectus* fossils are always found in the context of this volatile geology and rugged geography suggesting that *Homo erectus* thrived in these unstable landscapes during its Africa to Asia migration. Still, much like the clusters of hominin sites near lakes in northern China, we cannot rule out the possibility that environmental conditions leading to differential fossil preservation may account for the numerous fossil finds in this migration zone.

However, if there was a preference for volatile environments it may explain *Homo erectus*' departure from Africa (or Georgia) slightly before 1.8 Ma and arrival in extreme Southeast Asia not long thereafter. During a rather brief period (1.98–1.79 Ma) of low sea level, produced by northern hemisphere glaciation, the Tethys corridor became intermittently accessible. Considering the age of *H. erectus* fossils on Java and the sea level record, it is most likely that hominins made their first crossing into Sunda between 1.8–1.74 Ma during glaciations corresponding to either Marine Isotope Stage 62, 60, or 58, when sea level was at least 50 m lower than present (Shackelton 1997). Dry glacial periods offered not only land bridges, but also a greater extent of savanna and open woodland. These heterogeneous vegetation mosaics may have supported a more diverse fauna than previously found, providing ideal conditions for faunal and human migration (van den Bergh et al. 2001; Yap 2002; Antón and Swisher 2004; Wynn 2004; Bettis et al. 2009).

It appears that *Homo erectus* and companion mammals took advantage of these open landscapes to migrate eastward from the East African Rift Valley across the Bab-el-Mandab isthmus (approximately the Afar passage, which links present-day Djibouti and Yemen) to the Arabian Peninsula, then on through the strait of Hormuz or present day Iraq and Iran (see Nikitas and Nikita 2005), before following the coastal plain around India to present day Java. It is most likely that *Homo erectus* did not navigate any bodies of water to reach present Java, but rather, over a series of generations, walked the length of the emergent Sunda continental shelf off East Asia's present south coast. At present, the Indonesian archipelago's 14,000 tropical islands represent the highlands of the previously emergent Sunda landmass.

Two recent studies support a coastal route for this transcontinental migration from Africa to Asia. Peters and Vogel (2005) have analyzed the carbon isotope ratio of tooth enamel of African *Homo erectus*, and found evidence that *Homo erectus* either was eating C_4 grasses, sedges and broad-leafed herbs, or was eating the small animals that had eaten these same plants. In either case, this links *Homo erectus* with wetland and marsh habitats, which commonly occur around coastal areas. As already stated, the most plausible departure of *Homo erectus* from Africa (or Georgia) occurred at a time when glacial ice accumulations in the Northern Hemisphere locked up large volumes of the planet's water causing significant drops in sea level across the globe. Faure et al. (2002) have proposed that freshwater springs were abundant on emergent continental shelves during these low sea stands. Thus, by following a coastal route along the newly emergent continental shelves, *Homo erectus* would have been able to disperse taking advantage of a constant fresh water supply, favoring habitats in the form of coastal oases (Faure et al. 2002). Evidence both in Africa and in Asia at Sangiran is consistent with this hypothesis, as there is a prolonged temporal and spatial association of *Homo erectus* with wetland, coastal habitats rich in C_4 plants. Our localized data on climate change in Sangiran indicate that when *Homo erectus* arrived in the Solo Basin during accumulation of the upper Sangiran Formation, there was an abundance of C_4 plant contribution to the soil organic matter, with a steady increase in the contribution of C_3 plants as one proceeds up-section (Bettis et al. 2009). This data strengthens the correlation of early C_4 plant availability and consumption (either directly or through the ingestion of herbivores) with dietary changes in the evolving *Homo erectus* lineage as they first colonized Asia.

Early Hominins in Central Java

Since 1998, I have been conducting joint research in the Sangiran Dome (07°27.460′S, 110°50.360′E) with the Institute of Technology, Bandung (ITB). Our field team includes Yahdi Zaim and Yan Rizal (from ITB); Frachroel Aziz, Sudijono, Suminto and Sutikno Bronto from the Geological Research and Development Centre, Bandung; Art Bettis, Scott Carpenter, Roy Larick, and Mark Reagan from the University of Iowa; and Andrew Wulff from Western Kentucky University. Our work has shown that the earliest groups of *Homo erectus* likely arrived at the coastal swamps of south-central Sunda between 1.8 and 1.6 Ma and that sometime during this period, they arrived at the Sangiran locale (Larick et al. 2000; Ciochon et al. 2005). About 1.5 Ma, fast-flowing streams began building and cutting beds of coarser sediment. The lower and middle Bapang Formation stream banks represent the landscapes on which *Homo erectus* actually lived. About 900 kya, the hominins and most other contemporary large mammals seem to have left the area. In the meantime, volcanic debris had continued to accumulate up to the present era.

In order to understand the associations of pumice clasts and human fossils, our Indonesian-American team has undertaken detailed study of the sedimentary framework for the Sangiran Dome. We are studying the sedimentary dynamics of fossils bones and pumice clasts in stream environments and are analyzing the petrographic variety in volcanic minerals throughout the Dome. The final step is to calibrate these findings with hornblende eruption ages at a number of stratigraphic levels (Larick et al. 2001).

The rich bone beds of the Bapang Formation provide abundant information, with the lowest portion holding highly fragmented human fossils within the coarse gravel sediments. The pumice clasts contain green hornblende that yields eruption ages between 1.51 and 1.47 Ma. Above the base, a second series of Bapang deposits holds less fragmented human cranial elements. Middle range pumice clasts have green and brown hornblende crystals with age from 1.33 to 1.24 Ma. The highest part of the Bapang Formation has yielded the youngest hominins in association with brown hornblende, with the eruption ages clustering around 1 Ma.

The lahars at the base of the Sangiran Formation contain pumice clasts and their hornblende gives eruption ages of 2–1.8 Ma (Bettis et al. 2004). Unfortunately, the Sangiran Formation above the lahars has no pumice. For the present, we are not able to include the most important early hominin fossil beds directly in our scheme. Nevertheless, the bracketing dated material indicates that *Homo erectus* arrived after 1.8 Ma and before 1.6 Ma. Our results give the first radiometrically calibrated scheme for the emergence of this part of Sunda (including present-day Java), as well as for the arrival, entrenchment and disappearance of *Homo erectus*. *Homo erectus* occupied the southern part of Sunda for nearly 700,000 years beginning about 1.6 Ma. With an occupation of this duration, and with the existence at Sangiran Dome of 43 *Homo erectus* fossils with currently known provenience (Fig. 8.3), we may speak of an evolutionary sequence for Javan *Homo erectus* supported by a sedimentary framework and $^{40}Ar/^{39}Ar$ chronology.

New Paleoecological Studies in the Sangiran Dome

The focus of our paleoanthropological research in the Sangiran Dome of central Java has recently been expanded to include a much greater emphasis on paleoenvironmental context. The long sedimentary sequence of the Dome, spanning a time range of 2.0–0.5 Ma makes it uniquely amenable to stable isotope analyses of an abundant fossil assemblage: from mammals to clams and snails, to fossil pollen and other plant remains. It is here, in these propitious circumstances, that more than 80 fossils of *Homo erectus* have been found, spanning a 700 kya time range from 1.6 to 0.9 Ma. This unique co-occurrence makes Sangiran an outstanding locality to test the effects of climate change on the evolving human lineage.

Deposits of the upper Sangiran Formation accumulated in a variety of brackish and lake-edge environments while the overlying Bapang Formation accumulated in a braided stream environment (Bettis et al. 2004). Both formations are rich in organic deposits, fossiliferous zones (both vertebrates and invertebrates), and paleosol horizons (Ciochon et al. 2003; Larick et al. 2004; Bettis et al. 2009). These are providing a wealth of information about climatic conditions and the environment that attracted and sustained *H. erectus* in central Java.

Coincident with the occurrence of *H. erectus* fossils are exceptionally high $\delta^{13}C$ values of both soil organic matter (−12%) and freshwater mollusks (as much as 14% higher than modern counterparts). These high earliest Pleistocene values are associated with the presence of C_4 grasses confirmed by preliminary pollen investigations. Unlike C_4 savannah grasses of temperate latitudes, these tropical grasses prefer wet soil conditions (Johnston 1996). As mentioned earlier, in the overlying Bapang Formation (a *H. erectus*-bearing braided stream deposit), $\delta^{13}C$ values progressively shift toward C_3 plant compositions suggesting drier conditions. This long-term trend is confirmed by both the changing nature of rooting structures (an indicator of the type of vegetation cover) and the up-section increase in soil carbonate accumulation (signaling seasonal moisture deficits) (Bettis et al. 2009). The correlation of several proxies (soil organic matter, soil carbonate nodules, and freshwater mollusks) offers a unique opportunity for understanding the timing of climate-driven vegetation changes in this area.

Analysis of the organic matter from the upper Sangiran Formation, interpreted as a marsh and lake-margin deposit, suggests 70–80% contribution from C_4 plants, grasses that grew in wet, swampy conditions, which is confirmed by the physical characteristics of the paleosols that mark former stable land surfaces throughout the Upper Sangiran and Bapang formations (Bettis et al. 2009). These carbon isotope ratios are dramatically higher than those of their modern counterparts from landscapes that are dominated by C_3 plants.

The paleosols are grouped into eight distinct pedotypes, which are described in detail in Bettis et al. (2009). The emerging picture from the pedomorphs, combined with all other lines of evidence, illustrates that when *Homo erectus* arrived in the Solo Basin during accumulation of the upper Sangiran Formation (at least 1.6 Ma), the area was a low relief landscape along the upper reaches of a shallow estuary (Fig. 8.4a). Freshwater marshes and marsh-edge environments supported sedges, water-tolerant grasses, ferns, and water-tolerant trees and a variety of aquatic and semi-aquatic vertebrates such as small hippos, various cervids and crocodiles. However, by 1.3 Ma, during the accumulation of the middle Bapang Formation, local environmental conditions had changed (Fig. 8.4b). Braided streams draining nearby volcanic highlands provided intermittent floods of sandy, silty and clayey sediment, forming a dynamic and diverse riverine landscape characterized by open woodland, savanna and tree lined channels. These environments attracted stegodons, suids, various bovid species, and *Homo erectus*.

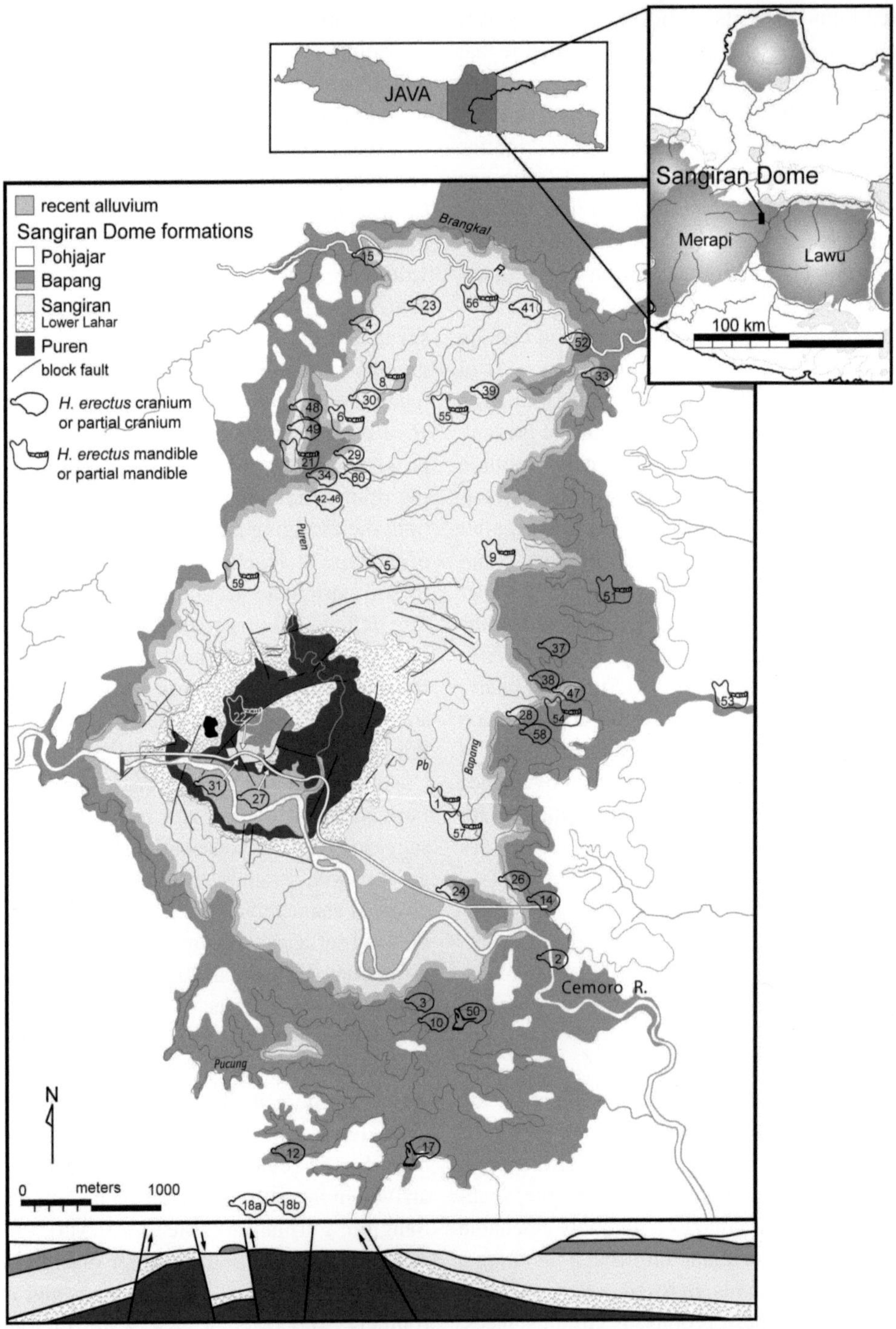

Fig. 8.3 The location of the Sangiran Dome in Central Java and a plan view showing the geographical extend of formations with precise placement of hominin find spots. Its so-called "domed-shape" arose as mud volcanoes pushed up older sediments from below. After subsequent erosion, a sequence of Pliocene and Early Pleistocene sediments outcrop in concentric rings, the oldest at the middle of the Dome. The exposed sequence consists of four principal formations (from oldest to youngest): Puren, Sangiran, Bapang, and Pohjajar. The *Homo erectus* find spots, depicted here by skull and mandible symbols, are found in the Upper Sangiran and Bapang Formations, shown here in *light gray* and *medium gray*, respectively. The small numbers inside the hominin cranium and mandible symbols generally refer to the "Sangiran" or "S" numbers used in the older, colloquial numbering system for the hominins of the Sangiran Dome (for a discussion of new numbering systems, see Larick et al. 2000:732–733; and for a comparison of numbering systems, see Fig. 11 in Larick et al. 2004)

a Upper Sangiran

b Middle Bapang

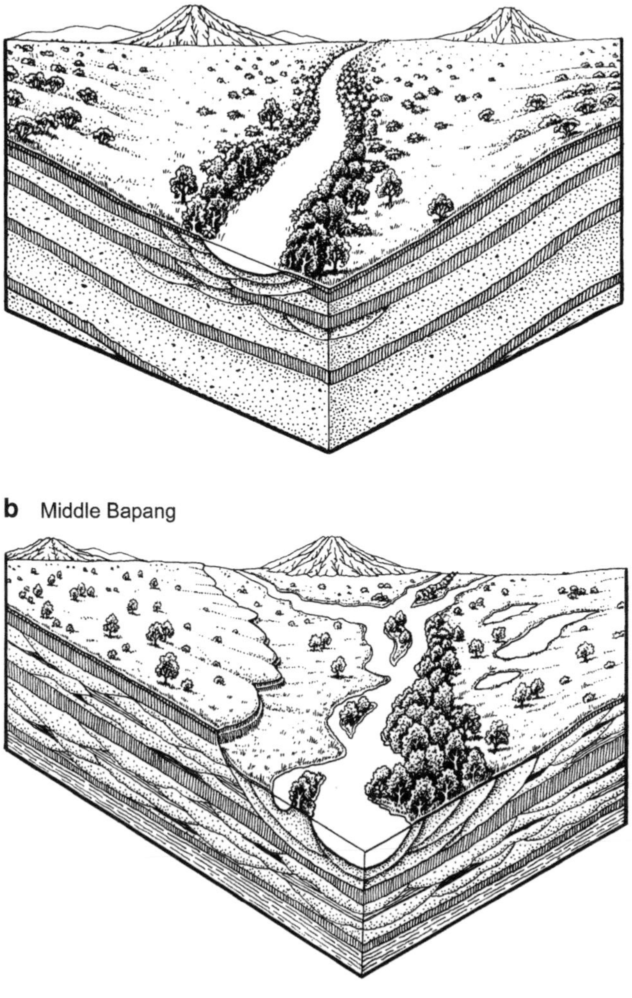

Fig. 8.4 An artist's depiction of the landscapes of the Solo Basin illustrating the differences between the paleoecological regimes of (**a**) Upper Sangiran Formation (1.6 Ma) when hominins first arrived, which were characterized by freshwater marshes and marsh-edge environments; and (**b**) Middle Bapang Formation sequences (1.3 Ma), that featured braided streams and open woodlands when *Homo erectus* flourished

Discussion and Conclusions

Based on the review of proposed hominin cave sites presented in this paper, I now suggest that the *Homo erectus* should *not* be included as a member of the *Stegodon-Ailuropoda* fauna (Fig. 8.1 graphs the geographical range of the *Stegodon-Ailuropoda* fauna). It is interesting to note that *Homo erectus* sites with significant cranial remains, such as Gongwangling (Shanxi province; 34°11′N, 109°29′E), Hexian (Anhui province; 31°53′N, 118°12′E) and Tangshan (Jiangsu province; 32°03′N, 119°03′E), are all found north of the *Stegodon-Ailuropoda* fauna. These *H. erectus* sites are also found north of the Qinling Mountains (Fig. 8.1). It is possible that early hominins may have inhabited parts of southern China, such as in river valleys or areas devoid of forest, but these hominins were not part of the heavily forested, humid-climate adapted mammalian fauna of the region. A good example of this phenomenon is the transient hominin occupation discovered in the Bose Basin of Guangxi (Hou et al. 2000). Hou and colleagues report large Acheulean-like tools associated with tektites and charcoal in laterite deposits that once represented fluvial terraces. They surmise that the tektite fall, dated to 803 Ka, caused an episode of woody-plant burning that resulted in widespread deforestation. With the forests gone, exposed cobble outcrops became a source of raw materials for hominins. Because of the intensely degrading laterite soil, no mammalian fossil remains have been found in these deposits, though I suspect if faunal remains *are* ever found, it will not be of elements of the *Stegodon-Ailuropoda* fauna. Hominins were present in the Bose Basin during this early Middle Pleistocene time interval because unique events brought about a change in the regional habitat.

From a different view, if we examine the historical tradition of identifying "hominin" presence in southern China on the basis of isolated teeth, the story of "*Hemanthropus*" exemplifies some of the problems that have plagued Asian paleoanthropology – specifically, fragmentary material and lack of context. This hominin taxon was created by von Koenigswald (1957a, b) on the basis of a small number of teeth collected in apothecary shops across southern China and island Southeast Asia, as there has long been a tradition in Chinese medicine to use fossil teeth for medicinal purposes. Von Koenigswald discovered this source of fossils in the early 1930s and collected many thousands of teeth over several decades. The fossil teeth, mostly excavated from Pleistocene caves in southern China, represented the elements of the *Stegodon-Ailuropoda* fauna and could be sorted into the natural hominoid groups of *Pongo* and *Gigantopithecus*. However, a small number of teeth did not fit either of those categories, leading von Koenigswald (1957a, b) to erect a new genus, "*Hemanthropus*," for this group, which he believed to be a distant relative of the Australopithecinae. Four decades ago it was believed that these teeth may be evidence of *Australopithecus* in Asia, though further work by researchers (e.g., Wolpoff 1982: 507) showed that the teeth of "*Hemanthropus*" were more likely worn or atypical members of the genus *Pongo*. Interestingly, with the identification of a new species of fossil ape at Mohui Cave in Guangxi, the "*Hemanthropus*" teeth once again become relevant – not because of their purported hominin status, but rather for their documentation of Pleistocene ape diversity. To investigate this further, in October 2005, I traveled to the Senckenberg Museum in Frankfurt to see the "*Hemanthropus*" collection. Among the many worn orangutan teeth I found several small ape teeth that very much resembled the small ape teeth from Mohui. Perhaps, von Koenigswald was the first to recognize the true diversity of Pleistocene apes in China.

In conclusion, this paper presents a summary of early hominin research at many sites in China and peninsular Southeast Asia during the last two decades, and attempts to tackle two separate, though related subjects, that are critical to our understanding of hominin occupation and movements in Asia during the Pleistocene (see also Ciochon 2009). The first is the problematic identification of hominins in the *Stegodon-Ailuropoda* fauna of southern China and peninsular Southeast Asia, and the second focuses on the hominin record of island Southeast Asia, specifically the Sangiran Dome of Java. The research in the Sangiran Dome is my current, ongoing field project, while the research on the "hominin" sites of southern China and Southeast Asia represents work undertaken in the 1980s and 1990s. I have written about both subjects here because each impacts our understanding of when hominins first arrive in Asia on their long dispersal out of Africa/Georgia. Based upon the information presented in this paper, I argue that comparison of putative hominin remains, such as those at Longgupo, with the unidentified hominoid teeth found at Mohui, will help elucidate ape diversity in this region during Pleistocene. Further hominin research in the Nihewan Basin of northeastern China, and especially in central Java, can provide valuable data about the movement of hominins into Asia, by providing information on their preferred habitats and dispersal patterns. The continued critical reevaluation of fossil materials from Pleistocene sites throughout Asia is of vital importance to elucidate the vagaries of hominin occupation within this dynamic and significant region.

Acknowledgments I thank my colleagues, Dr. Yahdi Zaim of the Institute of Technology Bandung, Java and Dr. Wang Wei of the Natural History Museum of Guangxi Zhuang Autonomous Region, China for access to specimens, for field collaborations and for many valuable discussions. For financial support, I thank the L.S.B. Leakey Foundation, the Collaborative Interdisciplinary Project (CIP) award program in the University of Iowa (UI), the Human Evolution Research Fund at the UI Foundation, the Center for Global and Regional Environmental Research at the UI, the Dean of the UI College of Liberal Arts and Sciences, and UI Vice President for Research. I thank Terry Harrison for useful discussions on *Lufengpithecus*. Art Bettis and Scott Carpenter of the University of Iowa, Department of Geosciences provided detailed analyses on the paleoecology of the Sangiran Dome. Invaluable editorial and research assistance was provided by K. Lindsay Eaves-Johnson. Anna Waterman assisted in proof-reading, while James Rogers, Audrey Cropp and Nathan Totten helped with figures.

References

Antón, S. C., & Swisher, C. C., III. (2004). Early dispersals of Homo from Africa. *Annual Review of Anthropology, 33*, 271–296.

Bao, T.V., & Kha, L.T. (1966). Report on paleontological and paleoanthropological research at Tham Khuyen, Binh Gia District (Lang Son Province). Documents of the Institute of Archaeology, Hanoi (in Vietnamese).

Barbour, G. B., Licent, E., & Teilhard de Chardin, P. (1926). Geological study of the deposits of the San Kan Ho Basin. *Bulletin of the Geological Sciences of China, 5*, 263–278.

Bettis III, E.A., Zaim, Y., Larick, R.R., Ciochon, R.L., Suminto, Rizal, Y., Reagan, M., et al. (2004). Landscape development preceding *Homo erectus* immigration into Central Java, Indonesia: the Sangiran formation lower lahar. *Palaeogeography, Palaeoclimatology, Palaeoecology 206*, 115–131.

Bettis III, E.A., Milius, A.K., Carpenter, S.J., Larick, R.R., Zaim, Y., Rizal, Y., et al. (2009). Way out of Africa: Early Pleistocene paleoenvironments inhabited by *Homo erectus* in Sangiran, Java. *Journal of Human Evolution 56*, 11–24.

Cai, B., & Li, Q. (2004). Human remains and the environment of Early Pleistocene in the Nihewan Basin. *Science in China Series D-Earth Sciences, 47*, 437–444.

Chaimanee, Y., Jolly, D., Benammi, M., Tafforeau, P., Duzer, D., Moussa, I., et al. (2003). A Middle Miocene hominoid from Thailand and orangutan species. *Nature, 422*, 61–65.

Chaimanee, Y., Suteethorn, V., Jintasakui, P., Vidthayanon, C., Marandat, B., Jaeger, J-J. (2004). A new orang-utan relative from the late Miocene of Thailand. *Nature 427*, 439–441.

Chaimanee, Y., Yamee, C., Tian, P., Khoawiset, K., Marandat, B., Tafforeau, P., et al. (2006). *Khoratpithecus piriyai*, a Late Miocene Hominoid of Thailand. *American Journal of Physical Anthropology, 131*, 311–323.

Ciochon, R. L. (2009). The mystery ape of Pleistocene Asia. *Nature, 458*, 910–911.

Ciochon, R. L., & Bettis, E. A., III. (2009). Asian *Homo erectus* converges in time. *Nature, 458*, 153–154.

Ciochon, R. L., & Eaves-Johnson, K. L. (2004). China: archaeological caves. In J. Gunn (Ed.), *Encyclopedia of cave and karst science* (pp. 221–225). New York/London: Fitzroy Dearborn.

Ciochon, R. L., & Olsen, J. W. (1986). Paleoanthropological and archaeological research in the Socialist Republic of Vietnam. *Journal of Human Evolution, 15*, 623–633.

Ciochon, R. L., Olsen, J. W., & James, J. (1990). *Other Origins: The search for the giant ape in human prehistory*. New York: Bantam Books. 262 pp.

Ciochon, R. L., Long, V. T., Larick, R., González, L., Grün, R., de Vos, J., et al. (1996). Dated co-occurrence of *Homo erectus* and *Gigantopithecus* from Tham Khuyen Cave, Vietnam. *Proceedings of the National Academy of Sciences of the United States of America, 93*, 3016–3020.

Ciochon, R.L., Bettis III, A., Larick, R., Zaim, Y., Suminto, Rizal, Y., Reagan, M., et al. (2003). *Homo erectus* landscapes: Paleosols in the Bapang and upper Sangiran Formations, Solo Basin, Central Java. *American Journal of Physical Anthropology* (Suppl 36), p. 77.

Ciochon, R.L., Zaim, Y., Grine, F., Rizal, Y., Larick, R., Bettis III, E.A., et al. (2005). A new *Homo erectus* maxilla from the Bapang basal unit, Sangiran Dome, Central Java. *American Journal of Physical Anthropology* (Suppl 40), p. 86.

Colbert, E. H., & Hooijer, D. R. (1953). Pleistocene mammals from the limestone fissures of Szechwan, China. *Bulletin of the American Museum of Natural History, 102*, 1–134.

Cuong, N. L. (1971). On the excavation of the caves Hang Hum, Tham Khuyen and Kao Leng. *Khao Co Hoc, 11/12*, 7–11 (in Vietnamese).

Cuong, N. L. (1985). Fossile menschenfunde aus Nordvietnam. In J. Hermann & H. Ullrich (Eds.), *Meschwerdund – Biotischer und Gesellschaftlicher Entwicklungsprozess* (pp. 96–102). Berlin: Akademis Verlag.

Etler, D. A., Crummett, T. L., & Wolpoff, M. H. (2001). Longgupo: Early *Homo* colonizer or Late Pliocene *Lufengpithecus* survivor in South China? *Human Evolution, 16*, 1–12.

Faure, H., Walter, R., & Grant, D. (2002). The coastal oasis: ice age springs on emerged continental shelves. *Global and Planetary Change, 33*, 47–56.

Gao, J. (1975). Australopithecine teeth associated with *Gigantopithecus*. *Vertebrata PalAsiatica, 13*, 81–88 (in Chinese with English abstract).

Harrison, T. (2000). Archeological and ecological implications of the primate fauna from prehistoric sites in Borneo. *Bulletin of the Indo-Pacific Prehistory Association, 14*, 90–100.

Harrison, T., Ji, X., & Su, D. (2002). On the systematic status of the late Neogene hominoids from Yunnan Province, China. *Journal of Human Evolution, 43*, 207–227.

He, Z., & Jia, L. (Eds.). (1997). *Yuanmou Hominoid fauna*. Kunming: Yunnan Science and Technology Press. 270 pp.

Hou, Y., Potts, R., Yuan, B., Guo, Z., Deino, A., Wang, W., et al. (2000). Mid-Pleistocene Acheulean-like stone technology of the Bose Basin, South China. *Science, 287*, 1622–1626.

Hou, Y. M., & Zhao, L. X. (2009). An archeological view for the presence of early humans in China. *Quaternary International*, doi: 10.1016/j.quaint.2009.09.025.

Huang, W. (1985). On the stone industry of Xiaochangliang. *Acta Anthropologica Sinica, 4*, 301–307.

Huang, W. (1991). *Wushan hominid site*. Beijing: Ocean Press. 230 pp.

Huang, W., & Zheng, S. (1999). Comprehensive study of Longgupo Hominid site. *Longgupo Prehistoric Culture, 1*, 57–63.

Huang, W., Ciochon, R. L., Gu, Y., Larick, R., Fang, Q., Yonge, C., et al. (1995). Early *Homo* and associated artefacts from Asia. *Nature, 378*, 275–278.

Huang, W., Gu, Y., Ciochon, R., Larick, R., Fang, Q., de Vos, J., et al. (1996). Whose teeth? – reply. *Nature, 381*, 202.

Huffman, O. F., Zaim, Y., Kappelman, J., Ruez, D. R., Jr., de Vos, J., Rizal, Y., et al. (2006). Relocation of the 1936 Mojokerto skull discovery site near Perning, East Java. *Journal of Human Evolution, 50*, 431–451.

Hyodo, M., Nakaya, H., Urabe, A., Saegusa, H., Shunrong, X., Jiyun, Y., et al. (2002). Paleomagnetic dates of hominid remains from Yuanmou, China, and other Asian sites. *Journal of Human Evolution, 43*, 27–41.

Jin, C.-Z., Qin, D.-G., Pan, W.-S., Tang, Z.-L., Liu, J.-Y., Wang, Y., et al. (2009). A newly discovered *Gigantopithecus* fauna from Sanhe Cave, Chongzuo, Guangxi, South China. *Chinese Science Bulletin, 54*, 788–797.

Johnston, W. H. (1996). The place of C_4 grasses in temperate pastures in Australia. *New Zealand Journal of Agricultural Research, 39*, 527–540.

Keates, S. 1995. *The Significance of the older Palaeolithic occurrences in the Nihewan Basin, northern China*. Unpublished Ph.D. dissertation. Oxford: University of Oxford.

Kha, L. T., & Long, V. T. (1976). Ten years of paleontological research. *Khao Co Hoc, 18*, 9–16 (in Vietnamese).

Kunimatsu, Y., Ratanasthien, B., Nakaya, H., Saegusa, H., & Nagaoka, S. (2004). Earliest Miocene hominoid from Southeast Asia. *American Journal of Physical Anthropology, 124*, 99–108.

Kunimatsu, Y., Ratanasthien, B., Saegusa, H., Nakaya, H., Nagaoka, S., Suganuma, Y., et al. (2005). An additional specimen of a large-bodied Miocene hominoid from Chiang Muan, northern Thailand. *Primates, 46*, 65–69.

Larick, R., & Ciochon, R. L. (1996). The African emergence and early Asian dispersals of the genus *Homo*. *American Scientist, 84*, 538–551.

Larick, R., Ciochon, R. L., Zaim, Y., Sudijono, S., Rizal, Y., & Aziz, F. (2000). Lithostratigraphic context for Kln-1993.05-SNJ, a fossil colobine maxilla from Jokotingkir, Sangiran Dome. *International Journal of Primatology, 21*, 731–759.

Larick, R., Ciochon, R. L., Zaim, Y., Sudijono, S., Rizal, Y., Aziz, F., et al. (2001). Early Pleistocene $^{40}Ar/^{39}Ar$ ages for Bapang Formation hominins, Central Jawa, Indonesia. *Proceedings of the National Academy of Sciences of the United States of America, 98*, 4866–4871.

Larick, R., Ciochon, R. L., & Zaim, Y. (2004). *Homo erectus* and the emergence of Sunda in the Tethys realm. *Athena Review, 4*, 32–39.

Li, H. M., & Wang, J. D. (1982). Magnetostratigraphic study of several typical geologic sections in North China. In T. S. Liu (Ed.), *Quaternary geology and environment of China* (pp. 33–38). Beijing: China Ocean Press.

Li, P., Qian, F., Ma, X., Pu, Q., Xing, L., & Ju, S. (1976). Preliminary study on the age of the Yuanmou Man by paleomagnetic technique. *Scientia Sinica, 6*, 579–591.

Li, Y., Yang, J., Xia, Z., & Mo, D. (1998). Tectonic geomorphology in the Shanxi Graben System, northern China. *Geomorphology, 23*, 77–89.

Li, H., Yang, X., Heller, F., & Li, H. (2008). High resolution magnetostratigraphy and deposition cycles in the Nihewan Basin (North China) and their significance for stone artifact dating. *Quaternary Research, 69*, 250–262.

Long, V. T., & Du, H. V. (1981). Zoological species belonging to the Pleistocene and geochronology of sediments containing them in caves and grottos in the northern part of Viet Nam. *Khao Co Hoc (new series), 1*, 16–19 (in Vietnamese).

Long, V. T., de Vos, J., & Ciochon, R. L. (1996). The fossil mammal fauna of the Lang Trang Caves, Vietnam, compared with Southeast Asian fossil and recent mammal faunas: the geographic implications. *Indo-Pacific Prehistory Association Bulletin, 14*, 38–56.

Matthew, W. D., & Granger, W. (1923). New fossil mammals from the Pliocene of Sze-chuan, China. *Bulletin of the American Museum of Natural History, 48*, 563–598.

Merceron, G., Taylor, S., Scott, R., Chaimanee, Y., & Jaeger, J.-J. (2006). Dietary characterization of the hominoid *Khoratpithecus* (Miocene of Thailand): evidence from dental topographic and microwear analyses. *Naturwissenschaften, 93*, 329–333.

Nikitas, P., & Nikita, P. (2005). A study of hominin dispersal out of Africa using computer simulations. *Journal of Human Evolution, 49*, 602–617.

Olsen, J. W., & Ciochon, R. L. (1990). A review of evidence for postulated Middle Pleistocene occupations in Viet Nam. *Journal of Human Evolution, 19*, 761–788.

Pei, W.-C. (1957). The zoogeographical divisions of Quaternary mammalian faunas in China. *Vertebrata Palasiatica, 1*, 9–24.

Pei, W. (1987). Carnivora, Proboscidea, and Rodentia from Liucheng *Gigantopithecus* Cave and other caves in Guangxi. *Memoirs of the Institute of Vertebrate Palaeontology and Palaeoanthropology, Academia Sinica, Beijing, Ser. A, 18*, 1–134.

Peters, C., & Vogel, J. (2005). Africa's wild C_4 plant foods and possible early hominid diets. *Journal of Human Evolution, 48*, 219–236.

Qi, G., Dong, W., Zheng, L., Zhao, L., Gao, F., Yue, L., et al. (2006). Taxonomy, age and environment status of the Yuanmou hominids. *Chinese Science Bulletin, 51*, 1001–6538 (in Chinese with English abstract).

Rink, W. J., Wei, W., & Bekken, D. (2008). Geochronology of *Ailuropoda-Stegodon* fauna and *Gigantopithecus* in Guangxi Province, southern China. *Quaternary Research, 69*, 377–387.

Schick, K. D., & Dong, Z. (1993). Early Paleolithic of China and Eastern Asia. *Evolutionary Anthropology, 2*, 22–35.

Schick, K. D., Toth, N., Qi, W., Desmond Clark, J., & Etler, D. (1991). Archaeological perspectives in the Nihewan Basin, China. *Journal of Human Evolution, 21*, 13–26.

Schwartz, J. H., & Tattersall, I. (1996). Whose teeth? *Nature, 381*, 201–202.

Schwartz, J. H., Long, V. T., Cuong, N. L., Kha, L. T., & Tattersall, I. (1994). A diverse hominoid fauna from the late middle Pleistocene breccia cave of Tham Khuyen, Socialist Republic of Vietnam (excluding Hylobatidae). *Anthropological Papers of the American Museum of Natural History, 73*, 1–11.

Schwartz, J. H., Long, V. T., Cuong, N. L., Kha, L. T., & Tattersall, I. (1995). A review of the Pleistocene hominoid fauna of the Socialist Republic of Vietnam (excluding Hylobatidae). *Anthropological Papers of the American Museum of Natural History, 76*, 1–24.

Shackelton, N. J. (1997). The deep-sea sediment record and the Pliocene-Pleistocene boundary. *Quaternary International, 40*, 33–35.

Shen, G., Gao, X., Gao, B., & Granger, D. E. (2009). Age of Zhoukoudian *Homo erectus* determined with $^{26}Al/^{10}Be$ burial dating. *Nature, 458*, 198–200.

Swisher, C.C., Curtis, G.H., Jacob, T., Getty, A.G., Suprijo, A., Widiasmoro (1994). Age of the earliest known hominids in Java, Indonesia. *Science, 263*, 1118–1121.

Tougard, C., Chaimanee, Y., Suteethorn, V., Triamwichanon, S., & Jaeger, J.-J. (1996). Extension of the geographic distribution of the giant panda (*Ailuropoda*) and search for the reasons for its progressive disappearance in Southeast Asia during the latest Middle Pleistocene. *Comptes Rendus de l'Académie des Sciences, Paris, Série II, 323*, 973–979.

van den Bergh, G. D., de Vos, J., & Sondaar, P. Y. (2001). The late Quaternary palaeogeography of mammal evolution in the Indonesian Archipelago. *Palaeogeography, Palaeoclimatology, Palaeoecology, 171*, 385–408.

von Koenigswald, G.H.R. (1957a). Remarks on *Gigantopithecus* and other hominoid remains from southern China. *Proc. Koninkl. Nederl. Akademie van Wetenschappen – Amsterdam, Ser. B, 60*, 153–159.

von Koenigswald, G.H.R. (1957b). *Hemanthropus* n.g. not *Hemianthropus*. *Proc. Koninkl. Nederl. Akademie van Wetenschappen – Amsterdam, Ser. B, 60*, 416.

Wang, Q. (1996). Assignment of Longgupo Cave hominid incisor from Wushan. *Acta Anthropologica Sinica, 15*, 320–323.

Wang, H., Deng, C., Zhu, R., Wei, Q., Hou, Y., & Boëda, E. (2005). Magnetostratigraphic dating of the Donggutuo and Maliang Paleolithic sites in the Nihewan Basin, North China. *Magnetostratigraphic dating of the Donggutuo and Maliang Paleolithic sites in the Nihewan Basin, North China. Quaternary Research, 64*, 1–11.

Wang, H., Deng, C., Zhu, R., & Xie, F. (2006). Paleomagnetic dating of the Cenjiawan Paleolithic site in the Nihewan Basin, northern China. *Science in China, Ser. D, 49*, 295–303.

Wang, W., Potts, R., Hou, Y., Chen, Y., Wu, H., Yuan, B., et al. (2005). Early Pleistocene hominin teeth recovered in Mohui Cave in Bubing Basin, Guangxi, South China. *Chinese Science Bulletin, 50*, 2777–2782.

Wang, W., Potts, R., Baoyin, Y., Huang, W., Cheng, H., Lawrence Edwards, R., et al. (2007). Sequence of mammalian fossils, including hominoid teeth, from the Bubing Basin caves, South China. *Journal of Human Evolution, 52*, 370–379.

Wei, Q., Meng, H., & Cheng, S. (1985). New Paleolithic site from the Nihewan beds. *Acta Anthropologica Sinica, 4*, 223–232.

Wolpoff, M. H. (1982). *Ramapithecus* and hominid origins. *Current Anthropology, 23*, 501–522.

Woo, R. K. (1962). The mandibles and dentition of *Gigantopithecus*. *Palaeontologia Sinica, New Ser. D, 11*, 1–94.

Wu, X. (2000). Longgupo hominoid mandible belongs to ape. *Acta Anthropologica Sinica, 19*, 1–10.

Wu, X., & Poirier, F. E. (1995). *Human Evolution in China*. In *A metric description of the fossils and a review of the sites*. New York: Oxford University Press. 303 pp.

Wynn, J. G. (2004). Influence of Plio-Pleistocene aridification on human evolution: evidence from paleosols of the Turkana Basin, Kenya. *American Journal of Physical Anthropology, 123*, 106–118.

Yap, S. Y. (2002). On the distribution patterns of Southeast-Asia freshwater fish and their history. *Journal of Biogeography, 29*, 1187–1199.

Zaim, Y., Ciochon, R. L., Polanski, J. M., Grine, F. E., Bettis III, E. A., Rizal, Y., et al. (2010). New *Homo erectus* maxilla from the base of the Bapang Formation Central Java (Indonesia). *Journal of Human Evolution*.

Zhang, Y. (1982). Variability and evolutionary trends in tooth size of *Gigantopithecus blacki*. *American Journal of Physical Anthropology, 59*, 21–32.

Zhang, Y. (1984). The "*Australopithecus*" of West Hupei and some early Pleistocene hominids of Indonesia. *Acta Anthropologica Sinica, 3*, 85–92 (in Chinese with English abstract).

Zheng, S. H. (2004). *Jianshi hominid site*. Beijing: Science Press. 412 pp. (in Chinese).

Zhu, R., An, Z., Potts, R., & Hoffman, K. A. (2003). Magnetostratigraphic dating of early humans in China. *Earth-Science Reviews, 61*, 341–359.

Zhu, R. X., Hoffman, K. A., Potts, R., Deng, C. L., Pan, Y. X., Guo, B., et al. (2001). Earliest presence of humans in northeast Asia. *Nature, 413*, 413–417.

Zhu, R. X., Potts, R., Xie, F., Hoffman, K. A., Deng, C. L., Shi, C. D., et al. (2004). New evidence on the earliest human presence at high northern latitudes in northeast Asia. *Nature, 431*, 559–562.

Zhu, R. X., Potts, R., Pan, Y. X., Yao, H. T., Lü, L. Q., Zhao, X., et al. (2008). Early evidence of the genus *Homo* in East Asia. *Journal of Human Evolution, 55*, 1075–1085.

Part III
South Asia

Introduction

In contrast with the paleontological and archeological evidence of an Early Pleistocene presence of hominins in parts of both China and Indonesia, evidence for hominins prior to the Middle Pleistocene in Southern Asia is rare and controversial. This is despite a geographical position intermediate between Africa and eastern Asia and an abundance of deposits of an appropriate age, the Siwalik Hills in the northern part of the region. The papers in this section review the paleontological and archeological record of the region in the context of early hominin dispersal, compare the southern Asia record with that of surrounding parts of Asia and discuss prospects for future research.

In "Early Pleistocene faunas of India and evidence of connections with other parts of the world", Rajeev Patnaik and Avinash Nanda review the rich paleontological record of the Plio-Pleistocene of northern India. Recent work has clarified the chronology of many parts of the Siwaliks. The Early Pleistocene Pinjor Formation contains several taxa that have been associated with early hominins in other parts of Eurasia, including *Megantereon* and *Theropithecus*. In addition, there are micromammals shared between northern India and Africa during this same time period, documenting a faunal connection. However, various claims of early hominin fossils from these deposits have proved to be unreliable.

In "The Indian subcontinent and 'Out of Africa I'", Parth Chauhan reviews the evidence for pre-Acheulean (Mode I) stone artifacts in the Plio-Pleistocene of the Indian subcontinent. In contrast with many other parts of the world, the Plio-Pleistocene of the Indian subcontinent does not contain any excavated sites with artifacts in primary contexts. He finds the lithic remains from Riwat and the Paabi Hills of Pakistan to be the best-studied and most convincing, but still controversial, evidence for the presence of hominins in the region prior to the Middle Pleistocene.

Michael Petraglia also addresses the archeological record of the region in "The Lower Paleolithic of the Indian subcontinent: hominin colonization, dispersals and occupation history". Petraglia discusses the Soan industries that have long been regarded as indicating an early hominin presence in South Asia, and concludes that they are most likely to be contemporary with Middle Paleolithic industries, rather than being pre-Acheulean (Mode I) assemblages dating to the Plio-Pleistocene boundary. He then addresses the later dispersal of Acheulean tools into Asia, noting that the earliest Acheulean assemblage in India, the Isampur Quarry, dated to 1.2 Ma, is similar in age to the earliest Acheulean artifacts in the Levant, while other assemblages date to much later. He also finds significant differences between the Acheulean assemblages of South Asia and those of East Asia, questioning the likelihood of a broad dispersal between the two regions.

The Editors

Chapter 9
Early Pleistocene Mammalian Faunas of India and Evidence of Connections with Other Parts of the World

Rajeev Patnaik and Avinash C. Nanda

Abstract In India, Early Pleistocene mammals have been found in Karewa Group of Kashmir Valley and the Pinjor Formation (Upper Siwalik Subgroup) of the Himalayan foothills. An attempt has been made here to integrate published fossil mammal data from the well dated sections in India to provide a Plio-Pleistocene biostratigraphic scheme. Occurrence of common mammals in Early Pleistocene hominin sites in Ethiopia, Algeria, Israel, Georgia, Indonesia, China as well as the Pinjor Formation of India and Pakistan indicates similar paleoenvironmental conditions and some faunal exchange. The beginning of the Pleistocene (1.8 Ma) saw an upsurge in tectonic activity all along the Himalayan foothills and this led to a change in depositional environment, slope and overall landscape. These factors, combined with overall global trend towards cool and dry conditions, were probably responsible for the spread of grasslands and shrinkage of forests, eventually aiding decline of the Siwalik mammalian faunal diversity.

Keywords Siwalik • Karewas • Early Pleistocene • Mammalian faunas • Biostratigraphy

Introduction

Way back in 1837, Falconer and Cautley reported fossil primates from the Siwalik Hills of the Indian subcontinent. This was one of the first records of fossil primates anywhere in the world. Since then, progress in South Asian paleoanthropological research has been rather slow. However, South Asia's geographic position between the well-known East African and East Asian hominin yielding localities suggests it should be further explored to search for Late Neogene and Quaternary hominins. Recent findings of hominins from Central Asia and Indonesia have further increased the importance of the Indian subcontinent because it is positioned at the cross-roads of human evolution and dispersion. Therefore, a better understanding of the South Asian Plio-Pleistocene mammalian fauna, their origins, and paths of dispersal, in the context of climate change and hominin evolution, is of immediate relevance.

Early Pleistocene deposits of India are exposed in the Siwalik foreland and Kashmir inter-montane basins. The Siwalik Group of sediments are exposed all along the southernmost foothills of the Himalayas, from Pakistan in the west to Myanmar in the east. These sediments range in age from 18.3 Ma (Johnson et al. 1985) to 0.22 Ma (Ranga Rao et al. 1988) and are well known for their wealth of fossil vertebrates. In the Upper Siwaliks, Early Pleistocene (1.8–0.78 Ma) deposits form part of the Pinjor Formation. The lower contact of the Pinjor Formation (with the underlying Tatrot Formation) is placed at 2.58 Ma, and the upper contact (with the overlying Boulder Conglomerate Formation) varies in age from 1.79 Ma to 0.6 Ma (Ranga Rao et al. 1988, 1995; Nanda 2002; Kumaravel et al. 2005). Pilgrim (1913) first proposed a division of the Siwaliks based on lithology and faunal composition. Because of the time-transgressive nature of various lithological boundaries, present day researchers prefer to use biostratigraphic interval-zones based on the first and last appearances of large mammals that have been tied to sections dated by magnetostratigraphy and tephrochronology in both Pakistan and India (Keller et al. 1978; Opdyke et al. 1979; Barry et al. 1982; Hussain et al. 1992; Ranga Rao et al. 1988, 1995; Nanda 2002).

Barry et al. (1982) proposed two interval-zones for the Upper Siwaliks of Pakistan: the *Hexaprotodon sivalensis* Interval-Zone (5.3–2.9 Ma) and the *Elephas planifrons* Interval-Zone (2.9–1.5 Ma). Later, Hussain et al. (1992) modified the range of the *Elephas planifrons* Interval-Zone from 2.9–1.5 Ma to 3.4–2.7 Ma, and proposed a third Upper Siwalik zone termed as the *Elephas hysudricus* Range-Zone (2.7–?Ma). As the upper part of the Upper Siwalik deposits are better exposed in India than Pakistan, Nanda (1997a) proposed 0.6 Ma as the upper limit of the *Elephas hysudricus* Range-Zone. Agarwal et al. (1993, p. 235) extended the

R. Patnaik (✉)
Center of Advanced Study in Geology, Punjab University, Chandigarh, 160014, India
e-mail: rajeevpatnaik@gmail.com

A.C. Nanda
Wadia Institute of Himalayan Geology, Dehra Dun 248 001, India
e-mail: nandaac@rediffmail.com

J.G. Fleagle et al. (eds.), *Out of Africa I: The First Hominin Colonization of Eurasia*, Vertebrate Paleobiology and Paleoanthropology,
DOI 10.1007/978-90-481-9036-2_9,

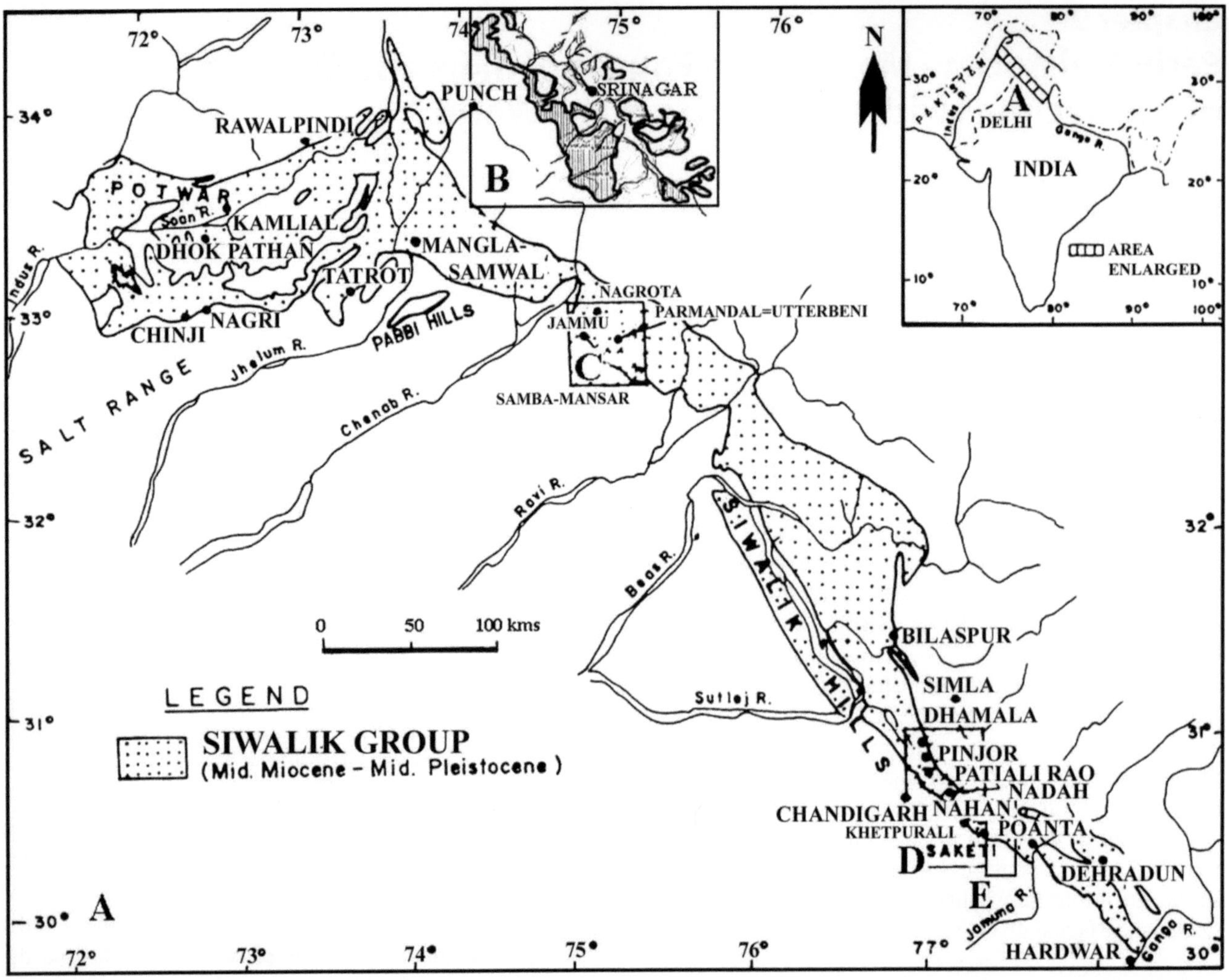

Fig. 9.1 Locality map showing Karewa deposits (**A**), and well-studied Siwalik areas (**B**); these include, Jammu (**C**), Chandigarh (**D**), and Haripur (**E**). Modified after Nanda (2002)

lower limit of the *E. planifrons* from 2.9 to 3.6 Ma. The younger *Elephas hysudricus* Interval-Zone, has been renamed as the *Equus sivalensis* Interval-Zone (2.6–0.6 Ma) by Nanda (2002). Recently, based on mammals recovered from well-dated Upper Siwalik sections in the Pabbi Hills, Pakistan, Dennell et al. (2006) have suggested further division of the *E. hysudricus* faunal zone. According to them, the *E. hysudricus-Sivatherium* part ranges from 2.7 to 1.7 Ma and is defined by the local extinction of *Sivatherium giganteum*, hippopotamids, anthracotheres and a large canid; the *E. hysudricus-Crocuta-Ursus-Panthera* zone represents the time period between 1.7 and 0.9 Ma. However, in the present paper we will use the *Equus sivalensis* Interval-Zone; the most common taxa occurring in this zone include *Elephas hysudricus, Stegodon insignis, Equus sivalensis, Rhinoceros palaeindicus, R. sivalensis, Coelodonta platyrhinus*, species of *Cervus* including *Rucervus, Sus* spp., *Sivatherium giganteum, Hemibos* spp. and *Bos acutifrons.*

The Karewa inter-montane basin, with around 1,000 m of Plio-Pleistocene glaciofluvial and lacustrine sediments (Burbank and Johnson 1982; Kusumgar et al. 1986; Kotlia 1990, Figs. 1–3), lies in the Kashmir Valley, between the northern Main Himalayan range and southern Pir Panjal Range. Karewa sediments range in age from ~ 4 Ma to 200 Ka (Kusumgar et al. 1986; Agrawal et al. 1987) and are commonly divided into the Lower and Upper Karewa Formations. On the Pir Panjal flanks, the Lower Karewas are gently inclined and overlain by nearly horizontal Upper Karewa deposits. The Upper Karewas are better exposed on the northern part of Himalayan flank. The mammalian taxa recorded from the Karewas include *Bos, Sus, Felis, Elephas hysudricus, Equus sivalensis, Rhinoceros, Sivatherium giganteum, Hexaprotodon sivalensis, Cervus punjabiensis, Cervus sivalensis, Canis vitastensis*, and the small mammals *Mus jacobsi, Ragapodemus debruijni, Kilarcola, Golunda* and *Episoriculus* (De Terra and Patterson 1939; Badam 1968; Tewari and Kachroo 1977; Kotlia 1990, 1992, 1998; Sahni and Kotlia 1985).

In this paper, we discuss the record of Early Pleistocene mammals from well dated sections in India and attempt an intra-regional correlation with similar deposits in Pakistan. In the

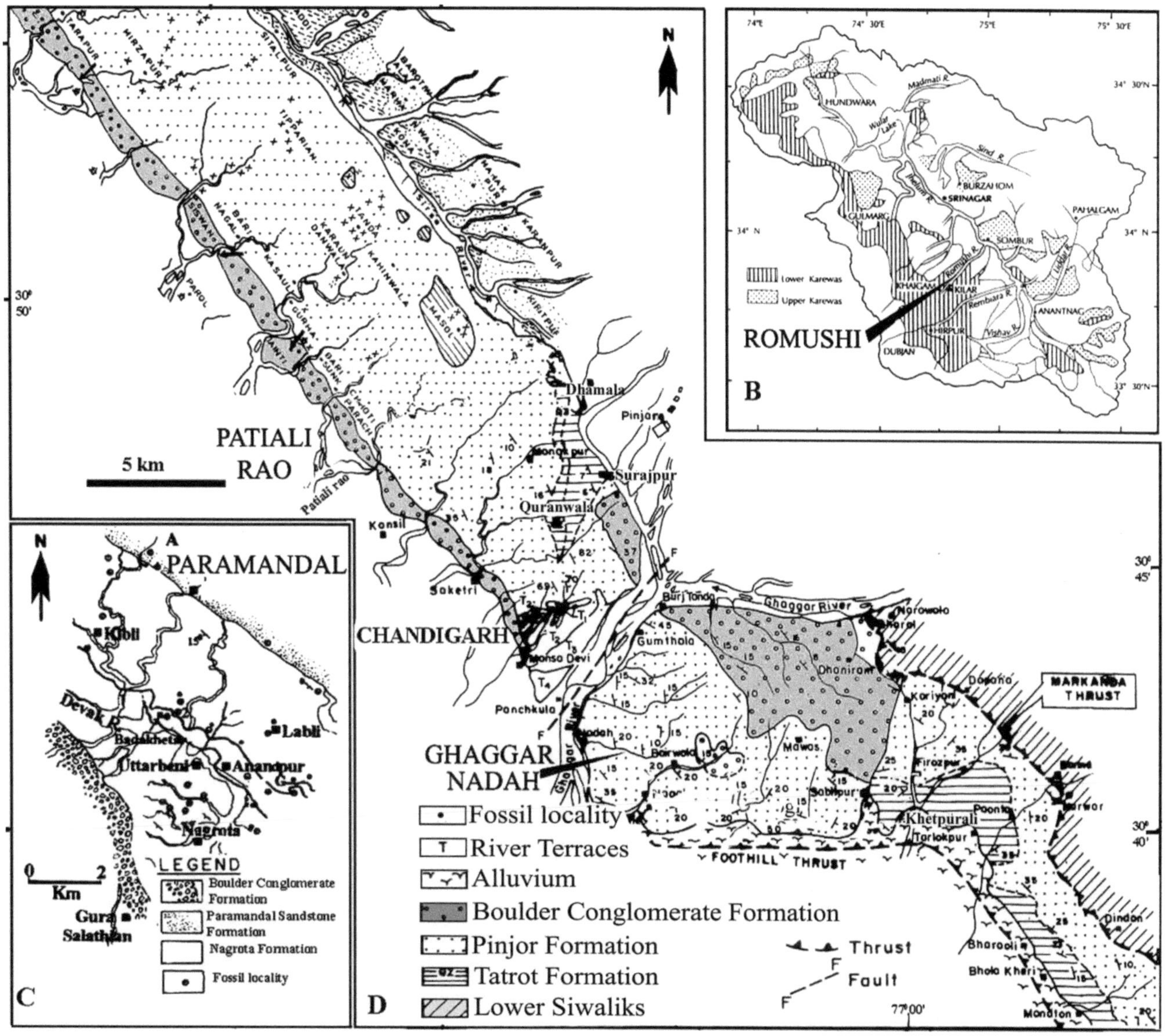

Fig. 9.2 Geological map of the (**B**) Karewa deposits of Kashmir (After Kotlia 1990), (**C**) the Jammu region (Modified after Basu 2004), and (**D**) the Chandigarh region (Modified after Sahni and Khan 1964; Nanda 2002). Locations of these regions are shown in Fig. 9.1a

light of the available fossil, isotopic, rock magnetic and pedogenic data, we discuss the Early Pleistocene paleoclimatic conditions in India with reference to human occupation. Furthermore, we compare the occurrence of common taxa in other parts of the world (including those from where fossil hominins have been recorded) such as East Africa, Israel, Georgia, Myanmar, Indonesia and China.

Plio-Pleistocene Deposits of Indo-Pakistan and the Mammalian Fauna

Siwalik sediments have yielded a large number of diverse fossil vertebrates since the first half of nineteenth century (see Colbert 1935; Pilgrim 1939; Sahni and Khan 1964). Attempts to place these fossils in a geochronological framework began in the late 1970s (Keller et al. 1978; Opdyke et al. 1979; Azzaroli and Napoleone 1982; Johnson et al. 1985). The magnetostratigraphy of fossiliferous strata in the Upper Siwaliks of Pakistan led Opdyke et al. (1979) to propose that *Equus, Elephas, Bos* and *Cervus* (i.e., cervids with antlers) marked the beginning of the 'Pinjor Fauna', coinciding with the Gauss – Matuyama (Chron C2An\C2r) magnetic boundary at 2.48 Ma (2.58 Ma, Cande and Kent 1995). The type area of the Pinjor Formation lies near Chandigarh and has been dated from 2.58 to 0.63 Ma (Ranga Rao et al. 1995). It is almost impossible to correlate the Pinjor Formation of the Chandigarh region to the Upper Siwalik sequences of Jammu or the Potwar Plateau of Pakistan by strike mapping. It has been proposed that these sequences in Jammu or Potwar, lying 300 or 400 km northwest of Chandigarh, should not be classified as part of

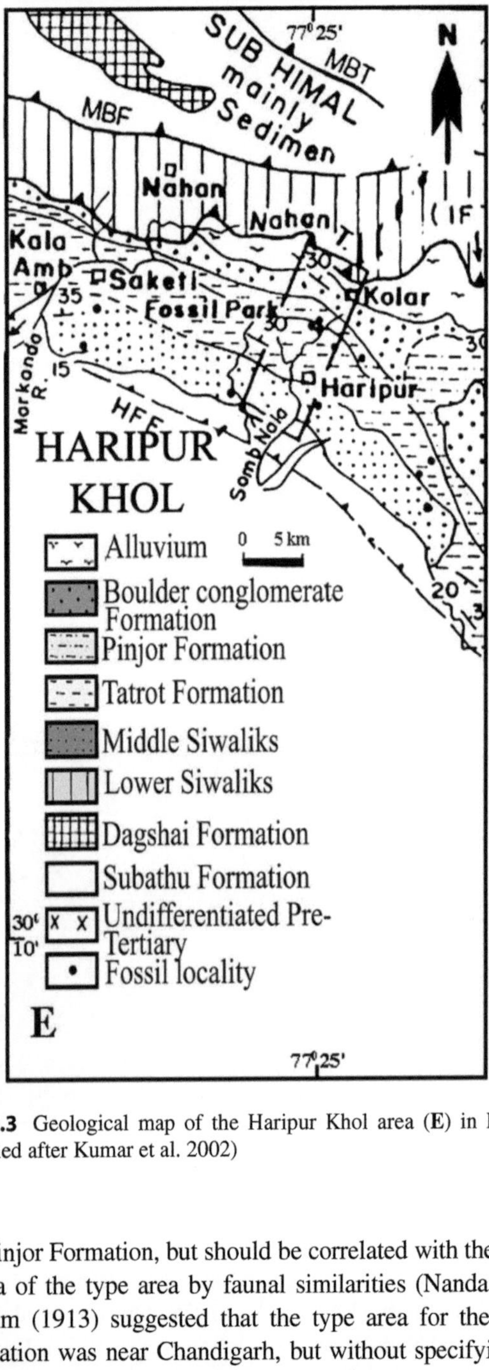

Fig. 9.3 Geological map of the Haripur Khol area (E) in Fig. 9.1, modified after Kumar et al. 2002)

the Pinjor Formation, but should be correlated with the Pinjor Fauna of the type area by faunal similarities (Nanda 2002). Pilgrim (1913) suggested that the type area for the Pinjor Formation was near Chandigarh, but without specifying any type section. The name 'Pinjor' is derived after the township of 'Pinjore' (also spelled as 'Pinjaur'), which lies on alluvium about 15 km northeast of Chandigarh.

In the Chandigarh region, the three well dated sections containing Early Pleistocene mammals are Patiali Rao, Ghaggar and Nadah (Figs. 9.1 and 9.2). The Haripur Khol section lies around 100 km east of Chandigarh (Figs. 9.1 and 9.2). In the Jammu region, the Upper Siwalik succession is divided into three formations – the Parmandal Sandstone, Nagrota Formation and Boulder Conglomerate. The Nagrota Formation has two bentonitic tuff beds, which have been dated as 2.8 ± 0.56 Ma and 2.31 ± 0.54 Ma (Ranga Rao et al. 1988). As mentioned in the introduction, well-dated Pinjor sediments exposed near Jammu, Chandigarh and Haripur have yielded fossil mammals that can be assigned to the Early Pleistocene (Figs. 9.1–9.3). These taxa include *Canis pinjorensis*, *Elephas hysudricus*, *Equus sivalensis*, *Rhinoceros palaeindicus*, *R. sivalensis*, *Coelodonta platyrhinus* (= *Punjabitherium platyrhinus*), *Cervus punjabiensis*, and *Leptobos falconeri*. In Jammu and Chandigarh sections (Figs. 9.4 and 9.5) *Elephas planifrons* and *Sivatherium* occur at younger levels than defined by the various mammalian zones developed in the Upper Siwaliks. Therefore, it appears that these sections need to be restudied and the taxonomic status of these mammals needs to be re-evaluated. On the other hand, if these younger occurrences of *Elephas planifrons* and *Sivatherium* are found to be consistent in other sections as well, then the mammalian zones have to be redefined. Mammalian taxa from Karewa deposits of Kashmir that can be safely placed in the Early Pleistocene are *Elephas hysudricus, Equus sivalensis, Cervus sivalensis* and *Canis*.

Dennell (2004) provided a list of mammals from the Pabbi Hills, Pakistan. Those that can be assigned to the Early Pleistocene include *Hemibos triquetricornis*, *Damalops palaeindicus*, cf. *Cervus*, cf. *Dama*, *Sivatherium giganteum*, Suid (*Sus*, Opdyke et al. 1979), *Equus* cf. *sivalensis*, *Rhinoceros sivalensis*, *Elephas hysudricus*, *Stegodon*, a Hippopotamid (probably *Hexaprotodon sivalensis*), *Crocuta crocuta*, *Pachycrocuta brevirostris*, *Felis*, cf. *Panthera uncia*, *Megantereon cultridens*, cf. *Canis cautleyi*, an herpestid and an ursid. Recently, Dennell et al. (2006, Figs. 10 and 11) have revised the biostratigraphic scheme provided by Opdyke et al. (1979). There appear to be regional differences within the Upper Siwalik deposits of India and Pakistan. *Theropithecus* and *Camelus* are absent in the Siwaliks of Pakistan, whereas they are found in the Indian part. Conversely, *Megantereon, Pachycrocuta, Ursus* and anthracotheres are absent in India, but occur in the Siwaliks of Pakistan (Dennell et al. 2006). It may be noted here that in fact *Megantereon falconeri, Pachycrocuta* and the anthracothere *Merycopotamus dissimilis* have been reported from the Indian part of Pinjor Formation exposed near Chandigarh (Colbert 1935; Barry 1981; Gaur 1987; see Lihoreau et al. 2007, for a review). Fossils of *Ursus* are known from the Indian part of the Kangra Upper Siwaliks (Colbert 1935), Narmada Valley, and Kurnool Caves (see Prasad 1996, for a review). The macaque *Procynocephalus subhimalayanus* (Verma 1969; Szalay and Delson 1979) that has been recorded from the Pinjor deposits of India has not been found in Pakistan. *Theropithecus* is represented by just one specimen and *Procynocephalus* by very few specimens (Szalay and Delson 1979). Therefore, it appears that the regional difference in fauna between Plio-Pleistocene of India and Pakistan is not really a major one and is likely to decrease as more and more areas are explored.

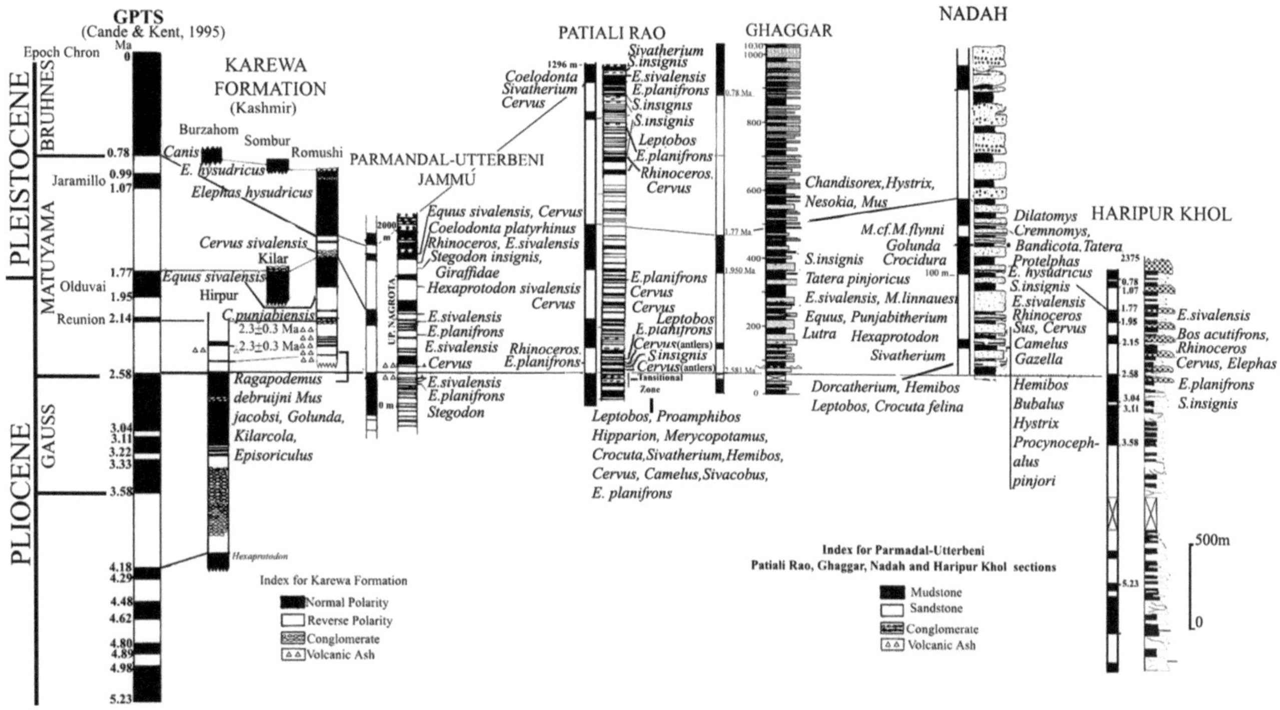

Fig. 9.4 Various mammal bearing Plio-Pleistocene sections in India which are well-dated using either magnetostratigraphy or tephrachronology or both. Data compiled from Azzaroli and Napoleone (1982), Tandon et al. (1984), Ranga Rao et al. (1988, 1995), Kotlia (1990), Agarwal et al. (1993), Sangode et al. (1996) and Kumaravel et al. (2005)

Plio-Pleistocene Paleoecology and Climate in the Indian Subcontinent

Although data based on various proxies indicate that the Indian monsoon system started intensifying by the Late Miocene, 10–8 Ma ago (Quade et al. 1989; Dettman et al. 2001; Harrison et al. 1993; Derry and France-Lanord 1997; Sanyal et al. 2004; Patnaik 2003), the East Asian monsoon was probably earlier at the Oligo-Miocene Boundary (Clift 2006; Ramstein et al. 1997; Sun and Wang 2005; Guo et al. 2001). Patnaik (2003; in press) observed that the distribution and abundance of modern grade murids in the Pliocene and those of at present tend to overlap, suggesting that a monsoonal climate similar to that of today was probably established by the Early Pliocene. By the Late Pliocene a significant diversification of murids might have been caused by the intensification of the monsoon system (Patnaik 2003; in press).

In the Siwaliks, it has been a common practice to reconstruct paleoecology based on vertebrate assemblages ranging in age by over millions of years and distributed across hundreds of kilometers. Rao and Patnaik (2001) recovered a diverse palynoassemblage from the Pinjor sediments at the Nadah section (Fig. 9.4), dated to just before the beginning of Pleistocene (at ~1.8 Ma). If we consider the minimum sedimentation rate of Pinjor sediments (i.e., 10 cm/kyr Sangode and Kumar 2003), this 2 m grey swamp deposit would represent ca. 20,000 years. This assemblage contained a variety of algal and fungal remains, pteridophytic spores, gymnospermous and angiospermous pollen. This assemblage was comparable to 13 modern families, of which three families are tropical to subtropical, three are tropical to temperate, three are temperate, and four are cosmopolitan. The pteridophytic spores indicate moist and shady habitats, and the fungal spores are indicative of warm and humid conditions. The overall vegetation pattern indicates a tropical-subtropical humid climate. The pollen of the temperate flora belonging to the Magnoliaceae (*Magnolia*) was likely to have been transported from the upland montane areas in the north. The lower part of the section was also full of carbonized wood and grass suggesting forest fires during dry periods. The montane elements belonging to the Pinaceae are predominant at the top of the section. The presence of zygospores of the Zygnemataceae is indicative of stagnant shallow and more or less mesotrophic (clear with moderate amount of nutrients) freshwater habitats (Fig. 9.5). This suggestion is supported by the presence of the charophyte genus *Lamprothamnium*, represented by *L. papulosum* and *L. succintium* from this locality, and indicates an oligo-mesohaline environment (Bhatia 1999). This evidence, along with the presence of murid-gerbillid rodents and lizards, indicates that Nadah sediments were deposited under an unpredictable seasonal monsoonal climate (Patnaik 2003). A hypothetical

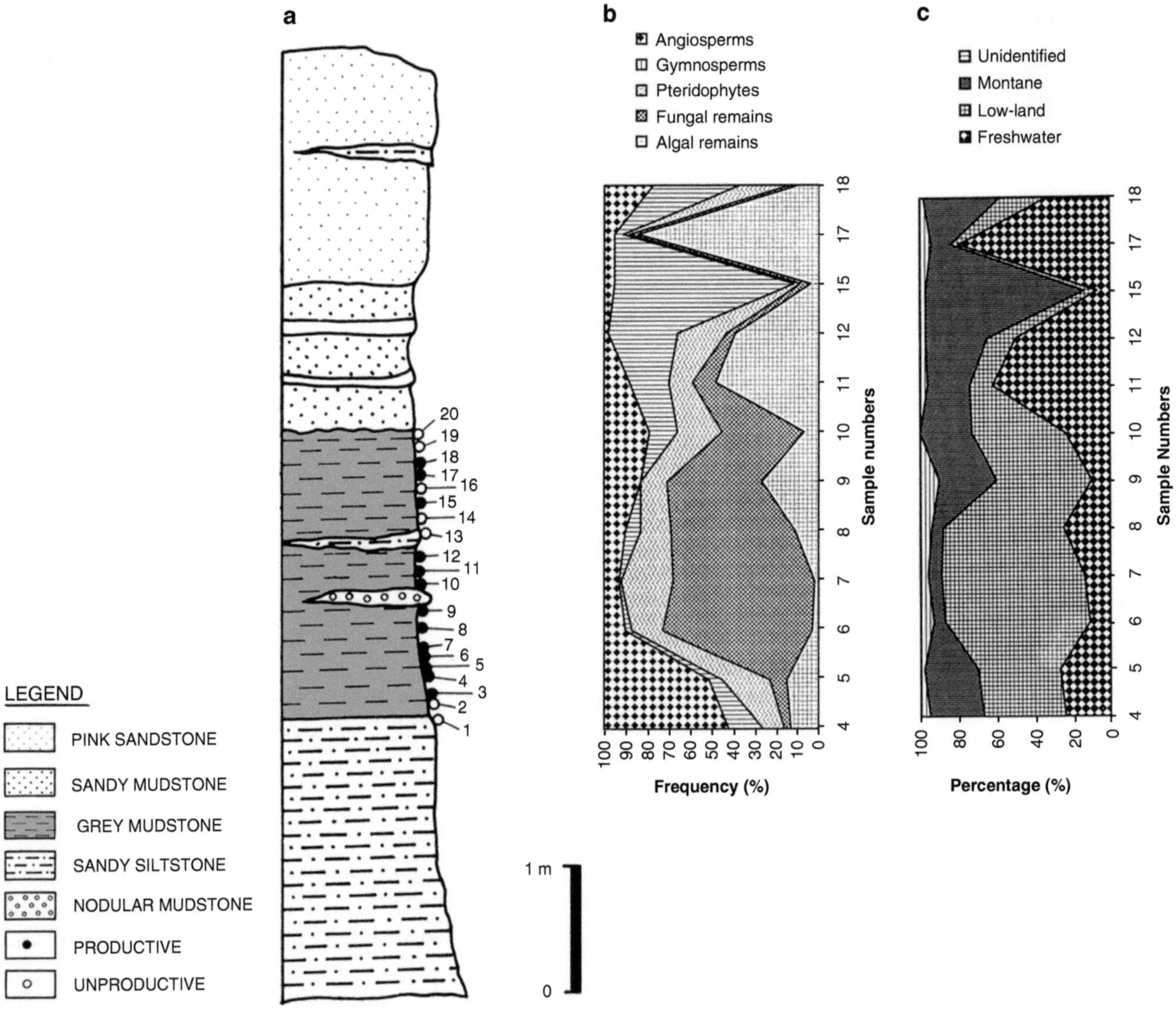

Fig. 9.5 (**a**) Lithocolumn at the Nadah locality (Figs. 9.1 and 9.2d) with positions of samples collected for pollen analysis, (**b**) representation of different plant groups, and (**c**) percentage of various palynotaxa belonging to various ecological groups (Modified after Rao and Patnaik 2001)

reconstruction of the Early Pleistocene based on the Nadah microfossil assemblage has been presented (Figs. 9.6 and 9.7). Phadatre et al. (1994) studied Siwalik pollens and spores ranging from 4–1 Ma in the Haripur Khol section and concluded that 4–3.5 Ma had dry grasslands, 3.5–2.7 Ma had muddy and marshy conditions, between 2.7 and 2.5 Ma there were well developed ponding conditions, and 2.5–1 Ma was rather cool and dry.

In general, the large mammals of the Pinjor Formation suggest the presence of grasslands and a well-watered landscape (Gaur 1987; Dennell 2003). This scenario holds true for the early part of the Pinjor Formation (the latest Pliocene). As we move into the Early Pleistocene browsers such as *Sivatherium* and hippopotamids became locally extinct in the Pabbi Hills, and the overall fauna indicates the presence of dry-season grasslands (Dennell et al. 2006). Paleosols are abundant in the Siwalik deposits. Based on color, micromorphology, maturity, and magnetic properties of paleosols, the Late Pliocene has been considered as warm and humid whereas Early Pleistocene was drier, less oxygenated (reducing) and cooler (Thomas et al. 2002; Sangode et al. 2001; Sangode and Bloemendal 2004). A similar view is presented by Sanyal et al. (2004), who, based on variation in oxidation, hydroxylation, and humification, inferred the presence of a warm-humid climate during the Early Pliocene, an intermediate phase during early Middle Pliocene, and warm oxidative phase during the Middle to Late Pliocene. Thomas et al. (2005: 341) studied stable isotopes from calcretes embedded in paleosols and estimated that the formation temperature and precipitation for Tatrots was ca. 25°C and

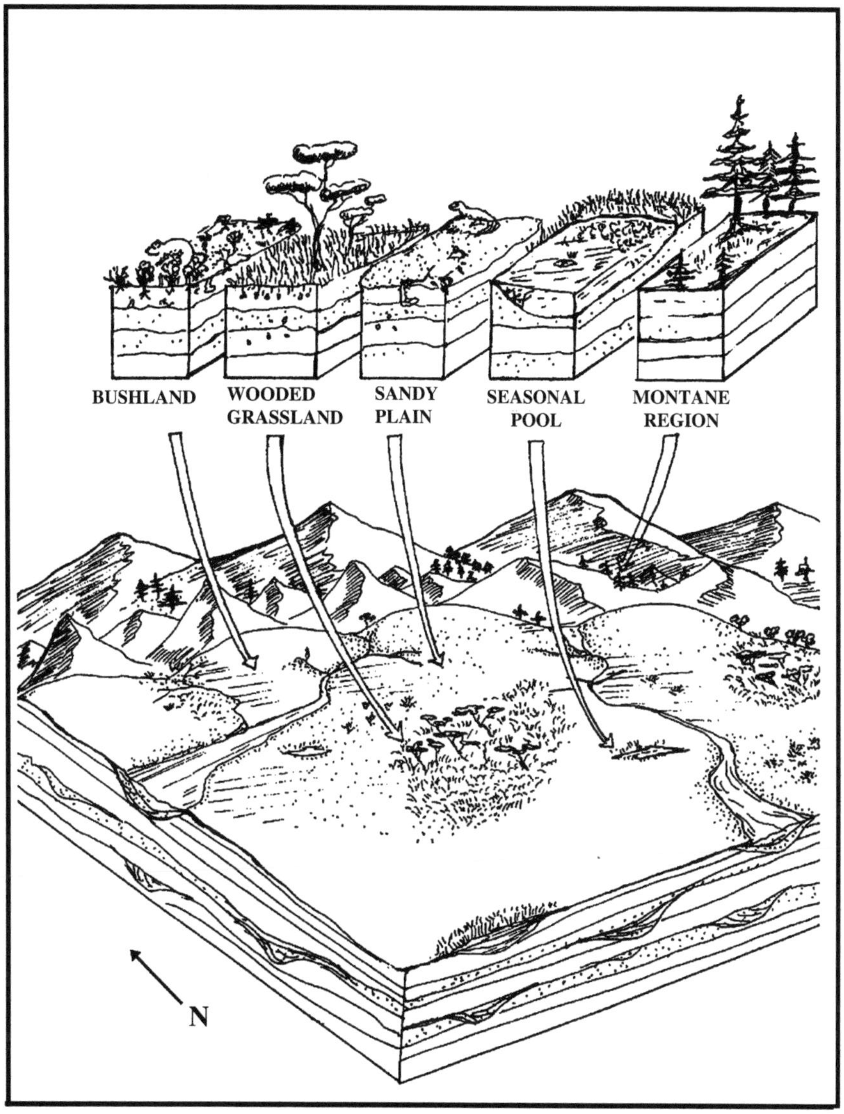

Fig. 9.6 Microfossil based paleoecological reconstruction of the Early Pleistocene Nadah locality in the Pinjor Formation (Patnaik 2003)

1,100 mm and those for Pinjors were 21°C and 1,000 mm, respectively.

Dennell (2003) and Dennell and Roebroeks (2005) reviewed Early Pleistocene deposits and their mammalian fauna with special reference to paleoecological conditions. They pointed out that grasslands covered a large area from West Africa to China in the Plio-Pleistocene and conditions were conducive for *Homo erectus* and its associated mammalian fauna to disperse. Occurrence of murids in Ethiopia, Kenya, Tanzania and Indo-Pakistan during the Late Pliocene also indicates that there were no physical barriers between the Indian sub-continent and East Africa to prevent faunal migrations to and from these regions (Patnaik 2000, 2001).

Connections with Other Parts of the World

It appears from the above discussion that the conditions in South Asia during the Plio-Pleistocene were indeed adequate for early hominins to either colonize or simply use as a route to Southeast Asia. In this context, it is necessary to analyze how similar the Pinjor mammalian taxa were to those of the Early Pleistocene in Africa, Europe, Central Asia, and East and Southeast Asia (Table 9.1). East African genera that are also known from the Pinjor Formation are *Golunda*, *Mus*, *Crocuta*, *Canis*, *Panthera*, *Mellivora*, *Megantereon*, *Pachycrocuta*, *Elephas*, *Equus*, *Hexaprotodon*, *Camelus*, *Sivatherium*, *Gazella*, *Oryx* and *Theropithecus* (Tchernov 1992; Patnaik 2000; Leakey and Werdelin 2010). At the species

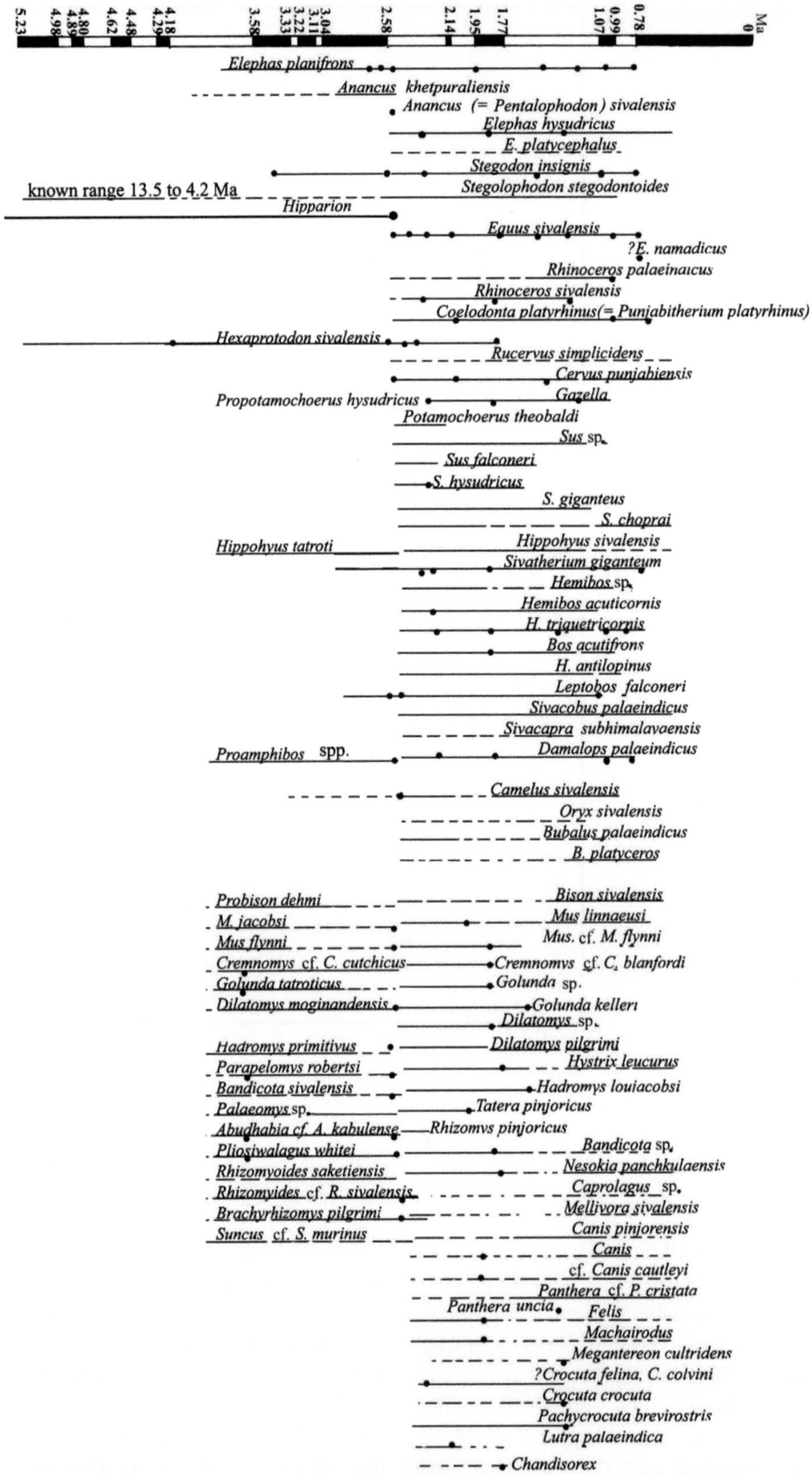

Fig. 9.7 Preliminary temporal ranges of various Plio-Pleistocene Siwalik mammals of Indo-Pakistan (Data from Nanda 2002; Dennell 2004; Dennell et al. 2006). *Filled dots* represent known sites, *complete lines* represent known ranges and *dotted lines* represent uncertain ranges

Table 9.1 Comparison of large mammal fauna from Early Pleistocene Hominin sites from Eurasia and the Pinjor formation of Indo-Pakistan. Only well dated mammals from Pinjor have been included

Dmanisi, Georgia	'Ubeidiya, Israel	Sangiran Dome, Java	Indo-Pakistan
Units A and B	Fi Member	Sangiran Formation	Pinjor Formation
(Gabunia et al. 2000)	(Tchernov 1987)	Upper Black Clays	(Nanda 2002; Dennell 2004)
Ursus etruscus	*Ursus etruscus*	(Aziz 2001)	Ursidae
Canis etruscus	*Canis cf. arnensis*	*Stegodon elephantoides*	*Canis cautleyi*
Martes sp.	*Canis* sp.	*Bubalus palaeokerabau*	*Canis pinjorensis*
Megantereon megantereon	*Vulpes* sp.	*Bibos palaeosondaicus*	*Megantereon cultridens*
Homotherium crenatidens	*Lutra* sp.	*Homo erectus*	*Panthera uncia*
Panthera gombaszoegensis	*Vormela cf. peregusna*		*Pachycrocuta brevirostris*
Pachycrocuta perrieri	*Megantereon cf. cutridens*	From equivalent aged faunas elsewhere on Java	*Felis* sp.
		(de Vos 1985; Sondaar et al. 1996)	
Archidiskodon meriodionalis	*Panthera gombaszoegensis*	*Panthera* sp.	*Crocuta crocuta*
Dicerorhinus etruscus etruscus	*Lynx* sp.	*Stegodon trigonocephalus*	Herpestidae
Equus stenonsis	*Felis* sp.	*Hexaprotodon sivalensis*	*Stegodon* sp.
Gazella borbonica	*Crocuta crocuta*	*Sus stremmi*	*Elephas hysudricus*
Soergelia sp.	*Herpestes* sp.	Bovidae (several)	*Hexaprotodon sivalensis*
Dmanisibos georgicus	*Mammuthus meridionalis*	Boselaphini	Hippopotamidae
Eucladoceros aff. senezensis	*Dicerorhinus e. etruscus*	Cervidae (several)	*Sus* sp.
Cervidae cf. arvernoceros	*Equus cf. tabeti*		*Equus sivalensis*
Cervus perrieri	*Equus cf. caballus*		*Rhinoceros sivalensis*
Dama nesti	*Kolpochoerus olduvaiensis*		*Rhinoceros palaeindicus*
Paleotragus sp.	*Sus strozzii*		*Sivatherium giganteum*
Homo ex. gr. erectus	*Hippopotamus behemoth*		*Damalops palaeindicus*
Hippopotamus gorgops	*Camelus* sp.		*Hemibos triqueticonis*
	Giraffidae		*Cervus sivalensis*
	Praemegaceros verticornis		*Cervus punjabiensis*
	Cervidae		*Gazella* sp.
	Bos sp.		*Leptobos falconeri*
	Gazella cf. gazella		*Coelodonta platyrhinus*
	Gasellospira torticornis		
	Macaca sylvanus		
	Homo sp.		

level, only *Theropithecus oswaldi* and *Pachycrocuta brevirostris* are found in both Pinjor and East Africa. The North African Early Pleistocene hominin site of Ain Hanech (Algeria) has yielded *Equus*, *Gazelle* and *Hippopotamus* (see Dennell 2003). The faunal composition of the hominin sites of Ubeidiya, Israel, and Dmanisi, Georgia are given in Table 1 (after Antón and Swisher 2004). Fossil remains from three Early Pleistocene localities in the An Nafud desert, Saudi Arabia, include those of *Crocuta crocuta*, *Pelorovis* cf. *oldowayensis*, *Oryx*, a large felid *Panthera* cf. *gombaszoegensis*, fox (*Vulpes* cf. *vulpes*), an elephant (probably *Elephas recki*), *Equus*, the hippopotamid *Hexaprotodon*, as well as various types of alcelaphines, bovids and camelids (Thomas et al. 1998; Dennell 2003). Central Asian latest Pliocene (1.77–1.95 Ma) locality, Kuruksay has yielded *Canis etruscus*, *Ursus* cf. *etruscus*, *Pliocrocuta*, *Lynx*, *Acinonyx*, *Megantereon*, *Homotherium*, *Archidiskodon*, *Dicerorhinus*, *Equus stenonis*, *Paracamelus*, *Sivatherium*, and *Damalops palaeindicus* (Sotnikova et al. 1997). *Sivatherium* and *Damalops palaeindicus* are also known from Pinjors (Dennell 2010). The Indian genus *Hemibos* has recently been recorded from the Early-Middle Pleistocene transition in Italy (Martinez-Navarro and Palombo 2004). Early Pleistocene Upper Irrawaddy fauna of Myanmar has several taxa in common with Pinjors (Takai et al. 2006). They include *Potamochoerus*, *Merycopotamus dissimilis*, *Hexaprotodon palaeindicus*, *Cervus* sp., *Hemibos triquetricornis*, *Rhinoceras sivalensis*, *Stegodon insignis* and *Elephas hysudricus*. The faunal remains from the Early Pleistocene at Xiaochangliang, China include *Cervus* and *Gazella* (Zhu et al. 2001). Other forms from the Pleistocene of China that are also known from Pinjor include, *Megantereon*, *Panthera*, *Pachycrocuta brevirostris*, *Crocuta*, *Canis*, *Equus*, *Rhinoceras*, *Stegodon*, *Coelodonta*, *Gazella*, *Leptobos*, *Hemibos*, *Potamochoerus* and *Sus* (Colbert 1940; Tang 1980; Han 1987; He 1997; Zhu et al. 2003; Qiu et al. 2004). *Procynocephalus wimani* from China (Schlosser 1924) is very similar to Pinjor *Procynocephalus subhimalayanus*

(Verma 1969), but the former probably came from Pliocene deposits. Hominin occupation in southern Europe during the Early Pleistocene is still debated, but large carnivores such as the large sabertooth felid *Megantereon* and the giant hyena *Pachycrocuta* (both present in Pinjor deposits of the Pabbi Hills, Pakistan and India) were present, though not common (Turner 1992; Rook et al. 2004). Another taxon with a widespread distribution is the cercopithecid *Theropithecus oswaldi*, which is represented in Plio-Pleistocene deposits of South Africa, East Africa, North Africa, Israel, Spain, Italy and India (Delson 1993; Delson et al. 1993; Gibert et al. 1995; Belmaker 2002; Rook et al. 2004; Gupta 1977; Gupta and Sahni 1981). The age of the Indian fossil *Theropithecus* is unclear as the only specimen (found at Mirzapur) does not come from a well-dated section (Delson 1993), but is likely to be ca. 1 Ma.

Tectonic Activity, Climate Change and Migration of the Siwalik Mammals at the Plio-Pleistocene Boundary

The Plio-Pleistocene boundary is placed at the top of the Olduvai subchron, which is now dated at 1.77 Ma by Cande and Kent (1995). The contact of the Pinjor and Boulder Conglomerate formations in the Jammu – Chandigarh – Haripur regions has been found to be time transgressive, and ranges from 1.79 to 0.6 Ma (Fig. 9.8). A similar situation prevails in Pakistan (see Opdyke et al. 1979: 31). This time transgressive nature of the contact between the Boulder Conglomerate and the succession yielding the Pinjor Fauna (called the Pinjor Formation in Chandigarh or the upper part of the Nagrota Formation in Jammu) is very significant, and sheds light on the various aspects concerning the Plio-Pleistocene boundary, sedimentation pattern, the extinction and migration of the Pinjor Fauna, and the intensification of the last phase of the Himalayan orogeny (Nanda 2002; Kumar et al. 2002).

Kumar et al. (2002) argue that major tectonic activity occurred at ca. 1.77 Ma along the Intra-Foreland thrust at the Haripur section. This caused the partial to complete replacement of transverse trunk drainage by a piedmont drainage system that brought in a huge amount of conglomerates. In the Ghaggar river section, considerable accumulation of boulder conglomerates started at 1.79 Ma, indicating a predominance of braided channel and alluvial fan environments (Kumaravel et al. 2005). It is interesting to note that exactly at this time conglomerates also occur in the Karewa section of Kashmir (Fig. 9.4), though strong influx of conglomerates in these deposits is known to have occurred around 3.5 Ma (Burbank and Johnson 1982). Today, the Karewa lacustrine and fluvio-deltaic sediments, representing the last 4 million years, are situated at 1,700–1,800 m above sea level, between the Zanskar (4,500–6,100 m) and the Pir Panjal (3,600–4,600 m) ranges. However, the presence of murid rodents similar to those of the Siwaliks suggests that the sediments at Khaigam might have been deposited at a low altitude. The lens-shaped fossil plant remains in the Lower Karewas (Gupta 1992) also point towards a low altitude status of the basin. Large mammals from Karewas such as *Hexaprotodon*, *Rhinoceros*, *Sivatherium* and *Elephas* also indicate low altitudinal and warm-humid conditions. Sahni (1936), based on paleobotanical, archeological and structural evidence, suggested that Paleolithic groups were able to cross the Himalayas (there is no hard evidence to support this idea). The rise of the Pir Panjal terminating free migration between

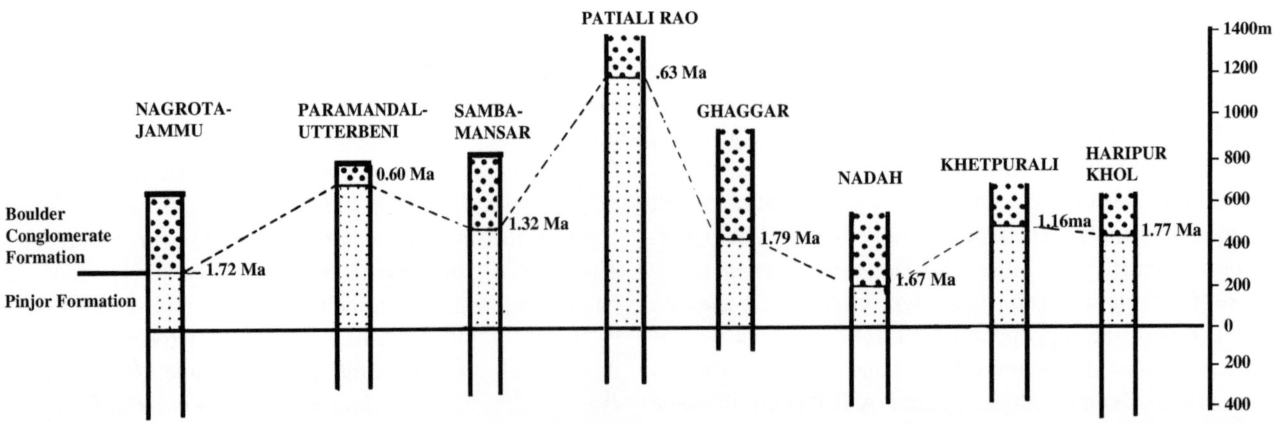

Fig. 9.8 Position of the base of the Boulder Conglomerate Formation or the upper limit of the successions, which has yielded Pinjor Fauna in various sections of the Jammu, Chandigarh and Haripur regions. Magnetostratigraphic dates for the Jammu region are based on the work of Ranga Rao (1986), Ranga Rao et al. (1988) and Agarwal et al. (1993); those of Patiali Rao, Ghaggar, Nadah, Khetpurali and Haripur Khol sections are after Ranga Rao et al. (1995), Kumaravel et al. (2005), Azzaroli and Napoleone (1982); Tandon et al. (1984) and Sangode et al. (1996) respectively. Various localities are marked in Figs. 9.1–9.3

the North Indian Plate and Tibet is a fairly recent phenomenon and it has risen by 2,700–3,000 m since the Middle Pleistocene (Valdiya 1991). Sakai et al. (2006) are of the opinion that Siwaliks might have been uplifted in a major way along the Main Boundary thrust around 1 Ma ago.

In both the Nagrota – Jammu and Nadah sections, the Plio-Pleistocene boundary almost coincides with the base of the Boulder Conglomerate Formation. Near the Plio-Pleistocene boundary, a dramatic change is indicated in the sedimentation pattern from alternating mudstone – sandstone (e.g., Pinjor Formation or Nagrota Formation) to the conglomeratic intervals (e.g., the Boulder Conglomerate). The presence of thick succession of conglomerates (clasts mainly of boulder or cobble size) indicates an intensification of the Himalayan orogeny within the provenance area, and its signature was evident in the foothills.

Tectonic activity led to an increase in gradient and appearance of distinctive Tertiary clast-bearing conglomerates deposited in a distal alluvial fan setting (Kumar et al. 2002; Opdyke et al. 1979). Today, such debris flow is caused by steep slope, abundant supply of suitable regolith, high pore water pressure and sparse vegetation (Innes 1983; Costa 1984; van Stein et al. 1988; Kumar et al. 2002). Due to slope instability, trees are not well supported and only grasses can survive (Hoorn et al. 2000). The inception of glaciation in the Northern Hemisphere during the Late Pliocene (Shackleton et al. 1984) caused an intensification of the monsoon climate system (Gupta and Melice 2003; Gupta and Thomas 2004). The monsoon climate, which is characterized by long dry winters and short wet summers, supports grasslands.

The work carried out by Nanda (1997b, 2002) reveals that in only two sections, those at Parmandal-Utterbeni and Patiali Rao, were vertebrate fossils found in strata younger than the Olduvai subchron. In these two sections, the Pinjor Fauna continued uninterrupted without any change (Figs. 9.4 and 9.7). However, in the Jammu – Nagrota, Nadah, and Haripur sections, the fauna does not continue upward as boulder conglomerates start appearing in these sections. Fairly thick alternating clast and mudstone deposits occur in the Lower Boulder Conglomerate Formation. Though one would not expect faunal preservation in such large sized clastic deposits of alluvial fan origin, there is hardly any evidence of vertebrate remains in the fine-grained mudstone facies of this formation either. The fact remains that many elements of the earlier Siwalik faunas are not known from this region after 0.6 Ma. However they are still present elsewhere in deposits in other parts of the region such as the Indo-Gangetic plain and Peninsular India, because some of these taxa, including *Crocuta* cf. *C. sivalensis, Stegodon insignis, Elephas hysudricus, Potamochoerus theobaldi* and *Bubalus palaeindicus*, have been found in the Middle and Late Pleistocene sediments of Narmada and Godavari River Valleys.

Discussion and Conclusions

The Pinjor fauna has been found to be very similar to the Early Pleistocene Upper Irrawady fauna at the specific level, indicating an extended paleobiogeographic province. There seems to be some faunal interchange between South Asia and China in the Early Pleistocene as indicated by the common presence of *Pachycrocuta brevirostris* and *Procynocephalus* in these regions. Common occurrence of *Damalops palaeindicus* and *Sivatherium* in the Pinjors and Central Asia may indicate dispersal of these two taxa from South to Central Asia at the Plio-Pleistocene boundary. A comparison of the Early Pleistocene Pinjor mammalian fauna with those of hominin sites in Eurasia shows the existence of several common genera but few common species (Table 9.1), probably suggesting similar ecological conditions but not much exchange of fauna. African bovids such as *Damalops*, *Oryx*, *Hippotragus* and *Kobus* are represented in the Siwaliks around 3–2.5 Ma (Turner and O'Regan 2007). However, exchange between East Africa and South Asia slowed down by the Early Pleistocene (Turner and O'Regan 2007). A particularly interesting observation is the association of the large sabertooth felid *Megantereon* and *Homo erectus*. Rook et al. (2004; see also Martínez - Navarro 2004) are of the opinion that *Homo*, *Theropithecus oswaldi*, *Megantereon whitei*, and *Hippopotamus* were among the first mammals to disperse out of Africa in the Early Pleistocene. With *Megantereon*, Hippopotamids and possible paleolithic evidence already recorded from the Pinjor sediments of Pakistan (Dennell 2003, 2004), and *Theropithecus* reported from Boulder Conglomerate deposits of India (Gupta 1977; Gupta and Sahni 1981), it would be worthwhile to carry out further field work in the region to search for *Homo erectus*. Having said this, it seems surprising that after 150 years of fieldwork in the Siwaliks we still don't have hominin fossils. The meager evidence of early hominin presence in South Asia could be due to the presence of large competing predators and a dearth of raw materials for making tools in the Late Pliocene and Early Pleistocene (Dennell 2007). However, it has been found that the Indian part of the Upper Siwalik deposits do contain large amounts of quartzite (a preferred raw material used by early hominins in making Soanian tools, although probably at a later stage), with pebble/cobble bearing boulder conglomerate horizons at ca. 5.5 Ma in the Ranital-Kangra section (Sangode et al. 2003; Sanyal et al. 2004) and at ca. 3.36 Ma in Haripur Khol section (Kumar et al. 2002). Further, it is interesting to note that although post-Siwalik deposits are rich in quartzite clasts, there is hardly any occurrence of quartzite clasts in the Early to Middle Pleistocene Boulder Conglomerate Formation, which primarily contain clasts of sandstones derived from Tertiary, Subathu, Dagshai and Kasauli Formations.

Tectonic activity intensified along the Himalayan foothills by the beginning of the Pleistocene. This was caused by movement along the Main Boundary Thrust and Intra Foreland Thrust. This activity greatly changed the riverine system by increasing fan deposits in the piedmont regions and decreasing the flood plain deposits. Inter-fluve flood plains are the areas where flora and fauna thrived and fan deposits restricted the preferred habitat of large mammals. Increases in gradient meant that mainly grasses could survive on the steep slopes, which in turn would have further adversely affected the browsers. The conditions may have deteriorated further with the global trend towards cool and dry conditions during the Pleistocene, evidence of which has also been recorded in the Siwalik paleosols. For over 18 million years the diverse Siwalik fauna underwent a series of extinctions, migrations and introductions as the "archaic" elements went extinct and "modern" elements appeared such that an essentially modern fauna was in place by the Middle Pleistocene. Only 24 of 49 taxa belonging to the Plio-Pleistocene Pinjor Formation (Nanda 2002) survived into the Early Pleistocene, and only around six survived into the Middle Pleistocene of Peninsular India and the Indo-Gangetic Plain.

Faunal turnover around 2.5 Ma is well documented in the East African terrestrial record (Behrensmeyer et al. 1997). Although the Plio-Pleistocene boundary is demarcated mainly in marine deposits, it is not very clear in terrestrial sequences from a paleoclimatic and faunal perspective. It appears that the tectonic and faunal events associated with well-dated Pinjor sediments of the Himalayan foothills offer potential for studies related to Plio-Pleistocene boundary events in a terrestrial record. It is near the Plio-Pleistocene boundary that the process of decline and possible migration of the Pinjor Fauna started and it is also marked in some drainage basins by the change in sedimentation pattern and the intensification of Himalayan orogeny. Such studies may be particularly rewarding if microfossils are also taken into account.

Acknowledgments We would like to express our sincere thanks to Profs. John Fleagle, Richard Leakey, Frederick Grine, Lawrence Martin and John Shea for the opportunity to participate and contribute to the workshop on "Out of Africa I: Who, Where and When?" Rajeev Patnaik thanks DST, New Delhi (PURSE;SR/S4/ES-171/2005) for the financial support for ongoing research in the Siwaliks. We would like to express our thanks to Dr. R. Potts, R. Teague and anonymous reviewers for their critical evaluation of the manuscript and constructive comments. One of us (ACN) is also thankful to Dr. B. R. Arora, Director, Wadia Institute of Himalayan Geology, Dehra Dun for providing facilities.

References

Agarwal, R. P., Nanda, A. C., Prasad, D. N., & Dey, B. K. (1993). Geology and biostratigraphy of the Upper Siwalik of Samba area, Jammu foothills. *Journal of Himalayan Geology, 4*, 227–236.

Agrawal, D. P., Kotlia, B. S., Kusumgar, S., & Gupta, S. K. (1987). Quaternary palaeoenvironmental changes in northwest India. In A. Sahni & R. Gaur (Eds.), *Perspectives in the human evolution* (pp. 223–260). Delhi: Renaissance Publishing House.

Antón, S. C., & Swisher, C. C. (2004). Early dispersals of *Homo* from Africa. *Annual Review of Anthropology, 33*, 271–296.

Aziz, F. (2001). New insight on the Pleistocene fauna of Sangiran and other hominid sites in Java. In T. Simanjuntak, B. Prasetyo, & R. Handini (Eds.), *Sangiran: Man, culture and environment in Pleistocene times* (pp. 260–271). Jakarta: Yayosan Obor Indonesia.

Azzaroli, A., & Napoleone, G. (1982). Magnetostratigraphic investigation of the Upper Sivaliks near Pinjor, India. *Rivista Italiana di Paleontologia, 87*, 739–762.

Badam, G. L. (1968). Note on the occurrence of fossil vertebrates in the Karewas of Kashmir. *Research Bulletin Punjab University, 19*, 453–455.

Barry, J. C. (1981). *The Hyaenids Pachycrocuta and Crocuta in the Villafrancian of East Africa and Southern Asia*. Proceedings of field conference on Neogene/Quaternary Boundary, India, pp. 9–10.

Barry, J. C., Lindsay, E. H., & Jacobs, L. L. (1982). A biostratigraphic zonation of the Middle and Upper Siwaliks of the Potwar Plateau of northern Pakistan. *Palaeogeography, Palaeoclimatology, Palaeoecology, 37*, 95–130.

Basu, P. K. (2004). Siwalik mammals of the Jammu Sub-Himalaya, India: An appraisal of their diversity and habitats. *Quaternary International, 117*, 105–118.

Behrensmeyer, A. K., Todd, N. E., Potts, R., & McBrinn, G. E. (1997). Late Pliocene faunal turnover in the Turkana Basin, Kenya and Ethiopia. *Science, 278*, 1589–1594.

Belmaker, M. (2002). First evidence of the presence of *Theropithecus* sp. in the Southern Levant. *Israel Journal of Zoology, 48*, 165.

Bhatia, S. B. (1999). Revision of the Charophyte Flora of the Siwalik Group (Neogene-Quaternary) of the Lesser Himalaya, India. *Australian Journal of Botany, 47*, 49–74.

Burbank, D. W., & Johnson, G. D. (1982). Intermontane basin development in the past 4 Myr in the northwest Himalaya. *Nature, 298*, 432–436.

Cande, S. C., & Kent, D. V. (1995). Revised calibration of the geomagnetic polarity timescale for the Late Cretaceous and Cenozoic. *Journal of Geophysical Research, 100*(B4), 6093–6095.

Clift, P. D. (2006). Controls on the erosion of Cenozoic Asia and the flux of clastic sediment to the Ocean. *Earth and Planetary Science Letters, 241*, 571–580.

Colbert, E. H. (1935). Siwalik mammals in American Museum of Natural History. *Transactions of the American Philosophical Society, New Series, 26*, 1–401.

Colbert, E. H. (1940). Pleistocene mammals from the Ma Kai valley of Northern Yunnan, China. *American Museum Novitates, 1099*, 1–10.

Costa, J. E. (1984). Physical geomorphology of debris flow. In J. E. Costa & P. J. Fleisher (Eds.), *Developments and applications of geomorphology* (pp. 268–317). Berlin: Springer.

De Terra, H., & Patterson, T. T. (1939). *Studies on Ice Age in India and associated human cultures* (vol. 493, pp. 1–453). Washington, DC: Carnegie Institution.

De Vos, J. (1985). Faunal stratigraphy and correlationof the Indonesian hominid sites. In E. Delson (Ed.), *Ancestors: The hard evidence* (pp. 215–220). New York: A.R. Liss.

Delson, E. (1993). *Theropithecus* fossils from Africa and India and the taxonomy of the genus. In N. G. Jablonski (Ed.), *Theropithecus: The rise and fall of a primate genus* (pp. 157–189). Cambridge: Cambridge University Press.

Delson, E., Eck, G. G., Leakey, M. G., & Jablonski, N. G. (1993). A partial catalogue of fossil remains of *Theropithecus*. In N. G. Jablonski (Ed.), *Theropithecus: The rise and fall of a primate genus* (pp. 499–525). Cambridge: Cambridge University Press.

Dennell, R. W. (2003). Dispersal and colonisation, long and short chronologies: how continuous is the Early Pleistocene record for hominids outside East Africa? *Journal of Human Evolution, 45*, 421–440.

Dennell, R. W. (2004). Early hominin landscape in the Northern Pakistan. Investigations in the Pabbi Hills. *BAR International Series, 1265*, 1–454.

Dennell, R. W. (2007). "Resource-rich, stone-poor": Early hominin land use in large river systems of northern India and Pakistan. In M. D. Petraglia & B. Allchin (Eds.), *The evolution and history of human populations in South Asia* (pp. 41–68). The Netherlands: Springer.

Dennell, R. W. (2010). Out of Africa I: Current problems and future prospects. In: J. G. Fleagle, J. J. Shea, F. E. Grine, A. L. Baden, & R. E. Leakey (Eds.), *Out of Africa I: The first hominin colonization of Eurasia* (pp. 247–273). Dordrecht, The Netherlands: Springer.

Dennell, R. W., & Roebroeks, W. (2005). Out of Africa: An Asian perspective on early human dispersal from Africa. *Nature, 438*, 1099–1104.

Dennell, R. W., Coard, R., & Turner, A. (2006). The biostratigraphy and magnetic polarity zonation of the Pabbi Hills, northern Pakistan: An Upper Siwalik (Pinjor Stage) Upper Pliocene-Lower Pleistocene fluvial sequence. *Palaeoecology, Palaeoclimatology and Palaeogeography, 234*, 168–185.

Derry, L. A., & France-Lanord, C. (1997). Himalayan weathering and erosion fluxes: Climate and tectonic controls. In W. F. Ruddiman (Ed.), *Tectonic uplift and climate change* (pp. 289–312). New York/London: Plenum.

Dettman, D. L., Kohn, M., Quade, J., Ryerson, F. J., Ojha, T. P., & Hammidullah, S. (2001). Seasonal stable isotope evidence for a strong Asian monsoon throughout the past 10.7 m.y. *Geology, 29*, 31–34.

Falconer, H., & Cautley, P. T. (1837). On some fossil remains of the order Quadrumana from the Siwalik Hills. *Journal Asiatic Society of Bengal, 6*, 354–360.

Gabunia, L., Vekua, A., Lordkipanidze, D., Swisher, C. C., Ferring, R., Justus, A., et al. (2000). Earliest Pleistocene hominid cranial remains from Dmanisi, Republic of Georgia: Taxonomy, geological setting, and age. *Science, 288*, 1019–1025.

Gaur, R. (1987). *Environment and ecology of early man in northwest India* (p. 252). Delhi: B. R. Publishing Corporation.

Gibert, J., Ribot, F., Gibert, L., Leakey, M., Arribas, A., & Martinez-Navarro, B. (1995). Presence of the cercopithecid genus *Theropithecus* in Cueva Victoria (Murcia, Spain). *Journal of Human Evolution, 28*, 487–493.

Guo, Z. T., Ruddiman, W. F., Hao, Q. Z., Wu, H. B., Qiao, Y. S., Zhu, R. X., et al. (2001). Onset of Asian desertification by 22 Myr ago inferred from loess deposits in China. *Nature, 416*, 159–163.

Gupta, H. P. (1992). Changing pattern of vegetation in the intermontane basin of Kashmir since 4 Ma: A palynological approach. *Palaeobotanist, 40*, 354–375.

Gupta, A. K., & Melice, J.-L. (2003). Orbital forcing of the Plio–Pleistocene Indian monsoons: Benthic foraminiferal proxies from ODP Site 758. *Current Science, 85*, 179–184.

Gupta, A. K., & Thomas, E. (2004). Initiation of Northern Hemisphere glaciation and strengthening of the northeast Indian monsoon: Ocean Drilling Program Site 758, eastern equatorial Indian Ocean. *Geology* (January 2003), *31*(1), 47–50.

Gupta, V. J. (1977). Fossil cercopithecoid from the Lower boulder Conglomerate Formation (Middle Pleistocene) of Mirzapur, Tehsil Kharar, District Ropar, Punjab. Recent Researches in *Geology, 3*, 450–452.

Gupta, V. J. & Sahni, A. (1981). Theropithecus delsoni, a new cercopithecine species from the upper Siwaliks of India. Bull Indian geol. Assoc. 14 (1981), 69–71.

Han, D. (1987). Artiodactyla fossils from Liucheng Gigantopithecus Cave in Guangxi. *Memoirs of Institute of Vertebrate Palaeontology and Palaeoanthropology, 18*, 135–208.

Harrison, T. M., Copeland, P., Hall, S. A., Quade, J., Burner, S., Ojha, T. P., et al. (1993). Isotopic preservation of Himalyan/Tibetan uplift, denudation, and climatic histories of two molasse deposits. *Journal of Geology, 101*, 157–175.

He, Z. (1997). *Yuanmou hominoid fauna*. Kunming, China: Yunana Science and Technology Press.

Hoorn, C., Ojha, T., & Quade, J. (2000). Palynological evidence for vegetation development and climatic change in the sub-Himalayan zone (Neogene, Central Nepal). *Palaeogeography, Palaeoclimatology, Palaeoecology, 163*, 133–161.

Hussain, S. T., Van Den Bergh, G. D., Steensma, K. J., De Visser, J. A., De Vos, J., Arif, M., et al. (1992). Biostratigraphy of the Plio-Pleistocene continental sediments (Upper Siwaliks) of the Mangla-Samwal Anticline, Azad Kashmir, Pakistan. *Proceedings Koninklijke Nederlandse Akademie van Wetenschappenn, Amsterdam, 95*, 65–80.

Innes, J. L. (1983). Debris flow. *Progress in Physical Geography, 21*, 663–677.

Johnson, N. M., Stix, J., Tauxe, L., Cerveny, P. F., & Tahirkheli, R. A. K. (1985). Paleomagnetic chronology, fluvial processes, and tectonic implications of the Siwalik deposits near Chinji village, Pakistan. *Journal of Geology, 93*, 27–40.

Keller, H. M., Tahirkheli, R. A. K., Mirza, M. A., & Opdyke, N. D. (1978). Magnetic polarity stratigraphy of the Upper Siwalik deposits, Pabbi Hills, Pakistan. *Earth and Planetary Science Letters, 36*, 187–201.

Kotlia, B. S. (1990). Large mammals from the Plio-Pleistocene of Kashmir Intermontane Basin, India, with reference to their status in Magnetic Polarity time Scale. *Eiszeitalter und Gegenwart, 40*, 38–52.

Kotlia, B. S. (1992). Pliocene murids (Rodentia, Mammalia) from Kashmir Basin, northwestern India. *Neus Jahrbuch fur Geologie und Palaeontologie, 184*, 339–357.

Kotlia, B. S. (1998). *Rhagapodemus debruijni* from the Pliocene of Kashmir Basin: First record of *Rhagapodemus* from India (Mammalia: Rodentia). *Senkenbergiana Lethaea, 78*, 213–216.

Kumar, R., Ghosh, S. K., Sangode, S. J., & Thakur, V. C. (2002). Manifestation of Intra-Foreland thrusting in the Neogene Himalayan Foreland Basin fill. *Journal of Geological Society of India, 59*, 547–560.

Kumaravel, V., Sangode, S. J., Kumar, R., & Siddaiah, N. S. (2005). Magnetic polarity stratigraphy of Plio-Pleistocene Pinjor Formation (type locality), Siwalik Group, NW Himalaya, India. *Current Science, 88*, 1453–1461.

Kusumgar, S., Kotlia, B. S., Agarwal, D. P., & Sahni, A. (1986). Biochronologie des fossils de vertebras des formations des Karewas du Cachmire, Inde. *L'Anthropologie, 90*, 151–164.

Leakey, M., Werdelin, L. (2010). Early Pleistocene mammals of Africa: Background to dispersal. In J. G. Fleagle, J. J. Shea, F. E. Grine, A. L. Baden, & R. E. Leakey (Eds.), *Out of Africa I: The first hominin colonization of Eurasia* (pp. 3–11). Dordrecht, The Netherlands: Springer.

Lihoreau, F., Barry, J., Blondel, C., Chaimanees, Y., Jaeger, J. J., & Brunet, M. (2007). Anatomical revision of the genus *Merycopotamus* (Artiodactyla; Anthracotheriidae): Its significance for Late Miocene mammal dispersal in Asia. *Palaeontolgy, 50*, 503–524.

Martínez-Navarro, B. (2004). Hippos, pigs, bovids, sabertoothed tigers, monkeys and hominids dispersals during Late Pliocene and Early Pleistocene times through the Levantine Corridor. In N. Goren-Inbar & J. D. Speth (Eds.), *Proceedings of the conference "Human Paleoecology in the Levantine Corridor"* (pp. 37–51). Oxford: Oxbow Books.

Martínez-Navarro, B., & Palombo, M. (2004). Occurrence of the Indian genus *Hemibos* (Bovini, Bovidae, Mammalia) at the Early-Middle Pleistocene transition in Italy. *Quaternary Research, 61*, 314–317.

Martínez-Navarro, B., & Rita, P. M. (2004). Occurrence of the Indian genus *Hemibos* (Bovini, Bovidae, Mammalia) at the Early-Middle Pleistocene transition in Italy. *Quaternary Research, 61*, 314–317.

Nanda, A. C. (1997a). Some biostratigraphic observations based on the Upper Siwalik faunas of the Siwalik Group of India and Nepal. In N. P. Wijayananda, P. G. Cooray, & P. Mosley (Eds.), *Geology in South Asia – II* (pp. 171–189). Dehiwala, Sri Lanka: Geological Survey & Mines Bureau.

Nanda, A. C. (1997b). Comments on Neogene/Quaternary Boundary and associated faunas in the Upper Siwalik of Chandigarh and Jammu. In: *Proceedings of First Regional GEOSAS Workshop on Quaternary geology of South Asia* (pp. 120–136). Madras, India: Anna University, pp. 120–136.

Nanda, A. C. (2002). Upper Siwalik mammalian faunas of India and associated events. *Journal of Asian Earth Science, 21*, 47–58.

Opdyke, N. D., Lindsay, E., Johnson, G. D., Johnson, N., Tahirkheli, R. A. K., & Mirza, M. A. (1979). Magnetic polarity stratigraphy and vertebrate paleontology of the Upper Siwalik Subgroup of northern Pakistan. *Palaeogeography, Palaeoclimatology, Palaeoecology, 27*, 1–34.

Patnaik, R. (2000). Siwalik murid rodents: Origin and dispersal. *Himalayan Geology, 21*, 145–151.

Patnaik, R. (2001). Late Pliocene micromammals from Tatrot formation (Upper Siwalik) exposed near village Saketi, Himachal Pradesh, India. *Palaeontographica, 261*, 55–81.

Patnaik, R. (2003). Reconstruction of Upper Siwalik palaeocology and palaeoclimatology using microfossil palaeocommunities. *Palaeoecology, Palaeoclimatology and Palaeogeography, 197*, 133–150.

Patnaik R. (2010). Fossil murine rodents as ancient monsoon indicators of the Indian subcontinent. *Quaternary International 192*, 43–51.

Phadatre, N. R., Kumar, R., & Ghosh, S. K. (1994). Stratigraphic palynology, floristic succession and the Tatrot/Pinjor boundary in Upper Siwalik sediments of Haripur Khol Area District Sirmuar (H.P.), India. *Himalayan Geology, 15*, 69–82.

Pilgrim, G. E. (1913). The correlation of the Siwaliks with mammal horizons of Europe. *Records of the Geological Survey of India, 43*, 264–326.

Pilgrim, G. E. (1939). The fossil Bovidae of India. *Palaeontologia Indica, New Series, 26*, 1–356.

Prasad, K. N. (1996). Pleistocene cave Fauna, Kurnool from Peninsular India. *Journal of Caves and Karst Studies, 58*(1), 30–34.

Qiu Zhanxiang Deng Tao, & Wang Banyue (2004). Early Pleistocene mammalian fauna from Longdan, Dongxiang, Gansu, China. *Palaeontologia Sinica, 191*(27), 1–198.

Quade, J., Cerling, T. E., & Bowman, J. R. (1989). Development of Asian monsoon revealed by marked ecological shift during the latest Miocene in northern Pakistan. *Nature, 342*, 163–166.

Ramstein, G., Fluteau, F., Besse, J., & Jousamme, S. (1997). Effect of orogeny, plate motion and land-sea distribution on Eurasian climate change over the past 30 million years. *Nature, 386*, 788–795.

Ranga Rao, A. (1986). North-west Himalayan foothills: Its stratigraphical record and tectonic phases. *Bulletin Oil and Natural Gas Commission, 23*, 109–128.

Ranga Rao, A., Agarwal, R. P., Sharma, U. N., Bhalla, M. S., & Nanda, A. C. (1988). Magnetic polarity stratigraphy and vertebrate palaeontology of the Upper Siwalik Subgroup of Jammu Hills, India. *Journal Geological Society of India, 31*(4), 361–385.

Ranga Rao, A., Nanda, A. C., Sharma, U. N., & Bhalla, M. S. (1995). Magnetic polarity stratigraphy of the Pinjor Formation (Upper Siwalik) near Pinjore, Haryana. *Current Science, 68*, 1231–1236.

Rao, M. R., & Patnaik, R. (2001). Palynology of Pliocene sediments of Pinjor formation, Haryana, India. *Palaeobotanist, 50*, 267–286.

Rook, L., Martínez-Navarro, B., & Howell, F. C. (2004). Occurrence of *Theropithecus* sp. in the Late Villafranchian of Southern Italy and implication for Early Pleistocene "Out of Africa" dispersals. *Journal of Human Evolution, 47*, 267–277.

Sahni, B. (1936). The Karewas of Kashmir. *Current Science, 5*, 10–16.

Sahni, M. R., & Khan, E. (1964). Stratigraphy, structure and correlation of the Upper Shivaliks east of Chandigarh. *Journal of the Palaeontological Society of India, 4*, 61–74.

Sahni, A., & Kotlia, B. S. (1985). Kaewa microvertebrates: Biostratigraphical and palaeoecological Implications. *Current Trends in Geology, 6*, 29–43.

Sakai, H., Sakai, H., Yahagi, W., Fuji, R., Hayashi, T., & Upreti, B. N. (2006). Pleistocene rapid uplift of the Himalayan frontal ranges recorded in the Kathmandu and Siwalik basins. *Palaeoecology, Palaeoclimatology and Palaeogeography, 242*, 16–27.

Sangode, S. J., & Bloemendal, J. (2004). Pedogenic transformation of magnetic minerals in Plio-Pleistocene palaeosols of the Siwalik Group, NW Himalaya, India. *Palaeoecology, Palaeoclimatology, and Palaeogeography, 212*, 95–118.

Sangode, S. J. & Kumar, R. (2003). Magnetostratigraphic correlation of the Late Cenozoic fluvial sequences from NW Himamalaya. Current Science, 84(8),1014–1024.

Sangode, S. J., Kumar, R., & Ghosh, S. K. (1996). Magnetic polarity of Siwalik sequences of Haripur area (HP), NW Himalaya. *Journal of Geological Society of India, 47*, 683–704.

Sangode, S. J., Bloemendal, J., Kumar, R., & Ghosh, S. K. (2001). Plio-Pleistocene pedogenic changes in the Siwalik palaeosols: A rock magnetic approach. *Current Science, 81*, 387–392.

Sangode, S. J., Kumar, R., & Ghosh, S. K. (2003). Magnetic Polarity Stratigraphy the Late Miocene Siwalik Group sediments from Kangra Re-entrant, H.P., India. *Himalayan Geology, 24*, 47–61.

Sanyal, P., Bhattacharya, S. K., Kumar, R., Ghosh, S. K., & Sangode, S. J. (2004). Mio-Pliocene monsoonal record from Himalayan foreland basin (Indian Siwalik) and its relation to vegetational change. *Palaeogeography, Palaeoclimatology, Palaeoecology, 205*, 23–41.

Schlosser, M. (1924). Fossil primates of China. *Palaeontologica Sinica Series D, 1*, 1–16.

Shackleton, N. J., Backman, H., Zimmerman, D. V., Kent, M. A., Hall, D. G., Roberts, D., et al. (1984). Oxygen isotope calibration of the onset of ice-rafting and history of glaciation in the North Atlantic region. *Nature, 307*, 620–623.

Sondaar, P., Aziz, F., van den Bergh, G. D., & de Vos, J. (1996). Faunal change and hominid evolution during Quaternary of Jawa. *Geological Research and Development Centre Paleontology Series, 8*, 1–10.

Sotnikova, M. V., Dodonov, A. E., & Pen'kov, A. V. (1997). Upper Cenozoic bio-magnetic stratigraphy of central Asian mammalian localities. *Palaeoecology, Palaeoclimatology and Palaeogeography, 133*, 243–258.

Sun, Xiangjun, Wang, & Pinxian. (2005). How old is the Asian monsoon system? – Palaeobotanical records from China. *Palaeogeography, Palaeoclimatology and Palaeoecology, 222*, 181–222.

Szalay, F. S., & Delson, E. (1979). *Evolutionary history of the primates* (p. 580). New York: Academic.

Takai, Masanaru, Saegusa, H., Thaung-Htike, Zin-Maung-Maung-Thein. (2006). Neogene mammalian fauna in Myanmar. *Asian Paleoprimatology, 4*, 143–172.

Tandon, S. K., Kumar, R., Koyama, M., & Niitsuma, N. (1984). Magnetic polarity stratigraphy of the Upper Siwalik Subgroup, east of Chandigarh, Punjab Sub-Himalaya, India. *Journal Geological Society of India, 25*, 45–55.

Tang, Y. (1980). An early Pleistocene mammalian fauna from Yuxian, Hebei Province and its significance towards stratigraphic subdivision. *Vertebrate PalAsiatica, 18*, 256–267.

Tchernov, E. (1987). The age of the 'Ubeidiya formation, an early Pleistocene hominid site in the Jordan Valley, Israel. *Israel Journal of Earth Sciences, 36*, 3–30.

Tchernov, E. (1992). Eurasian-African biotic exchanges through the Levantine corridor during the Neogene and Quaternary. In W. von Koenigswald & L. Werdelin (Eds.), *Mammalian migration and dispersal events in the European Quaternary* (vol. 153, pp. 103–123). Frankfurt a. M.: Courier Forschungsinstitut Senckenberg.

Tewari, B. S., & Kachroo, P. K. (1977). On the occurrence of *Equus sivalensis* from Karewas of Shupiyan, Kashmir Valley. *Recent Research in Geology, 3*, 468–477.

Thomas, H., Geraads, D., Janjou, D., Vaslet, D., Memseh, A., Billiou, D., et al. (1998). Découverte des premières faunes pléistocene de la peninsula Arabique dans le désert du Nafoud (Arabie Saoudite). *Compte Rendu de l'Academie des Sciences Paris, 326*, 145–152.

Thomas, J. V., Prakash, B., & Mohindra, R. (2002). Lithofacies and paleosol analysis of the Middle and Upper Siwalik Groups (Plio-Pleistocene), Haripur-Kolar section, Himachal Pradesh, India. *Sedimentary Geology, 150*, 343–366.

Thomas, J. V., Prakash, B., Mohindra, R., Bhattacharya, S. K., & Jani, R. A. (2005). Syndepositional tectonics and Paleoclimatic implications of Paleosols of the Plio-Pleistocene Siwalik groups, Haripur-Kolar section, H. P., India. *Himalayan Geology, 26*, 327–345.

Turner, A. (1992). Large carnivores and earliest European hominids: changing determinants of resource availability during the Lower and Middle Pleistocene. *Journal of Human Evolution, 22*, 109–126.

Turner, A., & O'Regan, H. J. (2007). Afro-Eurasian mammalian fauna and early hominin dispersals. In M. D. Petraglia & B. Allchin (Eds.), *The evolution and history of human populations in South Asia* (pp. 23–39). The Netherlands: Springer.

Valdiya, K. S. (1991). Quaternary tectonic history of northwest Himalaya. *Current Science, 61*, 664–668.

Van Stein, H., De Ruig, J., & Hoozemans, F. (1988). Morphological and mechanical aspects of debris flows in parts of the French Alps. *Zeitschrift fur Geomorphologie, 32*, 143–161.

Verma, B. C. (1969). *Procynocephalus pinjori* sp. nov. a new fossil primate from the Pinjor beds (Lower Pleistocene) east of Chandigarh. *Journal of Paleontological Society of India, 13*, 53–57.

Zhu, R. X., Hoffman, K. A., Potts, R., Deng, C. L., Pan, Y. X., Guo, B., et al. (2001). Earliest presence of humans in northeast Asia. *Nature, 413*, 413–417.

Zhu, R. X., An, Z., Potts, R., & Hoffman, K. A. (2003). Magnetostratigraphic dating of early humans in China. *Earth Science Reviews, 61*, 341–359.

Chapter 10
The Indian Subcontinent and 'Out of Africa I'

Parth R. Chauhan

Abstract The last few decades of paleoanthropological research has raised important issues about the rate and chrono-geographical extent of early hominin dispersals from Africa into Eurasia. Owing to its geographic position, the Indian subcontinent has a pivotal role to play in addressing such issues. This ecologically diverse landmass critically lies between the three sources of the oldest *Homo* fossils in the Old World and a southern route of expansion from Africa to Southeast Asia, through this region, has often been inferred. Claims of Plio-Pleistocene Oldowan assemblages have been made since the 1960s and come from the Narmada Valley in central India and from the Siwalik Hills in northern Pakistan and northern India. This paper critically reviews each of these claims and broadly discusses associated Plio-Pleistocene environments and geographic routes of entry. A large majority of these reported occurrences represents unsubstantiated claims and require further scientific verification through additional evidence. Tentative scenarios for the current absence of paleoanthropological evidence older than the Middle Pleistocene are also briefly discussed. This current lack of Oldowan assemblages, however, does not reflect an unquestionable absence of hominin occupation in the region. Ecologically conducive environments in the form of open grasslands, a seasonal monsoon regime, diverse fauna and eco-habitats (i.e., diverse hunting/scavenging opportunities), and an abundance of water and stone resources suggest the possible earlier presence of hominins in South Asia. Obviously, much more field research is required to test and confirm their early presence/absence in this geographically important region of the Old World.

Keywords Dispersal routes • Early *Homo* • Lower Karewas • Narmada Valley • Paleoenvironments • Plio-Pleistocene • Siwalik Hills • South Asia

P.R. Chauhan (✉)
The Stone Age Institute & CRAFT Research Center,
1392 W. Dittemore Road, Bloomington (Gosport), IN 47433
email: pchauhan@indiana.edu

Introduction

The last few decades of paleoanthropological research in Eurasia on the earliest dispersals from Africa, has raised important issues about the rate and chrono-geographical extents of early hominin occupations (Rendell et al. 1989; Swisher et al. 1994; Larick and Ciochon 1996; Gabunia et al. 2000; Bar-Yosef and Belfer-Cohen 2001; Anton and Swisher 2004; Dennell 2004; Langbroek 2004; Dennell and Roebroeks 2005). Owing to its geographic position, the Indian subcontinent has a pivotal role to play in addressing such issues and subsequent colonization events (Petraglia 2010). This landmass critically lies between the three sources of probably the earliest *Homo* fossils in the Old World (Woldegabriel et al. 2000; Gabunia et al. 2000; Swisher et al. 1994) and a southern route of expansion, from Africa to Southeast Asia, through this region has often been inferred (e.g., Larick and Ciochon 1996; Dennell 2004; Petraglia 2005). The subcontinent is relatively well-known for its long and rich record of Paleolithic and related Quaternary evidence in the form of stone tools, vertebrate and invertebrate fossils, and dynamic environmental and climatic signatures at a regional level (Misra 2001; Petraglia 2001). The Early Paleolithic record of the Indian subcontinent has been traditionally divided into Mode 1 (pre-Acheulean) and Mode 2 (Acheulean) industries (Misra 1987, 1994; Petraglia 1998, 2001; Gaillard and Mishra 2001; Chauhan 2009a). Most of these localities have been dated through the Thorium-Uranium method and mostly include Middle and Upper Pleistocene Acheulean sites (Mishra 1995; Pappu 2001).

Following the early impact of the Clactonian evidence in England (see Dennell and Hurcombe 1992), a pre-Acheulean technology based on pebbles and cobbles was also proposed for the Indian subcontinent in the form of the Soanian industry in what is now northern Pakistan (de Terra and Paterson 1939). Later work by the British Archaeological Mission to Pakistan (BAMP) in the 1980s resulted in a major revision of de Terra and Paterson's interpretations of the Soanian evidence (see Rendell et al. 1989). Subsequently, multiple lines of evidence including a comparison of Soanian and Acheulean technology

J.G. Fleagle et al. (eds.), *Out of Africa I: The First Hominin Colonization of Eurasia*, Vertebrate Paleobiology and Paleoanthropology,
DOI 10.1007/978-90-481-9036-2_10, © Springer Science+Business Media B.V. 2010

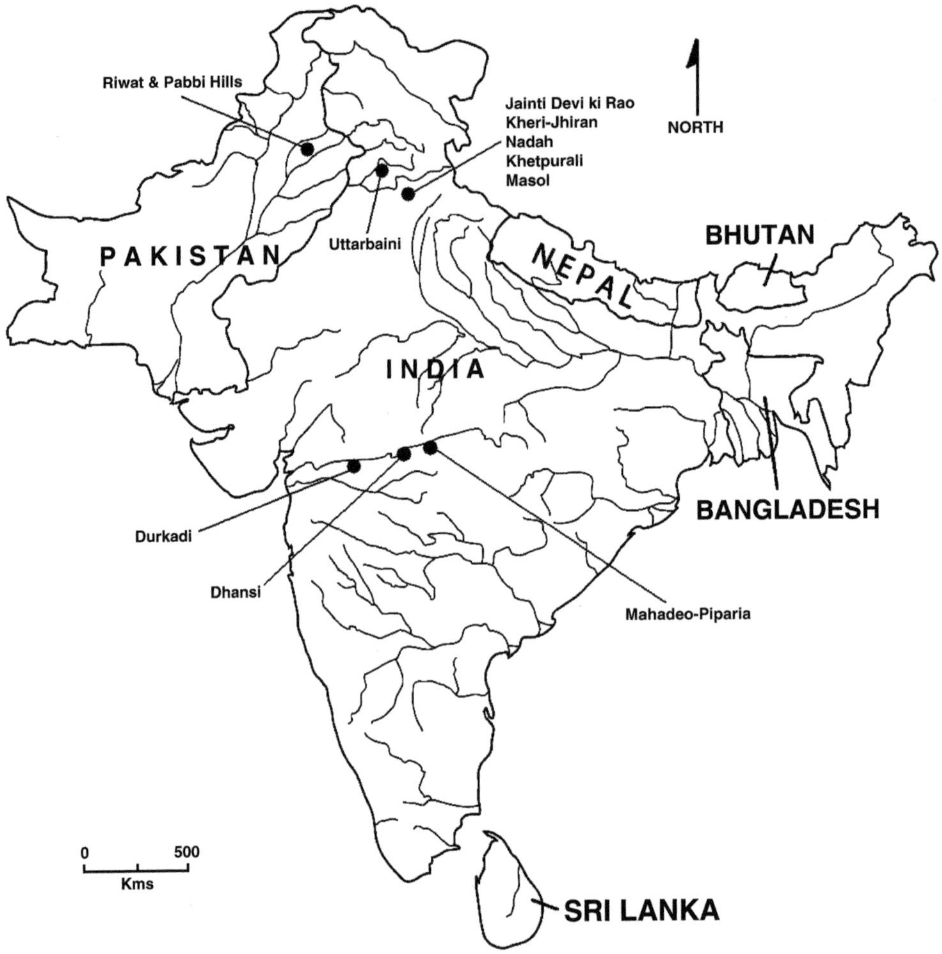

Fig. 10.1 General locations of paleoanthropological sites discussed in the text

(Gaillard 1995), landscape geoarcheology (Chauhan 2008a), surveys of dated geological features (Soni and Soni 2005) and a comparative morphometric analysis (Lycett 2007) clearly revealed that the majority of Soanian assemblages, if not all, represents a Mode 3 technology and relatively post-dates the Acheulean (Gaillard and Mishra 2001; Chauhan 2003). Subsequent claims for a pre-Acheulean occupation have come from the Narmada Valley of central India and from the Siwalik Hills in the northern zones of Pakistan and India (Fig. 10.1 and Table 10.1). This paper critically reviews these claims and broadly discusses associated Plio-Pleistocene environments and geographic routes of entry. Possible explanations for the absence of paleoanthropological evidence older than the Middle Pleistocene are also briefly discussed.

The Biogeography of South Asia

The Indian subcontinent covers an area of over four million square kilometers and essentially comprises the regions of India, Pakistan, Nepal, Sri Lanka, Bhutan, and Bangladesh. To the west of peninsular India is the Arabian Sea, to the east, the Bay of Bengal, and to the south is the Indian Ocean. The entire region comprises a diverse spectrum of ecological and topographical zones combined with a complex geological history. The north is dominated by the Greater and Lesser Himalaya and the Siwalik Hills, all almost geographically parallel and temporally successive to each other. This mountainous terrain includes northern Pakistan, northern India, most parts of Nepal, and Bhutan. South of these mountain and hill ranges are the Indo-Gangetic plains located in all South Asian countries except Sri Lanka and Bhutan. The plains are followed to the south by the great Thar Desert (in eastern Pakistan and northwestern India), and the Aravalli and Vindhyan range of hills. These hills are located north of the Deccan Plateau, a prominent landscape of peninsular India, which includes the Western and Eastern Ghats (ranges of hills). Although most parts of India are recognized as being tropical or sub-tropical (Mohapatra 1985), such rugged landscapes (the Ghats) are especially prominent along the coasts of peninsular India, southeastern India (Kerala), and northeastern India or east of Bangladesh. The subcontinent is also interspersed with complex drainage systems and

Table 10.1 The South Asian paleoanthropological sites as discussed in the text and associated attributes.

Site	Reported Age [dating method]	Stratigraphic context	Material reported	References	Comments
Riwat	2.0–2.2 Ma? [PM, GS]	Stratified in gritstone	3–23 cores and flakes (based on numerical ranking)	Rendell et al. 1987	Requires corroboration through in situ fine-grained contexts
Pabbi Hills a	2.2 – 1.7 Ma? [PM, GS]	Surface of fine-grained sediments	198 cores (various types), flakes, flake blades, scrapers, knife	Dennell 2004	Requires corroboration through in situ contexts
Pabbi Hills b	1.4–1.2 Ma? [PM, GS]	Surface of fine-grained sediments	307 (same as above)	Dennell 2004	Requires corroboration through in situ contexts
Pabbi Hills c	1.2–0.9 Ma? [PM, GS]	Surface of fine-grained sediments	102 (same as above)	Dennell 2004	Requires corroboration through in situ contexts
Uttarbaini	>1.6/>2.8 Ma? [FS]	Stratified below dated ash horizon	Not reported	Verma 1989, 1991	Requires confirmation, re-dating of ash
Jainti Devi ki Rao	EP-MP? [GS, BS]	Stratified within Lower Boulder Conglomerate Fm.	150 Acheulian handaxes, cleavers, choppers, large flakes	Sharma 1977	Requires confirmation, dating through *in situ* fine-grained contexts
Kheri-Jhiran	LP-EP? [GS, BS?]	Stratified within Pinjor Fm.	>45 'Abbevillian' handaxes, choppers, scrapers, 'rounded pebble tools', a discoid	Verma 1975	Requires corroboration
Nadah	2.2 – 2.0 Ma? [GS, BS]	Stratified (?) within Pinjor Fm.	*H. erectus* maxillary incisor	Singh et al. 1988	Ambiguous; require diagnostic specimen(s)
Khetpurali & Masol	ca. 3.4 Ma? [GS, BS]	Eroded out from Tatrot Fm.?	*Hominid fossils*: (mandibular, proximal femur, distal femur, patella, & post-cranial fragments, *stone tools*: (choppers, flakes?), *other vertebrate fossils*	Singh 2003	Represents false claims and requires substantiation
Durkadi	1 Ma? [GS, AT]	Stratified in surface of conglomerate	650 artifacts (see Table 10.2)	Armand 1979, 1983	Requires dating and corroboration through fine-grained contexts
Mahadeo Piparia	Early MP? [GS, AT]	Stratified in surface of conglomerate	>1,215 but not all Early Paleolithic	Multiple papers of A.P. Khatri; Supekar 1985	Requires dating and confirmation of context

PM: paleomagnetism; GS: geo-stratigraphy; FS: fission track; AT: artifact typology; BS: biostratigraphy; LP: Late Pliocene; EP: Early Pleistocene; MP: Middle Pleistocene.

associated numerous ecological and geographic features such as deciduous woodlands, tropical evergreen forests, savanna landscapes, semi-arid and arid scrub lands, arid sand deserts, and periglacial loessic landforms (Korisettar and Rajaguru 2002); caves, canyons, rockshelters, lakes, pools, and springs are also found in high numbers. With the exception of the Narmada and Tapi Rivers, most rivers in central and peninsular India flow from west to east and all exhibit unique fluvio-sedimentary regimes (Gupta 1995).

In the past as well as today, these different ecozones provided access to different types and shapes of lithic raw material, usually conditional upon factors of sedimentation and associated fluvial, tectonic, and erosional mechanisms. For example, in the Siwalik region, rounded and sub-rounded quartzite and sandstone clasts dominated the landscape during various phases of hominin occupation. In parts of north-central India, Acheulean assemblages were made on pink granite. In the Western and Eastern Ghats, however, the primary raw material was basalt and doleritic dykes, occurring as fragments from bedrock outcrops and water-worn clasts belonging to the Deccan volcanic Traps. Further south, in parts of Karnataka, limestone bedrock was the dominant raw material type available in the form of tabular slabs. The bi-annual Indian monsoon system may have also played a major role in the seasonal availability, in some areas, of raw materials in river or stream beds during the dry periods (Dennell 2007).

South Asian Plio-Pleistocene Environments

At the moment, the best source of evidence regarding paleoenvironmental conditions during the Plio-Pleistocene in South Asia comes from the Upper Siwalik sequence and the Lower Karewa deposits. This data from South Asia is in general accordance with the currently-known global evidence (Petraglia 2005; also see Patnaik and Nanda 2010) and each is discussed briefly in turn below. During the Late Miocene-Early Pliocene time, there was a marked change from a generally humid and warm climate to drier, cooler conditions (Wang 1994). On the Indian peninsula as well as in southern China, this change resulted in the gradual replacement of evergreen tropical rainforest by deciduous forests (see Morley 2000 for an overview). At this time and also during the Early Pleistocene, the presence of open grasslands or C_4 vegetation has been hypothesized for the Siwalik region (see Quade et al. 1989). The cooling period during the Early Pliocene was followed by a Middle Pliocene warmer period (*c.* 3.5–3.0 Ma) (see Meijaard and Groves 2004) and a warm oxidative phase during the terminal Pliocene (Sangode et al. 2001). General climatic conditions at this time may have been more humid than in younger contexts, evidence indirectly supported by the extinction of herbivore browsers at the Plio-Pleistocene boundary indicating a possible reduction in woodland environments (Dennell 2004). Sanyal et al. (2004) examined sections at Haripur Khol in the Siwaliks of northern India and recognized four major stages of floral successions spanning from 4.0 to 1.0 myr. Stage IV (2.5–1.0 myr) is marked by the presence of a volcanic ash bed and plant evidence indicating cool and dry conditions and a possible change in the concentration of CO_2 (p. 34). In southern and central Peninsular India, this change is broadly supported by a shift from tropical humid to semi-arid conditions during the Pliocene-Pleistocene transition, as broadly inferred from the mineral composition of certain palaeosols (Pal et al. 1989).

After the Upper Siwalik Formations, the next best source of Plio-Pleistocene climatic change in South Asia comes from the Lower Karewa ("flat topped plateaus") deposits in Kashmir. These sediments were deposited when the Pir Panjal range near the Kashmir Valley was uplifted and formed a large lake during the mid-Pliocene or around 4 myr (de Terra and Paterson 1939; Burbank and Johnson 1983). This was followed by extensive fluvial erosion leading to the exposure of the lacustrine sediments which have been dated through a large number of radiometric dating methods such as ^{14}C, thermoluminescence, paleomagnetism, fission track, and so forth (Agrawal et al. 1989, 1990). The loess deposits above the lacustrine sediments represent the Upper Karewas and are basically estimated to be younger than 300 ka (Singhvi et al. 1987). Agrawal et al. (1990:67) describe the following climatic sequence for the various lithostratigraphic zones of Late Pliocene age:

> Krachipatra locality is partly covered by Remzone 6 and spans about 2.4–2.2 myr. It shows a cool temperate climate with little variation in precipitation….

Prior to that time or

> …up to 3.8 myr the climate seems to be warm temperate…

and

> …From 3.7 to 2.6 myr, there is a transition from a sub-tropical type of climate to a cool temperate type (p.69).

Biostratigraphic comparisons further established that certain mammalian species inhabited the region only after 2.4 myr – *Cervus panjabiensis*, *C. kashmiriensis*, *Elephas hysudricus*, and *Equus sivalensis*, to name a few examples. Occurrences of specific micromammal species at 2.4, 1.8–1.6, and 0.4 Ma are all associated with cold oscillations and further supported by pollen and microvertebrate assemblages and stable isotope results (see Agrawal et al. 1990 for specific references to these data). A high ratio of C/N suggests a cold phase at the timing of the Olduvai event in the region, followed by another cold phase at about 0.73 myr (Agrawal 1988; Ganjoo 1990; Sangode et al. 2001). As Dennell (2007) points out – the majority of well-preserved Late Pliocene-Early Pleistocene paleoanthropological sites in both Africa and Asia are in lacustrine contexts. Therefore, the Lower Karewas represent a promising source of evidence for some of the earliest hominin occupation in the region, and deserve focused paleoanthropological investigations.

In contrast to the Himalayan and sub-Himalayan regions described above, our knowledge of the Mio-Pliocene and Plio-Pleistocene terrestrial records in peninsular India is severely limited, possibly a consequence of the low spatial profiles of such time-constrained sedimentary exposures. In other words, Quaternary sediments of peninsular India are restricted to the Middle Pleistocene and younger fluvial and occasionally lacustrine deposits found in spatial association with extant drainage systems (Gupta 1995). Sediments older than the Middle Pleistocene in peninsular India, or south of the Siwalik Hills, are not laterally extensive or may have been misidentified. A good example of this comes from the recent investigations by Pappu et al. (2003) at Attirampakkam, where they proved that sediments previously thought to be of Cretaceous age for many decades, are actually of Pleistocene age. Similar geological situations may exist elsewhere in peninsular India where older Plio-Pleistocene sediments (even if laterally restricted) have not yet been properly identified or promising sediments have been misidentified. For instance, some appropriate exposures (i.e., Early Pleistocene) are known to occur as vertical sections, fluvial paleo-channels, or as calcium carbonate deposits in parts of Rajasthan in northwest India (Dhir et al. 2004), within the Dhansi Formation of the central Narmada Basin (Tiwari and Bhai 1997), and possibly in central Maharashtra in association with Early Acheulean sites (see Gaillard and Mishra 2001; Deo et al. 2007; Sangode et al. 2007). Biostratigraphic chronology and research has been conducted extensively in India and shows clear evidence of constant taxonomic change throughout the Plio-Pleistocene, although no index fossils of large mammals have been identified for subdividing the Pleistocene (Badam 1979; A.C. Nanda, personal communication, 2006). The only known 'pre-Middle Pleistocene' Neogene vertebrate fossil locality in peninsular India is Piram Island off the coast of western India (Prasad 1974) but remains undated and has been compared with the Early Pliocene Siwalik fauna (Tatrot zone) (Chauhan 2008b). Despite this paucity of older faunal material, there is even some evidence of faunal migrations during the Pleistocene *from* South Asia into other parts of Eurasia (e.g., Martínez-Navarro and Palombo 2004).

The Two Routes of Entry into the Indian Subcontinent

The most obvious geographic areas of entry into the subcontinent (and subsequent population movements *from* India) are Afghanistan and Pakistan to the northwest and Myanmar in the northeast. Both zones offered coastal routes as well inland mountain passes (e.g., Khyber in the northwest) and ecologically-rich tropical evergreen and deciduous forests (in the northeast), without any major mountain ranges or deserts impeding potential movements (e.g., Dowsett et al. 1994). From the lack of appropriate evidence, the high-altitude Himalayan range was probably not penetrated during Plio-Pleistocene times; the earliest Paleolithic evidence in the Himalayan region of Kashmir and Ladakh is probably of Middle Pleistocene age (Sankalia 1971; Joshi et al. 1974; Sharma 1995). The NW corridor probably offered more suitable raw material than the NE corridor. Although hominins in northern Pakistan (Rendell et al. 1989; Dennell 2004) may have been restricted to that area due to a lack of abundant raw material in the Indo-Gangetic zone and on the Pinjor paleo-landscape (Dennell 2007), other hominin groups (if any) entering the subcontinent closer to the Arabian Sea through southern Afghanistan and southern Pakistan would probably not have encountered similar challenges. This is inferred from the presumably-available raw material sources in the intermediate zones such as the Sulaiman Mountains in central Pakistan and the Rohri Hills in southern Pakistan. The Thar Desert may have acted as a temporary ecological barrier but also contained numerous isolated raw material sources such as the Jayal Gravel Ridge and similar occurrences (e.g., Rajaguru et al. 1996), and once crossed, ample outcrops of Aravalli quartzites were available on the landscape. Indeed, small or moderate sized zones without any or minimal raw materials may have played an important role in increasing short distance ranging patterns for small hominin groups (when searching for raw material) invading new territories for the first time.

If early *Homo* arrived in Southeast Asia by about 1.8 Mya (Swisher et al. 1994) or no earlier than 1.7 myr (Sémah et al. 2000) through East Asia, then that species must have become well-adapted to forested/tropical environments by the time it reached those zones. There is currently no explanation why there may not have been movement westward into peninsular India or even gene flow between South Asia and Southeast Asia if early hominin groups occupied both regions simultaneously. Quaternary deposits in NE India are represented by ferruginous conglomerates with occasional Paleolithic material and this pattern appears to increase as one moves east and northeast of the subcontinent. One reason for the dearth of Early Paleolithic archeological evidence in this region may be poor visibility for survey purposes or a limited occurrence of time-specific deposits in this thickly-vegetated zone. This region has a rich record of younger non-biface assemblages, most of which may be related culturally to contemporaneous lithic industries from Southeast Asia. Except for the Ganges and Brahmaputra Rivers, the drainage system was not as extensive or braided as the NW region. Similar to the Indo-Gangetic landscape in the north and northwest (Dennell 2007), however, the Bay of Bengal (Alam et al. 2003) and the Ganges-Brahmaputra drainage system may have also acted as temporary barriers (Field and Lahr 2005) during the Late Pliocene-Early Pleistocene, due to a low occurrence of raw materials in the region and general delta environment. Cordaux et al. (2004) also recognize the NE region as a barrier during the last few millennia (from modern Y chromosome and mtDNA gene pools) and possibly during the Pleistocene as well. Nonetheless, this unique area may hold greater potential for Early Pleistocene sites reflecting a major, but hitherto unknown, human and faunal corridor between peninsular India and southeastern Asia (e.g., Mishra et al. 2010).

Central India

Narmada Valley

Central India is dominated by the Narmada River which flows through Madhya Pradesh and Gujarat from Amarkantak in the east to the Arabian Sea in the west – a total of about 1,300 km.

Numerous Quaternary geological and archeological investigations have been conducted in the entire valley since the nineteenth century (see Kennedy 2003). Prehistoric hominin occupation associated with the Narmada River appears to have occurred since at least the Middle Pleistocene, or potentially earlier (discussed later). Direct evidence of repeated human occupation is reflected by numerous sites ranging from the Early Paleolithic to the Chalcolithic Periods (Misra 1997). The region is most famous for yielding the oldest fossil hominin evidence in the subcontinent at Hathnora, which is represented by an incomplete calvarium and two clavicles and a possible rib fragment (Sonakia 1984; Sankhyan 1997, 2005). The cranium remains to be accurately dated but has been variably attributed by different investigators as *Homo erectus*, archaic *H. sapiens* or *H. heidelbergensis* (Sonakia 1984; Kennedy and Chiment 1991; Sankhyan 1997, 2005; Cameron et al. 2004). Because it possesses diverse but undiagnostic anatomical traits, Athreya (2007) has suggested that it be provisionally classified as *Homo* sp. indet. Initial paleomagnetic studies led by Agrawal et al. (1988) suggested that the Narmada Quaternary deposits fall within the Brunhes Chron (<0.73 Myr), while studies by the Geological Survey of India indicated an Early Pleistocene age for the oldest Quaternary sediments (see below). However, their respective studies were not applied to the same stratigraphic sections and because the Quaternary geology varies significantly across the entire basin, associated Early Paleolithic sites are yet to be securely dated including the ones discussed below. Based on the stratigraphic relationships, erosional unconformities, sedimentary mineralogy, granulometry, and structures, pedogenic characteristics, and tephra deposits, and paleomagnetic signatures, the central Narmada Quaternary sequence has been divided into seven formations listed here from oldest to youngest: Pilikarar, Dhansi, Surajkund, Baneta, Hirdepur, Bauras and Ramnagar (Tiwari and Bhai 1997). However, the author and his colleagues have recently demonstrated that the Pilikarar stratigraphic sequence cannot be defined as a geological formation (Patnaik et al. 2009), thus qualifying Dhansi as currently the oldest Quaternary formation in the central Narmada Basin. This is provisionally based on Rao et al.'s (1997) paleomagnetic results which demonstrate that the latter formation at the Dhansi type-site belongs to the Matuyama Chron. Claims of pre-Acheulean evidence have come from two sites in the valley: Mahadeo Piparia and Durkadi.

Mahadeo Piparia and Durkadi

In the 1960s, a pre-Acheulean lithic occurrence was reported in the form of the Mahadevian industry in the eastern part of the Narmada Valley. This industry was named after the site of Mahadeo-Piparia by Khatri (1962, 1966) who equated it to the Oldowan industry and interpreted it as a technological predecessor to the Indian Acheulean. Later excavations and stratigraphic observations by Sen and Ghosh (1963) and Supekar (1968) refuted Khatri's claim of a Mode 1 to Mode 2 transition here. A similar claim to that of Khatri's was made more systematically through controlled excavations by Armand (1979, 1983, 1985) who defined the Durkadian industry at the site of Durkadi from excavated contexts 2 km south of the lower Narmada channel. At both Durkadi and Mahadeo Piparia, a large amount of non-biface artifacts were recovered in stratified contexts and comprised of cores, choppers, flakes, "protobifaces", and other formal tool types. Both assemblages were recovered from within and over-lying the high-energy gravels of the Narmada River and many artifacts at both sites are relatively in fresh condition. This signified the use of the conglomerate surface through multiple visits for clast acquisition and stone tool production prior to the surface's burial by fine-grained sediments, a key geoarcheological feature at such sites associated with gravel or conglomerate contexts in the subcontinent. While Mahadeo-Piparia was discerned to be early Middle Pleistocene in age (Khatri 1962), Durkadi was interpreted to be about 1 Ma in relative age (Armand 1983) – mostly based on geo-stratigraphic and typological grounds.

It is also now generally accepted that the South Asian Acheulean is a result of early migrations of the genus *Homo* from Africa sometime in the Pleistocene rather than being an indigenous or regional technological development (Sankalia 1974). Therefore, neither the Mahadeo-Piparia and Durkadi, nor any other site in the Indian subcontinent, shows any convincing stratigraphic evidence for a technological transition from an Oldowan-type into the more sophisticated Acheulean technology (Jayaswal 1982). Nonetheless, both Durkadi and a part of Mahadeo-Piparia remain typo-morphological anomalies within the South Asian Early Paleolithic record and thus, merit reinvestigations to clarify their temporal and technological positions within the Indian Paleolithic framework. While Mahadeo-Piparia has yielded Acheulean bifaces and Middle Paleolithic elements in addition to the Mode 1 component (Supekar 1985), Durkadi continues to be ambiguous, despite Armand's (1983, 1985) report of one 'proto-cleaver', six 'proto-handaxes' and one 'Abbevillian' or evolved Durkadian handaxe. These eight specimens do not conform to the current typo-morphological definition of Acheulean bifaces as they lack bilateral and planform symmetry and adequate bifacial reduction. They also do not appear to resemble typical early developmental stages of the Acheulean as known from, for example, Olduvai Gorge, Konso-Gardula, Peninj and 'Ubeidiya (see Clark 1994). The remaining claims for pre-Middle Pleistocene paleoanthropological evidence come from multiple locations in the Siwalik Hills of northern Pakistan (2) and northern India (4).

Northern Pakistan

The Siwalik Hills

The Siwalik Hills or the Siwalik Foreland Basin consist of fluvial sediments (Fig. 10.2) deposited by hinterland rivers flowing southwards and southwestwards (Gill 1983) from the Lesser and Greater Himalayas, when the region south of these mountains was originally a vast depression or basin (referred to as the foredeep) (Brozović and Burbank 2000). They span from the western side of the Indus (northern Pakistan in the west) to the Bay of Bengal (Sikkim/Assam region in the east), covering a total length of approximately 2,400 km. The topography of the Siwalik Hills became a prominent feature on the landscape and reached its present elevation during Middle Pleistocene times (Kumar et al. 1994). The range is less than 13 km wide in places (average of 24 km), and it reaches an elevation between 900 and 1,200 m. Quartzite pebbles and cobbles were the main raw-material exploited by the hominin occupants of this eco-zone at multiple temporal intervals throughout the entire Siwalik range (Dennell 2007; Chauhan 2008a). In addition to being located within the Boulder Conglomerate Formation of the Upper Siwalik Subgroup (Johnson et al. 1982), these localized quartzite clasts also occur in streambeds, on Siwalik surfaces of varying ages, and in the terrace sections of intermontane valleys. Paleolithic sites in the Siwaliks are situated within a range of eco-geographic contexts and have been traditionally divided into two types – Acheulean and Soanian – and are found in the form of sites, site-complexes, find-spots and numerous surface scatters (e.g., de Terra and Paterson 1939; Stiles 1978; Mohapatra 1981; Chauhan 2007, 2008a). However, claims made of the earliest occupation in the Siwalik region are reported to be considerably older than both these industries and have not been classified as Soanian, despite some broad morphological similarities.

Fig. 10.2 A view of Upper Siwalik strata near Chandigarh in northern India

Northern India

Kheri-Jhiran (Northern India)

From paleontological investigations in the Solan District of Himachal Pradesh, Verma (1975:518) reported a rich vertebrate fossil locality from the Pinjor Formation as well as

> ...*closely associated* [emphasis mine] human artefacts – like crude handaxes, choppers, scrapers, light duty flakes and other pebble tools.

This locality, called GSI 107, is in the Kheri area and located about 30 km from the Pinjor type-section (Pinjore village). The Pinjor Formation here is thought to yield highly-fragmented but well-fossilized specimens of about a dozen different species within an area of 75 m^2. He reports *about* 45 fresh artifacts including 'Abbevillian type' handaxes, unifacial and bifacial choppers and scrapers, one discoid and several 'rounded pebble tools'. The raw materials are pebbles and cobbles of quartzite and chert. The artifacts are described as lacking retouch and being of 'crude typology'. Some of the artifacts were allegedly excavated in situ from the sandstone/conglomerate bed of the Pinjor Formation. The site also yielded about 50 unmodified quartzite pebbles (5–10 cm in diameter) in surface association with the vertebrate fossils. Twenty meters *below* GSI 107 and 45 m above, additional fossil material and 'pebble tools' are reported as well as a fourth occurrence 220 m above GSI 107 and near the interface between the Pinjor Formation and the overlying Boulder Conglomerate Formation. This fourth occurrence is described as yielding several 'crude' specimens similar in nature to those from GSI 107, as well as two bifacial scrapers/choppers and a 'multi-facetted' discoid. The evidence is collectively interpreted to represent a

> ...slow and gradual evolution in the culture and topology [typology?] of the artefacts through the long depositional history of Pinjors (Verma 1975:519).

Although the Kheri-Jhiran section from which the paleoanthropological material is thought to derive is not adequately described, a schematic figure by Verma shows the different occurrences of lithics and fossils within the section, all allegedly in situ. The photographs of four lithic specimens shown by Verma (1975:520) appear to be either pointed cores, core-fragments and/or pointed choppers. The flake scars are clearly visible and some of the illustrated artifacts morphologically resemble the pointed-core specimen from Riwat which may be of Late Pliocene age (discussed later). However, the current descriptions in the text and associated illustrations are inadequate, thus reflecting the ambiguous nature of the Kheri-Jhiran assemblages and their stratigraphic contexts. Its current status can be viewed as being typologically undiagnostic, as most Paleolithic surface scatters in the Siwalik region are. The Kheri-Jhiran occurrences most likely represent contexts where lithic specimens (some of which may even include naturally-flaked clasts) have fallen from the Boulder Conglomerate Formation and/or from above it, suggesting their possible post-Siwalik age. Additionally, the surface association of vertebrate fossils and lithic specimens is a common occurrence in the Siwalik region (Chauhan and Gill 2002; Chauhan 2008a, b) and not a single such locality has been proven to represent evidence of butchery or other types of hominin-modification. Nonetheless, the fact that Verma reports the material as being *in situ* and in association with vertebrate fossil material, this section and the surrounding area require further investigation.

Jainti Devi Ki Rao (Chandigarh Area, Northern India)

Near Chandigarh, Sharma (1977) reported an 'habitation site' from the Boulder Conglomerate Formation between Mullanpur and Parol on the northwestern bank of a seasonal stream called Jainti Devi ki Rao. From the geological and fossil vertebrate evidence, he assigned a Middle Pleistocene age to the material based on Upper Siwalik stratigraphy but also states that some or all of it may be Early Pleistocene. Interestingly, the investigator reports Acheulean bifaces in association with a prominent Mode 1 assemblage (pebble chopper/chopping-tool industry) as well as 'large flake tools'. From two vertical sections of the Lower Boulder Conglomerate about 20 ft in vertical thickness, the investigator reports 150 artifacts including many specimens in situ, although the majority appear to be rolled. The assemblage includes unifacial and bifacial choppers, massive 'borer-cum-choppers' (possibly just pointed choppers), bifacial handaxes and cleavers and large flakes (many flakes being made by the Clactonian technique). From the facetted platforms on some specimens, a prepared-core technology or the Levallois technique also appears to be present, with increased frequency in the younger context at the site. Based on the changing technology of the lithic industries and associated states of preservation, Sharma (1977:94) invokes broader environmental changes and recognizes an associated gradual evolution (similar to Verma 1975) of the behavioral evidence in this area from the Abbevilian to the Acheulean and beyond:

> Small flake-blade tools (along with the handaxe-cleaver and chopper industries) appear only in the later phases-the Upper Pleistocene and post-Pleistocene-indicating an Upper Palaeolithic culture in the region that is perhaps derived from the local Early Paleolithic cultures.

This scenario is presented using the Early Paleolithic evidence from the Boulder Conglomerate Formation combined with typologically younger lithic evidence from the nearby post-Siwalik terrace deposits (i.e., younger than the Boulder Conglomerate Formation) about 10–15 ft in height. From this younger context, Sharma reports 100 artifacts including flakes, blades and abundant debitage as well as a continuation of some Early Paleolithic types (choppers, handaxes, cleavers) but with decreased frequency, made on flakes and in fresh condition unlike those from the Boulder Conglomerate Formation. The post-Siwalik flakes are of various dimensions and morphology and their striking platforms show no signs of preparation, though secondary working and retouch appears to be prominent. The blades only occur in the post-Siwalik alluvium and are described to be technologically more refined (e.g., thin, elongated, prepared platforms but no retouch) and thought to belong to the Upper Paleolithic of Punjab.

Unfortunately, Sharma does not provide any photographs of the site or figures and a table for the lithics; a detailed stratigraphic description and figure are also lacking. The large amount of rolled specimens and the mixed nature of the lithic material may preclude the site as being in primary context or even a 'habitation site' as no evidence of refitting specimens is provided. Instead, the occurrence may be a result of frequent post-depositional processes such as seasonal fluvial activity from the monsoons, erosion through tectonic processes and colluvial action – all common in the Siwalik region (Chauhan and Gill 2002). In addition, it has long been proven that the rolled versus fresh condition of artifacts is not always a reliable indicator of the relative comparative age of the material. For example, older artifacts may have rolled less and thus be more fresh than younger artifacts from the same site that may have rolled over a longer distance or to a greater degree. Also, a linear model of lengthy *and continuous* technological evolution in the Siwaliks is highly unlikely. Rather, the Paleolithic record in the entire Siwalik region and the subcontinent in general is notably discontinuous (Dennell 2003) and the variable presence of Mode 1, Acheulean and Soanian assemblages clearly suggests intermittent occupation (Chauhan 2008a). Although there are significant deficiencies in the report by Sharma (1977), the area and particularly suitable outcrops of the Boulder Conglomerate Formation merit a survey for primary-context Paleolithic occurrences (preferably those capped by fine-grained sediments). Unlike the Pinjor Formation, the Boulder Conglomerate Formation contains a vast amount of quartzite clasts (along with the dominant sandstone clasts) for the production of stone tools which may have stimulated an increase in hominin occupation of the Siwalik frontal zone compared to Pinjor times (Chauhan 2008a). The location of sites in such contexts may yield valuable information regarding not only raw material exploitation and transport behaviors but also pinpoint the initial timing of colonization by incoming hominin groups.

Toka (Northern India)

In the Toka area in southern Himachal Pradesh, Verma and Srivastava (1984) reported Paleolithic artifacts eroding out from the Tatrot sediments of Pliocene age and in association with vertebrate fossils (see Gaillard and Mishra 2001 for a similar argument elsewhere in the Siwalik Hills). Despite the lack of excavations or *in situ* occurrences, they concluded that the artifacts on the Upper Siwalik slopes (but lacking on Lower Siwalik exposures) as well as some assemblages on the nearby Markanda terraces are eroding out from the ancient Siwalik surfaces. The investigators sought support for their observations from the previous work of Verma (1975) and Sharma (1977) (both discussed earlier):

> The tool types recovered from both these stratigraphic levels indicate the pre-existence of the culture and suggest the possibility that the artefacts occurring in the Siwalik outcrops in the Markanda Valley have their provenance in the Tatrot Formations (Verma and Srivastava 1984:17).

In conclusion, they state:

> The occurrences indirectly suggest that the toolmaker lived in this region during the Upper Pliocene times, contrary to the terrace deposits only and the early man appeared in the Siwalik region during the Middle Pleistocene (Verma and Srivastava 1984:19).

As a part of my doctoral research, basic geological attributes were examined at Toka to hypothesize about site formation processes and re-assess its stratigraphic context, partly in light of these claims. The resulting observations (Chauhan and Gill 2002) refuted Verma and Srivastava's claims through several types of evidence: (1) the concerned artifacts are Soanian (i.e., a combination of Modes 1 and 3) rather than being pre-Acheulean, (2) they derive from post-Siwalik Upper Pleistocene contexts capping the Tatrot sediments, (3) the post-Siwalik artifacts happened to be deflated on the underlying Tatrot sediments instead of eroding out of them, and (4) artifacts and vertebrate fossils are also mixed and thus not contemporaneous with each other. This was all confirmed through two test-trenches on a post-Siwalik terrace as well as a pre-existing water pipeline trench across a part of Toka through exposures of the Tatrot Formation (Chauhan 2008a). Where artifacts were found within Tatrot sediments or seemingly eroding out of them, they actually represented results of re-burial and/or re-exposure through colluvial action, monsoon-related surface runoff, or downslope displacement (Mohapatra and Singh 1979a; Chauhan and Gill 2002). In contrast, the contextual integrity of the artifacts appears to be associated with the post-Siwalik sedimentary layers above the Tatrot beds, implying a considerably younger age as Verma and Srivastava (1984:17) originally considered, but then negated, from observations at 75 localities (of only 5–15 artifacts at each location):

> Close association of stone artefacts and vertebrate fossils throughout the area under examination poses an intriguing problem as whether to accept them to be of a common stratigraphic level or taking one (fossils) as Pliocene in age and the artefacts of a later period, and accidental. This however, seems highly improbable.

From general observations by the author, this "close" but misleading association of stone artifacts and vertebrate fossils appears to be a result of winnowing and deflation from erosion and seasonal fluvial processes on the underlying Tatrot sediments (the source of the fossils) and post-Siwalik sediments (the source of the artifacts) in addition to the lack of post-Siwalik sedimentation (i.e., the lack of artifact burial) at different places on the site. Ultimately, it is presumed that the archeological material is not older than the associated post-Siwalik raw material source (i.e., Tirlokpur Nadi) since the Tatrot Formation exposures here and elsewhere does not contain any quartzite clasts for stone tool manufacture (Gill 1983).

Uttarbaini (Northern India)

A claim of an Early Pleistocene lithic occurrence has also been made from the Jammu-and-Kashmir region of northern India. Here, Verma (1989, 1991) reports artifacts *in situ* from below a tuffaceous layer which was initially dated to 1.6 ± 0.2 Ma and then re-dated to 2.8 ± 0.5 by Ranga Rao et al. (1988) using the fission-track method. However, this claim has not been verified through further detailed investigations including excavations and the application of other geochronological techniques. Not only does the ash require re-dating (R.K. Ganjoo, personal communication, 2008), but the context of the artifacts and their archeological integrity for that matter, also need to be confirmed. For example, no photographs of the section, the stratified ash or the artifacts are provided and a description of the assemblage composition is also lacking. Nonetheless, this alleged occurrence warrants a systematic survey of the area in light of the ash deposits whose inconsistent age may be more accurately constrained using the well-established Upper Siwalik biochronology (Nanda 2002; Dennell 2004) in addition to re-dating.

Riwat and the Pabbi Hills Assemblages (Northern Pakistan)

The best-studied but also the most controversial pre-Acheulean lithic evidence in South Asia is currently known from the Siwalik Hills of northern Pakistan and includes the ca. 2.0 myr site at Riwat (Rendell et al. 1989) and the 2.2–0.9 Mya old Mode 1 assemblages from the nearby Pabbi Hills (Hurcombe 2004). The work was carried out by the British Archaeological Mission to Pakistan (BAMP) which lasted almost two decades and also yielded the most securely-bracketed Acheulean find-spots in the subcontinent at Dina and Jalalpur (Rendell and Dennell 1985) and a unique post-Siwalik Paleolithic site dated to ~45 ka. (Rendell et al. 1989). The lithics at Riwat were first noticed in 1983 and then studied during subsequent visits over a decade. The site is a part of the Soan Syncline, a landscape-level geological feature that dips at an angle of about 10–15° on its southern edge (Rendell et al. 1989). Previous paleomagnetic applications (Burbank and Johnson 1983) had suggested that the Syncline formed in the Late Pliocene or between 1.9 and 2.1 Ma. Additionally, a volcanic tuff from the overlying horizontally-bedded fluvial sediments subsequently indicated an age of 1.6 ± 0.2 Ma through K/Ar. According to Dennell (2007), scientists have never questioned the age or the dating of the Soan Syncline stratigraphic sequence, which implies that the tilted artifact-bearing horizon was established prior to the folding and is considerably older than the overlying ca. 1.6 Ma old horizontal strata. Later work by Rendell et al. (1987, 1989) sought to confirm the Late Pliocene age of the concerned strata and also demonstrated that the artifact-bearing horizon was a prominent stratum of the Soan Syncline rather than being channel fill of a younger age. Through the collection of 280 samples from 71 sampling locations (with mean spacing of 1.7 m), they confirmed the magnetic polarity to belong to the Matuyama Chron (see Dennell 2007). The stratigraphic context of the Riwat assemblage is the lower gritstone/conglomerate horizon (LGC) in the syncline. An upper gritstone is 100–200 ka and the overlying loess yielded a TL date of 74.3±8.3 ka. A hemispherical disc core and a rolled handaxe were collected from this upper gritstone horizon.

Riwat

To distinguish between naturally-flaked clasts – a common feature in conglomeratic deposits – and genuine artifacts, the investigators developed a methodology based on experimentation and ranked the various archeological specimens based on length, breadth, thickness, flake features, number of directions of flake removal, percentage of remaining cortex, positive/negative scars, evidence of retouch, edge roundedness and post-depositional damage (Rendell et al. 1989). Ultimately, 23 flaked quartzite specimens were initially collected and ranked (see Table 7.2 in Rendell et al. 1989:110) after observing over 1,000 cobbles within the LGC, but only three specimens have been promoted as being the most convincing artifacts and thus have received the most attention (Fig. 10.3). Specimen R001 was first observed in 1983 to be imbedded in a gritstone/comglomerate horizon near the base of an erosional gully. This large core has eight or nine flake scars in three different directions and the size of the flake scars are thought to be comparable to Oldowan evidence from Bed I of Olduvai Gorge (Dennell 2007). It is worth noting however, that the overall dimensions and morphology of the actual core is rather large and pointed, unlike typical Oldowan cores which are generally smaller and amorphous. Except for the cortical butt, the remainder of the specimen exhibits flaking on its two faces, also not common in the Oldowan in general. Specimen R014 was extracted from a gritstone block that had fallen from the same horizon nearby and is represented by a large flake struck from a cobble and possesses a prominent bulb of percussion as well as ripple marks and again, eight flake scars from three different directions. The third specimen, R88/1, is a fresh Type-5 flake (Toth 1985) with a prominent bulb of percussion and with evidence of additional flaking from three directions and has positive and negative flake scars on each respective side. It was recovered from a freshly-eroding vertical section and 50 m from Specimen R001. Two additional specimens (R88/5 and R88/6) were originally counted as artifacts but were later discounted because they were thought to derive from post-Siwalik colluvial fill in the area. Overall, 1,264 clasts were plotted and studied in the LGC, however no additional artifacts were recovered. Therefore, in addition to the ca 2.0 Ma age requiring substantiation, the Riwat sample is meager and comes from a gravel horizon (i.e., secondary context) and thus, offers little behavioral information. On the other hand, the Pabbi Hills evidence offers paleoanthropologists greater behavioral and technological information, in spite of its surface context.

Pabbi Hills Assemblages

In comparison to the Riwat evidence, the Pabbi Hills assemblages (Hurcombe and Dennell 1993) are considerably richer but have often been overshadowed by the former. Following the work at Riwat, the main research objectives of BAMP in the Pabbi Hills were to pinpoint the context of the artifacts, confirm the Early Pleistocene age of much of the material, and distinguish between artifacts and naturally-fractured stone in the region. Only one find spot is reported and the rest of the specimens are surface occurrences distributed across the Siwalik landscape. The find spot is that of a stone tool on the surface of an escarpment of Sandstone 14. It fit in an

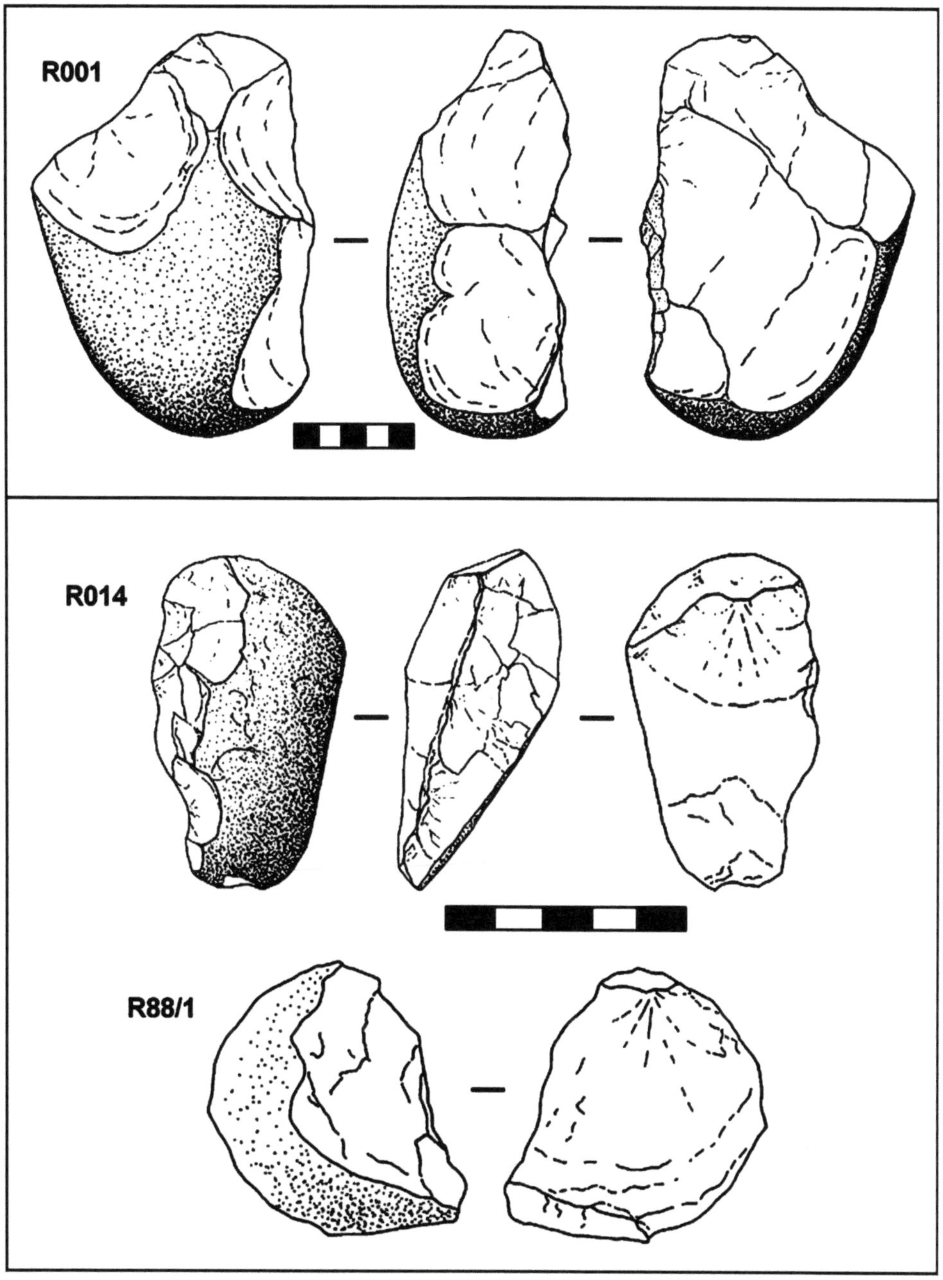

Fig. 10.3 The three main specimens from Riwat in northern Pakistan (Modified from Dennell 2007)

in situ socket located in a secondary channel context above the location from which it was recovered. A total of 607 specimens were interpreted as hominin-produced (Fig. 10.4 and Table 10.2) but their density of occurrence was observed to be very low: out of 211 locations where flaked-stone was recovered, isolated pieces occurred in 45% of the cases and no more than three specimens were found in 78% of the instances (Dennell 2007). The fluvial channel that deposited the sediments in the region was a part of a large floodplain environment and helped in establishing the depositional history of the area as well as possibly explained the absence of certain types of artifacts. Some of the possible debitage specimens collected weighed as little as 1 g. The investigators (see Dennell 2004) have also attempted to chronologically divide the entire assemblage based on the specimens' surface association with the underlying sediments and associated biochronology and stratigraphic correlation. For example, 102 specimens were distributed on sediments dated to 0.9–1.2 Ma; 307 specimens were collected from the surface of Sandstone 12, interpreted to be between 1.2 and 1.4 Ma; and 198 specimens were collected from a surface interpreted to be between 2.2 and 1.7 Ma. Approximately 41% are cores and 58% are

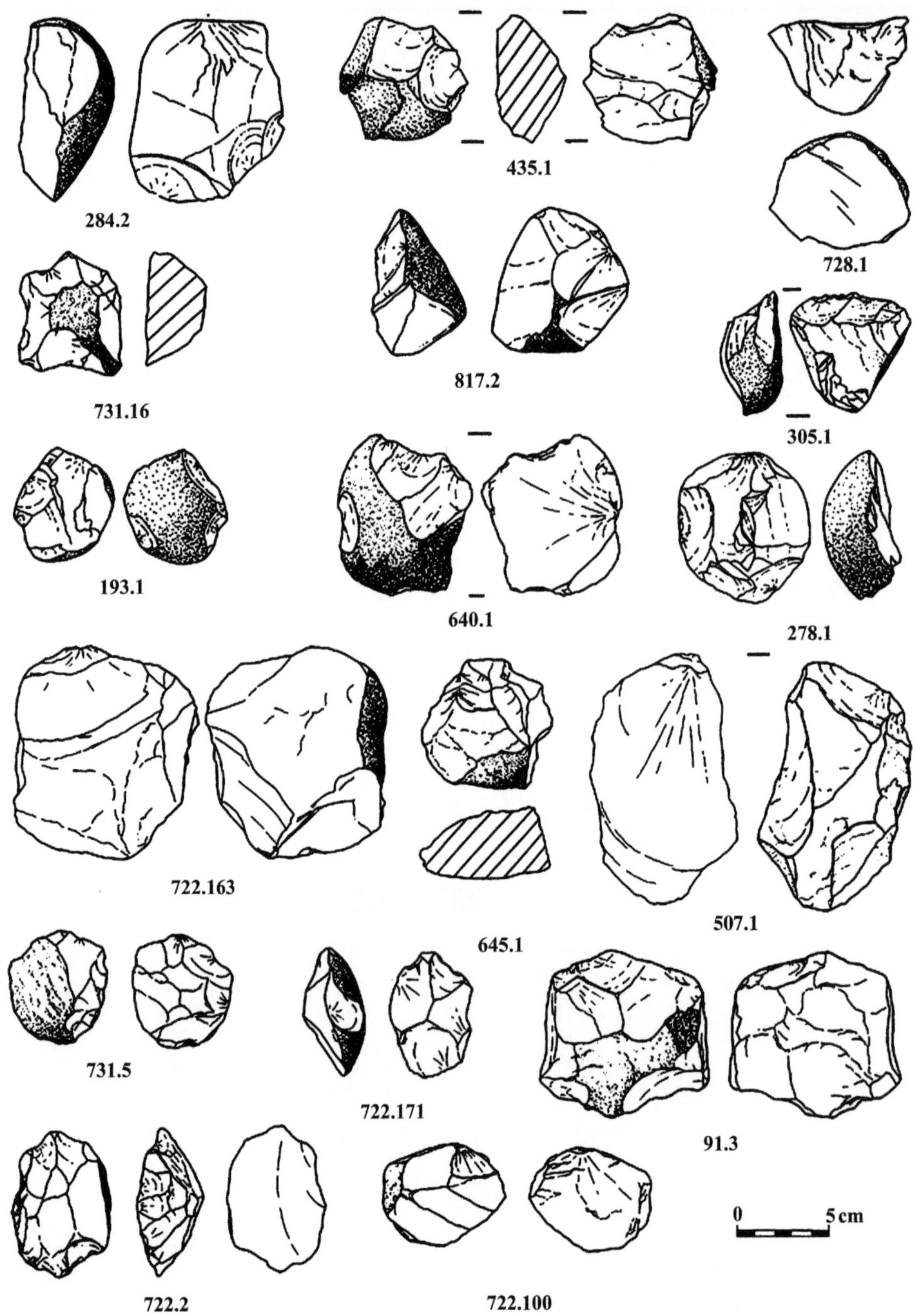

Fig. 10.4 Examples of lithic artifacts from the Pabbi Hills in northern Pakistan (After Hurcombe 2004: Fig. 8.12)

flakes and the majority of specimens (96%) were produced on quartzite with 2.8% of all specimens showing deliberate retouch (Hurcombe 2004). Six micro-cores, four hammerstones and six fragments of polished stone axes were made on flint and are thought to possibly belong to the Neolithic or a later phase.

Despite being typologically comparable to most typically African Oldowan assemblages, the investigators (see Dennell 2004, 2007) maintain their interpretations of the lithic artifacts and defend the contextual and behavioral integrities of the Pabbi Hills evidence using the following lines of argument: (a) except for the Neolithic-like specimens, there is a virtual lack in the region of artifacts from younger time periods such as the Acheulean and Middle or Upper Paleolithic, (b) the archeological evidence cannot be road or rail ballast because the latter is generally smashed, not flaked and the artifacts were found at higher elevations and several kilometers from the nearest road/railway, (c) there is no evidence of lithic or

Table 10.2 The Pabbi Hills assemblage composition (Modified from Hurcombe 2004)

Artifact type	Quantity	%	Retouched pieces	Quartzite	Flint	Other rock
Proto-biface	1	0.1	–	1	–	–
Straight edged core tool	5	0.8	3	5	–	–
Flake blade core	6	1.0	–	6	–	–
High action core tool (HACT)	14	2.3	–	14	–	–
Microcore	6	1.0	–	–	6	–
Squared core	10	1.6	–	10	–	–
Bipolar core	4	0.7	–	4	–	–
Shattered wedge core fragment	51	8.4	–	50	–	1
Irregular core	30	4.9	–	30	–	–
Split cobble	16	2.6	–	16	–	–
Unifacial core	16	2.6	–	16	–	–
Bidirectional/bifacial core	48	7.9	–	48	–	–
Simple disc core (discoid)	23	3.8	–	22	–	1
Complex disc core (discoid)	14	2.3	–	14	–	–
Polyhedron	3	0.5	–	3	–	–
TOTAL CORES/CORE-TOOLS	247	40.7	3	239	6	2
Shatter-split flake	2	0.3	–	2	–	–
Angular flake	11	1.8	–	11	–	–
Irregular flake	110	18.1	1	105	1	4
(Decortication flakes)	(21)	–	–	–	–	–
Split flake	64	10.5	–	60	3	1
(Decortication flakes)	(4)	–	–	–	–	–
Double bulb flake	5	0.8	–	5	–	–
Rounded simple flake	76	12.5	1	75	–	1
(Decortication flakes)	(39)	–	–	–	–	–
Rounded complex flake	13	2.1	–	12	–	1
Straight-edged flake	37	6.1	3	37	–	–
Flake blade	14	2.3	2	14	–	–
Scraper	6	1.0	6	5	1	–
HACT flake	6	1.0	–	6	–	–
Knife	1	0.2	1	1	–	–
TOTAL FLAKES/FLAKE TOOLS	345	56.8	14	333	6	6
TOTAL CHIPPED STONE	592	97.5	17	572	12	8
Ring stone	2	0.3	–	1	–	1
Rubbing stone	1	0.2	–	1	–	–
Rubbing stone flake	2	0.3	–	–	–	2
Polished stone fragment	6	1.0	–	2	–	4
TOTAL GROUND/POLISHED	11	1.8	0	4	0	7
Hammerstone flake	1	0.2	–	1	–	–
Hammerstone	3	0.5	–	3	–	–
TOTAL HAMMERSTONES	4	0.7	0	4	0	0
TOTAL SPECIMENS	607		17	580	12	15
% In assemblage			2.8	95.6	2.0	2.5

fossil material eroding/deflating from younger contexts and (d) it is unlikely (from the currently-observable erosional processes) that such old artifacts and fossils could have remained on the surface of these formations throughout the Middle and Upper Pleistocene; thus they have eroded out in recent decades from the underlying sediments. Dennell (2007:60) has further stated:

> As none of this material was found *in situ*, the case for dating it to the Early Pleistocene remains circumstantial. Nevertheless this type of field survey data forms an important part of the archaeological literature, and those readers who might reject this evidence on the grounds that it was found on the surface might reflect how much other data collected by field surveys elsewhere should also be rejected.

However, the most reliable chronological frameworks for early hominin occupation throughout the Old World have primarily come from well-excavated sites in fine-grained stratified contexts that were directly dated on an absolute scale.

Claims of Pre-Middle Pleistocene Hominin Fossils

The majority of human fossil material in the subcontinent comprises Late Pleistocene and Holocene specimens of *H. sapiens* from various parts of India and Sri Lanka (Kennedy 1999, 2001). The *Homo* calvaria from Hathnora in the central Narmada Valley in central India is thought to be of at least Late Pleistocene age (Patnaik et al. 2009) and currently represents the oldest and only pre-modern hominin fossil evidence in the Indian subcontinent (Sonakia and Biswas 1998). Two other finds alleged to be older have generally been ignored in published reviews, primarily because of their doubtful status as hominin fossils. For the sake of being comprehensive, they are formally included as a part of this critique. The two finds are thought to be of Early Pleistocene and Late Pliocene age respectively and both come from Nadah and Khetpurali in the Siwalik region near Chandigarh in northern India.

Nadah

The hominin specimen from Nadah is represented by a left maxillary central incisor recovered from the Pinjor Formation and attributed to *Homo erectus* (Singh et al. 1988). Although the lead author mentions his discovery of 'at least' three teeth thought to belong to three different individuals, the primary focus in their paper is on the one incisor. The specimen was discovered in 1985 from a buff-colored mudstone stratum 200 m from the base of the section and the basal portion of the Pinjor Formation and 0.5 km south of Nadah village. The other two incisors were recovered from 100 m away from the main specimen. The stratigraphic sequence here is approximately 275 m in total vertical height and consists of gray and greenish sandstones inter-bedded with brown and purple siltstone as well as cemented conglomerate and hard sandstone. Based on associated biostratigraphy and comparative stratigraphy of the Tatrot-Pinjor contact, the specimen is estimated to be 2.0–2.2 myr old. Singh and colleagues describe the tooth in great morphological detail and attempt to systematically prove the hominin classification of the specimen. They list several key features to invalidate it as a Siwalik hominoid but do not systematically compare the specimen with other mammalian species. Features thought to be characteristic to hominids are highlighted by the authors as: (a) a distinct occlusal abrasion pattern; (b) arched occlusal contour; (c) axial curvature of the longitudinal axis of the root and crown; (d) Pattern-3 prism of enamel structure; and (e) mesiodistal and labiolingual metric data falling in the then-known range of other hominid dental evidence (see Table I in Singh et al. 1988:570). Given the often ambiguous morphological overlap between certain hominin teeth and other mammalian species, the purported incisor(s) from Nadah remains circumstantial; it cannot be accepted as hominin until more diagnostic fossil specimens are recovered. At the very least, this specimen may represent a non-hominin primate incisor (A. Sahni, personal communication, 2003).

Khetpurali and Masol

Most recently, Singh (2003) has also reported hominid mandibular and post-cranial fragments in association with stone tools from the Tatrot Formation near Khetpurali Village. From available paleomagnetic dating results on the known Upper Siwalik Formations, he proposed an age of ca. 3.4 Ma for the evidence. All fossil and lithic specimens are reported to have eroded out from a brown siltstone bed approximately 10 m from the base of the Tatrot Formation which is thought to be 220 m thick here. An additional 50 vertebrate taxa were also recovered at and around this locality. The mandibular fragment comprises a lower right first molar (M1) and the alveoli of the third and fourth premolars and Singh highlights several features including (a) the low position of the mental foramen below the mesial root of M1; (b) pattern of worn enamel and dentine; and (c) the transversely thick horizontal rami as well as other related details such as a facet on the cusp and the width of the root. M1 is thought to be metrically double the size of that tooth in *Homo sapiens*. Later work allegedly yielded more hominid fossils including a similar mandibular ramus with the P3, P4, M1 and alveoli of the canine present, a proximal-end of a left femur, the distal-end of a left femur and a right patella (from a brown clay at the basal portion of the Tatrot section near Masol village). Unfortunately, the descriptions are not very helpful, but the fossil specimens are presumably classifiable as various large vertebrates rather than hominin (R. Patnaik, personal communication, 2003). The associated lithic illustrations are also of poor quality and lack a photographic scale. There are also no comparative tables or related data (i.e., geochronology, sedimentation, metric data for the fossils and lithics and so forth) that are normally found in the current literature regarding paleoanthropological finds of such significance. Early Paleolithic artifacts of a Mode 1 nature (including quartzite and ivory) are mentioned but the investigator does not provide any qualitative or quantitative details of the material except that they are made from quartzite pebbles and comprise unifacial and bifacial choppers. In that respect, they can easily be (and probably are) significantly younger Soanian assemblages that derived from younger contexts nearby. Overall, these localities reported by Singh (2003) do not appear to merit further scientific attention except for their geological and paleontological aspects.

Discussion and Conclusion

Although paleoanthropological evidence of pre-Middle Pleistocene age has been recovered in various parts of Western and Central Asia (Langbroek 2004), possibly the Arabian Peninsula (Petraglia 2003) and Southeast Asia (Swisher et al. 1994), the South Asian framework is not compatible with the current global paradigm. Even though there has been marginal topographical and ecological variation in specific regions within the subcontinent (e.g., Western and Easter Ghats or Deccan Trap) since the Late Pliocene, we currently have little information regarding climate and faunal dynamics and the potentially early *Homo* presence (if any) from the peninsular region. Well-preserved floral and behavioral faunal associations with Paleolithic assemblages are also rare in South Asia (Chauhan 2008b). As a consequence, we are not yet aware of the types of environmental and climatic contexts the majority of these assemblages were produced in. In contrast to Middle Pleistocene *Homo* populations, the earliest hominin groups that migrated out of Africa may not have been adequately successful in adapting to new landscapes and marginal lithic resources (Dennell 2003, 2007). This adaptive success is possibly reflected in the visible continuity of the South Asian archeological record from the Brunhes-Matuyama boundary onwards.

Despite the still controversial nature of the Riwat and Pabbi Hills lithic evidence from northern Pakistan, it remains the best-studied evidence among all claims reviewed in this paper. The Pabbi Hills material is morphologically and dimensionally more similar to classic Oldowan assemblages than is the Riwat evidence. However, because the Riwat assemblage is meager and comes from a gravel context and the Pabbi Hills evidence comes from surface contexts, both remain circumstantial and require corroboration through excavated sites in fine-grained primary contexts. Indeed, Dennell (2007:41) himself states:

> However, apart from a small amount of material *that remains controversial* [emphasis mine] from Riwat (Dennell et al. 1988) and the Pabbi Hills, Pakistan (Dennell 2004; Hurcombe 2004), there is no incontrovertible evidence that hominins were living in the northern part of the Indian subcontinent in the Early Pleistocene, even though it is the obvious corridor route between Southwest and Southeast Asia.

The other reports of pre-Acheulean occurrences in the Siwalik Hills (e.g., Uttarbaini) and Narmada Valley (e.g., Durkadi) of northern and central India require reinvestigations of their stratigraphic contexts and the precise age of the behavioral evidence. The sites of Kheri-Jhiran and Mahadeo Piparia appear to respectively represent a mixture of various lithic industries – Mode 1, Acheulean and younger assemblages. The claims of hominid fossils and stone tools from Late Pliocene and Late Miocene contexts at Nadah and Khetpurali (Singh et al. 1988; Singh 2003) respectively, require substantiation. It is important to note that the early Siwalik evidence does not necessarily suggest the presence or dispersal of Oldowan technology southwards into peninsular India. Although numerous Mode 1 assemblages are known from the Konkan coast, Karnataka, Uttar Pradesh, Bihar and West Bengal, Orissa, Andhra Pradesh and northeastern India (Jayaswal 1982), they are small and come from secondary contexts and remain typologically undiagnostic (Chauhan 2010). This inconspicuous absence of Oldowan occupation in the subcontinent, particularly peninsular India, may be attributable to a number of factors or scenarios, discussed below (see Chauhan 2006):

1. *Older evidence has not yet been properly recognized or was overlooked.* This is a possibility, given the virtual absence of contemporaneous vertebrate fossil material with which stone tools might be associated. In addition, goal-oriented and multidisciplinary investigations focusing on the recovery of Oldowan or Oldowan-age evidence has not yet been carried out systematically in the Indian Siwaliks or peninsular India. A few Mode 1 artifacts, especially specimens like primary flakes, can be easily overlooked if the age of the underlying sediments is unknown or if such evidence (i.e., pre-Acheulean) is not theoretically expected to exist in a given region. For example, if evidence comparable to the Pabbi Hills material was found on Pinjor sediments in the Indian Siwaliks or in peninsular India, it would probably be automatically interpreted as belonging to the Soanian industry, since the raw material and many tool types are identical. Indeed, Paleolithic archeologists in the Indian Siwaliks have rarely tested the stratigraphic integrity of the numerous 'seemingly Soanian' lithic scatters found on surfaces of Tatrot and Pinjor Formations.
2. *The hominin groups that initially occupied northern Pakistan did not continue into lower latitudes of peninsular India at that time and peninsular India was not colonized until the Brunhes-Matayuma boundary.* If systematic surveys and long-term research efforts fail to reveal older evidence in peninsular India, then early hominins may not have entered this region until about 800–700 ka, possibly due to various ecological or climatic factors, or other reasons currently unknown (Petraglia 1998). The lack of adaptive success has also been recently explored by Dennell (2003) and he also states the lack of raw materials in the region during the Late Pliocene-Early Pleistocene, resulting in a marginal hominin presence there (Dennell 2007). Instead, existing early hominin groups (e.g., northern Pakistan) may have dispersed along the Siwalik corridor to reach SE Asia or simply may have taken a northern route through Central and East Asia. This still raises one important issue: while the Indo-Gangetic plains may have acted

as a barrier between the Siwalik range and peninsular India, there is no explanation for why hominins may not have entered the peninsular region through routes closer to the Arabian Sea, areas which provided ample raw material unlike the Siwaliks and Indo-Gangetic floodplains.

3. *Such evidence has not been well-preserved and/or well-exposed and thus extremely rare and difficult to recover.* In a recent study on correlating collagen loss with fossil/lithic site location in Asia, Holmes et al. (2006) have demonstrated that the quality and mode of preservation of such material in southern Asia is comparable to the evidence from parts of Africa. More importantly, they state:

> Without a proper understanding of the limitations of the Asian data set it is not possible to truly evaluate the evidence for Palaeolithic site distribution across this vast area. (p. 15)

This holds true especially for the entire Indian subcontinent and almost all of peninsular India. A plausible reason for the current lack of paleoanthropological evidence older than the Middle Pleistocene in peninsular India may be that temporally-corresponding sedimentary exposures are not widespread and/or have not been extensively identified. In north-central India, suitable sediments lie deeply buried under Gangetic alluvium (Misra 2001) and in peninsular India, early Quaternary sedimentation was spatially restricted by bedrock topography and erosional and tectonic factors (see Mishra 2006–2007). The only sediments of Late Pliocene-Early Pleistocene age (in addition to the Upper Siwalik Sub-group and the Karewa deposits) in peninsular India are restricted to a few localities and even these are currently doubtful or unconfirmed through absolute dating. Furthermore, the most prominent evidence in South Asia, for a coastal route from East Africa to SE Asia, may be submerged from fluctuating sea levels (e.g., Flemming 2004). However, goal-oriented systematic surveys and detailed excavations of known sites will prove to be rewarding. In northern Pakistan, Dennell (2007) attributes the lack of fossil hominin material in the otherwise rich vertebrate fossil evidence to such possible factors as: (a) taphonomic bias towards the preservation of larger mammals (Dennell 2008), (b) seasonal flash floods, (c) water-borne infections and illnesses, and (d) episodic major flood events every 2–3 decades – hypothesized from modern analogs and thus having implications on raw material availability and procurement. Some of these factors may also be applicable in other regions of the subcontinent (e.g., Brahmaputra delta).

Although it is possible that peninsular India may have been bypassed by Late Pliocene/Early Pleistocene hominins, the current lack of Oldowan assemblages does not reflect an unquestionable absence of hominin occupation in the region. Obviously, much more research is required to test and confirm such hypotheses. Systematic surveys of known areas (e.g., Upper Siwalik Formations, Karewas) and new exposures (e.g., in peninsular India) are critically desired for confirming and modeling the earliest hominin dispersals between East Africa and Southeast Asia. The subcontinent's topo-geographic separation, by the Greater Himalayas and Tibetan Plateau, from the rest of the continent must have affected the movement and environmental adaptations of early *Homo*. Ecologically conducive environments, however, in the form of open grasslands, a seasonal monsoon regime, diverse fauna and eco-habitats (i.e., diverse hunting/scavenging opportunities), and an abundance of water and stone resources suggest the possible earlier presence of hominins in South Asia (Korisettar and Rajaguru 1998, 2002; Petraglia 2005). For example, Dennell (2004:213) points out that East African and South Asian climates were essentially similar between c. 2.0–1.5 Ma and an eastward dispersal "…can be regarded as simply a latitudinal dispersion…" Evidences gathered from the evolutionary histories of Siwalik murids (notably rats and mice), that have specific life histories and adapt to unpredictable climatic conditions, suggest that the monsoon system in the subcontinent started intensifying by the Early Pliocene (Patnaik 2003; also see Gupta and Mélice 2003). Many of these fossil murid sites geographically overlap with monsoon regimes as well as with hominid sites (Patnaik, 2010). Indirectly therefore, this relatively early development and a considerable longevity of the South Asian monsoon (Retallack 1995) also hints at the possibility of Early Pleistocene hominin presence in peninsular India.

The foregoing review and critique of the South Asian pre-Acheulean paleoanthropological evidence has included several claims of both lithic and hominid-fossil occurrences from northern Pakistan and two separate regions of India, respectively. All of these reported occurrences represent either unsubstantiated claims or require further scientific verification or corroboration through additional sites. Until we have absolute dates from multidisciplinary excavations in primary fine-grained sedimentary contexts, we need to be more cautious when interpreting currently insufficient and equivocal data. This is also provisionally supported from the lack of Late Pliocene/Early Pleistocene stone tools in SE Asia (Corvinus 2004) and the need for better chronological and contextual control at key site localities in Asia (e.g., Swisher et al. 1994; Zhu et al. 2004). Corresponding evidence that may support a Late Pliocene/Early Pleistocene occupation of Asia is also lacking at the moment in key intermediate regions such as the Iranian plateau and Afghanistan. Although Oldowan sites have been reported from Yemen and Djibouti (see Chauhan 2009b), they remain undated and associated Plio-Pleistocene land-bridge connections at the Strait of Bab-al-Mandab remain ambiguous (see Derricourt 2005). The earliest occupation of the Indian subcontinent currently appears to be by Early Acheulean hominins at the Brunhes-Matuyama polarity transition or slightly earlier. However, systematic multidisciplinary investigations in the

Siwalik Hills, Lower Karewa deposits and contemporaneous geological exposures in other parts of India may yet yield unequivocal evidence of the initial southern technological dispersal from East Africa to Southeast Asia.

Acknowledgements I am thankful to John Fleagle for the invitation to participate in this intellectually stimulating workshop and the Stony Brook students, John Shea and Richard Leakey for the excellent organization, making it a truly memorable event. John Fleagle and John Shea provided useful comments on an earlier draft of this paper. I thank Nicholas Toth and Kathy Schick of the Stone Age Institute for the continued support of my research. My paleoanthropological research in India has been funded by the Australian Research Council, National Geographic Society, Wenner-Gren Foundation for Anthropological Research, Fulbright Scholar Program and INQUA. This paper is a modified and updated version of Chauhan (2009c). Any errors or assumptions in this paper are entirely my own.

References

Agrawal, D. P. (1988). Palaeoclimatic data from Kashmir: A synthesis and some correlations. *Proceedings of the Indian National Science Academy, 54*(3), 333–342.

Agrawal, D. P., Kotlia, B. S., & Kusumgar, S. (1988). Chronology and significance of the Narmada formations. *Proceedings of the Indian National Science Academy, 54*(3), 418–424.

Agrawal, D. P., Dodia, R., Kotlia, B. S., Razdan, H., & Sahni, A. (1989). The Plio-Pleistocene geologic and climatic record of Kashmir Valley, India: A review and new data. *Palaeogeography, Palaeoclimitology, Palaeoecology, 73*, 267–286.

Agrawal, D. P., Dodia, R., Kotlia, B. S., Razdan, H., & Sahni, A. (1990). The Plio-Pleistocene geologic and climatic record of the Kashmir Valley, India: A review and new data. *Man and Environment, 15*(1), 57–72.

Alam, M., Alam, M. M., Curray, J. R., Chowdury, M. L. R., & Gani, M. R. (2003). An overview of the sedimentary geology of the Bengal Basin in relation to the regional tectonic framework and basin-fill history. *Sedimentary Geology, 155*, 179–208.

Antón, S., & Swisher, C. C., III. (2004). Early dispersals of homo from Africa. *Annual Reviews in Anthropology, 33*, 271–296.

Armand, J. (1979). The Middle Pleistocene pebble tool site of Durkadi in Central India. *Paleorient, 5*, 105–144.

Armand, J. (1983). *Archaeological excavations in the Durkadi Nala: An early Palaeolithic pebble-tool workshop in Central India.* Delhi: Munshiram Manoharlal Publishers.

Armand, J. (1985). The emergence of the handaxe tradition in Asia, with special reference to India. In V. N. Misra & P. Bellwood (Eds.), *Recent advances in Indo-Pacific prehistory* (pp. 3–8). New Delhi: Oxford/IBH.

Athreya, S. (2007). Was *Homo heidelbergensis* in South Asia? A test using the Narmada fossil from central India. In M. D. Petraglia & B. Allchin (Eds.), *The evolution and history of human populations in South Asia: Inter-disciplinary studies in archaeology, biological anthropology, linguistics and genetics* (pp. 137–170). Dordrecht, The Netherlands: Springer.

Badam, G. L. (1979). *Pleistocene fauna of India.* Pune: Deccan College Postgraduate and Research Institute.

Bar-Yosef, O., & Belfer-Cohen, A. (2001). From Africa to Eurasia – early dispersals. *Quaternary International, 75*, 19–28.

Brozovi , N., & Burbank, D. W. (2000). Dynamic fluvial systems and gravel progradation in the Himalayan foreland. *Geological Society of America Bulletin, 112*(3), 394–412.

Burbank, D. W., & Johnson, G. D. (1983). The late Cenozoic chronologic and stratigraphic development of the Kashmir Intermontane Basin, northwestern Himalaya. *Palaeogeography, Palaeoclimitology, Palaeoecology, 43*, 205–235.

Cameron, D., Patnaik, R., & Sahni, A. (2004). The phylogenetic significance of the Middle Pleistocene Narmada hominin. *Journal of Osteoarchaeology, 14*, 419–447.

Chauhan, P. R. (2003). An overview of the Siwalik Acheulian & reconsidering its chronological relationship with the Soanian – A theoretical perspective. Assemblage 7. Retrieved from http://www.assemblage.group.shef.ac.uk/issue7/chauhan.html

Chauhan, P. R. (2006). Human origins studies in India: Position, problems and prospects. Assemblage 9. Retrieved from http://www.assemblage.group.shef.ac.uk/issue9/chauhan.html

Chauhan, P. R. (2007). Soanian cores and core-tools from Toka, northern India: Towards a new typo-technological organization. *Journal of Anthropological Archaeology, 26*, 412–441.

Chauhan, P. R. (2008a). Soanian lithic occurrences and raw material exploitation in the Siwalik Frontal Zone, northern India: A geoarchaeological perspective. *Journal of Human Evolution, 54*, 591–614.

Chauhan, P. R. (2008b). Large mammal fossil occurrences and associated archaeological evidence in Pleistocene contexts of peninsular India and Sri Lanka. *Quaternary International, 192*, 21–42.

Chauhan, P. R. (2009a). The lower paleolithic of the Indian Subcontinent. *Evolutionary Anthropology, 18*, 62–78.

Chauhan, P. R. (2009b). Early *Homo* occupation near the gate of tears: Examining the paleoanthropological records of Djibouti and Yemen. In E. Hovers & D. Braun (Eds.), *Inter-disciplinary approaches to the Oldowan* (pp. 49–59) (Book Series: Vertebrate Paleobiology and Paleoanthropology). New York: Kluwer.

Chauhan, P. R. (2009c). Was there an Oldowan occupation in the Indian Subcontinent?: A critical appraisal of the earliest paleoanthropological evidence. In: *The Cutting Edge: New Approaches to the Archeology of Human Origins.* K. Schick and N. Toth (Eds.). Stone Age Institute Press. pp. 33–54.

Chauhan, P. R. (2010). Core-and-flake assemblages of central and peninsular India. In C. J. Norton & D. Braun (Eds.), *Asian paleoanthropology: From Africa to China and beyond.* Dordrecht, The Netherlands: Springer pp. 113–120.

Chauhan, P. R., & Gill, G. S. (2002). The impact of geological and anthropogenic processes on prehistoric sites on Siwalik slopes: A case study. *Bulletin of the Indian Geologists Association, 35*, 71–81.

Clark, J. D. (1994). The Acheulian industrial complex in Africa and elsewhere. In R. S. Corruccini & R. L. Ciochon (Eds.), *Integrative paths to the past- paleoanthropological advances in honor of F. Clark Howell* (pp. 451–469). Englewood Cliff, NJ: Prentice Hall.

Cordaux, R., Aunger, R., Bentley, G., Nasidze, I., Sirajuddin, S. M., & Stoneking, M. (2004). Independent origins of Indian caste and tribal paternal lineages. *Current Biology, 14*, 231–235.

Corvinus, G. (2004). *Homo erectus* in East and Southeast Asia, and the questions of the age of the species and its association with stone artifacts, with special attention to handaxe-like tools. *Quaternary International, 117*, 141–151.

De Terra, H., & Paterson, T. T. (1939). *Studies on the Ice Age in India and associated human cultures.* Carnegie Institute Publication 493. Washington, DC: Carnegie Institution of Washington.

Dennell, R. W. (2003). Dispersal and colonisation, long and short chronologies: How continuous is the Early Pleistocene record for hominids outside East Africa? *Journal of Human Evolution, 45*(6), 421–440.

Dennell, R. W. (2004). *Early hominin landscapes in Northern Pakistan: Investigations in the Pabbi hills.* BAR International Series 1265. Oxford: Archaeopress.

Dennell, R. (2007). "Resource-rich, stone-poor": Early hominin land use in large river systems of northern India and Pakistan. In M. D. Petraglia & B. Allchin (Eds.), *The evolution and history of human populations in South Asia – Interdisciplinary studies in archaeology, biological anthropology, linguistics and genetics* (Vertebrate Paleobiology and Paleoanthropology Series). Dordrecht: Springer.

Dennell, R. W. (2008). The taphonomic record of Upper Siwalik (Pinjor stage) landscapes in the Pabbi Hills, northern Pakistan, with consideration regarding the preservation of hominin remains. *Quaternary International, 192*, 62–77.

Dennell, R. W., & Hurcombe, L. (1992). Paterson, the British Clactonian and the soan flake industry: A re-evaluation of the Early Palaeolithic of Northern Pakistan. In C. Jarrige (Ed.), *South Asian archaeology* (pp. 69–72). Madison, WI: Prehistory Press.

Dennell, R. W., & Roebroeks, W. (2005). An Asian perspective on early human dispersal from Africa. *Nature, 438*, 1099–1104.

Dennell, R. W., Rendell, H. R., & Hailwood, E. (1988). Early tool-making in Asia: Two-million-year-old artefacts in Pakistan. *Antiquity, 62*, 98–106.

Deo, S. G., Mishra, S., Rajaguru, S. N., & Ghate, S. (2007). Antiquity of Acheulian culture in upland Maharasthra: A geoarchaeological approach. In A. R. Sankhyan (Ed.), *Genome and people of India* (pp. 292–308). Calcutta: Concept Publishers.

Derricourt, R. (2005). Getting "Out of Africa": Sea crossings, land crossings and culture in the hominin migrations. *Journal of World Prehistory, 19*, 119–132.

Dhir, R. P., Tandon, S. K., Sareen, B. K., Ramesh, R., Rao, T. K. G., Kailath, A. J., et al. (2004). Calcretes in the Thar desert: Genesis, chronology and palaeoenvironment. *Proceedings of the Indian Academy of Sciences – Earth and Planetary Sciences, 113*, 473–515.

Dowsett, H., Thompson, R., Barron, J., Cronin, T., Fleming, F., Ishman, S., et al. (1994). Joint investigations of the Middle Pliocene climate I: PRISM paleoenvironmental reconstructions. *Global and Planetary Change, 9*, 169–175.

Field, J. S., & Lahr, M. M. (2005). Assessment of the Southern dispersal: GIS-based analyses of potential routes at oxygen isotopic stage 4. *Journal of World Prehistory, 19*, 1–45.

Flemming, N. C. (2004). Submarine prehistoric archaeology of the Indian continental shelf: A potential resource. *Current Science, 86*(9), 1225–1230.

Gabunia, L., Vekua, A., Lordkipanidze, D., Swisher, C. C., Ferring, R., Justus, A., et al. (2000). Earliest Pleistocene cranial remains from Dmanisi, Republic of Georgia: Taxonomy, geological setting, and age. *Science, 288*, 1019–1025.

Gaillard, C., (1995). An Early Soan assemblage from the Siwaliks: a comparison of processing sequences between this assemblage and of an Acheulian assemblage from Rajasthan. In: Wadia, S., Korisettar, R., Kale, V.S. (Eds.), Quaternary Environments and Geoarchaeology of India—Essays in Honour of Professor S.N. Rajaguru. Geological Society of India, Bangalore, pp. 231–245.

Gaillard, C., & Mishra, S. (2001). The Lower Palaeolithic in South Asia. In Origin of Settlements and Chronology of the Palaeolithic Cultures in Southeast Asia. Colloque International de la Fondation Singer Polinac (Under the direction of F. Sémah, C. Falgueres, D. Grimaud-Herve, and A. Sémah). Paris, 3–5 June, 1998. Semenanjung, Paris, pp. 73–92.

Ganjoo, R. K. (1990). The Plio-Pleistocene boundary in India: A re-appraisal. *Man and Environment, 15*(2), 29–34.

Gill, G. S. (1983). Sedimentology of the Siwalik Group exposed between the rivers Ghaggar and Markanda- Northwestern Himalaya. *Publication of the Centre of Advanced Study in Geology, Punjab University, 13*, 274–312.

Gupta, A. (1995). Fluvial geomorphology and Indian rivers. In S. Wadia, R. Korisettar, & V. S. Kale (Eds.), *Quaternary environments and geoarchaeolgy of India* (pp. 52–60). Bangalore: Geological Society of India.

Gupta, A. K., & Mélice, J.-L. (2003). Orbital forcing of the Plio-Pleistocene Indian monsoon: Benthic foraminiferal proxies from ODP Site 758. *Current Science, 85*(2), 179–184.

Holmes, K., Robson-Brown, K. A., Oates, W. P., & Collins, M. J. (2006). Assessing the distribution of Asian Palaeolithic sites: A predictive model of collagen degradation. *Journal of Archaeological Science, 33*, 971–986.

Hurcombe, L. (2004). The stone artefacts from the Pabbi Hills. In R. W. Dennell (Ed.), *Early hominin landscapes in Northern Pakistan: Investigations in the Pabbi Hills* (pp. 222–292). British Archaeological Reports International Series 1265. Oxford, Archaeopress.

Hurcombe, L. M., & Dennell, R. W. (1993). A Pre-Acheulean in the Pabbi Hills, northern Pakistan? In C. Jarrige (Ed.), *South Asian Archaeology 1989* (pp. 133–136). Proceedings of the International Conference of South Asian Archaeologists in Western Europe, Paris, July 1989.

Jayaswal, V. (1982). *Chopper-chopping component of palaeolithic India*. New Delhi: Agam Kala Prakashan.

Johnson, N. M., Opdyke, N. D., Johnson, N. D., Lindsay, E. H., & Tahirkheli, R. A. K. (1982). Magnetic polarity stratigraphy and ages of Siwalik Group rocks of the Potwar Plateau, Pakistan. *Palaeogeography, Palaeoclimatology, Palaeoecology, 37*, 17–42.

Joshi, R. V., Rajaguru, S. N., Pappu, S. N., & Bopardikar, B. P. (1974). Quaternary glaciation and palaeolithic sites in the Liddar Valley (Jammu-Kashmir). *World Archaeology, 5*(3), 369–379.

Kennedy, K. A. R. (1999). Paleoanthropology of South Asia. *Evolutionary Anthropology, 8*(5), 165–185.

Kennedy, K. A. R. (2001). Middle and Late Pleistocene hominids of South Asia. In P. V. Tobias, M. A. Raath, J. Moggi-Cecci, & G. A. Doyle (Eds.), *Humanity from African naissance to coming millennia* (pp. 167–174). Firenze: University of Firenze.

Kennedy, K. A. R. (2003). *God-apes and fossil men: Paleoanthropology of South Asia*. Ann Arbor, MI: University of Michigan Press.

Kennedy, K. A. R., & Chiment, J. (1991). The fossil hominid from the Narmada valley, India: Homo erectus and Homo sapiens? In P. Bellwood (Ed.), *Indo – Pacific prehistory 1990* (vol. 1, pp. 42–58). Canberra: Indo – Pacific Prehistory Association.

Khatri, A. P. (1962). Mahadevian: An Oldowan pebble culture of India. *Asian Perspectives, 6*(1), 186–197.

Khatri, A. P. (1966). Origin and evolution of hand-axe culture in the Narmada Valley (Central India). In D. Sen & A. K. Ghosh (Eds.), *Studies in prehistory (Robert Bruce Foote Memorial Volume)* (pp. 96–121). Calcutta: Firma K.L. Mukhopadhyay.

Korisettar, R., & Rajaguru, S. N. (1998). Quaternary stratigraphy, palaeoclimate and the Lower Palaeolithic of India. In M. D. Petraglia & R. Korisettar (Eds.), *Early human behaviour in global context: The rise and diversity of the Lower Palaeolithic record* (pp. 304–342). London: Routledge.

Korisettar, R., & Rajaguru, S. N. (2002). The monsoon background and the evolution of prehistoric cultures of India. In K. Paddayya (Ed.), *Recent studies in Indian archaeology* (pp. 316–348). New Delhi: Munshiram Manoharlal Publishers.

Kumar, R., Ghosh, S. K., & Phadtare, N. R. (Eds.). (1994). *Siwalik Foreland basin of Himalaya*. New Delhi: Oxford/IBH.

Langbroek, M. (2004). *'Out of Africa': An investigation into the earliest occupation of the Old World* (BAR International Series 1244). Oxford: Archaeopress.

Larick, R., & Ciochon, R. L. (1996). The African Emergence and Early Asian Dispersals of the Genus Homo. *American Scientist, 84*, 538–552 (The Magazine of Sigma Xi, The Scientific Research Society). November–December.

Lycett, S. (2007). Is the Soanian techno-complex a Mode 1 or Mode 3 phenomenon? A morphometric assessment. *Journal of Archaeological Science, 34*, 1434–1440.

Martínez-Navarro, B., & Palombo, M. R. (2004). Occurrence of the Indian genus *Hemibos* (Bovini, Bovidae, Mammalia) at the Early–Middle Pleistocene transition in Italy. *Quarternary Research, 61*, 314–317.

Meijaard, E., & Groves, C. P. (2004). Morphometrical relationships between South-east Asian deer (Cervidae, tribe Cervini): Evolutionary and biogeographic implications. *Journal of Zoology, 263*, 179–196.

Mishra, S. (1994). The South Asian Lower Palaeolithic. *Man and Environment, 19*(1–2), 57–72.

Mishra, S. (1995). Chronology of the Indian Stone Age: The impact of recent absolute and relative dating attempts. *Man and Environment, 20*(2), 11–16.

Mishra, S. (2006–2007). The Indian Lower Palaeolithic. *Bulletin of the Deccan College Research Institute, 56/57*, 47–94.

Mishra, S., Gaillard, C., Hertler, C., Moigne, A-M, Simanjuntak, T. (2010). India and Java: Contrasting records, intimate connections, *Quaternary International.* doi:10.1016/j.quaint.2009.11.040

Misra, V. N. (1987). Middle Pleistocene adaptations in India. In O. Soffer (Ed.), *The Pleistocene old world- regional perspectives* (pp. 99–119). New York: Plenum.

Misra, V. N. (1997). Early man and his environment in Central India. *Journal of the Palaeontological Society of India, 42*, 1–18.

Misra, V. N. (2001). Prehistoric human colonization of India. *Journal of Bioscience, 26*(4), 491–531.

Mohapatra, G. C. (1981). Acheulian discoveries in the Siwalik frontal range. *Current Anthropology, 22*(4), 433–435.

Mohapatra, G. C. (1985). Cultural ecology of early man in India. In S. B. Deo & K. Paddayya (Eds.), *Recent advances in Indian archaeology (Proceedings of the Seminar held in Poona in 1983)* (pp. 23–28). Pune: Deccan College Post-graduate and Research Institute.

Mohapatra, G. C., & Singh, M. (1979a). Stratified occurrence of lithic artifacts in the Siwalik frontal range of western sub-Himalaya. *Punjab University Research Bulletin (Arts), 10*(1–2), 65–77.

Morley, R. J. (2000). *Origin and evolution of tropical rain forests*. Chichester: Wiley.

Nanda, A. C. (2002). Upper Siwalik mammalian faunas of India and associated events. *Journal of Asian Earth Sciences, 21*, 47–58.

Pal, D. K., Deshpande, S. B., Venugopal, K. R., & Kalbande, A. R. (1989). Formation of di- and trioctahedral Smectite as evidence for paleoclimatic changes in Southern and Central Peninsular India. *Geoderma, 45*, 175–184.

Pappu, R. S. (2001). *Acheulian culture in Peninsular India*. New Delhi: D.K. Printworld (P) Ltd.

Pappu, S., Gunnell, Y., Taieb, M., Brugal, J., & Touchard, Y. (2003). Excavations at the Palaeolithic Site of Attirampakkam, South India: Preliminary findings. *Current Anthropology, 44*, 591–597.

Patnaik, R. (2003). Reconstruction of Upper Siwalik palaeoecology and palaeoclimatology using microfossil palaeocommunities. *Palaeogeography, Palaeoclimitology, Palaeoecology, 197*, 133–150.

Patnaik, R. (2010). Fossil murine rodents as ancient monsoon indicators of the Indian subcontinent, *Quaternary International.* doi: 10.1016/j.quaint.2010.04.005

Patnaik, R., & Nanda, A. C. (2010). Early Pleistocene mammalian faunas of India and evidence of connections with other parts of the world. In J. G. Fleagle, J. J. Shea, F. E. Grine, A. L. Baden, & R. E. Leakey (Eds.), *Out of Africa I: The first hominin colonization of Eurasia* (pp. 129–143). Dordrecht, The Netherlands: Springer.

Patnaik, R., Chauhan, P. R., Rao, M. R., Blackwell, B. A. B., Skinner, A. R., Sahni, A., et al. (2009). New geochronological, palaeoclimatological and Palaeolithic data from the Narmada Valley hominin locality, central India. *Journal of Human Evolution, 56*, 114–133.

Petraglia, M. D. (1998). The Lower Palaeolithic of India and its bearing on the Asian record. In M. D. Petraglia & R. Korisettar (Eds.), *Early human behaviour in global context: The rise and diversity of the Lower Palaeolithic record* (pp. 343–390). London: Routledge.

Petraglia, M. D. (2001). The Lower Palaeolithic of India and its behavioural significance. In L. Barham & K. Robson-Brown (Eds.), *Human roots: Africa and Asia in the Middle Pleistocene* (pp. 217–233). Bristol: Western Academic & Specialist Press.

Petraglia, M. D. (2003). The Lower Paleolithic of the Arabian Peninsula: Occupations, adaptations, and dispersals. *Journal of World Prehistory, 17*(2), 141–179.

Petraglia, M. D. (2005). Hominin responses to Pleistocene environmental change in Arabia and South Asia. *Geological Society of London, Special Publications, 247*, 305–319.

Petraglia, M. D. (2010). The Early Paleolithic of the Indian Subcontinent: Hominin colonization, dispersals and occupational history. In J. G. Fleagle, J. J. Shea, F. E. Grine, A. L. Baden, & R. E. Leakey (Eds.), *Out of Africa I: The first hominin colonization of Eurasia* (pp. 165–179). Dordrecht, The Netherlands: Springer.

Prasad, K. V. (1974). The vertebrate fauna from Piram Island, Gujarat, India. *Palaeontologica Indica* (Memoirs of the Geological survey of India XLI) *41*, 1–23.

Quade, J., Cerling, T. E., & Bowman, J. R. (1989). Development of Asian monsoon revealed by marked ecological shift during the latest Miocene in northern Pakistan. *Nature, 402*, 163–166.

Rajaguru, S. N., Mishra, S., & Ghate, S. (1996). A re-interpretation of gravel spreads in the Jaisalmer region, Thar Desert, India. *Journal of Arid Environments, 432*, 53–58.

Ranga Rao, A., Agrawal, R. P., Sharma, U. N., Bhalla, M. S., & Nanda, A. C. (1988). Magnetic polarity stratigraphy and vertebrate palaeontology of the Upper Siwalik Subgroup of Jammu Hills, India. *Journal of the Geological Society of India, 32*, 109–128.

Rao, K. V., Chakraborti, S., Rao, K. J., Ramani, M. S. V., Marathe, S. D., Borkar, B. T. (1997). Magnetostratigraphy of the Quaternary Fluvial sediments and Tephra of Narmada valley, Central India. In S. G. Udhoji & M. P. Tiwar (Eds.), *Quaternary geology of the Narmada valley- A multidisciplinary approach* (pp. 65–78). Special Publication No. 46 of the Geological Survey of India. Calcutta: Geological Survey of India.

Rendell, H. M., & Dennell, R. W. (1985). Dated Lower Palaeolithic artifacts from Northern Pakistan. *Current Anthropology, 26*, 93.

Rendell, H. M., Hailwood, E. A., & Dennell, R. W. (1987). Magnetic polarity stratigraphy of Upper Siwalik Sub-Group, Soan valley, Pakistan: Implications for early human occupation of Asia. *Earth and Planetary Science Letters, 85*, 488–496.

Rendell, H. M., Dennell, R. W., & Halim, M. A. (1989). Pleistocene and Palaeolithic investigations in the Soan valley, Northern Pakistan. B.A.R. International Series 544. Oxford: Archaeopress.

Retallack, G. J. (1995). Palaeosols of the Siwalik Group as a 15 Myr record of South Asian Palaeoclimate. *Memoirs Geological Society of India, 32*, 36–51.

Sangode, S., Mishra, S., Naik, S., Deo, S. (2007). Magnetostratigraphy of the Quaternary sediments associated with some Toba tephra and Acheulian artefact bearing localities in the western and central India. *Gondwana Magazine*, Special Volume 10, 111–121.

Sangode, S. J., Bloemendal, J., Kumar, R., & Ghosh, S. K. (2001). Plio-Pleistocene pedogenic changes in the Siwalik palaeosols: A rock magnetic approach. *Current Science, 81*(4), 387–392.

Sankalia, H. D. (1971). New evidence for early man in Kashmir. *Current Anthropology, 12*(4–5), 538–561.

Sankalia, H. D. (1974). *The prehistory and protohistory of India and Pakistan*. Pune: Deccan College and Postgraduate Research Institute.

Sankhyan, A. R. (1997). A new human fossil from the Central Narmada Basin and its chronology. *Current Science, 73*, 1110–1111.

Sankhyan, A. R. (2005). New fossils of Early Stone Age man from Central Narmada valley. *Current Science, 88*, 704–707.

Sanyal, P., Bhattacharya, S. K., Kumar, R., Ghosh, S. K., & Sangode, S. J. (2004). Mio-Pliocene monsoonal record from Himalayan foreland basin (Indian Siwalik) and its relation to vegetational change. *Palaeogeography, Palaeoclimitology, Palaeoecology, 205*, 23–41.

Sémah, F., Saleki, H., & Falguères, C. (2000). Did Early Man reach Java during the Late Pliocene? *Journal of Archaeological Science, 27*, 763–769.

Sen, D., & Ghosh, A. K. (1963). Lithic culture – complex in the Pleistocene sequence of the Narmada Valley, Central India. *Rivista Di Scienze Prestoriche, 18*, 3–23.

Sharma, J. C. (1977). Palaeolithic tools from Pleistocene deposits in Punjab, India. *Current Anthropology, 18*(1), 94–95.

Sharma, K. K. (1995). Quaternary stratigraphy and prehistory of the Upper Indus Valley, Ladakh. In S. Wadia, R. Korisettar, & V. S. Kale (Eds.), *Quaternary environments and geoarchaeology of India-Essays in honour of Professor S.N. Rajaguru* (pp. 98–108). Bangalore: Geological Society of India.

Singh, M. P. (2003). First record of a Middle Pliocene hominid from the Siwalik Hills of South Asia. *Human Evolution, 18*, 213–228.

Singh, M. P., Sahni, A., Kaul, S., & Sharma, S. K. (1988). Further evidence of hominid remains from the Pinjor Formation, India. *Proceedings of the Indian National Science Academy, 54*, 564–573.

Singhvi, A. K., Bronger, A., Pant, R. K., & Sauer, W. (1987). Thermoluminescence dating and its implications for the chronostratigraphy of loess-paleosol sequences in the Kashmir Valley (India). *Chemical geology (Isotopic Geoscience Section), 65*, 45–56.

Sonakia, A. (1984). The skull cap of early man and associated mammalian fauna from Narmada Valley alluvium, Hoshangabad area, Madhya Pradesh, India. *Records of the Geological Survey of India, 113*, 159–172.

Sonakia, A., & Biswas, S. (1998). Antiquity of the Narmada *Homo erectus*, the early man of India. *Current Science, 75*(4), 391–393.

Soni, A. S., Soni, V. S. (2005). Paleolithic tools from the surface of optically stimulated luminescence dated alluvial fan deposits of Pinjaur Dun in NW sub-Himalayas. *Current Science* 88(6), 867–871.

Stiles, D. (1978). Palaeolithic artifacts in Siwalik and post-Siwalik deposits. *Kroeber Anthropology Society Papers, 53–54*, 129–148.

Supekar, S. G. (1968). *Pleistocene stratigraphy and prehistoric archaeology of the Central Narmada Basin*. Ph.D. Dissertation, Deccan College Postgraduate and Research Institute, Pune.

Supekar, S. G. (1985). Some observations on the Quaternary stratigraphy of the central Narmada valley. In V. N. Misra & P. Bellwood (Eds.), *Recent advances in Indo-Pacific prehistory* (pp. 19–27). New Delhi: Oxford/IBH.

Swisher, C. C. III., Curtis, G. H., Jacob, T., Getty, A. G., Suprijo, A., Widiasmoro. (1994). Age of earliest known hominids in Java, Indonesia. *Science, 263*, 1118–1121.

Tiwari, M. P., & Bhai, H. Y. (1997). Quaternary stratigarphy of the Narmada valley. In S. G. Udhoji & M. P. Tiwari (Eds.), *Quaternary geology of the Narmada valley- A multidisciplinary approach* (pp. 33–63). Special Publication No. 46 of the Geological Survey of India. Calcutta: Geological Survey of India.

Toth, N. (1985). The Oldowan reassessed: A close look at early stone artifacts. *Journal of Archaeological Science, 12*, 101–120.

Verma, B. C. (1975). Occurrence of Lower Palaeolithic artefacts in the Pinjor member (Lower Pleistocene) of Himachal Pradesh. *Journal of the Geological Society of India, 16*(4), 518–521.

Verma, B. C. (1989). Time stratigraphic position of the Early Palaeolithic culture in the Siwaliks of northwest India based on fission-track age of the associated sediments. *Current Science, 58*, 242–244.

Verma, B. C. (1991). Siwalik Stone Age culture. *Current Science, 61*(8), 496.

Verma, B. C., & Srivastava, J. P. (1984). Early Palaeolithic sites of the outer Siwalik belt in Sirmur District, Himachal Pradesh – their age and stratigraphic position. *Man and Environment, 8*, 13–19.

Wang, L. J. (1994). Sea surface temperature history of the low latitude western Pacific during the last 5.3 million years. *Palaeogeogragpy, Palaeoclimatology, Palaeoecology, 108*, 379–436.

Woldegabriel, G., Heiken, G., White, T. D., Asfaw, B., Hart, W. K., & Renne, P. R. (2000). Volcanism, tectonism, sedimentation, and the paleoanthropological record in the Ethiopian Rift System. *Geological Society of America, Special Paper, 345*, 83–99.

Zhu, R. X., Potts, R., Xie, F., Hoffman, K. A., Deng, C. L., Shi, C. D., et al. (2004). New evidence on the earliest human presence at high northern latitudes in northeast Asia. *Nature, 431*, 559–562.

Chapter 11
The Early Paleolithic of the Indian Subcontinent: Hominin Colonization, Dispersals and Occupation History

Michael D. Petraglia

Abstract The Indian subcontinent contains a multitude of Early Paleolithic sites which are important for understanding dispersal processes and the paleodemography of early humans. Mode I sites appear to be scarce whereas Acheulean occurrences are found in some abundance, particularly within basins that provided a perennial water supply, high biomass, and raw material sources. Though the record of Acheulean habitation appears to extend over the long term in South Asia, site distributions and site counts do not necessarily imply that populations were large and permanent in any particular region. Acheulean tool-making traditions in South Asia contrast considerably with contemporaneous tool assemblages in East Asia, though there are intriguing technological similarities with the Chinese stone tool assemblages from the Luonan Basin, suggesting a more complicated scenario for the evolution of populations in Asia.

Keywords Indian subcontinent • Dispersals • Basin model • Mode I • Soan • Acheulean

Introduction

The dispersal of early hominins out of Africa and their colonization of Asia is a significant topic in human evolutionary studies. The remarkable finds at Dmanisi and the dating of the Indonesian localities have provided new insights into hominin dispersals at the Plio-Pleistocene boundary (e.g., Swisher et al. 1994; Gabunia et al. 2000). Given that the earliest finds for the presence of hominins outside of Africa are relatively scarce, researchers have grappled with the nature of the evidence and its implications with respect to 'long' and 'short' chronologies (e.g., Rolland 1998, 2001; Dennell 2003). Although some have argued that the earliest Plio-Pleistocene dispersals into Eurasia indicate relatively continuous occupation as part of multiple dispersal events (e.g., Mithen and Reed 2002; Antón and Swisher 2004), others contend that early dispersals were temporally and spatially discontinuous, implying short-term success after colonizations (e.g., Dennell 2003). In Arabia and the Levant there is ample evidence for Acheulean dispersals by 1.4 Ma and later (Bar-Yosef 1998; Bar-Yosef and Belfer-Cohen 2001; Petraglia 2003, 2005), while in East Asia the sample of sites increases substantially after ca. 1 Ma ago (e.g., Schick 1994; Zhu et al. 2001). The Middle Pleistocene localities in Eurasia indicate a more permanent record of hominin presence in comparison with earlier periods. Important questions remain about the geographic distribution of hominin populations and how such groups differentiated through range expansions and subsequent isolation (Lahr and Foley 1998; Foley 1999).

South Asia plays a pivotal role in any discussion of Out of Africa dispersals given its central geographic position between western and eastern Asia and its low latitude position. Indeed, the Indian subcontinent contains Early Paleolithic assemblages that should be informative about hominin colonization and dispersal processes as well as the relative success of the populations in adapting to new environments (e.g., Petraglia 1998, 2001) (Fig. 11.1). The goal of this chapter is to explore the role of South Asia in the processes of hominin expansions along the southern dispersal route.

Geography and Paleoenvironments

Early Paleolithic occupation of the subcontinent varied according to geographic and environmental conditions. The subcontinent contains a number of distinct regions and ecozones, including the Greater and Lesser Himalayan mountain range, the sub-Himalayan Siwalik Hills, the Indo-Gangetic plain and the peninsula. The Himalayas and their associated mountain ranges to the west and east mark a dramatic northern boundary to the Indian subcontinent. The uplifted and folded deposits of the Siwalik Hills of the sub-Himalayas

M.D. Petraglia (✉)
Research Laboratory for Archaeology and the History of Art, University of Oxford, Dyson Perrins Building, South Parks Road, OX1 3QY Oxford, UK
e-mail: michael.petraglia@rlaha.ox.ac.uk

J.G. Fleagle et al. (eds.), *Out of Africa I: The First Hominin Colonization of Eurasia*, Vertebrate Paleobiology and Paleoanthropology,
DOI 10.1007/978-90-481-9036-2_11, © Springer Science+Business Media B.V. 2010

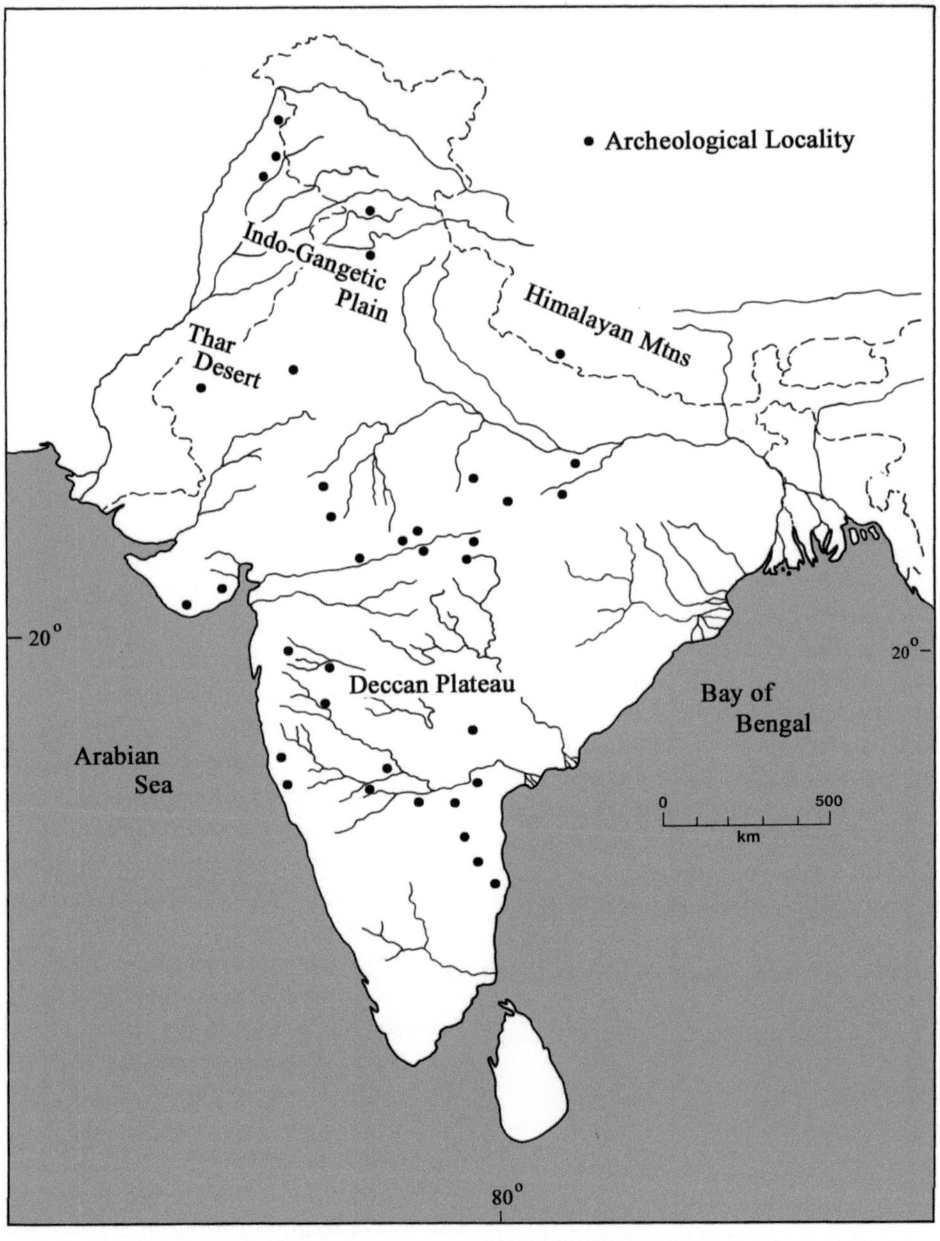

Fig. 11.1 The Indian subcontinent showing key Early Paleolithic localities and site complexes

contain Pleistocene fauna and Paleolithic localities (e.g., Rendell et al. 1989; Dennell 2004). In the south, the peninsula represents one of the oldest and least geologically disturbed large land masses in the world, often referred to as the Deccan Plateau. The peninsula and its basins and river valleys have yielded a large number of Acheulean sites in a variety of ecological and topographic settings (e.g., Pappu 1995; Dennell 1998; Korisettar and Rajaguru 1998; Petraglia 1998, 2001, 2006; Korisettar 2002). As will be described below, the basins of the subcontinent have particular spatial boundaries, and these would have influenced Paleolithic occupation of the region (see Fig. 11.2). Between the Himalayas and the peninsula lies the Indo-Gangetic plain, which has deep alluvial deposits stretching from the Indus River of Pakistan in the west to the eastern deltaic lowlands of Bangladesh. These large rivers and deltas may have acted as barriers, thereby limiting dispersals of mammals and hominins (Kretzoi 1961–1964; Field and Lahr 2005; Field et al. 2007). During the Early Pleistocene, the relative paucity of available lithic raw materials in the Indo-Gangetic plains may have also prevented successful colonization of the region (Dennell 2007).

The Asian–Indian monsoon is one of the most important features of planetary atmospheric circulation and it must have played a role in changing hominin adaptations

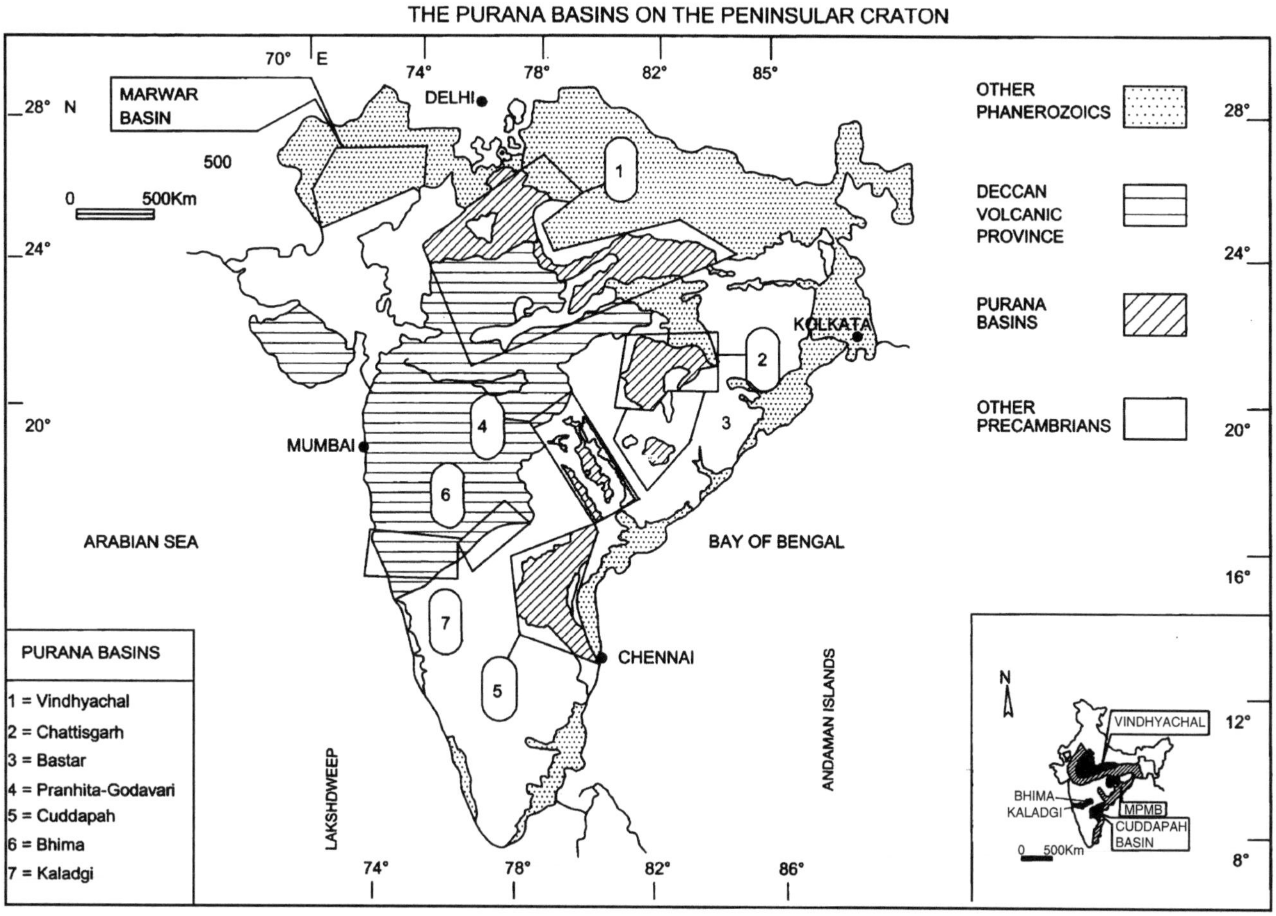

Fig. 11.2 Map showing major Purana basins of peninsula India (After Korisettar 2007: Fig. 3). The smaller Gondwana basins are at the margins of the Purana basins. Korisettar has made the argument that the central focus of Paleolithic settlement was in the basins, and these would have strongly influenced dispersals in the subcontinent

(e.g., Baker et al. 1995; Wasson 1995; Bowler et al. 1995; Korisettar and Ramesh 2002). The major part of the Indian subcontinent has been a monsoonal environment since the Miocene, although fluctuations and shifts in its intensity are registered through time, which may be related to Himalayan–Tibetan uplifts (Retallack 1995; An et al. 2001). Paleoclimatic research has shown that during the Quaternary, the intensity of the monsoon fluctuated in accordance with global oscillations, and the prevalence of wet and dry cycles controlled floral communities (e.g., Joshi 1970; Agrawal 1992; Bowler et al. 1995; Clemens et al. 1996). Deep-sea cores from the central Indian Ocean record changes in radiolarian plankton during the last 1.4 Ma, with cyclical changes over the last 500 Ka suggesting monsoonal oscillations due to orbital forcing (Gupta and Fernandes 1997; Gupta 1999). Indian riverine systems are distinguished by their strong seasonality in flow as governed by the monsoon (Gupta 1995; Baker et al. 1995). From the distribution of rainfall patterns, a variety of ecosystems are formed, including deciduous woodlands, tropical evergreen forests, savannas, semi-arid to arid scrub lands and deserts.

Hydrological changes are recorded in the Thar Desert, indicating change from wet to dry environments and decrease in stream flow (Misra 1987, 1995). Forests in the mountains along the western border of the country fluctuated between woodland and savanna ecosystems, in turn affecting habitats across the Deccan Plateau (Korisettar 1994; Korisettar and Rajaguru 1998). Given the long-term operation of the fluctuating monsoon over the Pliocene and Pleistocene, sedimentary records show some uniformity in geomorphic processes, phases of deposition and erosion, and formation of laterites and ferricretes (Korisettar 1994; Ollier 1995; Korisettar and Rajaguru 1998; Pappu 1999). Paddayya (1982) has surmised that hominins had to adapt to seasonal changes in the monsoon, suggesting that settlement was marked by wet season dispersal of groups and dry season aggregation of groups near spatially limited water supplies. High resolution ecological information is limited, although soil carbonate analysis of Early Pleistocene deposits in the Upper Siwaliks indicate a grassland setting (Quade et al. 1993) with *Equus*, *Rhinoceros*, *Elephas* and *Stegodon* (Turner 2004). On the peninsula, associations of fossil ungulates with archeological localities attest to the prevalence of savanna habitats with ponded facies and riverine gallery forests (Korisettar and Rajaguru 1998). Faunal remains include *Bos*, *Bubalus*, *Cervus*, *Elephas*, *Equus*, *Gazella*, *Hexaprotodon*, *Leptobos*, *Stegodon* and *Sus* (Badam 1984; Misra 1989). Whereas mammalian faunas have been recovered in several areas, only one early human fossil has been found to date, a calvaria from the middle reaches of the Narmada River Valley (Sonakia 1985), classified as belonging to a large-brained form variably attributed to *Homo erectus* or *Homo heidelbergensis* (e.g., Kennedy 1999; Cameron et al. 2004; Athreya 2007).

Earliest Occupation of the Subcontinent

The Upper Siwaliks and the Earliest Colonization

If the long chronology for hominin occupations of Eurasia is correct, there should be an expectation that Mode I assemblages of Late Pliocene or Early Pleistocene age would be recovered in South Asia. Indeed, one might expect that the Plio-Pleistocene environments of peninsular India would be particularly attractive, with their high biomass in grassland and savanna settings (Dennell 1998, 2010). Yet, the evidence for early Mode I occupations of the subcontinent is meager to non-existent.

The Riwat locality in the Soan Valley represents one of the best potential cases for a Late Pliocene presence of hominins in the subcontinent. The Riwat locality contains a small number of flaked pieces that are from a boulder conglomerate context dating to ca. 1.9 Ma or more (Burbank and Reynolds 1984; Dennell et al. 1988; Rendell et al. 1989). Two pieces that are considered to be definitive cultural objects were obtained from the conglomerate. Other pieces thought to be cultural objects were found in an associated conglomerate horizon. Great effort has been put on establishing the contexts of the finds and the cultural nature of the flaked pieces. In a second important locality in the Upper Siwaliks, archeological investigations in the Pabbi Hills has yielded stone artifacts on erosional surfaces of fossiliferous deposits (Dennell 2004). Overall, 607 pieces of flaked stone were found that were considered to be made by hominins. The artifacts consisted of simple cores and flakes. The pieces were found in very low densities across exposures. Flaked pieces were found in 211 locales, mostly consisting of isolated pieces where no more than three objects were found at a single place (Hurcombe 2004). Approximately half (n = 307) of the stone pieces were found on exposures dating to 1.4–1.2 Ma; 102 were found on exposures dating to 0.9–1.2 Ma; and 198 pieces on exposures of deposits that belonged to, or that were earlier than, the Olduvai Subchron, ca. 1.7–2.2 Ma.

If such information can be further supported with additional evidence, the Upper Siwaliks would certainly take center stage in an investigation of the early colonization of the subcontinent. While much effort has been expended on documenting the age and context of the finds and evaluating the artifactual nature of the stone pieces, additional information is needed to bolster the current findings and dispel lingering doubts (e.g., Petraglia 1998; Klein 1999). Other than in the Upper Siwaliks, there is no place on the Indian subcontinent where analysts can convincingly claim an early hominin presence.

The Soan Industry and Its Antiquity

Archeologists have assumed that the discovery of simply flaked stone tools would provide evidence for the earliest hominin occupation in the subcontinent. Unifacially and bifacially flaked cobbles and pebbles found on surfaces of tributaries of the Indus system, the Himalayan foothills, and the Potwar plateau were potential candidates for an early presence of hominins. The identification of Pre-Soan, Early Soan and Late Soan tool types were thought to form an evolutionary sequence and a separate and distinct tradition from the handaxe and cleaver industry (DeTerra and Paterson 1939). After their initial discovery, the identification of these cobble and pebble tools was assumed to be evidence for the antiquity of early hominins in the region, though this has been a contentious and confusing subject given the lack of stratigraphic control and chronometric ages.

Archeologists working in South Asia have recognized and maintained the typological and technological distinction between Soan and Acheulean industries (e.g., Sankalia 1974; Paddayya 1984; Misra 1987, 1989). Soan industries typically consist of unifacial tools on cores and rarer bifacial pieces, whereas the Acheulean is usually defined on the basis of standardized bifaces (i.e., handaxes, cleavers, picks) on cores and flakes and unifacial tool types on cores and retouched flakes. While there is a generally recognized technological and typological difference between these industries, the Soan assemblages occasionally contain bifacially flaked cores and choppers (Gaillard 2006; Chauhan 2007) and the Achuelean sometimes contains unifacial large cutting tools, suggesting that the Soan–Acheulean tool-type dichotomy is not as distinct as traditionally portrayed (Petraglia 1998). Indeed, a comparison of Soan artifacts from the Siwaliks with Acheulean materials from Singi Talav concluded that there were close overlaps and parallels in the processing sequences and forms of the two industries (Gaillard 1994, 1995). In this view, Soan assemblages contained pebbles and cobbles that were trimmed along their edges both unifacially and bifacially, some forms reminiscent of tools associated with Acheulean assemblages, including so-called "cleavers" and "handaxes". These observations were supported by examination of Acheulean assemblages that contained bifaces that were roughly trimmed, thus retaining characteristics of the original clast. From a processing viewpoint, it was argued that the Soan and Acheulean forms show convergences in striking and trimming of pieces (Gaillard 1995).

Assemblage variation in reduction techniques and tool forms is tied, in part, to raw material form. The Soan industry is the result of tool manufacture on pebbles and cobbles; as a consequence, core-flake assemblages may be more commonly found in areas such as the Himalayan foothills and outwash of northern India where extensive gravel deposits occur (Petraglia 1998). Recent work on Soan and Acheulean industries in the Siwaliks confirms that available clast size is an important factor which conditions tool morphology (Chauhan 2007). Within Acheulean assemblages it appears that hominins selected various clast forms to manufacture different items. Thus, spatial variations in tool type frequencies may reflect raw material availability and type; for instance, the frequency of choppers in Acheulean assemblages varies from 43% in north-central India to only 11% in southeastern India (Ghosh 1985). Within sites, variations in tool types can be found, such as at Adamgarh Rockshelter, where choppers were often made on cobbles and handaxes and cleavers were manufactured on large clasts and flakes (Joshi 1978; Petraglia 2006).

It has been pointed out that some Soan and unifacial chopper tool assemblages contain Levallois elements indicating closer approximation to Middle Paleolithic industries than to Mode I occurrences (Corvinus 1998; Chauhan 2007; Lycett 2007a). As a result, it has been argued that some Soan localities are not contemporaneous with older Mode I assemblages or Acheulean sites. In such a view, Soan assemblages are not useful for discussions about the earliest dispersals into the subcontinent or for comparisons with older Mode I occurrences east of the Movius Line (Lycett 2007b). Indeed, current interpretation is that the early stratigraphic observations for the Soan series was incorrect (Dennell 1995) and that the great majority of the familiar Soan localities (Jayaswal 1982) should not be considered to be early hallmarks of hominin occupation in the subcontinent. Although it is possible that some Soan localities may be slightly older than Acheulean assemblages (Dennell 2000–2001), most Soan sites in the Himalyan foothills are likely to date to the late Middle and Late Pleistocene (e.g., Stiles 1978; Soni and Soni 2005; Gaillard 2006; Chauhan 2007), and are thus contemporaneous with, or younger than, the Late Acheulean.

The current situation therefore indicates that Soan assemblages are a distinct reduction strategy which is based on the availability and size of certain clast forms. These assemblages may contain occasional bifacial forms that share some technological convergences with Acheulean implements, though this is a minor tool component with the industry. From a chronological perspective, Soan localities do not necessarily precede the Acheulean industry, but rather they are likely to be contemporary with Middle Paleolithic industries in the subcontinent. Certain Soan assemblages appear to be a Late Pleistocene industry that incorporates both simple core-flake forms and Levallois elements. Thus, based on currently available evidence, many identified Soan assemblages should no longer be considered as potential representatives of the earliest occupations of the subcontinent.

So, Are There Occupations at the Plio-Pleistocene Boundary?

If hominins reached East Asia by 1.8 Ma, it has been hypothesized that South Asia should contain assemblages at least that old (Dennell 1998; Petraglia 1998). Such a dispersal event presumes movement of hominins from South Asia to East Asia, following a southern or low latitude route. Although such an hypothesis is plausible, the evidence to support such a dispersal is not clear. The only evidence in South Asia to indicate an early exit from Africa is the material from the Soan Valley and the Pabbi Hills of Pakistan. If these localities are accepted as valid archeological occurrences, the sites would certainly provide a central geographic point between West Asia and East Asia. However, if the Upper Siwalik localities are not accepted as genuine occurrences, the absence of evidence would be a sign that hominins were not present in the region, despite the presence of environments that were similar to those found in East Africa. If hominins were not present in the Indian subcontinent by the Pliocene, it would indicate that dispersal routes were not along a southern route, but were instead focused in northern Eurasia. Geographic barriers that limited or prevented a southerly migration may have included the large river systems outflowing from the Himalayas or the arid landscapes in the northwest during glacial periods. Comparisons between environmental settings and faunas of northern India and southern China would be instructive, as it would help to explain if corridors existed for hominin dispersal.

Whether or not researchers consider the Upper Siwalik localities to be legitimate archeological occurrences, it is clear that Pliocene and Early Pleistocene sites are rare to absent. Indeed, although pedestrian survey efforts have been undertaken for decades, most surveys have failed to locate Pliocene and Early Pleistocene localities. It is possible that some of the early localities have been missed or overlooked owing to various factors, including complexity of depositional and postdepositional processes, light artifact density, and lack of the application of chronometric dating techniques.

Although such limitations exist, the scarcity of Mode I sites in peninsular India is probably a pattern that will hold up even if additional surveys are conducted as numerous surveys within the basins of peninsular India (Korisettar 2004, 2007) have failed to locate convincing Mode I sites. Thus, the weak and contentious archeological evidence of South Asia indicates that either Late Pliocene and Early Pleistocene sites are absent, or that hominins were present, but such populations were small or not permanent in the region. While it is entirely conceivable that Mode I sites will be found with more intensive field work, current indications are, in agreement with Dennell's (2003) argument for Eurasia, that the South Asian evidence is unlikely to indicate a long-term, permanent record of early hominin populations.

Acheulean Colonization and Dispersal

Acheulean Dispersals

The subcontinent contains a large cutting tool industry which shares close technological characteristics with Acheulean assemblages in the West (Clark 1994; Schick 1994; Petraglia 2003). Like its western counterparts, the South Asian Acheulean contains relatively large bifacially flaked forms such as handaxes and cleavers, often consisting of a high proportion of the tools found within localities. The large cutting tools show systematic production of bifacially flaked cores and flake blanks.

Comparative study of the Olorgesailie and Hunsgi–Baichbal assemblages indicated technological parallels in large cutting tool assemblages (Noll and Petraglia 2003). In both East Africa and South Asia, large clasts of various raw materials were selected to fashion large cutting tools. While biface size at certain localities varied as a consequence of particular raw material type and flaking intensity, there was a general tendency to manufacture large tools (57% of the bifaces range between 150 and 160 mm). The relatively low degree of resharpening indicated that there was a high discard rate over small, localized areas. The East African and Indian biface assemblages therefore shared parallels in tool morphology as well as in transport patterns.

Given that Africa contains the earliest Acheulean assemblages, large cutting tool assemblages in South Asia must be viewed as a product of an Out of Africa dispersal event. The earliest claim for Acheulean occupation in India is at the Isampur Quarry, where mammalian teeth associated with stone tools produced age estimates of ca. 1.2 Ma by ESR (Blackwell et al. 2001; Paddayya et al. 2002). If such an early dispersal event occurred, it would be consistent with the earliest movements of Acheulean hominins into the Levant, presumably by *Homo erectus* using Acheulean toolkits (Bar-Yosef 1998). Otherwise all available chronometric dates so far achieved in the subcontinent indicate younger ages for the Acheulean, dated to 700–400 ka in Pakistan (Rendell and Dennell 1985; Dennell 1998) and to >350–250 Ka in India (e.g., Petraglia 1998). This later age range would incorporate many Achuelean assemblages that contain well-made bifacial implements and cores showing preparatory techniques (Fig. 11.3). Later Acheulean dispersals are thought to have occurred throughout Eurasia by 800 ka (Rolland 1998, 2001; Saragusti and Goren-Inbar 2001; Dennell 2003). It is possible that the presence of abundant Late Acheulean industries may signal a later dispersal event into the subcontinent by *Homo heidelbergensis* populations. At the present time, the precise timing of the entry of Acheulean hominins into the subcontinent cannot be resolved, though we can be confident that such populations were certainly present by 700 ka.

Dispersals and the Basin Model

One of the most important geographic models to emerge in recent years is the Basin Model, which hypothesizes that Paleolithic site distribution in the subcontinent is linked with the Purnana and Gondwana basins (Figs. 11.2 and 11.4) (Korisettar 2004, 2007). There are 14 main structural basins that are defined by their characteristic lithosequences and formation. The basins vary in size and have diverse shapes and areal extents, measuring from 700 to 60,000 km^2. The basins share some common features, such as plentiful water supplies (perennial springs and ponds), a high biomass, favorable geomorphic settings (e.g., caves and rockshelters) and an abundant supply of lithic resources for stone tool manufacture. As indicated by the Early Paleolithic records of these basins, hominins were attracted to the basins and their margins (Fig. 11.5).

A specific geographic distribution of the Purana–Gondwana basins has been observed that may have influenced ranging patterns of hominins. The basins occur in the northwest, the peninsula and along the eastern half of India and near the coast. Thus, it is concluded that the routes of dispersal followed the interior basins and not the coastlines, contrary to hypotheses which argue for littoral zone expansions (e.g., Turner and O'Regan 2007) (Fig. 11. 4). Indeed, in South Asia there is little information that Acheulean populations were using coastal environments, though there are near coastal sites in both western and eastern parts of the subcontinent. However, these sites are within or close to basin margins, thus Acheulean populations may have been adaptations to use of interior terrestrial zones and not coastal resources. Of course, it is possible that Acheulean sites are

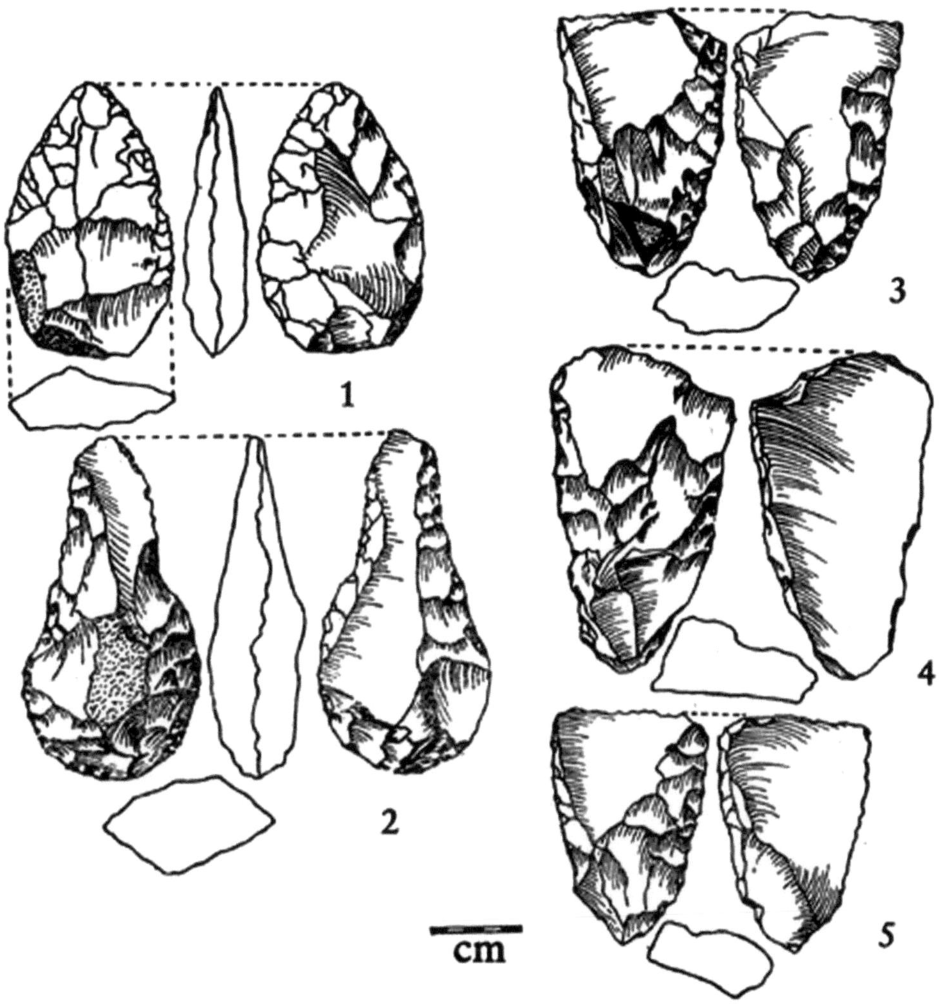

Fig. 11.3 Handaxes and cleavers from Bhimbetka rockshelter excavation (III-F-23). Note the fine invasive trimming scars along the lateral margins and the use of flakes to produce cleavers, often using preparatory core techniques (After Misra 1985)

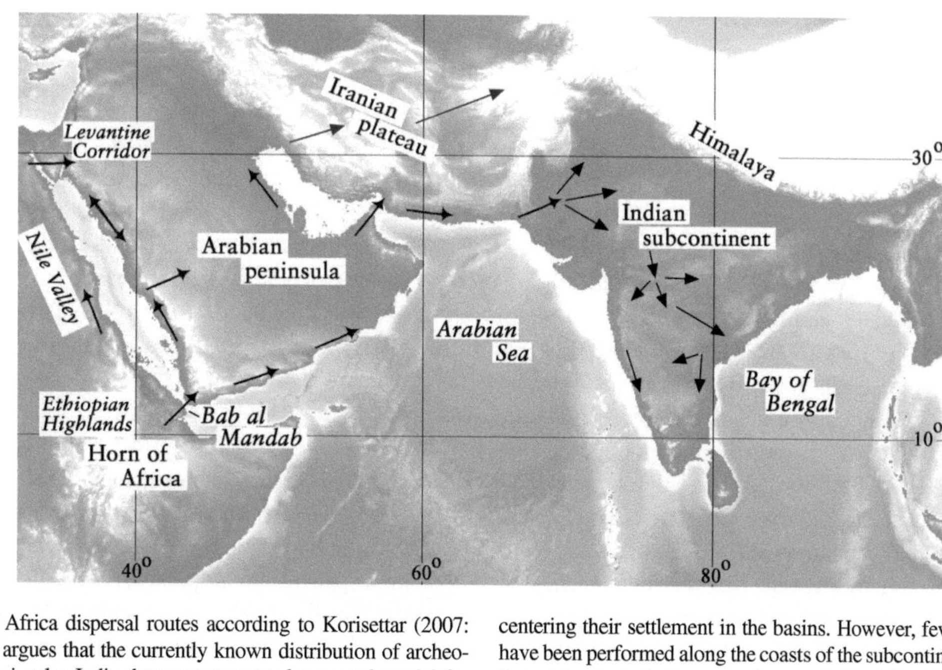

Fig. 11.4 Out of Africa dispersal routes according to Korisettar (2007: Fig. 2). Korisettar argues that the currently known distribution of archeological sites in peninsular India does not support the coastal model for hominin expansion across the subcontinent and into Southeast Asia. Rather, Korisettar argues that hominins would have dispersed in terrestrial zones, centering their settlement in the basins. However, few systematic surveys have been performed along the coasts of the subcontinent and it is probable that archeological sites are submerged. Petraglia (2003) has argued that Acheulean hominins utilized the coasts in Arabia, hence it is possible that hominins may have been present on the coasts of the subcontinent

Fig. 11.5 The Hunsgi–Baichbal Valley showing the plentiful Acheulean occurrences in this area as identified by Paddayya during surface survey. Korisettar (2007) points out that the dense network of low order streams dates to the Holocene, exposing many Acheulean localities. The Hunsgi–Baichbal Valley is in the Bhima basin. The basin morphology and its geological structure favored spring activity under Pleistocene high water tables and spring-fed pools and ponds in topographic lows. The valley would have contained a high biomass of plants and animals, thus making it attractive to Acheulean hominins

buried along the continental shelf (Flemming 2004) and in certain places along the southern dispersal route, Acheulean sites have been found along the coast (Petraglia 2003) (Figs. 11.1 and 11.4).

Occupation Intensity During the Acheulean

The Acheulean record of the subcontinent may be described as very rich based on the presence of hundreds (if not thousands) of Acheulean localities and the large number of artifacts present in sites (e.g., Misra 1987; Paddayya 1984; Pappu 2001; Petraglia 1998, 2006). However, such characterizations, while warranted in a general sense, tell us little about temporal and spatial records from a population perspective. To understand the nature of hominin occupation, it may be instructive to examine site counts to ascertain potential habitation length. Such interpretations are of course crude estimates, and we are well aware that site counts may be influenced by many behavioral and postdepositional factors.

If the overall number of Paleolithic sites in valleys is examined relative to time, it is clear that continuous occupations in areas cannot be documented (Petraglia 2007). For instance, although the identification of 196 Acheulean locales in the Hunsgi–Baichbal Valley may appear to indicate significant occupation, an evaluation by time indicates otherwise. In a long chronology model (i.e., ca. 750,000 years in duration), there is an average of one site produced every 3,826 years, whereas in a short chronology (i.e., ca. 250,000 years in duration) a total of one site is produced every 1,275 years. The record of habitation is even coarser for the Malaprabha Valley, the 27 Acheulean sites representing the production of one site every 9,259 years in a short chronology. Although simple calculations, such data indicate that the valleys were not consistently occupied through time, even if alterations in preservation are taken into account. Moreover, if site density is a sign of permanency, then the accumulation of Paleolithic sites in each valley indicates little intensive use (220 Early and Middle Paleolithic sites distributed over 500 km^2 in the Hunsgi–Baichbal Valley; 105 Early and Middle Paleolithic sites distributed over a 6,250 km^2 of the Malaprabha Valley). This locality information, together with the identification of Acheulean sites in many basins of peninsular India, indicates that populations were probably small and did not live their group existence in any one place. If this is the case, it is very likely that populations were shifting from one basin to another through time, thereby accounting for sites in many places, though not in the counts that we would expect if occupation was continuous in particular areas.

The South Asian Acheulean and 'Acheulean-Like' Assemblages to the East

In appraisals of the geographic distribution of Early Paleolithic industries in Asia versus those in the West, some analysts continue to support the notion of the Movius Line (e.g., Schick 1994; Petraglia 1998, 2006; Norton et al. 2006; Lycett 2007b). Yet, the presence of Mode II bifaces in East Asia has been construed as a breakdown of the geographic divide between the Mode I and Acheulean industries (Yi and Clark 1983; Huang 1989). Indeed, a close relationship was demonstrated between tool types and raw material types, shapes, and availability among Mode I and Mode II sites in China and India (Leng 1992, 1998). Additionally, it has been surmised that certain Indian assemblages with few bifaces may share characteristics with Chinese assemblages classified as Mode II (Petraglia 1998).

With respect to site frequency, East Asia does not evidence a systematically produced Acheulean Industrial Complex, having just a handful of sites with claims of Mode II technology. Indeed, even within localities with Mode II forms, it is

Fig. 11.6 Examples of cleavers from the Baichbal Valley. These tool forms are common in some localities, making up a large percentage of large cutting tools. Most of the cleavers are manufactured using preparatory core techniques, and these are made of end- and side-struck flake blanks. The blanks have particular shapes, with attention placed on producing a lateral bit. The flake blanks are then trimmed along dorsal and lateral edges to produce a distinctive shape (see Petraglia et al. 2005 for a discussion of this reduction method)

usually the case that unifacial forms dominate the assemblage. In contrast, India contains a rich Acheulean heritage, and although no systematic count of identified sites has been performed, it is clear that the total number of localities runs into the many hundreds. For example, surveys in the Raisen District, the Hunsgi and Baichbal Valleys, and the Kaladgi Basin have shown that more than 370 genuine Acheulean localities are present in these areas alone (Jacobson 1985; Paddayya 2001; Petraglia 2006).

In terms of counts of large cutting tools, there are large differences between South Asia and East Asia. For example, although Bose is a 'biface rich' assemblage for East Asia, only small numbers have been tabulated, e.g., 35 bifacial large cutting tools were reported in an inter-regional comparison (Hou et al. 2000). In contrast, the frequency of large cutting tools in India runs into the tens of thousands, and single localities have produced more bifaces than the total combined count of bifaces found throughout China. To demonstrate the point, more than 500 large cutting tools were collected from only ten sites in the Hunsgi and Baichbal Valleys (Paddayya and Petraglia 1993). Single sites have produced high counts in excavations, such as at Bhimbetka rockshelter where over 300 large cutting tools were recovered and at Chirki–Nevasa where more than 500 large cutting tools were uncovered (Misra 1985; Corvinus 1983).

Analysts have pointed out that there are some technological parallels between East Asian tool forms and those found in the Acheulean. Study of large cutting tools from Bose, China, indicates that the tools were "Acheulean-like" due to aspects of manufacture, bold flaking patterns, and high flake scar counts (Hou et al. 2000). Appraisal of handaxe sites east and west of the Movius Line indicates that there were metric similarities in lengths and widths of tools found in Korea, India and Kenya (Norton et al. 2006). However, significant differences in handaxe thickness were found between the Acheulean assemblages from Olorgesailie and Hunsgi–Baichbal and from those of the Imjin/Hantan River Basins (IHRB) in Korea. In a more comprehensive appraisal using a wider range of sites, it was found that the IHRB handaxes shared similarity in refinement with those found in Acheulean assemblages from Africa and India (Petraglia and Shipton 2008; Shipton and Petraglia 2008). Indeed, the comparative data shows that the IHRB handaxes fall within range of the Acheulean handaxe thicknesses and refinement, though most similar to the thickest and least refined tool assemblages. In consideration of the differences in tool forms, the rarity of bifacial forms, and the potentially younger age of the Korean localities, it appears that the IHRB sites are not affiliated with the Acheulean tradition but have more in common with other Asian industries dominated by core and flake tools.

The similarities between the IHRB assemblages and those of the Acheulean are thought to reflect technological convergence. This may be the case for other LCT assemblages in East Asia, such as at Bose, which shows rare bifacial flaking, a high frequency in the manufacture of unifacial implements, the use of heavy tools, and predominance of core and flake industries.

Acheulean assemblages from South Asia contrast in other key ways when compared to those from East Asia. Quarrying behavior at the Acheulean site of Isampur, India, demonstrates that stone tool manufacture was part of a set of procedures in tool making, showing that there were different reduction techniques being used to manufacture handaxes and flake-cleavers (Petraglia et al. 1999, 2005; Petraglia and Paddayya 2006). Perhaps the most distinctive difference between the East and South Asian tool artifact types is the heavy use of cleavers in India (Figs. 11.3 and 11.6). Cleavers are plentiful in sites (e.g., 274 from Chirki–Nevasa; 215 from Bhimbetka; 53 from Hunsgi V), and they sometimes form up to 60% of the tool assemblages (Ghosh 1985). Examination of manufacturing procedures at the Isampur quarry source indicates that cleavers are part of a planned core strategy, which requires predetermination in order to strike a large side-struck flake with a predicted outline shape (Petraglia et al. 2005). Such strategies appear to be precursors of prepared core technologies inclusive of the Levallois technique on smaller cores, such as found at Bhimbetka and in the Malaprabha Valley (Misra 1985; Petraglia et al. 2003). Though there is some regularity in the production of large cutting tools in India, they can range from boldly struck or they can be quite refined in their flaking patterns. For instance, at the Bhimbetka rockshelter and the Gunjana Valley, many of the bifaces show small and shallow trimming scars on lateral margins to regularize shape, likely indicating some intentionality in their manufactured form (Misra 1985; Raju 1988). All of these technological factors, taken together, indicate significant differences between East and South Asian tool assemblages.

In sum, Mode II implements in East Asia are rare compared to those found at localities in the West, biface forms are not a persistent tool form in the East Asia industry, and the shape characteristics, symmetry and morphology of the Mode II artifacts are not equivalent to the Acheulean industries of India. While the geographic dichotomy for biface production may not be as well defined as originally observed, the presence of a common "Acheulean" industry in East Asia has not been adequately established, thus many assemblages with bifacial forms may be considered to be technological convergences with the Acheulean.

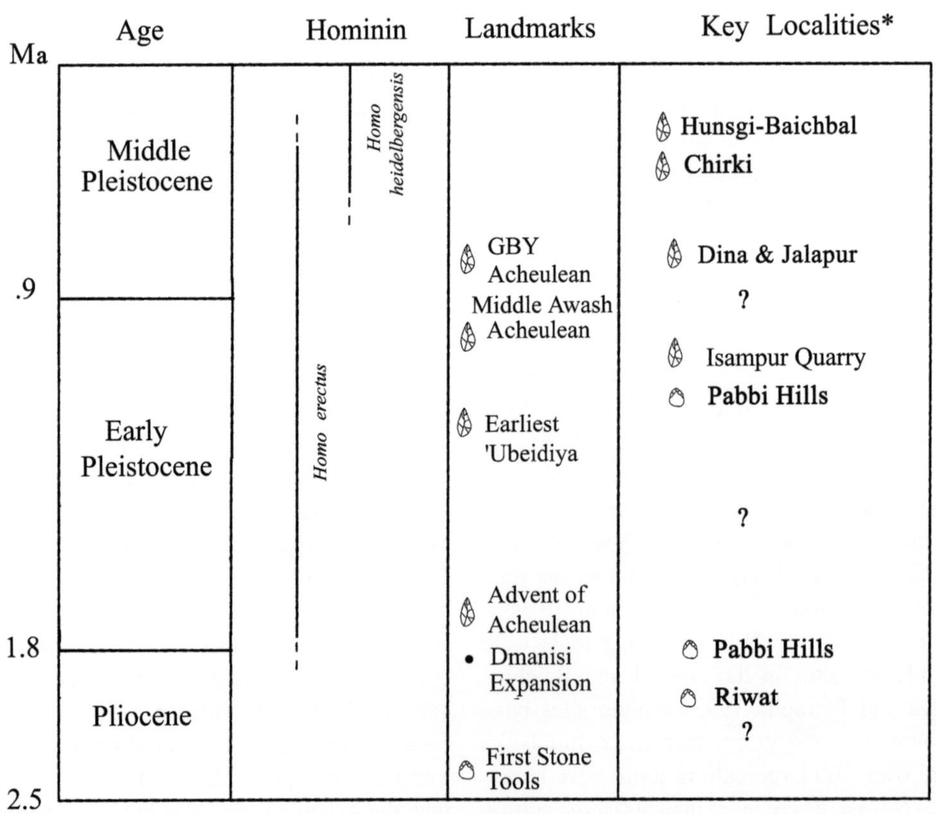

Fig. 11.7 General outline of key events in hominin evolution and important localities in the Indian subcontinent

The Luonan Basin: An Acheulean Sortie into China?

Though various claims have been made that Mode II assemblages are present in East Asia, few places have provided plentiful and convincing evidence for an Acheulean industry. Recent work in the Luonan Basin of China, however, is particularly intriguing as investigators have shown that handaxes and cleavers occur in numerous localities (Wang 2005, 2007). A comparison between the Luonan Basin assemblages and large cutting tool industries indicates that the Chinese tools comfortably fit within the Acheulean range of variation (Petraglia and Shipton 2008). The Luonan handaxes, however, tend to be slightly thicker and less refined than most Acheulean assemblages, thereby overlapping with the least refined Acheulean localities. The identification of 119 cleavers from the Luonan localities is significant as no other area in East Asia has produced such a large number of such artifact forms. The recovery of the bifacial artifacts from surface contexts remains a problem as the age of the lithic assemblages has not been firmly established. Nevertheless, the investigators have reported the recovery of lithic artifacts from the stratified loess deposits (Lu et al. 2007: Fig. 2). Recent chronometric dating of the Luonan sequence and comparison with the central Chinese Loess Plateau indicates an age range from at least 144 ka to 0.78 Ma (Lu et al. 2007). Though it appears that lithic artifacts are derived from Middle Pleistocene deposits, uncertainty remains about the stratigraphic provenience of the surface finds, which comprise the majority of the tool assemblages. If the Luonan localities and their LCT assemblages hold up to further scrutiny, it may provide the first potential demonstration of an Acheulean presence in East Asia, though this needs to be validated through future excavations, artifact analyses and dating.

Should the Luonan evidence be supported through further field work, two possible interpretations for the presence of an Acheulean in East Asia may be put forward. One possibility is that Pleistocene hominins at Luonan independently replicated Acheulean technology, in accordance with an argument that the large bifacial blanks were sought, and that final forms are the product of raw materials and reduction intensity (e.g., Davidson and Noble 1993; McPherron 2000). Another possibility is that hominins with knowledge of Acheulean toolmaking strategies dispersed into East Asia from a western source. The overall paucity of Acheulean localities in East Asia suggests that if such a dispersal event occurred, it was short-lived and not prolonged, or that once present in China, there was an abandonment of Acheulean manufacturing methods. If there was a spread of Acheulean toolmakers into East Asia, we would expect that populations were employing routes to avoid the Himalayas and the Tibetan plateau, either along a northern route, at about 40°N, or along a southerly route, at about 25°N. Future survey in regions poorly known to paleoanthropology may reveal additional Acheulean localities along either of these routes.

Conclusion

Although there has been much discussion of colonization and dispersal events in hominin evolution, treatments in the literature have been geographically biased. Most discussion has centered on the connections between East Africa, the Levant and East Asia, with little discussion of other major areas, including those situated along the Indian Ocean Rim, particularly Arabia and South Asia. Unfortunately, the archeological record of South Asia has played little role in Out of Africa models, despite the central geographic position of the subcontinent in any potential southern dispersal route. In this regard, the goal of this chapter was to indicate the potential importance of the South Asian record through some working hypotheses (Fig. 11.7), though the paucity of hominin remains and the chronological limitations of this record is recognized.

In agreement with Dennell (2003), it may be argued that the rarity of Mode I occurrences in South Asia indicates a sporadic or non-permanent occupation of the subcontinent. Although further work may detail the presence of Mode I sites, it is unlikely that the evidence will ever be construed as a permanent and continuous occupation by Pliocene and Early Pleistocene hominins. The reasons for the scarcity of Mode I sites to emerge in the subcontinent is not at all clear, as early hominins probably had the ability to penetrate the region and favorable environments were present. One possibility is that the Indo-Gangetic drainage system acted as a barrier and the lack of workable stone was detrimental to tool-dependent dispersing populations (Dennell 2007).

On the other hand, Acheulean occupation of South Asia is certain and it is linked to at least one dispersal event from the West. Based on chronometric evidence, it is difficult to determine if Acheulean hominins were present in the subcontinent prior to 1 Ma. More secure evidence indicates a dispersal into the subcontinent after 700 Ka, probably accounting for the high frequency of well-made Acheulean tool forms and advanced preparatory techniques in toolmaking. Such chronological and technological evidence would coincide with the presence of *Homo heidelbergensis* in India. However, given that the Narmada specimen shares many attributes with *Homo erectus* (Athreya 2007), the presence of this species, or more than one archaic hominin population in South Asia cannot be discounted.

Based on the known distribution of Acheulean sites in South Asia, hominin populations appear to have occupied many basins within peninsular India and in valleys of the sub Himalayan zone. While the Acheulean evidence in South

Asia supports the survival of hominin populations in the region over hundreds of thousands of years, the record does not necessarily indicate that populations were large and permanent in any one region. Rather, Acheulean populations may have been small and dispersing between various basins within the subcontinent.

On the whole, Acheulean tool-making traditions in South Asia contrast considerably with contemporaneous tool assemblages in East Asia. If this is the case, it is possible that variations in cultural traditions is accounted for by relative isolation between South Asian and East Asian populations, though not necessarily a consequence of differences in cognitive or adaptive abilities. Such evidence implies that there were geographic barriers between populations, and that the source of populations in East Asia may have been part of a dispersal event that did not emanate from the subcontinent. However, potential Acheulean assemblages in the Luonan Basin of China indicate that short-lived penetrations into East Asia may have occurred. The precise source of this Acheulean expansion is uncertain, but given the geographic position of Luonan, in Central China, it is unlikely that populations dispersed from India. Although much high-quality field research is needed to examine many of the issues raised here, it is clear that South Asia provides an exciting frontier to explore Out of Africa models and microevolutionary processes of population expansion and contraction.

References

Athreya, S. (2007). Was *Homo heidelbergensis* in South Asia? A test using the Narmada fossil from central India. In M. D. Petraglia & B. Allchin (Eds.), *The evolution and history of human populations in South Asia* (pp. 137–170). Dordrecht: Springer.

Agrawal, D. P. (1992). *Man and environment in India through ages.* New Delhi: Books & Books.

An, Z., Kutzbach, J. E., Prell, W. L., & Porter, S. C. (2001). Evolution of Asian monsoons and phased uplift of the Himalaya-Tibetan plateau since Late Miocene times. *Nature, 411*, 62–66.

Antón, S. C., & Swisher, C. C., III. (2004). Early dispersals of *Homo* from Africa. *Annual Review of Anthropology, 33*, 271–296.

Badam, G. L. (1984). Pleistocene faunal succession of India. In R. O. Whyte (Ed.), *The evolution of the East Asian environment, Vol. II, palaeobotany, palaeogeology and palaeoanthropology* (pp. 746–775). Hong Kong: University of Hong Kong.

Baker, V. R., Ely, L. L., Enzel, Y., & Kale, V. S. (1995). Understanding India's rivers: Late Quaternary palaeofloods, hazard assessment and global change. In S. Wadia, R. Korisettar, & V. S. Kale (Eds.), *Quaternary environments and geoarchaeology of India* (pp. 61–77). Bangalore: Geological Society of India.

Bar-Yosef, O. (1998). Early colonizations and cultural continuities in the Lower Palaeolithic of western Asia. In M. D. Petraglia & R. Korisettar (Eds.), *Early human behaviour in global context: the rise and diversity of the Lower Palaeolithic record* (pp. 221–279). London: Routledge.

Bar-Yosef, O., & Belfer-Cohen, A. (2001). From Africa to Eurasia – early dispersals. *Quaternary International, 75*, 19–28.

Blackwell, B. A. B., Fevrier, S., Blickstein, J. I. B., Paddayya, K., Petraglia, M., Jhaldiyal, R., et al. (2001). ESR dating of an Acheulean quarry site at Isampur, India. *Journal of Human Evolution, 40*, A3.

Bowler, J. M., Jones, R., & An, Z. (1995). Climatic change in Asia and the western Pacific: its significance for greenhouse scenarios. In S. Wadia, R. Korisettar, & V. S. Kale (Eds.), *Quaternary environments and geoarchaeology of India* (pp. 1–21). Bangalore: Geological Society of India.

Burbank, D. W., & Reynolds, R. G. (1984). Sequential late Cenozoic disruption of the northern Himalayan foredeep. *Nature, 311*, 114–118.

Cameron, D., Patnaik, R., & Sahni, A. (2004). The phylogenetic signifance of the Middle Pleistocene Narmada hominin cranium from central India. *International Journal of Osteoarcheology, 14*, 419–447.

Chauhan, P. (2007). Soanian cores and core-tools from Toka, northern India: towards a new typo-technological organization. *Journal of Anthropological Archaeology, 26*, 412–441.

Clark, J. D. (1994). The Acheulean industrial complex in Africa and elsewhere. In R. S. Corruccini & R. L. Ciochon (Eds.), *Integrative paths to the past* (pp. 451–469). Upper Saddle River, NJ: Prentice-Hall.

Clemens, S. C., Murray, D. W., & Prell, W. L. (1996). Nonstationary phase of the Plio-Pleistocene Asian monsoon. *Science, 274*, 943–948.

Corvinus, G. (1983). *A survey of the Pravara River system in Western Maharashtra, Vol. 2: The excavations of the Acheulean site of Chirki-on-Pravara, India.* Tubingen: Institut fur Urgeschichte.

Corvinus, G. (1998). Lower Palaeolithic occupations in Nepal in relation to South Asia. In M. D. Petraglia & R. Korisettar (Eds.), *Early human behaviour in global context: The rise and diversity of the Lower Palaeolithic record* (pp. 391–417). London: Routledge.

Davidson, I., & Noble, W. (1993). Tools and language in human evolution. In K. Gibson & T. Ingold (Eds.), *Tools, language and cognition in human evolution* (pp. 363–388). Cambridge: Cambridge University Press.

Dennell, R. W. (1995). The Early Stone Age of Pakistan: a methodological review. *Man and Environment, 20*(1), 21–28.

Dennell, R. W. (1998). Grasslands, tool making and the hominid colonization of southern Asia: a reconsideration. In M. D. Petraglia & R. Korisettar (Eds.), *Early human behaviour in global context: The rise and diversity of the Lower Palaeolithic record* (pp. 280–303). London: Routledge.

Dennell, R. W. (2000–2001). Palaeolithic studies in India since Independence: from an outsider's point of view. Bulletin of the Deccan College Post-Graduate and Research Institute 60–61, 175–187.

Dennell, R. W. (2003). Dispersal and colonisation, long and short chronologies: how continuous is the Early Pleistocene record for hominids outside East Africa? *Journal of Human Evolution, 45*, 421–440.

Dennell, R. (2004). *Early Hominin Landscapes in Northern Pakistan: Investigations in the Pabbi Hills.* Oxford: BAR International Series, No 1265.

Dennell, R. (2007). "Resource-rich, stone-poor": early hominin land use in large river systems of northern India and Pakistan. In M. D. Petraglia & B. Allchin (Eds.), *The evolution and history of human populations in South Asia* (pp. 41–68). Dordrecht: Springer.

Dennell, R. (2010). "Out of Africa 1": current problems and future prospects. In: J. G. Fleagle, J. J. Shea, F. E. Grine, A. L. Baden, & R. E. Leakey (Eds.), *Out of Africa I: the first hominin colonization of Eurasia* (pp. 247–273). Dordrecht: Springer.

Dennell, R. W., Rendell, H., & Hailwood, E. (1988). Early tool-making in Asia: two-million-year-old artefacts in Pakistan. *Antiquity, 62*, 98–106.

DeTerra, H. & Paterson, T.T. (1939). Studies on the Ice Age in India and associated human cultures (no. 393). Washington, D.C.: Carnegie Institute of Washington.

Field, J. S., & Lahr, M. M. (2005). Assessment of the southern dispersal: GIS-based analyses of potential routes at Oxygen Isotopic Stage 4. *Journal of World Prehistory, 19*, 1–45.

Field, J., Petraglia, M. D., & Lahr, M. M. (2007). The southern dispersal hypothesis and the South Asian archaeological record: examination of dispersal routes through GIS analysis. *Journal of Anthropological Archaeology, 26*, 88–108.

Flemming, N. C. (2004). Submarine prehistoric archeology of the Indian continental shelf: a potential resource. *Current Science, 86*, 1225–1230.

Foley, R. A. (1999). The evolutionary geography of Pliocene hominids. In T. Bromage & F. Schrenk (Eds.), *African biogeography, climatic change, and hominid evolution* (pp. 328–348). Oxford: Oxford University Press.

Gabunia, L., Vekua, A., Lordkipanidze, D., Swisher, C. C., III, Ferring, R., Justus, A., et al. (2000). Earliest Pleistocene hominid cranial remains from Dmanisi, Republic of Georgia: taxonomy, geological setting, and age. *Science, 288*, 1019–1025.

Gaillard, C. (1994) Processing sequences in Indian Lower Palaeolithic: examples from Acheulean and Soanian. Paper presented at the 15th Indo-Pacific Prehistory Association Congress, Chiang Mai, Thailand.

Gaillard, C. (1995). An Early Soan assemblage from the Siwaliks: a comparison of processing sequences between this assemblage and of an Acheulean assemblage from Rajasthan. In S. Wadia, R. Korisettar, & V. S. Kale (Eds.), *Quaternary environments and geoarchaeology of India* (pp. 231–245). Bangalore: Geological Society of India.

Gaillard, C. (2006). Les premiers peuplements d'Asie du Sud: vestiges culturels. *Comptes Rendus Paleovol, 5*, 359–369.

Ghosh, A. K. (1985). An attempt to understand the Acheulian succession in India from spatial distributions. In V. N. Misra & P. Bellwood (Eds.), *Recent advances in Indo-Pacific prehistory* (pp. 29–34). New Delhi: Oxford–IBH.

Gupta, A. (1995). Fluvial geomorphology and Indian rivers. In S. Wadia, R. Korisettar, & V. S. Kale (Eds.), *Quaternary environments and geoarchaeology of India* (pp. 52–60). Bangalore: Geological Society of India.

Gupta, S. M. (1999). Radiolarian monsoon index Pyloniid group responds to astronomical forcing in the last 500,000 years: evidence from the central Indian Ocean. *Man and Environment, 24*, 99–107.

Gupta, S. M., & Fernandes, A. A. (1997). Quaternary radiolarian faunal changes in the central Indian Basin: inferences to paleomonsoonal oscillation of the 10° S hydrographic front. *Current Science, 72*, 965–972.

Hou, Y., Potts, R., Yuan, B., Guo, Z., Deino, A., Wang, W., et al. (2000). Mid-Pleistocene Acheulean-like stone technology of the Bose Basin, South China. *Science, 287*, 1622–1626.

Huang, W. (1989). Bifaces in China. *Human Evolution, 4*, 87–92.

Hurcombe, L. M. (2004). The stone artifacts from the Pabbi Hills. In R. W. Dennell (Ed.), *Early hominin landscapes in Northern Pakistan: Investigations in the Pabbi Hills* (pp. 222–292). Oxford: BAR International Series, No 1265.

Jacobson, J. (1985). Acheulean surface sites in Central India. In V. N. Misra & P. Bellwood (Eds.), *Recent advances in Indo-Pacific prehistory* (pp. 49–57). New Delhi: Oxford–IBH.

Jayaswal, V. (1982). *Chopper-chopping component of palaeolithic India*. Delhi: Agam Kala Prakashan.

Joshi, R. V. (1970). The characteristics of the Pleistocene climatic events in the Indian subcontinent - land of monsoon climate. In S. B. Deo & M. K. Dhavilikar (Eds.), *Indian antiquary* (pp. 53–63). Bombay: Popular Prakashan.

Joshi, R. V. (1978). *Stone Age cultures of central India: Report on the excavations of rock-shelters at Adamgarh, Madhya Pradesh*. Poona: Deccan College Post-Graduate and Research Institute.

Kennedy, K. A. R. (1999). Paleoanthropology of South Asia. *Evolutionary Anthropology, 8*, 165–185.

Klein, R. (1999). *The human career*. Chicago: Chicago University Press.

Korisettar, R. (1994). Quaternary alluvial stratigraphy and sedimentation in the Upland Deccan Region, Western India. *Man and Environment, 19*, 29–41.

Korisettar, R. (2002). The archaeology of the South Asian Lower Palaeolithic: history and current status. In S. Settar & R. Korisettar (Eds.), *Indian archaeology in retrospect: Prehistory* (pp. 1–65). New Delhi: Indian Council of Historical Research and Manohar Publishers.

Korisettar, R. (2004). Geoarchaeology of the Purana and Gondwana basins of peninsular India: peripheral or paramount. Presidential address, Sixty-fifth session of the Indian History Congress. Bareilly, Uttar Pradesh (off-print).

Korisettar, R. (2007). Toward developing a basin model for Paleolithic settlement of the Indian subcontinent: geodynamics, monsoon dynamics, habitat diversity and dispersal routes. In M. D. Petraglia & B. Allchin (Eds.), *The evolution and history of human populations in South Asia* (pp. 69–96). Dordrecht: Springer.

Korisettar, R., & Rajaguru, S. N. (1998). Quaternary stratigraphy, palaeoclimate and the Lower Palaeolithic of India. In M. D. Petraglia & R. Korisettar (Eds.), *Early human behaviour in global context: The rise and diversity of the Lower Palaeolithic record* (pp. 304–342). London: Routledge.

Korisettar, R., & Ramesh, R. (2002). The Indian monsoon: roots, relations and relevance. In S. Settar & R. Korisettar (Eds.), *Indian archeology in retrospect: Archeology and interactive disciplines* (pp. 23–59). New Delhi: Indian Council of Historical Research and Manohar Publishers.

Kretzoi, M. (1961–1964). Mammal faunae and the continental geology of India. *Acta Geologiica Academiae Scientiarum Hungarica, 7–8*, 310–312.

Lahr, M. M., & Foley, R. A. (1998). Towards a theory of modern human origins: geography, demography, and diversity in recent human evolution. *Yearbook of Physical Anthropology, 41*, 137–176.

Leng, J. (1992). *Early Palaeolithic technology in China and India*. Unpublished Ph.D. dissertation, University Microfilms International, Ann Arbor, MI.

Leng, J. (1998). Early Palaeolithic quartz industries in China. In M. D. Petraglia & R. Korisettar (Eds.), *Early human behaviour in global context: The rise and diversity of the Lower Palaeolithic record* (pp. 418–436). London: Routledge.

Lu, H., Zhang, H., Wang, S., Cosgrove, R., Zhao, C., Stevens, T., et al. (2007). A preliminary survey on loess deposit in Eastern Qinling Mountains (Central China) and its implication for estimating age of the Pleistocene lithic artifacts. *Quaternary Sciences 27*(4), 1–9 (in Chinese).

Lycett, S. J. (2007a). Is the Soanian techno-complex a Mode 1 or Mode 3 phenomenon? a morphometric assessment. *Journal of Archeological Science, 34*, 1434–1440.

Lycett, S. J. (2007b). Why is there a lack of Mode 3 Levallois technologies in East Asia? A phylogenetic test of the Movius-Schick hypothesis. *Journal of Anthropological Archaeology, 26*, 541–575.

McPherron, S. (2000). Handaxes as a measure of the mental capabilities of early hominids. *Journal of Archeological Science, 27*, 655–663.

Misra, V. N. (1985). The Acheulean succession at Bhimbetka, Central India. In V. N. Misra & P. Bellwood (Eds.), *Recent advances in Indo-Pacific prehistory* (pp. 35–47). New Delhi: Oxford–IBH.

Misra, V. N. (1987). Middle Pleistocene adaptations in India. In O. Soffer (Ed.), *The Pleistocene old world* (pp. 99–119). New York: Plenum.

Misra, V. N. (1989). Stone Age India: an ecological perspective. *Man and Environment, 14*, 17–64.

Misra, V. N. (1995). Geoarchaeology of the Thar Desert, Northwest India. In S. Wadia, R. Korisettar, & V. S. Kale (Eds.), *Quaternary environments and geoarchaeology of India* (pp. 210–230). Bangalore: Geological Society of India.

Mithen, S., & Reed, M. (2002). Stepping out: a computer simulation of hominid dispersal from Africa. *Journal of Human Evolution, 43*, 433–462.

Noll, M. P., & Petraglia, M. D. (2003). Acheulean bifaces and early human behavioral patterns in East Africa and South India. In M. Soressi & H. L. Dibble (Eds.), *Multiple approaches to the study of bifacial technologies* (pp. 31–53). Philadelphia: University of Pennsylvania Press.

Norton, C. J., Bae, K., Harris, J. W. K., & Lee, H. (2006). Middle Pleistocene handaxes from the Korean peninsula. *Journal of Human Evolution, 51*, 527–536.

Ollier, C. D. (1995). New concepts of laterite formation. In S. Wadia, R. Korisettar, & V. S. Kale (Eds.), *Quaternary environments and geo-archaeology of India* (pp. 309–323). Bangalore: Geological Society of India.

Paddayya, K. (1982). *The Acheulean culture of the Hunsgi Valley (Peninsular India): A settlement system perspective*. Poona: Deccan College Postgraduate and Research Institute.

Paddayya, K. (1984). India. In O. Bar-Yosef et al. (Eds.), *Neue Forschungen zur Altsteinzeit* (pp. 345–403). Verlag: CH Beck.

Paddayya, K. (2001). The Acheulian culture project of the Hunsgi and Baichbal Valleys, Peninsular India. In L. Barham & K. Robson-Brown (Eds.), *Human roots: Africa and Asia in the Middle Pleistocene* (pp. 235–258). Bristol: Western Academic and Specialist Press.

Paddayya, K., & Petraglia, M. D. (1993). Formation processes of Acheulean localities in the Hunsgi and Baichbal Valleys, Peninsular India. In P. Goldberg, D. T. Nash, & M. D. Petraglia (Eds.), *Formation processes in archaeological context* (pp. 61–82). Madison: Prehistory Press.

Paddayya, K., Blackwell, B. A. B., Jhaldiyal, R., Petraglia, M. D., Fevrier, S., Chaderton, D. A., II, et al. (2002). Recent findings on the Acheulian of the Hunsgi and Baichbal Valleys, Karnataka, with special reference to the Isampur excavation and its dating. *Current Science, 83*, 641–647.

Pappu, R. S. (1995). The contribution of the earth sciences to the development of Indian archaeology. In S. Wadia, R. Korisettar, & V. S. Kale (Eds.), *Quaternary environments and geoarchaeology of India* (pp. 414–434). Bangalore: Geological Society of India.

Pappu, R. S. (2001). *Acheulian culture in peninsular India*. New Delhi: D.K. Printworld.

Pappu, S. (1999). A study of natural site formation processes in the Kortallayar Basin, Tamil Nadu, South India. *Geoarcheology, 14*, 127–150.

Petraglia, M. D. (1998). The Lower Paleolithic of India and its bearing on the Asian record. In M. D. Petraglia & R. Korisettar (Eds.), *Early human behaviour in global context: The rise and diversity of the Lower Paleolithic record* (pp. 343–390). New York: Routledge.

Petraglia, M. D. (2001). The Lower Palaeolithic of India and its behavioural significance. In L. Barham & K. Robson-Brown (Eds.), *Human roots: Africa and Asia in the Middle Pleistocene* (pp. 217–233). Bristol: Western Academic and Specialist Press.

Petraglia, M. D. (2003). The Lower Paleolithic of the Arabian peninsula: occupations, adaptations, and dispersals. *Journal of World Prehistory, 17*, 141–179.

Petraglia, M. D. (2005). Hominin responses to Pleistocene environmental change in Arabia and South Asia. In M. J. Head & P. L. Gibbard (Eds.), *Early-Middle Pleistocene transitions: The land-ocean evidence, Special Publications, 247* (pp. 305–319). London: Geological Society.

Petraglia, M. D. (2006). The Indian Acheulean in global perspective. In N. Goren-Inbar & G. Sharon (Eds.), *Axe Age: Acheulian toolmaking from quarry to discard* (pp. 389–414). London: Equinox Publishing.

Petraglia, M. D. (2007). Palaeolandscapes and Palaeolithic locales of southern India: a consideration of hominin dispersals and adaptations, *Pragdhara. Journal of Uttar Pradesh State Department of Archeology, 17*, 27–44.

Petraglia, M. D., & Paddayya, K. (2006). New insights on the Acheulean: a case study from southern India. *Humankind, 1*, 61–75.

Petraglia, M., LaPorta, P., & Paddayya, K. (1999). The first Acheulean quarry in India: stone tool manufacture, biface morphology, and behaviors. *Journal of Anthropological Research, 55*, 39–70.

Petraglia, M. D., & Shipton, C. (2008). Large cutting tool variation West and East of the Movius Line. *Journal of Human Evolution, 55*, 952–956.

Petraglia, M. D., Schuldenrein, J., & Korisettar, R. (2003). Landscapes, activity, and the Acheulean to Middle Paleolithic transition in the Kaladgi Basin, India. *Eurasian Prehistory, 1*, 3–24.

Petraglia, M. D., Shipton, C., & Paddayya, K. (2005). Life and mind in the Acheulean: a case study from India. In C. Gamble & M. Porr (Eds.), *The Hominid Individual in Context* (pp. 197–219). London: Routledge.

Quade, J., Cerling, T. E., Bowman, J. R., & Jah, A. (1993). Paleoecological reconstruction of floodplain environments using palaeosols from Upper Siwalik Group sediments, northern Pakistan. In J. F. Schroder (Ed.), *Himalaya to the sea: Geology, geomorphology and the quaternary* (pp. 213–226). London: Routledge.

Raju, D. R. (1988). *Stone Age hunter-gatherers: An ethno-archaeology of Cuddapah Region, South-east India*. Pune: Ravish Publishers.

Rendell, H., & Dennell, R. W. (1985). Dated Lower Palaeolithic artefacts from northern Pakistan. *Current Anthropology, 26*, 393.

Rendell, H., Dennell, R., & Halim, M. A. (1989). *Pleistocene and Palaeolithic investigations in the Soan Valley, Northern Pakistan*. Oxford: BAR International Series, No 544.

Retallack, G. J. (1995). Palaeosols of the Siwalik Group as a 15 Myr record of South Asia palaeoclimate. In S. Wadia, R. Korisettar, & V. S. Kale (Eds.), *Quaternary Environments and Geoarchaeology of India* (pp. 36–51). Bangalore: Geological Society of India.

Rolland, N. (1998). The Lower Palaeolithic settlement of Eurasia, with special reference to Europe. In M. D. Petraglia & R. Korisettar (Eds.), *Early human behaviour in global context* (pp. 187–220). London: Routledge.

Rolland, N. (2001). The initial peopling of Eurasia and the early occupation of Europe in its Afro-Asian context: major issues and current perspectives. In S. Milliken & J. Cook (Eds.), *A very remote period indeed* (pp. 78–94). Oxford: Oxbow Books.

Sankalia, H. D. (1974). *The prehistory and protohistory of India and Pakistan*. Poona: Deccan College Postgraduate and Research Institute.

Saragusti, I., & Goren-Inbar, N. (2001). The biface assemblages from Gesher Benot Ya'aqov, Israel: illuminating patterns of "Out of Africa" dispersal. *Quaternary International, 75*, 85–89.

Schick, K. D. (1994). The Movius Line reconsidered: perspectives on the earlier Paleolithic of Eastern Asia. In R. S. Corruccini & R. L. Ciochon (Eds.), *Integrative paths to the past* (pp. 569–596). Upper Saddle River, NJ: Prentice-Hall.

Shipton, C., & Petraglia, M. D. (2008). Inter-continental variation in Acheulean bifaces. In C. J. Norton & D. Braun (Eds.), *Asian Paleoanthropology: From Africa to China and beyond*. Dordrecht: Springer 45–55.

Sonakia, A. (1985). Skull cap of an early man from the Narmada valley alluvium (Pleistocene) of central India. *American Anthropologist, 87*, 612–616.

Soni, A. S., & Soni, V. S. (2005). Palaeolithic tools form the surface of optically stimulated luminescence dated alluvial fan deposits of Pinjaur Dun in NW sub-Himalayas. *Current Science, 88*, 867–871.

Stiles, D. (1978). Palaeolithic artifacts in Siwalik and post-Siwalik deposits. *Kroeber Anthropology Society Papers, 53–54*, 129–148.

Swisher, C., III, Curtis, G. H., Jacob, T., Getty, A. G., Suprijo, A., & Widiasmoro, A. (1994). Age of the earliest known hominids in Java, Indonesia. *Science, 263*, 1118–1121.

Turner, A. (2004). Vertebrate fossils: Carnivora. In R. Dennell (Ed.), *Early hominin landscapes in northern Pakistan: Investigations in the Pabbi Hills* (pp. 404–411). BAR International Series: Oxford.

Turner, A., & O'Regan, H. J. (2007). Afro-Eurasian mammalian fauna and early hominin dispersals. In M. D. Petraglia & B. Allchin (Eds.), *The evolution and history of human populations in South Asia* (pp. 23–39). Dordrecht: Springer.

Wang, S. (2005). *Perspectives on hominid behaviour and settlement patterns: a study of the Lower Palaeolithic sites in the Luonan Basin, China*. Oxford: Archeopress.

Wang, S. (2007). The Paleolithic open-air sites in the Luonan Basin, China. Shaanxi Provincial Institute of Archeology, Cultural Relics Administrative Committee of Shangluo District, Museum of Luonan County. Beijing: Science Press (in Chinese).

Wasson, R. J. (1995). The Asian monsoon during the Late Quaternary: a test of orbital forcing and palaeoanalogue forecasting. In S. Wadia, R. Korisettar, & V. S. Kale (Eds.), *Quaternary environments and geoarchaeology of India* (pp. 22–35). Bangalore: Geological Society of India.

Yi, S., & Clark, G. A. (1983). Observations on the Lower Palaeolithic of Northeast Asia. *Current Anthropology, 24*, 181–202.

Zhu, R. X., Hoffman, R. A., Potts, R., Deng, C. L., Pan, Y. X., Guo, B., et al. (2001). Earliest presence of humans in northeast Asia. *Nature, 413*, 413–417.

Part IV
Europe and Western Asia

Introduction

In "Early Pleistocene faunal connections between Africa and Eurasia: an ecological perspective", Miriam Belmaker uses a variety of faunal analyses of the Early Pleistocene site of 'Ubeidiya in Israel to examine whether the initial dispersal of hominins from Africa to Eurasia was primarily driven by external factors, specifically the extension of African savannahs into the Levant due to climate change, or by intrinsic factors that permitted long range dispersal and subsequent adaptation to a new environment. Through a series of multivariate analyses, she finds that although Early Pleistocene faunas of the Levant contain a few African mammals, these are generally rare and that the overall composition of the mammalian fauna is one of Mediterranean woodland rather than African savannah. In analyzing the ecology of the individual taxa that dispersed from Africa she concludes that the dispersal of hominins and other African taxa in the Early Pleistocene is the result of the adaptations and dispersal abilities of individual species rather than a wholesale transposition of an African savannah fauna into Eurasia.

In "Early Pleistocene faunas of Eurasia and hominin dispersals", Bienvenido Martínez-Navarro reviews the abundant evidence for faunal dispersal between Africa and Eurasia throughout the Pliocene and Pleistocene. He then reviews the systematics and biogeography of the taxa that are most commonly found in association with the dispersal of early hominins at the base of the Pleistocene, noting the importance of correct identification of the taxa at individual sites. In contrast with Belmaker, he argues that the dispersal of hominins and other African mammals into Eurasia at the base of the Pleistocene was part of a northern extension of African habitats, both woodland and savannah. In his view, the carnivorous diet of the genus *Homo* was critical to its ability to disperse into other continents with different climate and different faunas.

In "Fossil skulls from Dmanisi: a paleodeme representing earliest *Homo* in Eurasia" Philip Rightmire and David Lordkipanidze describe the remarkable cranial material from the site of Dmanisi in the Republic of Georgia. They review the geology of the site and demonstrate that the Dmanisi fossils date to slightly younger than 1.77 Ma. After a detailed summary of the morphology of the crania and mandibles from the site, they conclude that all of the material is best attributed to a single paleodeme, and that the difference among specimens are attributable to normal intrapopulational variation, including difference in sex and age. In size, all of the Dmanisi crania fall at the lower limit of other fossils attributed to *Homo erectus*. While the Dmanisi fossils are placed in *Homo erectus*, they show some similarities to more primitive hominins from East Africa placed in *Homo habilis*. The Dmanisi cranial material could support either an African or Asian origin of *Homo erectus* and, concomitantly, arguments that the earliest hominins to leave Africa were more primitive than *Homo erectus*.

The Editors

Chapter 12
Early Pleistocene Faunal Connections Between Africa and Eurasia: An Ecological Perspective

Miriam Belmaker

Abstract During the Early Pleistocene, hominins dispersed from Africa to Eurasia via the Levant. The presence of African taxa in higher latitudes has been interpreted as indicative of an Africa savanna habitat in the Early Pleistocene of the Levant and as support for the hypothesis that the spread of grasslands into higher latitudes promoted the dispersal of hominins into Eurasia. To infer the paleoenvironment of the Levant during the Early Pleistocene, a multivariate analysis based on abundance and presence-absence data of mesoherbivores was used. Results indicate that in the Early Pleistocene, the Levant was situated in a Mediterranean woodland habitat. The presence of African taxa in novel environments may be explained using a long distance dispersal model and suggests that the spatial and temporal abundance pattern of African taxa in the Levant is consistent with the early stages of dispersal and is not indicative of the presence of savanna habitats in higher latitudes. The ecological success of hominins in higher latitudes as attested to by their widespread geographic distribution is consistent with their adaptation to novel environments and suggests that they may have been pre-adapted to variable environments.

Keywords Biogeography • Long distance dispersal • Mediterranean woodland • Savanna • Southern Levant • 'Ubeidiya

Introduction

The causes and forces which influenced hominin dispersal from Africa into Eurasia during the Late Pliocene and Early Pleistocene are some of the most enigmatic phenomena in human evolution (Schick and Zhuan 1993; Gabunia and Vekua 1995; Larick and Ciochon 1996; Bar-Yosef 1998; Potts 1998; Arribas and Palmqvist 1999; Bar-Yosef and Belfer-Cohen 2001; Antón et al. 2002; Potts 2002; Antón and Swisher 2004; Martínez-Navarro 2004). The large biogeographic range, from the Iberian peninsula (Fuente Nueve-3 and Barranco León ca. 1.3 Ma; Agustí et al. 2000) to China (Yuanmou ca. 1.7 Ma; Zhu et al. 2003) attained by hominins less than 0.5 Ma after the initial dispersal ca. 1.8–1.9 Ma (Bar-Yosef and Belfer-Cohen 2001) attests to the ecological success of the species. Two main types of hypotheses have been proposed to explain the occurrence, and especially the timing of this major event in human evolution.

Extrinsic hypotheses have suggested that the expansion of grassland habitats into higher latitudes (≥30°N) during the Late Pliocene and Early Pleistocene provided an ecological means by which early hominins, adapted to such environments, were able to follow their preferred habitats northward and eastward towards India and China. If hominins were adapted to a savanna environment (Vrba 1988, 1995; Wynn 2004), the success of their dispersal was dependent on the presence of such habitats in the new region, specifically in the early stages. Thus, during periods of climatic change, the spatial extent of the savanna may have increased, resulting in a range increase of many savanna adapted species.

Dennell (2003:422) has suggested that one of the basic assumptions we should consider when discussing hominin dispersal events is that "the earliest Eurasians preferentially occupied grasslands and open scrub and woodlands, as in East Africa" as they were adapted to hot and dry conditions and open grasslands. The Aullan dispersal event (ca. 1.8 Ma) associated with the new appearance of African faunas in Eurasia included the African genus *Homo* (Martínez-Navarro 2010; Martínez-Navarro and Palmqvist 1995).

The alternative hypotheses emphasize that the dispersal from Africa is an event unique to hominins and that intrinsic factors may have provided the impetus for their dispersal as well as for their ecological success in higher latitudes. The initial dispersal of hominins has been attributed to both morphological and behavioral characteristics such as the capacity for long distance walking (Steudel 1994), endurance running (Bramble and Lieberman 2004), heat adaptation (Walker and Leakey 1993), greater brain capacity (Aiello 1993; Aiello and Wheeler 1995) and social structure (Kroll 1994).

M. Belmaker (✉)
Department of Anthropology, Harvard University, Peabody Museum, 11 Divinity Ave, Cambridge, MA, 02138, USA
e-mail: belmaker@fas.harvard.edu

J.G. Fleagle et al. (eds.), *Out of Africa I: The First Hominin Colonization of Eurasia*, Vertebrate Paleobiology and Paleoanthropology,
DOI 10.1007/978-90-481-9036-2_12, © Springer Science+Business Media B.V. 2010

However, other forces may have been important in the subsequent ecological success of hominins at higher latitudes. Bar-Yosef and Belfer-Cohen (2001) have suggested that the release from tropical diseases may have allowed for an increase in population size in higher latitudes. The Variability Selection Hypothesis states that the adaptability of hominins to a wide range of habitats and specifically to a variable climate in Africa may have provided a pre-adaptation to novel environments (Potts and Teague 2010; Potts 1998, 2002). Such a pre-adaptation may be biological as suggested by Potts (1998, 2002) or behavioral such as technological innovations (Larick and Ciochon 1996).

Another set of hypotheses emphasizes the importance of meat in the diet for both the initial dispersal from Africa as well as ecological success. The increase in body and brain size required a better quality diet (Aiello and Wheeler 1995; Cordain et al. 2000, 2001; Aiello et al. 2001; Aiello and Wells 2002). In other words, hominins evolved to become active hunters and hence needed larger territories (Walker and Shipman 1996). Taphonomic analyses of faunal remains have suggested that Plio-Pleistocene hominins had primary access to carcasses and a high proportion of meat in the diet (Brantingham 1998; Hemmer 2000; Domínguez-Rodrigo et al. 2002; Domínguez-Rodrigo 2002, 2003). The scavenging hypothesis suggests that the increased reliance of hominins on meat of scavenged carcasses (rather than hunting) led to hominins following large African predators into Eurasia (Turner 1984, 1992; Arribas and Palmqvist 1999; Turner 1999). This is supported by evidence suggesting that hominins in East Africa may have only had secondary access to carcasses (Blumenschine 1986, 1987, 1988; Blumenschine and Cavallo 1992; Blumenschine et al. 1994, 2003).

The hypotheses discussed above encompass a wide scope of both biological and cultural phenomena and rigorous testing is needed to tease apart those that may have had an impact on hominin dispersals from Africa. Testing the environmental hypothesis requires a robust paleoecological reconstruction of Early Pleistocene "Out of Africa" sites. The presence of African savanna habitats in higher latitudes during the Early Pleistocene would lend support to the environmental hypotheses. On the other hand, the absence of such habitats would favor the intrinsic hypotheses.

Testing the Environmental Hypothesis: The Case of the Southern Levant

The Southern Levant is one of the major dispersal routes between Africa and Eurasia used throughout the Neogene (Tchernov 1988b). The Southern Levant constitutes a unique province. In the west, it is bordered by a fertile Mediterranean region, which rapidly subsides eastwards to the semiarid steppes of the Syria Arabia desert. In the north it is bounded by the Zagros and Taurus mountains and in the south by the isthmus of the Suez (Yom-Tov et al. 1988; Tchernov and Belmaker 2004). Since the onset of the Neogene, the Southern Levant was geographically a corridor between Africa and Eurasia. During different time periods, geological changes (i.e., tectonics, climate and sea level changes) altered the "permeability" of this land bridge. Thus, during certain times, the region allowed for animals to disperse from one region to another as opposed to other periods in which it served as a barrier (Yom-Tov et al. 1988; Tchernov and Belmaker 2004).

Several routes have been suggested to connect Africa and Eurasia via the Southern Levant: along the Nile valley, the Red Sea coast, via the Sinai and the Negev or across the Bab el Mandeb straits (Dennell 2010; Potts and Teague 2010). Absence of securely dated archeological sites along any of the routes precludes their definite identification. Dispersal via the first three routes would be terrestrial, following fresh to brackish lakes present throughout (e.g., as can be noted by the Nahal Zihor handaxe assemblage in the Arava valley; Ginat et al. 2003). This route has been shown to be the most probable based on vegetation and climate models (Holmes 2007). In addition, it has been hypothesized that hominins may have preferred dispersal routes via hilly and rough terrains generated by tectonic and volcanic movements such as those present in the African rift valley, Sinai, Red Sea coast and Arava valley (King and Bailey 2006). Crossing the Bab el Mandeb straits, which would have required sea crossing, would appear less probable as there is no evidence for a land bridge during this time period (Derricourt 2006; Fernandes et al. 2006).

In order to analyze the ecological success of hominins in higher latitudes, it is necessary to examine the earliest sites out-of-Africa with unequivocal evidence for hominin remains or activities. 'Ubeidiya, in the central Jordan Valley, Israel, is located 3 km south of the Sea of Galilee (Figs. 12.1 and 12.2). The site exhibits human remains (Tobias 1966a, b; Belmaker et al. 2002), and rich lithic (Bar-Yosef and Goren-Inbar 1993) and faunal assemblages (Haas 1966, 1968; Tchernov 1986). Long-range biochronological correlation in conjunction with paleomagnetic dating has suggested a date of ca. 1.6–1.2 Ma for the most of the fossil bearing sequence (Tchernov 1987; Sagi et al. 2005; Sagi 2005). Thus, 'Ubeidiya is not only the earliest site in the Southern Levant with such evidence but also has a rich mammalian assemblage providing the abundance data necessary for paleoecological reconstructions (Fig. 12.3).

In this paper, I demonstrate that although African taxa are present in the Early Pleistocene site of 'Ubeidiya, a paleoecological reconstruction using multivariate analysis of the mesoherbivore community suggests that this locality should be assigned to a Mediterranean woodland habitat. The presence

Fig. 12.1 Location of Plio-Pleistocene sites in the Southern Levant mentioned in the text

of African taxa in such an environment will be explained using a long distance dispersal (LDD) model.

Multivariate Analysis of the Large Mammal Community of 'Ubeidiya

Large mammals have been widely used both for paleoecological reconstructions (Andrews et al. 1979; Andrews 1992, 1995a, b, c, 1996; Reed 1997, 1998) as well as for studies of community and environmental changes through time (Bobe and Eck 2001; Bobe et al. 2002; Alemseged 2003). The main premise underlying paleoecological reconstructions based on fauna is the unique niche requirements of the individual species. The presence of an indicator mammalian species and its relative abundance have been used as evidence for climatic change e.g., the finding of reindeer, present today in higher latitudes, in Plio-Pleistocene sites in the lower latitudes of Europe served to indicate the limits of glaciations during the Pleistocene (Delpech and Heintz 1976).

Despite being widely used, the indicator taxon method has several shortcomings (Andrews 1995c). When inferring single species' adaptations there is a strong dependency on precise taxonomic identification and the association of the species to a specific habitat. The presumed habitat associated with extinct species is usually based on similarities to extant related species. This premise, of resemblance in paleo and

Fig. 12.2 Overview of the 1971 'Ubeidiya excavation showing the excavation of strata I 26, I 15 and II 23 (© 'Ubeidiya archive, reprinted with permission)

Fig. 12.3 Close up of stratum I 26c at 'Ubeidiya showing scatter of bones from the 1971 excavation. *Arrows* denote the presence of bones (© 'Ubeidiya archive, reprinted with permission)

modern habitats between phylogentically close species, although probably true for most cases, should not be taken as an *a priori* assumption (Andrews 1995c). Another factor presents difficulties in the use of mammals as paleoecological indicators: relative species frequencies may be sensitive to taphonomic biases. These may be due to time and space averaging, depositional milieu, differential preservation, random effects or anthropogenic bias (Voorhies 1969; Behrensmeyer 1975, 1978, 1982, 1984, 1992, 2001; Behrensmeyer and Hill 1980; Brain 1981; Behrensmeyer and Cooke 1985; Badgley 1986; Behrensmeyer et al. 1986, 1992; Walker 1993; Lyman 1994; Andrews 1995a; Arribas and Palmqvist 1998; Flessa 2000).

To overcome these problems, "taxon free" or "phylogeny free" methods have been developed. Ecomorphological diversity was developed by Andrews and colleagues (1979).

This method associates ecomorphological characteristics within a specimen regardless of its species taxonomic identification. For instance, some specimens may be identified to species level while others (due to fragmentation or anatomical position) cannot. Specimens can be classified as one of four ecomorphological characters based on morphology (gross or micro) and measurements. Thus, a tooth fragment that may not be classified to species can be classified as a "grazer" based on the hypsodonty index. Size, locomotion and diet can thus be inferred from the specimens themselves, and can include specimens that may be only identified to higher level taxa rather than those identified to species only. This allows for ecological comparison of sites that have no species in common due to spatial and temporal distances, but have similar ecomorphological diversities.

Andrews and colleagues (1979) compared four modern communities from various habitats: lowland forest, montane forest, flood plain and woodland-bush-land. Taxonomy at the ordinal level, size, locomotion and diet distributions were found to differ significantly among these habitats. Moreover when fossil assemblages were studied, they could be assigned to modern communities in a similar manner. Such analyses have since been done for many Neogene and Quaternary sites (Kay 1977; Andrews and Nesbit Evans 1978; Andrews et al. 1979; Nesbit Evans et al. 1981; Andrews 1989, 1992, 1995b, c, 1996; Cerling et al. 1992; Gunnell 1995; Morgan et al. 1995; Reed 1996, 1997, 1998; Gagnon 1997; Fernández-Jalvo et al. 1998).

The cenogram graph was developed by Valverde (1967) and later by Legendre (1986, 1987a). This method examines mammalian species (all species except chiroptera and carnivora) ranked by abundance against log estimated body weight. The slopes produced can be correlated with open versus closed and humid versus dry environments. This method has been applied to paleontological sites worldwide (Valverde 1967; Legendre 1986, 1987a, b; Gunnell 1990; Legendre et al. 1991; Montuire 1994, 1995, 1998; Montuire and Desclaux 1997; Aguilar et al. 1998; Wilf et al. 1998).

In paleoecological reconstructions of hominin sites, these presence-absence models have been developed using primarily African biomes and habitats. On the other hand, application of the model to Plio-Pleistocene sites from the circum-Mediterranean region has lead to inconclusive results (Belmaker 2002). Application of Andrews' ecomorphological model to the Early Pleistocene site of 'Ubeidiya did not allow assignment of the fossil community to any of the modern African comparative sites.

A method based on abundance of individuals rather than presence-absence of species, was developed by Vrba (1980) and by Greenacre and Vrba (1984). Using multivariate analysis based on census data, they have shown that abundance of individuals of modern bovid tribes can be used to separate habitats based on vegetation cover (height and spacing of trees and bushes). Areas with higher proportions of bush and tree cover have a high abundance of Cephalopini, Tragelaphini and Reduncini, without any members of the Antilopini tribe. However, the use of individual abundance data, i.e., Number of Identified Specimens or NISP is particularly problematic in fossil analyses as it is mostly driven by taphonomic rather than paleoenvironmental factors (Behrensmeyer et al. 2000). Nonetheless, confining the analysis to taxa of a similar size range (in the broad sense), and a community of trophically similar and sympatric species (Hubbell 2001) reduces the effect of sampling of rare species (such as primates and carnivores) as well as collection bias of smaller and larger taxa. Moreover, in order to test the robustness of the analysis, a similar test based on the presence-absence of taxa may be performed as well.

Following the method developed by Vrba (1980), I applied a Correspondence Analysis (CA) (Legendre and Legendre 1998) to abundance values as well as presence–absence data of mesoherbivores (ruminants, suids and equids) from various Plio-Pleistocene and modern assemblages. CA is a method of ordination in a reduced space. It consists of describing the trend or order of variation of the objects (e.g., sites) with respect to all descriptors (e.g., taxon abundances). The *n* descriptors are reduced to the most important variables, which explain the most variance in the data. The major axis is the direction of the maximum variance of the scatter of the points (Legendre and Legendre 1998).

In her multivariate analysis, Vrba (1980), focused on bovid tribes. To facilitate comparison between a wide range of habitats, both African and Eurasian, I opted to include tribes and genera of mesoherbivores, i.e., Ruminantia, Suidae and Equidae following Alemseged (2003).

Due to common confusion between names of vegetation types, habitats and biomes, prior to any paleoecological reconstruction it is important to be explicit about what we are comparing or attempting to reconstruct (Belmaker, this paper). There is a clear distinction between vegetation forms (e.g., woodland, grassland or rain-forest) versus habitats or biomes (e.g., Mediterranean, savanna) which encompass both the spatial distribution of the vegetation forms as well as aspects of climate, temperature, precipitation and seasonality (Eiten 1992). Thus, one can analyze the presence of woodland, grassland or the mosaic of both in several distinct types of habitat such as African savanna or Mediterranean woodland. Although some habitats may have a similar distribution of vegetation types (woodland or grassland), the climate and habitat may differ (Eiten 1992).

This is of particular importance for the comparison between Africa and Mediterranean biomes presented here. Superficial similarities exist between woodland parks in East Africa and a park forest in the Mediterranean region. Both exhibit sparse trees with grassland in between. Both regions are mosaic environments, which include woodland and grassland. The extent of woodland is usually dependent on precipitation and natural fires.

African savanna grasslands contain widely scattered trees or shrubs. These may grade into "tree savanna," "shrub savanna," or "savanna woodland". This forms mosaic landscapes in which clumps, groves of woody plants are dispersed throughout a grassy matrix. Thus, savanna landscape denotes areas where savanna vegetation is dominant but may be interspersed with riparian or gallery forest, or patches of woodland, swamps, or marshes (Scholes and Archer 1997).

Mediterranean woodlands are dominated by sclerophyllous evergreen shrubs and trees. Different communities are recognized based on gradients of moisture and temperature that result in a range of ground coverage of various heights and density. Thus, woodland refers to regions with dense forest coverage (>80%) and maquis, garrigue and batha refer to decreasing levels of ground coverage as well as decreasing density and height of vegetation (Allen 2001). Tree density is also dependent on the presence of nutrients. For example, regions with high levels of phosphate in the soil may result in low tree density (0–20%) (Rabinovich-Vin 1986) producing is a "savanna-like" or "savannoid" habitat rather than a true savanna. Savanna grassland around the Mediterranean basin is regarded as anthropogenic (rather than climatic) in origin is considered to be the result of the clearing of trees and shrubs for grazing pasture (Allen 2001).

However, despite superficial similarities between savanna woodland and Mediterranean woodland in vegetation forms (i.e., low tree density with grassland between) there are differences in the precipitation regime and seasonality between the two biomes. The East African savanna is largely dominated by two wet periods (defined as one whose mean rainfall exceeds the mean for all months) and two dry periods (Delany and Happold 1979), while in the Mediterranean region, there is only one dry season which is long and is correlated with the hot season (Blondel and Aronson 1999).

In this study, I compare sub-tropical African biomes, both woodland and savanna habitats to Mediterranean woodlands and scrubland rather than the closed (woodland) versus open (grassland) vegetation types. Thus, similarity of the'Ubeidiya assemblages to one of the modern biomes will imply a climate, precipitation and seasonality together with the spatial distribution of vegetation types.

I compared the data of mesoherbivore tribes and genera from the site of 'Ubeidiya strata III 12, III 20, II 23–24, III 21–22, I 15–16, II 36 and I 26 (Belmaker 2006) to recent and Plio-Pleistocene sites from Africa and Europe as published in the literature. East African Plio-Pleistocene sites were: Shungura Formation Members A through G (Alemseged 2003) and Konso levels 4M, 4HA, 10L and 10M (Suwa et al. 2003). Recent African sites were: Serengeti grassland, Serengeti woodland, Manyara, Ngorongoro, Nairobi Park from East Africa and Kruger Park from South Africa (Schaller 1972). Eurasian Plio-Pleistocene sites were: Poggio Rosso, Italy (Mazza et al. 2004), Dmanisi, Georgia (Gabunia et al. 2000), Venta Micena, Spain (Palmqvist and Arribas 2001) and Untermassfeld, Germany (Kahlke 2000). Recent European sites were Rila National Park, Bulgaria (Spasov 1998) and the Province of Arezzo, Italy (Mattiolio et al. 2004) (Table 12.1).

Results for correspondence analysis based on presence-absence data (Fig. 12.4) indicate that the two first axes explain 62.93% of the variance and that total inertia (total CA variance) = 1.3712. Results for the abundance data (Fig. 12.5) indicate that the two first CA axes explain 42.26% of the variance and inertia (total CA variance) = 3.2961.

Overall, results are similar for both presence–absence and abundance distributions. Axis I distinguishes between the Eurasian fauna and African fauna (both recent and Plio-Pleistocene) for both presence–absence and abundance data. The mesoherbivore community of 'Ubeidiya is more similar to the recent European and Plio-Pleistocene communities than to any of the African communities (recent or Plio-Pleistocene) and have low negative values on axis I (<0 for the presence–absence data and <–0.5 for the abundance data) while all African communities (recent and Plio-Pleistocene) have positive values. All Eurasian communities, indicative of Mediterranean woodland, cluster together with the taxa of Palaearctic origins: cervids, caprines and the genus *Sus*. All African (recent and Plio-Pleistocene) communities cluster with taxa of African biogeographic origin: the bovid tribes-Antilopini, Alcelaphini, Hippotragini, Reduncini and Aepycerotini, and the suid genera *Metridiochoerus* and *Phacochoerus*. The position of *Equus* and Bovini tribe in the center of the diagram between European and African localities emphasizes their joint distribution on both continents.

Further distinction between assemblages based on Axis II cannot be achieved using presence-absence data only. Using abundance data, Axis II distinguishes between two African habitats: Serengeti Woodland, Manyara and Kruger Park (labeled Group A) all with negative values, while Ngorongoro, Savanna grassland and Nairobi Park (labeled Group B) have positive values of axis II. This is similar to the classification found by Greenacre and Vrba (1984) and which they interpreted as a difference in rainfall. Thus, Group B sites have a lower rainfall and thus attain a more open vegetation pattern than Group A. Group B can be classified based on the presence of Antilopini, Alcelaphini, Hippotragini and *Phacochoerus*. The Plio-Pleistocene sites of Konso 10 L and Konso 10 M may also be included in the classification. Group B can be classified based on the presence of *Metridiochoerus*, Reduncini and Aepycerotini. The Pliocene Shungura Formation sites and Konso 4M and 4HA can also be grouped here. The clustering of the European recent and Plio-Pleistocene sites and specifically that of 'Ubeidiya with similar values of CA axis II to African woodland (0.5 to –0.5), as opposed to the open grassland (1.00–1.50) strengthens the identification as a closed forest habitats within the woodland–scrubland continuum of the Mediterranean habitat.

Table 12.1 Mesoherbivore abundances for selected Modern and Plio Pleistocene sites in Eurasia and Africa

Site	Cerv	Alce	Anti	Redu	Bov	Capr	Trag	Aepy	Hippo	Kolp	Metr	Noto	Phac	Sus	Equu	Hippa
III 11-13	80.74	0	2.22	0	13.33	0	0	0	0.74	1.48	0	0	0	0	1.48	0
III 20	90.32	0	1.94	0	1.94	0	0	0	0.65	0	0	0	0	0	5.16	0
II 23-25	65.76	0	4.62	0	3.18	0	0	0	0.48	0.32	0	0	0	0.8	24.84	0
III 21-22	77.12	0	2.54	0	1.69	0	0	0	0.28	0.28	0	0	0	0	18.08	0
II 26-27	82.54	0	3.88	0	3.7	0	0	0	0.88	2.12	0	0	0	2.29	4.59	0
II 36	61.58	0	4.21	0	3.68	0	0	0	0	0.53	0	0	0	11.05	18.95	0
II 37	52	0	19	0	4	0	0	0	0	1	0	0	0	7	17	0
Konso 4M	0	29.4	2.7	30	6.76	0	3.86	2.5	0	16	0.96	1.54	0	0	4.26	1.74
Konso 4HA	0	34.14	0	42.68	6.09	0	4.87	0	0	7.31	0	0	0	0	3.65	3
Konso 10 L	0	63.33	1.25	14.58	0	0	0	0	5.83	5.83	0.48	0	0	0	5.23	1.66
Konso 10 M	0	63.97	3.13	9.15	0.84	0	0.84	0	0.24	3.86	8.43	0	0	0	9.63	1.32
Shungura Upper G	0	10.01	0	39.23	0.66	0	4	14.69	0	16.19	1.84	4	0	0	7.51	1.84
Shungura Lower G	0	1.51	0	45.99	1.65	0	17.23	13.2	0	9.83	3.14	5.12	0	0	1.25	1.01
Shungura F	0	6.01	0	19.08	6.36	0	16.96	10.6	0	10.6	5.3	16.96	0	0	0	8.13
Shungura E	0	0.84	0	16.46	4.22	0	23.63	10.13	0	31.65	0.84	9.28	0	0	0	2.95
Shungura D	0	0	0	21.05	0	0	10.53	10.53	0	10.53	0	36.84	0	0	0	10.53
Shungura C	0	1.6	0	17.95	8.65	0	15.38	18.91	0	20.51	1.28	7.37	0	0	0	8.33
Shungura B	0	1.11	0	26.94	9.17	0	7.22	11.11	0	5	2.22	33.61	0	0	0	3.61
Shungura A	0	0	0	5.31	3.54	0	1.77	3.54	0	8.85	0	73.45	0	0	0	3.54
Serengeti grassland	0	21.3	56.74	0.18	2.19	0	0.42	4.32	2.66	0	0	0	0.98	0	11.13	0
Serengeti woodland	0	0.01	3.35	1.16	35	0	1.65	33.04	22.67	0	0	0	1.23	0	1.81	0
Manyara	0	0	0	2.33	63.69	0	1.06	29.72	0	0	0	0	1.06	0	2.12	0
Ngorongoro	0	57.06	21.08	0.51	0.25	0	1.69	0	0.42	0	0	0	0	0	18.98	0
Nairobi Park	0	6.94	23.16	2.8	0	0	2.14	17.35	30.02	0	0	0	4.22	0	13.38	0
Kruger Park	0	5.32	0	2.05	4.11	0	3.25	79.06	0.61	0	0	0	0	0	5.57	0
Dmanisi	55.72	0	1.96	0	11.62	5.97	0	0	0	0	0	0	0	0	24.7	0
Untermassfeld	64.08	0	0	0	25.4	0	0	0	0	0	0	0	0	3.31	7.18	0
Poggo rosso	68.53	0	0	0	2.79	0	0	0	0	0	0	0	0	23	5.59	0
Venta Micena	25.71	0	0	0	14.45	12.04	0	0	0	0	0	0	0	0	47.78	0
Arezzo Italy	60.65	0	0	0	0	0.84	0	0	0	0	0	0	0	38.5	0	0
Rila	33.98	0	0	0	0	40.12	0	0	0	0	0	0	0	25.88	0	0

cerv. = cervidae, Alce. = Alcelphini, Anti. = Antilopini, Redu. = Reduncini, Bovi. = Bovini, Cape. = Caprini, Trag. = Tragelaphini, Aepy. = Aepycerotini, Hippo. = Hippotragini, Kolp. = Kolpochoerus, Metr. = Metridiochoerus, Noto. = Notochoerus, Phac. = Phacochoerus, Sus. = Sus, Equu. = Equus, Hippa. = Hipparion (Data for modern African habitats from Schaller (1972), data for Poggo Rossio from Mazza et al. (2004), data for Dmanisi from Gabunia et al. (2000), data for Venta Micena from Palmqvist and Arribas (2001), data for Untermassfeld from Kahlke (2000) (MNI quantification), data for Arezzo, Italy from Mattiolio et al. (2004) and data for Rila National park, Bulgaria from Spasov (1998), data for Shungura from Alemseged (2003) and for Konso from Suwa et al. (2003))

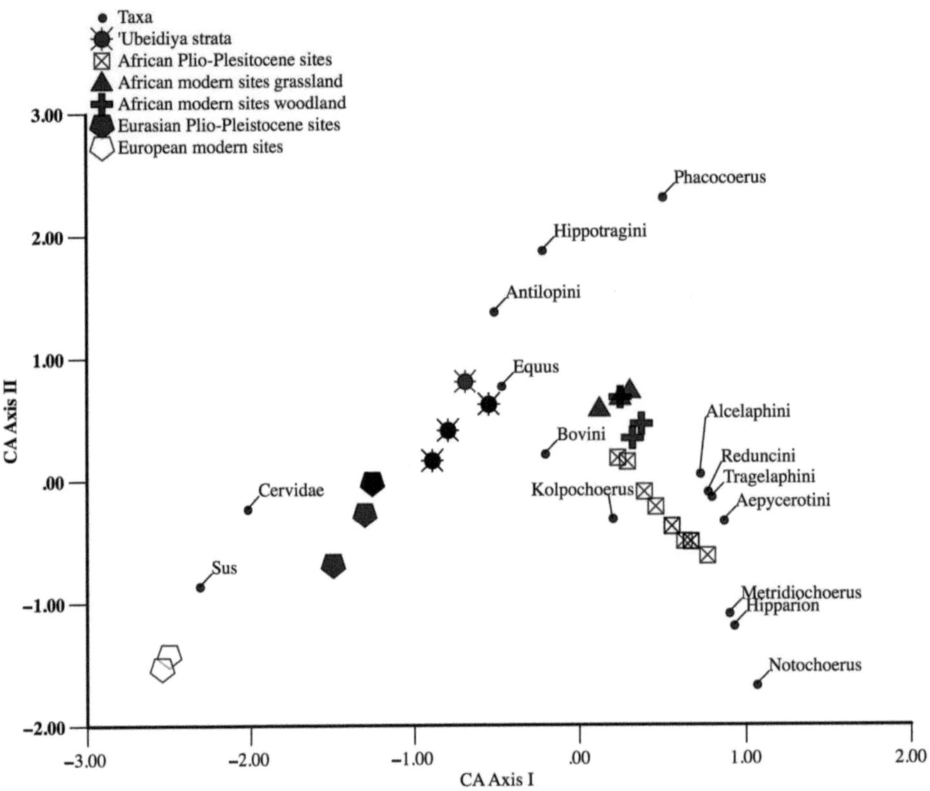

Fig. 12.4 Scattergram for correspondence analysis for presence-absence data for mesoherbivores from recent and Plio-Pleistocene African and Eurasian sites (Data for recent African habitats from Schaller 1972, data for Poggo Rossio from Mazza et al. 2004, data for Dmanisi from Gabunia et al. (2000), data for Venta Micena from Palmqvist and Arribas 2001, data for Untermassfeld from Kahlke 2000 (MNI quantification), data for Arezzo, Italy from Mattiolio et al. (2004) and data for Rila National park, Bulgaria from Spasov 1998, data for Shungura from Aelemseged 2003 and for Konso from Suwa and colleagues 2003)

Thus, while African taxa are present at 'Ubeidiya the overall signature of fauna at the site suggests that the community is more similar to a Mediterranean woodland than to an African savanna. While the fauna is certainly mixed and contains both Eurasian and African taxa, it does not identify the habitat in 'Ubeidiya as an African one but rather more as a Eurasian one.

How Can the Presence of African Taxa Be Explained in a Mediterranean Biome?

Contrary to the reconstruction presented here, several studies have suggested that the paleoecological reconstruction of 'Ubeidiya was open and semi-arid (Martínez-Navarro 2004, 2010; Klein 1999; Dennell 2004). This is based on the presence of African species and specifically *Oryx* sp., *Kolpochoerus olduvaiensis*, *Hippopotamus gorgops*, *Pelorovis oldowayensis* and *Crocuta crocuta*. Moreover, it has been suggested that the *Megantereon* remains found at the site should be identified as the African species *M. whitei*, further increasing the proportion of African species at the site (Martínez-Navarro 2004). Similarly, reconstruction of the paleoenvironment of the site of Dmanisi, Georgia, was interpreted by Palmqvist (2002:158) as "African savanna with tall grasses, shrubs and low bush/tree cover" (my emphasis). Moreover, ecomorphological analysis of Venta Micena has suggested that "Orce at Venta Micena was very similar to that represented in modern African savannas with tall grass and low bush/tree cover, suggesting that the countryside in the Guadix–Baza basin was relatively unforested during early Pleistocene times, as happens today" (Palmqvist et al. 2003:225). These reconstructions are inconsistent with conclusions presented here. Evidence from the correspondence analysis has shown that although African taxa are present in the southern Levant as well as in other circum-Mediterranean sites in the Early Pleistocene, the presence of taxa which require Mediterranean woodland habitats such as cervids and genus *Sus* suggest that the environment was more similar to a Mediterranean woodland rather than an African savanna.

Nonetheless, the faunal assemblages in 'Ubeidiya have a significant proportion of Africa taxa (Tchernov 1986) compared to other Eurasian sites such as Dmanisi (Vekua 1995; Palmqvist 2002), or the Italian sites of Selvella, Pietrafitta

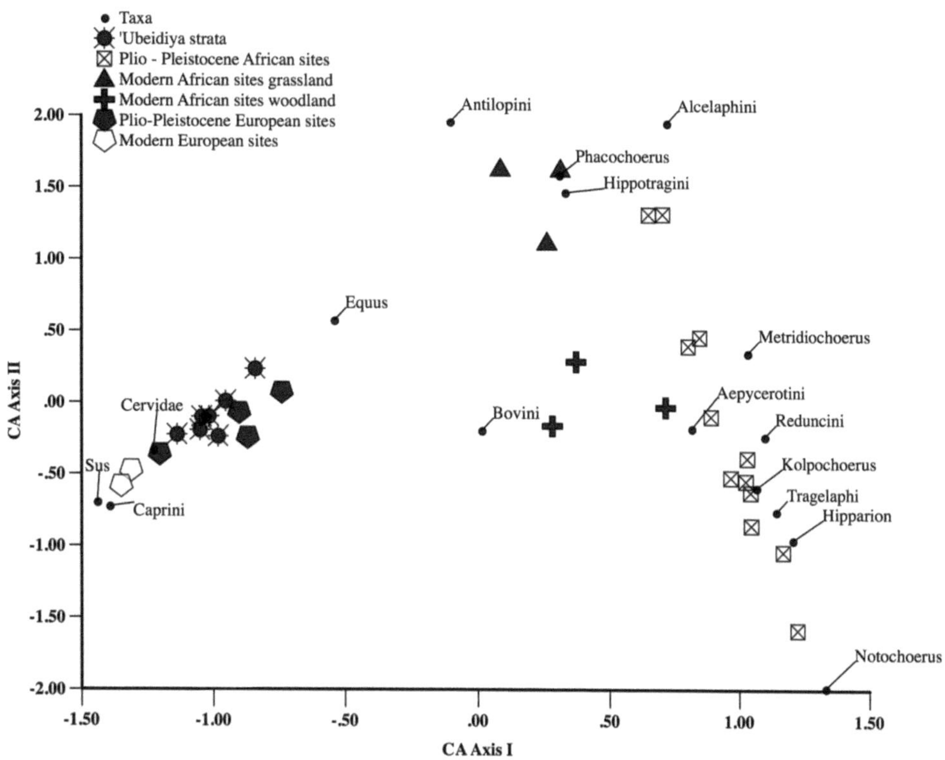

Fig. 12.5 Scattergram of correspondence analysis for abundance data for mesoherbivores from recent and Plio-Pleistocene African and Eurasian sites (Data for recent African habitats from Schaller 1972, data for Poggo Rossio from Mazza et al. 2004, data for Dmanisi from Gabunia et al. 2000, data for Venta Micena from Palmqvist and Arribas 2001, data for Untermassfeld from Kahlke 2000 (MNI quantification), data for Arezzo, Italy from Mattiolio et al. 2004 and data for Rila National park, Bulgaria from Spasov 1998, data for Shungura from Aelemseged 2003 and for Konso from Suwa et al. 2003)

and Pirro Nord (Caloi and Palombo 1997; Rook et al. 2004; Arzarello et al. 2007). I argue that the presence of African taxa in a Mediterranean woodland may be explained by a Long Distance Dispersal (LDD) model (Nathan 2001; Nathan et al. 2003) coupled with biotic interactions such as competition and predation and that the presence of African taxa in Plio-Pleistocene sites in the southern Levant is not inconsistent with the paleoecological reconstruction presented in this analysis. The abundance pattern (both within sites and across sites) of different ecomorphological groups of Africa taxa (grazers, browsers, aquatic taxa and carnivores) in the southern Levant during the Early Pleistocene is consistent with the expected from their adaptation to a novel environment and the presence of biotic competitors that they may have encountered.

The LDD Model

Dispersal is defined as the movements of organisms (in any stage of the life cycle) away from their parents' source location to a new location defined as an end location (Nathan 2001). Long distance dispersal (LDD) is the rare occasion in which species shift their range by moving over long distances. We define an exotic species as a species from a different biogeographic origin than the location in which it is present, suggesting long distance dispersal.

In order to expand its range, a species must be able to complete all of the following three processes:

1. Travel to a new area
2. Withstand potential unfavorable conditions during its passage
3. Establish a viable population upon its arrival (post depositional establishment)

If any one of the three processes is not completed, long-term colonization will not occur. Thus, an organism may travel successfully to a new area only to find it too harsh for its own existence or a superior competitor may already occupy the niche, resulting in the local extinction of the dispersing species (Nathan 2001). It might survive in the new region, but if there is not a large enough colonizing population of both sexes, a viable population will not be able to colonize the new region. Only if all three conditions occur, can colonization be observed in the paleontological record.

Diffusion (dispersal over several generations) is a complex yet distinctive pattern. Early waves of dispersal (invasion and range expansion) are slow, intermittent and may require repeated dispersal events. Following adaptation to the new environment and successful colonization, the geographic range increases at an exponential rate (Brown and Lomolino 1998). Post depositional establishment is dependent on the ecological condition at the site and traits specific to the biotic community i.e., the presence or absence of competitors or predators (Nathan 2001). Thus, if the dispersing species is adapted to the ecological conditions and is not excluded by biotic interaction such as competition with native taxa or predation, it will successfully establish itself, its abundance will increase, its range will expand or both. If either of these conditions will not occur, it will remain in low abundance and low geographic range until it will become extinct in the end location. According to this model, different ecomorphological groups will have different abundances both within any given site as well as across sites both in space and time depending on the habitat in the new region, time since their dispersal and local competitors.

Taxa which are abundant in the fossil record in the new region, and appear both in large numbers as well in many number of sites over a large geographic range are consistent with the third stage of colonization (expansion) and can be inferred to be ecologically good dispersers into the new region. On the other hand, taxa that are rare, or present in few sites are more consistent with the initial stages of dispersal and if they persist only during a short time span may be inferred to be poor dispersers into the new region. Following this model, this may be the result of a lack of adaptation to the abiotic conditions or competition with superior competitor taxa present in the region.

While the quantification of fossil abundances as a proxy for living community populations is problematic due to taphonomic biases, using several measures of biogeographic abundance (abundance within any given site as well as abundance across sites both in space and in time) reduces the biases of the fossil distribution in relation to the true biological one.

Testing the LDD Model in the Plio-Pleistocene of the Southern Levant

To test if African taxa in the southern Levant follow the pattern as expected from the model, the African taxa were assigned one of four ecomorphological groups: grazers (all orders), browsers (all orders), aquatic taxa and carnivores. The abundance of each taxon (by group) is observed both within sites and across sites (i.e., number of sites) in space and time.

Of the Early Pleistocene sites in the southern Levant with evidence for the presence of hominins, only a few have faunal remains that allow for a biogeographic comparison and have been used in this analysis (Fig. 12.1).

1. The mammalian fauna of Bethlehem, Palestine, is the oldest Plio-Pleistocene assemblage known from the Southern Levant. A bone cache was found in a well in the city of Bethlehem (Gardner and Bate 1937). The original finds suggested that lithics may be present but they have since been shown to be natural (Hooijer 1958). Based on the faunal assemblage it has been assigned to the Middle Villafranchian (Hooijer 1958).
2. The site of 'Ubeidiya, is located in the central Jordan Valley, Israel and is presented in detail in this study. The site exhibits human remains (Tobias 1966a, b; Belmaker et al. 2002), rich lithic (Bar-Yosef and Goren-Inbar 1993) and faunal assemblages (Haas 1966, 1968; Tchernov 1986). The site has been dated between 1.6–1.2 Ma (Tchernov 1987, 1988a; Sagi et al. 2005; Sagi 2005).
3. Evron Quarry (Tchernov et al. 1994; Ron et al. 2003) is located near Kibbutz Evron on the coastal plain of the Western Galilee, Israel. Paleomagentic studies have suggested a date ca. 1.0 Ma (below the 0.78 Ma Brunhes–Matuyama boundary) for the archeological bearing strata. The site has yielded *in situ* Acheulean deposits that include quartz/limestone pebbles and flint artifacts. Handaxes collected from the quarry were associated with the assemblage (Ronen 1991). A small faunal assemblage was retrieved ($n = 36$) (Tchernov et al. 1994).
4. Bitzat Ruhama is located in the eastern part of the southern coastal plain of Israel (Ronen et al. 1998). The site is estimated around 1.0 Ma based on magnetostratigraphy and RTL dating methods (Ron and Gvirtzman 2001). The site exhibits a large and highly variable lithic assemblage, with no bifaces, dominated by notches and denticulates (Zaidner et al. 2003; Zaidner 2003a, b). There is only a small faunal assemblage ($n = 36$) (Ronen et al. 1998).
5. Latamne is located ca. 40 km north of the Village of Hamma, Syria, on the Orontes River. It consists of a "living floor" with a large lithic assemblage and faunal assemblage (Clark 1967). The date of the site was estimated as ca. 0.7 Ma based on faunal correlations but the presence of the arvicolid *Lagurodon aranake* (Tchernov 1988b) and typo-technological affinities of the lithic assemblage suggest a data ca. 1.0 Ma (Tchernov 1988b).
6. Gesher Benot Ya'aqov is located in the northern Dead Sea rift, 4 km south of the Hula Valley, Israel. The site exhibits hominin paleoanthropological remains, a wealth of lithic remains, a large faunal assemblage and a unique botanical assemblage. The Acheulean industrial complex lithic assemblage represents a unique African technology of cleavers which suggests affinities to Africa and has

been interpreted as evidence for a second dispersal event (Goren-Inbar and Saragusti 1996). The site has been dated to the 0.78 Ma Brunhes-Matuyama boundary (Goren-Inbar et al. 2000).

Grazers

The presence of grazer taxa, and specifically the large grazer herbivores, has been interpreted as an indication for the spread of savanna habitats into higher latitudes during the Early Pleistocene (Martínez-Navarro 2004, 2010; Klein 1999; Dennell 2004). The grazer taxa in the Early Pleistocene in the Southern Levant are: *Pelorovis oldowayensis*, *Kolpochoerus olduvaiensis*, *Oryx* cf. *gazella*, *Equus tabeti* and *Theropithecus oswaldi*.

Pelorovis oldowayensis was first described in Olduvai from middle and upper Bed II and Bed III (Gentry 1978). *Pelorovis* was unknown outside of Africa until the first record in 'Ubeidiya (Geraads 1986), but since then it has been recorded as *P.* cf. *oldowayensis* at the site of Gesher Benot Ya'aqov (Martínez-Navarro 2004, 2010). Moreover, it was found in the Early Pleistocene deposits of the An Nafud desert in the north of Saudi Arabia (Thomas et al. 1998). In 'Ubeidiya, although it appears in all strata, it comprises only between 0.5% and 6% of the identified terrestrial large mammals specimens (Belmaker 2006).

Oryx cf. *gazella* has been identified in the southern Levant from only a few specimens in 'Ubeidiya and is not known from other Levantine sites. *Oryx* species from East Africa, North Africa and Arabia have been assigned different taxonomic names. It has been suggested by Martínez-Navarro (2004) that together with the *Oryx* from 'Ubeidiya, all should be assigned to the same species. These include the African *Oryx* cf. *gazella* from Ternifine, Olduvai, *Oryx* sp. from Koobi Fora and *Oryx eleulmensis* in Aïn Hanech.

Kolpochoerus olduvaiensis first begins a process of segregation from *K. limnetes*, as early as Olduvai Bed I. Although it is known from the Shungura Member G, it becomes common in Member K. In Olduvai, it ranged from Bed IV and appears in other Middle Pleistocene African sites (Harris 1983). In the southern Levant, *Kolpochoerus olduvaiensis* is known from 'Ubeidiya (Geraads et al. 1986), where it comprised only up to 2% of the assemblage. An endemic species *Kolpochoerus evronensis* (previously identified as *Metridiochoerus evronensis*) was found in Evron (Tchernov et al. 1994). Despite a brachydont dentition, stable isotope analysis on *K. limnetes* from Koobi Fora suggests that it was a grazer although it was dependent on water (Harris and Cerling 2002).

The equid at 'Ubeidiya was identified as *Equus* cf. *tabeti* by Eisenmann (1986) based on metapodial and tooth morphology. *Equus tabeti* has been found in Aïn Hanech (Arambourg 1970) and Koobi Fora (Eisenmann 1983). This species is very close to the African species *E. numidicus* and *E. oldowayensis* (Eisenmann 1983). Guérin and colleagues (1993) noted similarities between the 'Ubeidiya specimens and those from Latamne. In 'Ubeidiya, *Equus* cf. *tabeti* ranges from 5% to 24% among strata.

Multivariate analysis has suggested similarities between the equids from the Southern Levant, North African species (*E. numidicus* – *E. tabeti* lineage) and the specimens found in Venta Micena attributed to *Equus* cf. *altidens* and which have also been found in Spain (Orce, Cúllar de Baza, Cueva Victoria, Huèscar-1), Italy (Pirro Nord, Selvella), France (Sainzelles) and Germany (Süssenborn) (Arribas and Palmqvist 1999). This group is also similar to the recent *E. grevyi*. This lineage has been named "simplicidens" (1997) and includes *E. numidicus*, *E. tabeti* and *E. altidens*. The *E. numidicus* and *E. tabeti* appear in East Africa (Eisenmann 1983), while the first appearance of *E. altidens* is in Dmanisi (1.7 Ma) suggesting an African origin for the common ancestor of the lineage. In the southern Levant, the gracile equids were replaced in the Middle Pleistocene and probably as early as Gesher Benot Ya'aqov (0.78 Ma) by true equids *E. caballus* and by the Asiatic onager *E. hemiones* and *E. hydruntinus* (Eisenmann et al. 2002).

Both taxonomic schemes are consistent with the LDD model. If we accept a lumping taxonomy, i.e., that all circum-Mediterranean gracile equids belong to a single lineage, a "simplicidens" group (*sensu* Guerreo-Alba and Palmqvist 1997), which includes the sub populations identified as *E. numidicus*, *E. tabeti* and *E. altidens*, we can infer that upon dispersal from Africa ca. 2 Ma, this linage attained ecological success in Eurasia. The sub-population *E. tabeti* (Eisenmann 2004) represented a dispersal of small geographic range, which did not persist in the region past 1.0 Ma ('Ubeidiya and Latamne) and although they are recorded with moderate abundance (<25%) in 'Ubeidiya (Belmaker 2006), their spatial distribution pattern is similar to the pattern described for other grazers. This pattern is heightened, if we accept a splitting taxonomy, i.e., the presence of *E. tabeti sensu stricto* in 'Ubeidiya. In the latter taxonomy, the dispersal of *E. altidens* into Spain and Europe (either from Asia or Africa) would be viewed as a different dispersal event, perhaps via a different route. Nonetheless, in both taxonomic scenarios, Eurasian taxa replaced African ones as early as the onset of the Middle Pleistocene. However, the high abundance of equids does appear to represent an anomaly to the paleoecological reconstruction presented here. I suggest that this may result from erroneous *a priori* assumptions based on the ecomorphology of congeners and conspecifics.

Paleoecological reconstruction of *Equus tabeti* from Aïn Hanech based on analysis of limb bone measurements suggest that *E. tabeti* was adapted to an open country with a flat and hard ground (Eisenmann 1984). This adaptation was assumed to be representative of the 'Ubeidiya equid population as well.

Nonetheless, it is worth noting that the analysis of two subspecies of *E. stenosis* based on the same methodology indicates that they occupy a different habitat. Thus, the *E. stenosis* population from Saint-Vallier is inferred to occupy a not too open, humid and soft ground habitat while the *E. stenosis* population from La Puebla de Valverde occupied open, dry and hard ground (Eisenmann and Guérin 1984). Moreover, although equids as a family are classified as grazers, based on gross dental morphology, mesowear analysis has suggested that *E. capensis* from South Africa was a mixed feeder rather than a grazer (Kaiser and Franz-Odendaal 2004). This suggests that further analysis is needed to work out both the taxonomy as well as the ecomorphological characteristics to discern preferred habitat of local populations. Thus, further research is needed to determine the adaptation requirements of local populations, such as the *E. tabeti* in 'Ubeidiya, within the wide spread *Equus "simplicidens"* group as in the case of the variable *E. stenosis*.

The genus *Theropithecus* in general and the species *Theropithecus oswaldi* in particular was widely distributed in Africa during the Pleistocene (Jablonski 1993). North of the Sahara, specimens attributed to the genus are few yet point to dispersal of the genus from the Pliocene. In the Early Pleistocene, specimens as attributed to *T. oswaldi* have been found ranging from the Iberian Peninsula in the west and the Indian peninsula in the east.

A large cercopithecid calcaneus from the lowest stratum (III 12) in 'Ubeidiya can probably be assigned to *Theropithecus* cf. *oswaldi* (Belmaker 2010). This taxon has also been identified in Cueva Victoria, Spain (Gibert et al. 1995) and in Pirro Nord, Italy (Rook et al. 2004) although its presence in the latter site has been contested (Patel et al. 2007). The genus is also found in North Africa in the site of Ahl al Oughlam as the species *T. atlanticus* in the Pliocene and as *T. oswaldi* in Ternifine and Thomas Quarry in the Middle Pleistocene (Alemseged and Geraads 1998). The species, *T. delsoni* (identified by Delson (1993) as *T. oswaldi delsoni*), is also found in the Pleistocene of India, at the site Mirzapur (ca. 1.0 Ma) (Gupta and Sahni 1981; Delson 1993; Pickford 1993). Cranial and tooth morphology suggests similarities to *T. gelada* and suggestive of a graminivorous diet (Leakey 1993).

The geographic spread and dietary adaptation of this taxon suggests that its presence in the southern Levant may be consistent with the advance of grassland habitats. While this is plausible, it is not inconsistent with the reconstruction presented in the study. Ecological success of *Theropithecus* cf. *oswaldi* in the southern Levant may be partially explained by the absence of competitors in the new region. Moreover, evidence based on morphology of the limb bones has suggested that the locomotion of *Theropithecus oswaldi* was assigned to "open mixed" rather than "open terrestrial" as extant geladas. This is similar to extant *Papio anubis* a taxon that regularly uses some arboreal substrates (Elton 2002). Thus, they may have been adapted to utilize the woodlands present in the Mediterranean region. This further strengthens the reconstruction of the region as a wooded environment as opposed to an open one.

Overall, grazer taxa show a similar pattern and can be categorized as rare both in spatial distribution and in time. Their first appearance in the region is observed in 'Ubeidiya (ca. 1.6–1.2 Ma) and they are not present in the region later than Gesher Benot Ya'aqov (0.78 Ma). The distribution of African grazers in the southern Levant suggests a small geographic range which did not expand past their primary dispersal event. When a taxon (e.g., *Theropithecus oswaldi*) is more widely distributed, it may be attributed to lack of competition and/or a preference to a mixed and more woodland habitat. This generalized pattern is consistent with the first stage of dispersal; low level and intermittent. Moreover, since this group is rare in abundance, their distribution in out-of-African sites during the Early Pleistocene indicates that they were poor adaptors to the new habitats. This may have resulted from two phenomena; the climate and habitat were not suitable for the newly arriving taxa or the habitat included competing species or predators, which prevented successful establishment.

Despite superficial similarities between savanna woodland and Mediterranean woodland habitats (i.e., low tree density with grassland between) there are differences in the precipitation regime and seasonality between the two biomes as well as differences in geographic distribution of sub-habitats which affect the success of grazers in each biome. The East African savanna is largely dominated by two wet periods and two dry periods (Delany and Happold 1979). In the wet season there is a dietary overlap between grazer that is mitigated by the abundance of the vegetation growth. During the dry season and decrease in graze, competition is reduced primarily by "migration" between wet season range on the open plains and the dry season range in the woodland (Eltringham 1979; Maddock 1979). Migration patterns differ among taxa in order to minimize competition (Maddock 1979). In the Mediterranean region, there is only one dry season, which is long and is correlated with the hot season (Blondel and Aronson 1999). The Mediterranean landscape in constricted between the sea and the mountain ranges that encircle the basin and dissect it (Blondel and Aronson 1999). Due to the highly fragmented landscape, reduction of competition during the dry season is more difficult by migration and many taxa, specifically cervids, are mixed feeder which consume the browse of the sclerophyllous evergreen shrubs and trees during the dry seasons.

Poor adaptation of African grazers may have been caused by their inability to adapt to relative absence of graze in the dry season coupled with a high level of competition from the large local mixed feeders such as *Praemegaceros verticornis.*

Mesowear analysis on *Praemegaceros obscurus* specimens from Ceyssaguet suggest that it is a mixed-feeder closer to the browse dominated range of the browse-graze spectrum (Kaiser and Croitor 2004).

Thus, while African grazers are indeed present in the Southern Levant in the Early Pleistocene, their low abundances and restricted geographic range suggest that they are not indicative of the habitat in their region of origin i.e., African savanna, but rather point to a period of many small intermittent dispersal events, each of which may have ended in local extinction due to inadequate adaptation and poor competition with the local taxa. Thus, the presence of savanna and savanna woodland grazing herbivore taxa in the Plio-Pleistocene of the Southern Levant is not inconsistent with the multivariate analysis that suggests association with a Mediterranean woodland and scrubland.

Browsers

As opposed to the relatively high number of grazer taxa of African origin in the southern Levant, browsing taxa of African origin are relatively few. Their absence was used to further confirm the reconstruction of the southern Levant as an African savanna during the Early Pleistocene (Martínez-Navarro 2004, 2010). A single herbivore browser, *Giraffa* sp. is found among the African species in the southern Levant. This species first appears in the southern Levant in the Bethlehem fauna (Hooijer 1958), 'Ubeidiya (Haas 1966) and Latamne (Guérin et al. 1993). In 'Ubeidiya it is rare and only three individuals were found throughout the entire sequence (Belmaker 2006). The presence of this taxon in the Early Pleistocene is probably a straggler from earlier periods (Tchernov 1984), and as such has little bearing on the Early Pleistocene dispersal event.

Nonetheless, the relative low proportion of African browsing taxa is indicative of the paleoecological conditions present along the dispersal route between Africa and the southern Levant rather within the southern Levant itself. We can assume that if conditions along the route were unfavorable to such taxa, dispersal would be hindered. Thus, extended grasslands, as opposed to woodlands or presence of browse, may serve as a barrier to dispersal.

In northern Sahara and Arabia, the Pliocene grasslands that developed around 3 Ma were still present and formed the majority of the vegetation in Southwest and South Asia ca. 2 Ma during onset of hominid dispersal events. A case in point is the site of An Nafud in Arabia (Thomas et al. 1998). This site exhibits many of the African taxa present in the southern Levant (*Crocuta crocuta*; *Pelorovis* cf. *oldowayensis*; *Oryx* sp.) as well additional African alcelaphines and bovids whereas Palaearctic species such as cervids are notably absent. Stable isotope analyses of herbivore teeth indicate a diet of C_4 plants typical of open grassland (Thomas et al. 1998).

Thus, An Nafud is representative of the barrier or filter region between East Africa (from which the dispersal originated) and the southern Levant *sensu stricto* as the end location. The difference in the faunal composition between Arabia and the southern Levant during the Early Pleistocene, i.e., the presence of cervids and *Sus* in 'Ubeidiya and the higher proportion of bovids in An Nafud reiterate the paleoecological reconstruction of the southern Levant as Mediterranean rather than savanna.

The absence of browsing African taxa does not indicate that the southern Levant was a savanna habitat; rather it is indicative of the habitats along the dispersal routes. Although climatic shifts during the Late Pliocene would have supported the dispersal of grazing taxa through the Saharo–Arabian barrier, these habitats were probably not humid and wooded enough to be suited for browsing taxa. Thus dispersal routes (rather than the end location in the southern Levant) were in fact a filter that prevented browsing taxa from dispersing from Africa into higher latitudes during the Early Pleistocene, while allowing grazing or generalist taxa to do so. Moreover, the reconstruction based on presence-absence of African taxa only, as opposed to the multivariate analysis of the entire large mammal community presented here, fails to account for the Eurasian taxa indicative of woodlands such as multiple cervid species and *Sus*.

Aquatic Taxa

The presence of aquatic taxa, notably hippos, in higher latitudes has been cited as evidence for a large-scale dispersal event (Martínez-Navarro 2004, 2010). I argue that even though hippos graze one or two km from a water source (Eltringham 1999), they are primarily dependent on the existence of perennial waters and thus are less sensitive to the differences in seasonality and precipitation such as defined the differences between the African savanna and the Mediterranean region. The presence of hippos in the southern Levant in the Early Pleistocene is thus more indicative of the development of fresh water lake systems than of biome habitats (Mediterranean or savanna).

In the Pliocene-Early Pleistocene in Africa there were two *Hippopotamus* species: *H. gorgops* and *H. amphibius* (Boisserie 2005). During the Early and Middle Pleistocene a large form, *H. antiquus* (= *H major*) evolved in Europe (Mazza 1991). The origin of *H. antiquus* from an African *H. amphibius* (Kahlke 1997) or *H. gorgops* (Martínez-Navarro 2010) is debated. Unlike Europe, the Levant Hippos include only *H. gorgops*, *H. amphibius* and an endemic species *H. behemoth*.

Hippopotamus gorgops can be traced through the Early to Middle Pleistocene deposits of Olduvai Gorge (Coryndon and Coppens 1973; Coryndon 1976). It is present at 'Ubeidiya (Faure 1986), although it is very rare. This species had high crowned molars, small premolars, elevated orbits, and large and strongly ridged canines. It has been interpreted as adapted to aquatic life with a grazer habitat, which allowed it to exploit lacustrine conditions (Coryndon and Coppens 1973).

Hippopotamus behemoth was described by Faure (1986) as an endemic species and is one of the most common species (Belmaker 2006). This species has since been also identified in Latamne (Guérin et al. 1993). *Hippopotamus behemoth* falls with the size range of *H. gorgops* populations in Africa and it has been suggested that both may be a single species (Martínez-Navarro 2010; Martínez-Navarro et al. 2004), but further morphological studies are required to confirm or refute this hypothesis. The taxonomic relationship between this taxon and possible predecessors in Africa is unclear and may include forms related to *H. gorgorps* or *H. amphibius*.

Hipopotamus sp. has been described in the Pliocene deposits of Bethlehem (Gadner and Bate 1943; but see Hooijer 1958). *Hippopotamus amphibius* is present in all Southern Levantine sites from the Early Pleistocene: Evron (Tchernov et al. 1994), Bizat Ruhama (Ronen et al. 1998) and Gesher Benot Ya'aqov (Martínez-Navarro 2004), but due to the preservation of the material it could not be securely identified to species. However, given the size and morphology of the majority of the remains they are most probably *Hippopotamus amphibius*. *H. amphibius* is recorded from the Southern Levant as late as the end of the Middle Pleistocene (Tchernov 1984).

Hippopotamus amphibius is known from Pliocene deposits in Africa (Coryndon 1976). Thus, *Hippopotamus* dispersed into the Southern Levant from Africa in two main waves. The first included *H. amphibius* from ca. 1.2 Ma and perhaps as early as Bethlehem (Gardner and Bate 1937; Hooijer 1958). A second dispersal event during the Early Pleistocene included *H. gorgops*. Pending future analysis, *H. behemoth* would either be grouped with *H. gorgops*, or would have evolved from an in situ local evolution from either one of the two populations.

The dispersal of hippos in the Plio-Pleistocene into the Southern Levant was dependent on a chain of freshwater lakes. Three paleolakes: Kuntila, Zihor and Hiyyon were present in northeastern Sinai and along the margins of the Arava Valley in the Dead Sea Rift. Ostracod composition suggests freshwater lakes (Rosenfeld et al. 2002). In the Southern Jordan Valley, ostracod samples from the Mazar Formation (2.0–1.8 Ma) suggested a salinity of fresh to brackish waters (Almogi-Labin et al. 1995). In the Central Jordan Valley, ostracod analysis has suggested that the salinity of paleolake 'Ubeidiya was fresh to brackish (Almogi-Labin et al. 1995). The hydrogeographical connection between the Nile Valley and the Sea of Galilee is also evident by recent genetic studies of the cichlid *Astatotilapia flaviijosephi* that suggest that it separated from the other haplochromines during the Middle to Late Pliocene (2.5–3.3 Ma) and probably dispersed from Africa to the southern Levant via the Nile (Werner and Mokady 2004).

The absence of *H. gorgops* in Europe during the Early Pleistocene and the possible in situ evolution of the local Pliocene *Hippopotamus* population into *H. behemoth* during this time period in the southern Levant suggests a fragmentations of habitat into smaller lake basins (specifically between Europe and the southern Levant) promoting local speciation and hindering latitudinal dispersal. Moreover, this emphasizes the geographic position of the southern Levant as an evolutionary cul-de-sac. Dispersing species could not maintain sufficient gene flow with their ancestral populations. The Holocene presence of *H. amphibius* in the rivers of the coastal region indicates later dispersal events probably through the Nile (Tchernov 1988b).

Based on the morphology and distribution of *Hippopotamus* species in Plio-Pleistocene African lakes, it has been suggested that *H. gorgops* was a grazer that preferred lacustrine environments (Coryndon 1976). Stable isotope analysis on Pleistocene hippopotamids from Africa (Bocherens et al. 1996) and *H. antiquus* from Venta Micena (Palmqvist et al. 2003) suggests that they fed predominantly on aquatic vegetation and analysis of the diet of modern *H. amphibius* while indicative of a C_4 feeder includes a significant C_3 component in the diet (Boisserie et al. 2005). While the direct degree of competition between the *H. gorgops* and *H. amphibius* cannot be measured, it should be remembered that the overall carrying capacity in the tropical latitudes is larger than in higher ones (the latitudinal gradient) and is also a function of area (the species-area curve) (Rosenzweig 1995). Thus, large lake systems (as present in East Africa in the Plio-Pleistocene) may have been able to carry two closely related species of *Hippopotamus* while the smaller lake systems of the Southern Levant may not. Thus, if populations of *H. amphibius* inhabited the paleolakes of the southern Levant, the new dispersing population of *H. gorgops* would have encountered a higher level of competition, hindering its ecological success in the region.

The presence of abundant African aquatic taxa in the Plio-Pleistocene of the southern Levant is not inconsistent with results from the multivariate analysis, which suggests association with a Mediterranean woodland and scrubland, but is indicative of the local hippo population within the lake basin.

Carnivores

The carnivores have been cited as the main evidences for an African savanna habitat in the southern Levantine Early Pleistocene and for an open dispersal route between Africa and Eurasia during this time period. Specifically the presence of the hyaena and a dirk tooth felid are cited as indicative of such an environment (Martínez-Navarro 2004, 2010).

Four species of African carnivores are recorded in the southern Levant: *Crocuta crocuta*, cf. *Mellivora* sp., *Herpestes* cf. *ichenuemon* and *Megantereon whitei*. Since carnivores as secondary and tertiary consumers are high on the food chain, we expect them to have relatively low abundances compared to primary consumers (Krebs 2001) but their adaptation to the new habitat can be evaluated by an analysis of their continued habitation and range expansion.

Crocuta crocuta has been found in Africa as early as Member B in Shungura and at Olduvai (Petter 1973). 'Ubeidiya is the only Eurasian site, older than 1.0 Ma, in which *C. crocuta* is present rather than *Pachycrocuta brevirostris* (Martínez-Navarro 2004, Martínez-Navarro, et al 2009). It first appeared in Europe in the Late Early Pleistocene. It survived in Europe and the Southern Levant until the Late Pleistocene (Rabinovich 2002).

A single large and robust Mustelidae ulna was found in 'Ubeidiya. It exhibits morphology and size similar to the extant *Mellivora capensis* (Belmaker 2006). Petter (1987) described a specimen of *Mellivora* sp. of the size of the living species *M. capensis* from Laetoli and it has been suggested that all the material from Laetoli onwards may represent a single highly plastic lineage (Petter 1987; Werdelin and Lewis 2005). In North African Plio-Pleistocene sites *Mellivora* sp. has been found in Ahl al Oughlam (Geraads 1997), Ternifine, Bouknadel and Thomas Quarries (Michel 1988). In the southern Levant, it is present today throughout Israel, Saudi Arabia, Kuwait and Iraq (Harrison and Bates 1991). *Mellivora capensis* is rare in the Pleistocene and Holocene record. It has been recorded from the Late Pleistocene from 'Erq el-Ahmar (Vaufrey 1951) and in Nahal Hemar dated to 10000 BP (Dayan 1989).

Only a few specimens of *Herpestes* cf. *ichneumon* have been retrieved from 'Ubeidiya (Belmaker 2006). The genus *Herpestes* has a very patchy record in Africa (Werdelin and Lewis 2005). The Egyptian mongoose, *H. ichneumon*, is first recorded at Laetoli (Petter 1987), with a possible later occurrence at Olduvai (Petter 1973). In the southern Levant, it is present today in the northwestern Arabia peninsula, Israel and Southern Turkey (Harrison and Bates 1991).

The only fossil evidence of the dirk tooth felid, *Megantereon* sp. in the southern Levant is at 'Ubeidiya; an upper canine (UB 80) from stratum II 23 was assigned by Haas (1966) to *Megantereon* cf. *megantereon*. Ballesio (1986) reassigned the species to *Megantereon* cf. *cultridens*. Turner (1987) suggested that the Early Pleistocene marchidontids are a single taxon ranging through Eurasia and Africa, *Megantereon cultridens*. This species appears both in African and Eurasia as early as 3 Ma and survives as late as Untermassfeld around 1.2–0.9 Ma. Thus, it is not indicative of an African dispersal. Alternatively, it has been suggested that *M. cultridens* is a Eurasian species while *M. whitei* is an African species, which dispersed into Eurasia during the Early Pleistocene (Martínez-Navarro and Palmqvist 1995). *M. whitei* has been identified in the Early Pleistocene sites of Venta Micena, Spain, and Appolonia, Greece (Martínez-Navarro and Palmqvist 1995; Martínez-Navarro and Palmqvist 1996). Following this paradigm, it was suggested by Martínez-Navarro (2010, but see Lewis and Werdelin 2010) that as the 'Ubeidiya *Megantereon* is small of size it should also be identified as *M. whitei*. This species attains a wide geographic distribution and replaces the European Pliocene *M. cultridens*. Its last appearance is in Untermassfeld, Germany, around 1.2–0.9 Ma.

Three African carnivore taxa (*Herpestes ichneumon*, *Mellivora capensis* and *Crocuta crocuta*) have expanded their range in the southern Levant since the Early Pleistocene and continue to exist in the region until Late Pleistocene or even until recent times. *M. capensis* is recently present both in the Arava desert, the Galilee, Judea as well as the coastal plain (Mendelssohn and Yom-Tov 1999). *Herpestes ichenuemon* is common in the Mediterranean region but it is also found in the arid regions of the Negev and the Arava (Mendelssohn and Yom-Tov 1999). This recent distribution suggests that these species successfully adapted to the Mediterranean woodland, maquis and scrubland as well as more arid habitats (Harrison and Bates 1991). *Crocuta crocuta* originally occupied a wide range of open habitats in Africa including dry acacia plains, open savanna and rocky country (Kingdon 1974). It persisted in temperate and glacial Europe as late as Late Pleistocene period (Rabinovich 2002) suggesting an ecological tolerance to a wide range of habitats. A fourth carnivore, *Megantereon*, attained a wide geographic range but did not survive past the Middle Pleistocene. It has been suggested that this was due to the competition with the modern carnivore guild (Turner 1997).

To date, no taxa have been found in the Early Pleistocene or Pliocene sites in the southern Levant that may be inferred to occupy niches similar to those of the African carnivores. In Europe, *M. whitei* replaced *M. cultridens* and it has been suggested that it was perhaps a better competitor (Martínez-Navarro 2004). Thus, although no *M. cultridens* fossils were found in the southern Levant (probable due to the limited number of sites), a similar relationship between the taxa may be assumed. This suggests that the ecological success of the African carnivores in the southern Levant may be attributed to two facts. First, they are ecological generalists and had the ability to adapt to a wide range of environment. Second, the ecological niches occupied were presumably either not occupied by local ecologically-equivalent taxa, and thus the potential competition would have been reduced, or the new-comers were better competitors, as in the presumed case of *Megantereon whitei*. The presence of African carnivores in the Plio-Pleistocene of the southern Levant is not related to the presence of an African savanna as opposed to Mediterranean woodland and is not inconsistent with the results of the multivariate analysis presented in this study.

Discussion and Conclusions

Evidence from the large mammal communities in the southern Levant suggests that the environment during the Early Pleistocene was Mediterranean woodland–scrubland. Climatic change during this period facilitated the dispersal of large mammals and hominins across the Saharo–Arabian barrier. Other models based on presence or absence of indicator African taxa have suggested a paleoecological reconstruction of savanna grasslands for the southern Levant during the Early Pleistocene. The reconstruction of 'Ubeidiya as an Africa savanna is inconsistent with the reconstruction obtained from the multivariate analysis which suggests a greater similarity to a Mediterranean woodland. The presence of African taxa in a region to which they are not adapted to may be explained using ecological models of long distance dispersal and competition with local taxa.

African taxa in the southern Levant during the Early Pleistocene are mostly rare in abundance and restricted to a small geographic range. This is in accordance with the early stages of dispersal, which are characterized by intermittent dispersal events. Colonization of most taxa was not successful, probably due to a combination of environmental conditions and competition with native taxa. The taxa which did succeed in surviving in the new region were ecological generalists that could tolerate a wide range of environments and presumably did not have competitors from the local taxa. Thus, the presence of African taxa in the southern Levant, and specifically in 'Ubeidiya, is consistent with the reconstruction of the habitat as Mediterranean woodland–scrubland.

A possible explanation for the apparent discrepancies in reconstructions (presence–absence of indicative taxa versus community wide multivariate analysis) is in the differences between the savanna and Mediterranean woodlands. Thus, presence–absence data (which is often used) artificially increase the importance of rare species where in fact, their presence is ephemeral and due to unique ecological circumstances rather than indicative of a biome. The use of the entire community may help tease apart the differences between similar biomes.

Support for the reconstruction presented in this paper is offered by Tchernov's (1980) analysis of the avifauna of 'Ubeidiya indicating that the Palaearctic groups predominate the assemblages and only a few are tropical (Oriental or Ethiopian). The development of the Mediterranean elements from the Asian species took place shortly after the Messinian crisis but increased during the Pliocene and Early Pleistocene around the humid Mediterranean basin and resulted in a high proportion of endemic species (Tchernov 1980).

Further support can be obtained from pollen spectra obtained from Eurasian and Southern Levantine Plio-Pleistocene sites. Pollen analysis of the site of Dmanisi suggests a Mediterranean type climate analogous to recent Mediterranean woodlands (Gabunia et al. 2000). Macrofloral remains of fossilized leaves retrieved from 'Ubeidiya have been identified as *Pistacia lentiscus*, *Rhus tripratita* and *Myriophyllum* (Lorch 1966). A pollen spectrum was extracted from stratum III 12 and analyzed by A. Horwitz (in Bar-Yosef and Tchernov 1972). The analysis indicated 82% arboreal species of which the overwhelming majority are *Quercus* sp. followed by *Juniperus* sp. and *Olea* sp. Non-arboreal families include Gramineae, Cruciferae and Compositae. Cyperaceae comprise 8.4% of the pollen and attest to the water habitat present at the site. This pollen composition was interpreted as indicative of a pluvial environment. Both analyses confirm the presence of a Mediterranean park-forest with rocky and steppe terrain. The botanical remains of Gesher Benot Ya'aqov suggest the presence of Mediterranean wood and plant species and that the climate pattern in the Hula valley at the time of deposition resembled the seasonal Mediterranean pattern seen today (Goren-Inbar et al. 2004). While Mediterranean vegetation is dominated by evergreen sclerophyllous scrubland of oaks (*Quercus* spp.), olives (*Olea* spp.) and *Pistacia*, different communities occur depending on environmental gradients such as moisture, nutrients and temperatures (Allen 2001).

Differences in the large mammal community among different sites and specifically the mesoherbivore community may help distinguish between the sub-habitats within the Mediterranean biome. Thus, the small faunal assemblage in Bizat Ruhama (1.0 Ma) has revealed only bovid, equid and hippo remains (Ronen et al. 1998). In Evron (1.0 Ma), the cervid sample comprises only four specimens of a total of 36 (11%) identified specimens (all taxa), compared to bovids that are represented by 11 specimens (30%) (Tchernov et al. 1994). In 'Ubeidiya (1.6–1.2 Ma) cervids are 30–60% of each of the assemblages (Belmaker 2006). The absence or low proportion of cervids in Bizat Ruhama and Evron may be the results of the small sample size or may indicate a regional (or chronological) shift from more humid regions, where the forest cover would have been higher supporting a high cervid population to drier regions with a more open forest and scrubland habitat represented by a high equid frequency.

Hominin Dispersal

The dispersal of early hominins, as with other species, required three distinct phases; first, the impetus for dispersal and ability to disperse, second, physical access and crossing of the barrier and third, the post depositional establishment. The hominin dispersal pattern in the Early Pleistocene follows the pattern described for the large mammals but specifically for the carnivore guild: long range spatial distribution, long temporal distribution and ecological tolerance.

While early dispersal events for hominins were presumed to have been sporadic and intermittent (Dennell 2004), subsequently, they expanded their range and maintained an ongoing presence in higher latitudes. The long-range spatial distribution attained within the first 0.5 Ma of their dispersal suggests that they were most probably ecological generalists, rather than adapted to any specific environment, and had the ability to adapt to a wide range of environments.

Such ecological analogies between *Homo* and carnivores have been postulated since evidence for carnivory are present in the archaeological record ca. 2.5 Ma (Brantingham 1998). While a diet with a higher proportion of carnivory requires a larger home range to maintain (Bramble and Lieberman 2004), it is also more habitat tolerant (Foley 2002). Such ecological comparisons may be extended to behavioral ecology, genetic differentiation and species resilience such as suggested to exist between *Homo* and *Canis* (Arcadi 2006).

As the ecological success of dispersing hominins was not hindered by competition, their niche was probably not occupied by local ecological equivalent taxa. It is tempting to infer from this on the position of hominins within the predator guild as hunters or scavengers. However, because the southern Levantine carnivore guild is depleted in large scavengers ('Ubeidiya has the smaller *Crocuta* rather than *Pachycrocuta*) and large feline predators ('Ubeidiya has two large felines compared to five in East Africa (Lewis and Werdelin 2010), any such inference may be problematic using current models which focus on modern African carnivore guilds and ethnographic groups such as the Hadza as reference (Domínguez-Rodrigo 2002).

It has been suggested that the relative high proportion of African fauna at the site of 'Ubeidiya indicated that the hominin presence at the site should not be viewed as a dispersal event but rather as a range expansion into part of the greater African milieu (Klein 1999). This study has indicated that this is not the case and that the southern Levant was a Mediterranean woodland during the Plio-Pleistocene. Campbell (1972) has suggested that human dispersal consisted of two broad geographical expansions that were followed by biological and cultural adaptation: tropical to temperate dispersal and a much later temperate to arctic dispersal. These two phases represent an increased ability to cope and exploit harsh environments (Turner 1984; Dennell 2004). While the conditions in the Mediterranean are not as harsh, they are seasonal compared to African habitats and thus may represent "stepping stones" which facilitated further adaptation to the more temperate conditions at a later period. Such adaptation may include the controlled use of fire at Gesher Benot Ya'aqov (Goren-Inbar et al. 2004) and changes in subsistence patterns.

The success of hominins (as opposed to the large ungulates) in the Mediterranean region suggests that they were able to cope with the environmental differences imposed on them by the new environment. Thus, while the ability to cross the Saharo-Arabian barrier was mitigated by Late Pliocene and Early Pleistocene climatic change (Dennell 2004), the post dispersal establishment stems from processes that were unique to humans. If hominins were selected for variable environments as predicted by the Variability Selection Hypothesis (Potts 1998), they would have possessed the necessary pre-adaptation to facilitate the successful colonization in the Mediterranean biome which they would have encountered in their dispersal.

Acknowledgments I would like to thank the conference organizers for their kind invitation to participate in the stimulating and thought-provoking workshop. I am grateful to Ofer Bar-Yosef and the late Eitan Tchernov for their guidance, encouragement and continuing assistance throughout my thesis. My appreciation to Uzi Motro for his support. My thanks to D. Zohary and H. Leschner for insightful discussions. I appreciate the comments of the reviewers on earlier drafts of this manuscript. This research was funded by generous grants from the L.S.B. Leakey Foundation, Irene Levy Sala CARE Foundation and the Wenner Gren Richard Carley Hunt fellowship.

References

Aguilar, J.-P., Legendre, S., Michaux, J., & Montuire, S. (1998). Pliocene mammals and climatic reconstruction in the Western Mediterranean area. In J. H. Wrenn, J. P. Suc, & S. A. G. Leroy (Eds.), *The Pliocene: Time of change* (pp. 109–120). Utah: American Association of Stratigraphic Palynologists Foundation.

Agustí, J., Oms, O., Parés, J. M., Martínez-Navarro, B., & Turq, A. (2000). Dating and correlation of early human occupation in the Baza formation (Guadiz-Baza Basin, SE Spain). In D. Lordkipanidze, O. Bar-Yosef, & M. Otte (Eds.), Early humans at the gates of Europe (pp. 113–122). Liège: ERAUL 92.

Aiello, L. C. (1993). Neocortex size, group size, and the evolution of language. *Current Anthropology, 34*, 184–193.

Aiello, L. C., & Wells, J. C. K. (2002). Energetics and the evolution of the genus *Homo*. *Annual Review of Anthropology, 31*, 323–338.

Aiello, L. C., & Wheeler, P. (1995). Expensive-tissue hypothesis: the brain and digestive system in human and primate evolution. *Current Anthropology, 36*, 199–221.

Aiello, L. C., Bates, N., & Joffe, T. (2001). In defense of the expensive tissue hypothesis. In D. Falk & K. R. Gibson (Eds.), *Evolutionary anatomy of the primate cerebral cortex* (pp. 57–78). Cambridge: Cambridge University Press.

Alemseged, Z. (2003). An integrated approach to taphonomy and faunal change in the Shungura Formation (Ethiopia) and its implication for hominid evolution. *Journal of Human Evolution, 44*, 451–478.

Alemseged, Z., & Geraads, D. (1998). *Theropithecus atlanticus* (Thomas 1984) (Primates: Cercopithecidae) from the late Pliocene of Ahl al Oughlam, Casablanca, Morocco. *Journal of Human Evolution, 34*, 609–621.

Allen, H. D. (2001). *Mediterranean ecogeography*. London: Prentice Hall.

Almogi-Labin, A., Siman-Tov, R., Rosenfeld, A., & Debard, E. (1995). Occurrence and distribution of the foraminifer *Ammonia beccarii tepida* (Cushman) in water bodies, recent and Quaternary, of the Dead Sea Rift, Israel. *Marine Micropaleontology, 26*, 153–159.

Andrews, P. (1989). Palaeoecology of Laetoli. *Journal of Human Evolution, 18*, 173–181.

Andrews, P. (1992). Community evolution in forest habitats. *Journal of Human Evolution, 22*, 423–438.

Andrews, P. (1995a). Experiments in taphonomy. *Journal of Archaeological Science, 22*, 147–153.

Andrews, P. (1995b). Time resolution of the Miocene fauna from Paşalar. *Journal of Human Evolution, 28*, 343–358.

Andrews, P. (1995c). Mammals as paleoecological indicators. *Acta Zoologica Cracoviensa, 38*, 59–72.

Andrews, P. (1996). Palaeoecology and hominoid palaeoenvironments. *Biological Reviews, 71*, 257–300.

Andrews, P., & Nesbit Evans, E. (1978). The environment of *Ramapithecus* in Africa. *Journal of Human Evolution, 5*, 22–30.

Andrews, P., Lord, J. M., & Nesbit-Evans, E. M. (1979). Patterns of ecological diversity in fossil and modern mammalian faunas. *Biological Journal of the Linnean Society, 11*, 177–205.

Antón, S. C., Leonard, W. R., & Robertson, M. L. (2002). An ecomorphological model for the initial hominid dispersal from Africa. *Journal of Human Evolution, 43*, 773–785.

Antón, S. C., & Swisher, C. C. (2004). Early dispersals of *Homo* from Africa. *Annual Review of Anthropology, 33*, 271–296.

Arambourg, C. (1970). *Les Vertébrés Du Pleistocène De L'Afrrique Du Nord*. Paris: Éditions du Muséums.

Arcadi, A. C. (2006). Species resilience in Pleistocene hominids that travel far and ate widely: an analogy to the wolf like canids. *Journal of Human Evolution, 51*, 383–394.

Arribas, A., & Palmqvist, P. (1998). Taphonomy and paleoecology of an assemblage of large mammals: hyaenid activity in the Lower Pleistocene site at Venta Micena (Orce, Guadix - Baza Basin, Granada, Spain). *Geobios, 31*, 3–47.

Arribas, A., & Palmqvist, P. (1999). On the ecological connection between saber-tooths and hominids: faunal dispersal events in the Lower Pleistocene and a review of the evidence of the first human arrival in Europe. *Journal of Archaeological Science, 26*, 571–585.

Arzarello, A., Marcolini, F., Pavia, G., Pavia, M., Petronio, C., Petrucci, M., et al. (2007). Evidence for the earliest human occurrence in Europe: the site of Pirro Nord (Southern Italy). *Naturewissenaschaften, 94*, 107–112.

Badgley, C. (1986). Taphonomy of mammalian fossil remains from Siwalik rocks of Pakistan. *Paleobiology, 12*, 119–142.

Ballesio, R. (1986). Les Carnivores du gisement Pléistocène d'Oubéidiyeh. In E. Tchernov (Ed.), *Les Mammifères du Pléistocène Inférieur, de la Vallée du Jourdain a Oubéidiyeh* (pp. 63–92). Paris: Association Paléorient.

Bar-Yosef, O. (1998). Early colonizations and cultural continuities in the Lower Palaeolithic of Western Asia. In M. D. Petraglia & R. Korisettar (Eds.), *Early human behaviour in global context: The rise and diversity of the lower Palaeolithic record* (pp. 221–279). London: Routledge.

Bar-Yosef, O., & Belfer-Cohen, A. (2001). From Africa to Eurasia – early dispersals. *Quaternary International, 75*, 19–28.

Bar-Yosef, O., & Goren-Inbar, N. (1993). *The lithic assemblages of 'Ubeidiya: a lower Palaeolithic site in the Jordan Valley*. Jerusalem: The Hebrew University of Jerusalem.

Bar-Yosef, O., & Tchernov, E. (1972). *On the Palaeo-ecological history of the site of 'Ubeidiya*. Jerusalem: Israel Academy of Sciences and Humanities.

Behrensmeyer, A. K. (1975). The taphonomy and paleoecology of Plio-Pleistocene vertebrate assemblages east of Lake Rudolf, Kenya. *Bulletin of the Museum of Comparative Zoology, 146*, 473–578.

Behrensmeyer, A. K. (1978). Taphonomic and ecologic information from bone weathering. *Paleobiology, 4*, 150–162.

Behrensmeyer, A. K. (1982). Time resolution in fluvial vertebrate assemblages. *Paleobiology, 8*, 211–227.

Behrensmeyer, A. K. (1984). Taphonomy and the fossil record. *American Scientist, 75*, 558–566.

Behrensmeyer, A. K. (1992). Paleoenvironmental context and taphonomic modes. In A. K. Behrensmeyer, J. D. Damuth, W. A. DiMichele, R. Potts, H. D. Sues, & S. L. Wing (Eds.), *Terrestrial ecosystems through time* (pp. 15–93). Chicago: University of Chicago Press.

Behrensmeyer, A. K. (2001). Terrestrial vertebrates. In D. E. G. Briggs & P. R. Crowther (Eds.), *Paleobiology II* (pp. 318–320). Cornwall: Blackwell.

Behrensmeyer, A. K., & Cooke, H. (1985). Paleoenvironments, stratigraphy, and taphonomy in the African Pliocene and Early Pleistocene. In E. Delson (Ed.), *Ancestors: The hard evidence* (pp. 60–62). New York: Alan R. Liss.

Behrensmeyer, A. K., & Hill, A. (Eds.). (1980). *Fossils in the making*. Chicago: University of Chicago Press.

Behrensmeyer, A. K., Gordon, K. D., & Yanagi, G. T. (1986). Trampling as a cause of bone surface damage and pseudo cut marks. *Nature, 319*, 768–771.

Behrensmeyer, A. K., Damuth, J. D., DiMichele, W. A., Potts, R., Sues, H. D., & Wing, S. L. (Eds.). (1992). *Terrestrial ecosystems through time*. Chicago: University of Chicago Press.

Behrensmeyer, A. K., Kidwell, S. M., & Gastaldo, R. A. (2000). Taphonomy and paleobiology. *Paleobiology, 26*, 103–147.

Belmaker, M. (2002). Community structure changes through time – 'Ubeidiya as a case study. In H. Buitenhuis, A. M. Choyke, M. Mashkour, & A. H. Al-Shiyab (Eds.), *Proceedings of the 5th international conference of archaeozoology of the Southwest Asia and adjacent areas (62)* (pp. 9–21). Center for Archeological Research and Consultancy, Rijksuniversiteit, Groningen, The Netherlands: ARC publications.

Belmaker, M. (2006). *Community structure through time: 'Ubeidiya, a Lower Pleistocene site as a case study*. Ph.D. dissertation, The Hebrew University of Jerusalem, Israel.

Belmaker, M. (in press). On the road to China: the environmental landscape of the Lower Pleistocene in Western Eurasia and its implications for the dispersal of *Homo*. In C. J. Norton, D. R. Braun (Eds.), *Asian paleoanthropology: From Africa to China and beyond*. Dordrecht: Springer.

Belmaker, M. (2010). The presence of a large cercopithecine (cf. *Theropithecus* sp.) in the 'Ubeidiya Formation (Early Pleistocene, Israel). *Journal of Human Evolution, 58*, 43–56.

Belmaker, M., Tchernov, E., Condemi, S., & Bar-Yosef, O. (2002). New evidence of hominid presence in the Lower Pleistocene of the Southern Levant. *Journal of Human Evolution, 43*, 43–56.

Blondel, J., & Aronson, J. (1999). *Biology and wildlife of the Mediterranean Region*. Oxford: Oxford University Press.

Blumenschine, R. J. (1986). *Early hominid scavenging opportunities: Implications of carcass availability in the Serengeti and Ngorongoro ecosystems BAR International Series 283*. Oxford: B.A.R.

Blumenschine, R. J. (1987). Characteristics of an early hominid scavenging niche. *Current Anthropology, 28*, 383–407.

Blumenschine, R. J. (1988). Reinstating an early hominid scavenging niche: A reply to Potts. *Current Anthropology, 29*, 483–486.

Blumenschine, R. J., & Cavallo, J. A. (1992). Scavenging and human evolution. *Scientific American, 267*, 90–96.

Blumenschine, R. J., Cavallo, J. A., & Capaldo, S. D. (1994). Competition for carcasses and early hominid behavioral ecology: A case study and conceptual framework. *Journal of Human Evolution, 27*, 19–213.

Blumenschine, R. J., Peters, C. R., Masao, F. T., Clarke, R. J., Deino, A. L., Hay, R. L., et al. (2003). Late Pliocene *Homo* and hominid land use from Western Olduvai Gorge, Tanzania. *Science, 299*, 1217–1221.

Bobe, R., & Eck, G.G. (2001). Responses of African bovids to Pliocene climatic change. Paleobiology Memoirs 2 Supplement to 27, 1–47.

Bobe, R., Behrensmeyer, A. K., & Chapman, R. E. (2002). Faunal change, environment variability and late Pliocene hominin evolution. *Journal of Human Evolution, 42*, 475–497.

Bocherens, H., Koch, P. L., Mariotti, A., Geraads, D., & Jaeger, J. J. (1996). Isotopic biogeochemistry (^{13}C, ^{18}O) and mammalian enamel from African Pleistocene hominid sites. *Palaios, 11*, 306–318.

Boisserie, J.-R. (2005). The phylogeny and taxonomy of Hippopotamidae (Mammalia: Artiodactyla): a review based on morphology and cladistic analysis. *Zoological Journal of the Linnean Society, 143*, 1–26.

Boisserie, J.-R., Zazzo, A., Merceron, G., Blondel, C., Vignaud, P., Likius, A., et al. (2005). Diet of modern and late Miocene hippopotamids: Evidence from carbon isotope composition and microwear of tooth enamel. *Paleogeography, Paleoclimatology, Paleoecology, 221*, 153–174.

Brain, C. K. (1981). *The hunters or the hunted: An introduction to African cave taphonomy*. Chicago: University of Chicago Press.

Bramble, D. M., & Lieberman, D. E. (2004). Endurance running and the evolution of *Homo*. *Nature, 432*, 345–352.

Brantingham, J. P. (1998). Hominid-carnivore coevolution and invasion of the predatory guild. *Journal of Anthropological Archaeology, 17*, 327–353.

Brown, J. H., & Lomolino, M. V. (1998). *Biogeography*. Saunderland, MA: Sinauer.

Caloi, L., & Palombo, M. R. (1997). Biochronology of large mammals in the Early and Middle Pleistocene on the Italian peninsula. *Hystrix, 9*, 3–12.

Campbell, B. (1972). Man of all seasons. In B. Cambell (Ed.), *Sexual selection and the decent of man (1871–1971)* (pp. 40–58). Chicago: Aldine.

Cerling, T. E., Kappelman, J., Quade, J., Ambrose, S., Sikes, N., & Andrews, P. (1992). Paleoenvironment of *Kenyapithecus* at Fort Ternan. *Journal of Human Evolution, 23*, 371–377.

Clark, J. D. (1967). The Middle Acheulean occupation site at Latamne, northern Syria. *Quaternaria, 9*, 1–68.

Cordain, L., Miller, J. B., Eaton, S. B., Mann, N., Holt, S. H. A., & Speth, J. D. (2000). Plant-animal subsistence ratios and macronutrient energy estimations in worldwide hunter-gatherer diets. *American Journal of Clinical Nutrition, 71*, 682–692.

Cordain, L., Watkins, B. A., & Mann, N. J. (2001). Fatty acid composition and energy density of foods available to African hominids. Evolutionary implications for human brain development. In A. P. Simopoulos & K. N. Pavlou (Eds.), *Nutrition and fitness: Metabolic studies in health and disease* (pp. 114–161). Basel: Karger.

Coryndon, S. C. (1976). Fossil Hippoptamidae from Plio-Pleistocene successions of the Rudolf basin. In Y. Coppens, F. C. Howell, G. L. I. Issac, & R. E. F. Leakey (Eds.), *Earliest man and environment in the Lake Rodulf Basin: stratigraphy, paleoeoclogy and evolution* (pp. 238–250). Chicago: University of Chicago Press.

Coryndon, S. C., & Coppens, Y. (1973). Preliminary report on Hippopotamidae (Mammalia, Artiodactyla) from the Plio/Pleistocene of the Lower Omo Basin, Ethiopia. In L. S. B. Leakey, R. J. G. Savage, & S. C. Coryndon (Eds.), *Fossil vertebrates of Africa* (Vol. 3, pp. 139–158). London: Academic.

Dayan, T. (1989) *The succession and community structure of the carnivores of the Middle East in space and time*. Ph.D. dissertation, Tel Aviv University, Ramat Aviv, Israel.

Delany, M. J., & Happold, D. C. D. (1979). *Ecology of African mammals*. London: Longman.

Delpech, F., & Heintz, E. (1976). Les faunas de Paleolithique dans le sud-ouset de la France. In H. Lumley (Ed.), *Le Prehistoire Francaise I* (pp. 386–394). Bordeaux: Editions de Centre National de la Rechercher Scientifique.

Delson, E. (1993). *Theropithecus* fossils from Africa and India and the taxonomy of the genus. In N. G. Jablonski (Ed.), *Theropithecus: The rise and fall of a primate genus* (pp. 157–189). Cambridge: Cambridge University Press.

Dennell, R. W. (2003). Dispersal and colonisation, long and short chronologies: how continuous is the Early Pleistocene record for hominids outside East Africa? *Journal of Human Evolution, 45*, 421–440.

Dennell, R. W. (2004). Hominid dispersals and Asian biogeography during the Lower and Early Middle Pleistocene, c. 2.0–0.5 Mya. *Asian Perspectives, 43*, 205–226.

Dennell, R.W. (2010). "Out of Africa I": Current problems and future prospects. In J. G Fleagle, J. J. Shea, F. E. Grine, A. L. Baden, & R. E Leakey (Eds.), *Out of Africa I: the first hominin colonization of Eurasia* (pp. 247–273). Dordrecht: Springer.

Derricourt, R. (2006). Getting "Out of Africa": Sea crossing, land crossings and culture in the hominin migrations. *Journal of World Prehistory, 19*, 119–132.

Domínguez-Rodrigo, M. (2002). Hunting and scavenging by early humans: the state of the debate. *Journal of World Prehistory, 16*, 1–54.

Domínguez-Rodrigo, M. (2003). Bone surface modifications, power scavenging and the "display" model at early archaeological sites: a critical review. *Journal of Human Evolution, 45*, 411–415.

Domínguez-Rodrigo, M., De la Torre, I., De Luque, L., Alcala, L., Mora, R., Serrallonga, J., et al. (2002). The ST Site Complex at Peninj, West Lake Natron, Tanzania: implications for early hominid behavioural models. *Journal of Archaeological Science, 29*, 639–665.

Eisenmann, V. (1983). Family Equidae. In J. M. Harris (Ed.), *Koobi Fora Research Project* (The Fossil Ungulates: Proboscidae, Perissodactyla, and Suidae, Vol. 2, pp. 156–214). Oxford: Clarendon.

Eisenmann, V. (1984). Sur quelques caractères adaptifs de squelette d'*Equus* (Mammalia, Perissodactyla) et leurs implications paléoécolgiques. Bulletin du Muséum National d'historie Naturelle. 4E serie. section C, 185–195.

Eisenmann, V. (1986). Les Equidés d'Oubiédiyeh. In E. Tchernov (Ed.), *Les Mammifères du Pléistocène Inférieur, de la Vallée du Jourdain à Oubéidiyeh* (pp. 191–212). Paris: Association Paléorient.

Eisenmann, V. (2004). *Equus*: an evolution without lineages? In L. C. Maul & R.-D. Kahlke (Eds.), *Late Neogene and Quaternary biodiversity and evolution: regional developments and interregional correlations. VII International Palaeontological Colloquium in Weimar* (pp. 101–104). Weimar: Senckenberg.

Eisenmann, V., & Guérin, C. (1984). Morphologie functionnelle et environnement chez les Périssodactyles. *Geobios, Mém Spécial, 8*, 69–74.

Eisenmann, V., Helmer, D., & Sañia Segiu, M. (2002). The big equid from the Geometric Kebaran of Um el Tlel, Syria: *Equus valeriani, Equus capensis* or *Equus caballus*? In H. Buitenhuis, A. M. Choyke, M. Mashkour, & A. H. Al-Shiyab (Eds.), *Proceedings of the 5th International Conference of Archaeozoology of the Southwest Asia and Adjacent Areas (62)* (pp. 62–73). Center for Archeological Research and Consultancy, Rijksuniversiteit, Groningen, The Netherlands: ARC publications.

Eiten, G. (1992). How names are used for vegetation. *Journal of Vegetation Science, 3*, 419–424.

Elton, S. (2002). A reappraisal of the locomotion and habitat preference of *Theropithecus oswaldi*. *Folia Primatologica, 73*, 252–280.

Eltringham, S. K. (1979). *The ecology and conservation of large African mammals*. London: Macmillan.

Eltringham, S. K. (1999). *The Hippos. Natural history and conservation*. Cambridge: Cambridge University Press.

Faure, M. (1986). Les hippopotamides du Pleistocene ancien d'Oubeidiyeh (Israel). In E. Tchernov (Ed.), *Les Mammifères du Pléistocene Inférieur de la Vallée du Jourdain à Oubeidiyeh* (pp. 107–142). Paris: Association Paléorient.

Fernandes, C. A., Rohling, E. J., & Siddall, M. (2006). Absence of post-Miocene Red Sea land bridges: biogeographic implications. *Journal of Biogeography, 33*, 961–966.

Fernández-Jalvo, Y., Denys, C., Andrews, P., Terry-Williams, C., Daupin, Y., & Humprey, L. (1998). Taphonomy and paleoecology of Olduvai Bed I (Pleistocene, Tanzania). *Journal of Human Evolution, 34*, 137–172.

Flessa, K. W. (2000). Taphonomy – a process approach. *Science, 288*, 1971–1972.

Foley, R. (2002). Adaptive radiations and dispersals in hominin evolutionary ecology. *Evolutionary Anthropology Suppl, 1*, 32–37.

Gabunia, L., & Vekua, A. (1995). A Plio-Pleistocene hominid from Dmanisi, East Georgia, Caucasus. *Nature, 373*, 509–512.

Gabunia, L., Vekua, A., & Lordkipanidze, D. (2000). The environmental context of early human occupation of Georgia (Transcaucasia). *Journal of Human Evolution, 38*, 785–802.

Gagnon, M. (1997). Ecological diversity and community ecology in the Fayum sequence (Egypt). *Journal of Human Evolution, 32*, 133–160.

Gardner, E. W., & Bate, D. M. A. (1937). The bone-bearing beds of Bethlehem: their fauna and industry. *Nature, 140*, 43–433.

Gentry, A. W. (1978). Bovidae. In V. J. Maglio & H. B. S. Cooke (Eds.), *Evolution of African mammals* (pp. 540–572). Cambridge, MA: Harvard University Press.

Geraads, D. (1986). Ruminants Pléistocène d'Oubéidiyeh. In E. Tchernov (Ed.), *Les Mammiféres du Pléistocène Inférieur, de la Vallée du Jourdain a Oubéidiyeh* (pp. 143–182). Paris: Association Paléorient.

Geraads, D. (1997). Late Pliocene Carnivora from Ahl al Oughlam (Casablanca). *Geobios, 30*, 127–164.

Geraads, D., Guérin, C., & Faure, M. (1986). Suidae (Mammalia, Artiodacyla) du gisement Pléistocène ancien dù Oubiédiyeh. In E. Tchernov (Ed.), *Les Mammifères du Pléistocène Inférieur, de la Vallée du Jourdain a Oubéidiyeh* (pp. 93–106). Paris: Assciation Paléorient.

Gibert, J., Ribot, F., Gilbert, L., Leakey, M. G., Arribas, A., & Martínez-Navarro, B. (1995). Presence of the Cercopithecid genus *Theropithecus* in Cueva Victoria (Murcia, Spain). *Journal of Human Evolution, 28*, 487–493.

Ginat, H., Zilberman, E., & Saragusti, I. (2003). Early Pleistocene lake deposits and Lower Paleolithic finds in Nahal (wadi) Zihor, Southern Negev desert, Israel. *Quaternary Research, 59*, 445–458.

Goren-Inbar, N., & Saragusti, I. (1996). An Acheulian biface assemblage from Gesher Benot Ya'aqov, Israel: Indications of African Affinities. *Journal of Field Archaeology, 23*, 15–30.

Goren-Inbar, N., Feibel, C. S., Verosub, K. L., Melamed, Y., Kislev, M. E., Tchernov, E., et al. (2000). Pleistocene milestones on the out of Africa corridor at Gesher Benot Ya'aqov, Israel. *Science, 289*, 944–947.

Goren-Inbar, N., Alperson, N., Kislev, M. E., Simchoni, O., Melamed, Y., Ben-Nun, A., et al. (2004). Evidence of hominin control of fire at Gesher Benot Ya'aqov, Israel. *Science, 304*, 725–727.

Greenacre, M. J., & Vrba, E. S. (1984). A correspondence analysis of biological census data. *Ecology, 65*, 984–997.

Guérin, C., Eisenmann, V., & Faure, M. (1993). Les grands mammifères du gisement pléistocène moyen de Latamné (vallée de l'Oronte, Syrie). In P. Sanlaville, J. Besançon, L. Copeland, & S. Muhesen (Eds.), *Le Paléolithique de la vallée moyenne de l'Oronte (Syrie). BAR International Series 587* (pp. 169–178). Oxford: B.A.R.

Guerreo-Alba, S., & Palmqvist, P. (1997). Estudio morfomertrico del caballo de Venta Micena (Orce, Granada) y su comparacion con los equidos mordernos y del Plio-Pleistocene en el viejo y nuevo mundo. *Paleontologia i evolucio, 30–31*, 93–148.

Gunnell, G. F. (1990). Cenogram analysis of Bridger B mammalian fauna (Middle Eocene). *Journal of Vertebrate Paleontology, 10*, 26A.

Gunnell, G. (1995). Comparative paleoecology of Paleogene and Neogene mammalian faunas: Trophic structure and composition. *Palaeogeography, Palaeoclimatology, Palaeoecology, 115*, 265–286.

Gupta, V. L., & Sahni, A. (1981). *Theropithecus delsoni,* a new cercopithecine species from the Upper Siwaliks of India. *Bulletin of the Indian Geololgical Association, 14*, 69–71.

Haas, G. (1966). *On the vertebrate fauna of the Lower Pleistocene site of 'Ubeidiya.* Jerusalem: Israel Academy of Sciences and Humanities.

Haas, G. (1968). On the fauna of 'Ubeidiya. *Proceedings of the Israel Academy of Sciences and Humanities, 7*, 1–14.

Harris, J. M. (1983). Family Suidae. In J. M. Harris (Ed.), *Koobi Fora Research Project, vol 2. The fossil ungulates: Proboscidae, Perissodactyla and Suidae* (pp. 139–320). Oxford: Clarendon.

Harris, J. M., & Cerling, T. E. (2002). Dietary adaptations of extant and Neogene African suids. *Journal of Zoology, London, 256*, 45–54.

Harrison, D. L., & Bates, P. J. J. (1991). *The mammals of Arabia.* Sevenoaks, Kent: Harrison Zoological Museum.

Hemmer, H. (2000). Out of Asia: a paleoecological scenario of man and his carnivorous competitors in the European Lower Pleistocene. In D. Lordkipanidze, O. Bar-Yosef, & M. Otte (Eds.), *Early humans at the gates of Europe* (pp. 99–106). Liège: Etudes et Recherches Archéoloques de l'Univeristé de Liège.

Holmes, K. M. (2007). Using Pliocene paleoclimatic data to postulate dispersal pathways of early hominins. *Palaeogeography, Palaeoclimatology, Palaeoecology, 248*, 96–108.

Hooijer, D. A. (1958). An early Pleistocene mammalian fauna from Bethlehem. *Bulletin of the British Museum of Natural History, London, Geology, 3*, 265–292.

Hubbell, S. P. (2001). *The unified neutral theory of biodiversity and biogeography.* Princeton/Oxford: Princeton University Press.

Jablonski, N. G. (Ed.). (1993). Theropithecus. *The rise and fall of a primate genus.* Cambridge: Cambridge University Press.

Kahlke, R. D. (1997). Die Hippopotamus-Reste aus dem Unterpleistozän von Untermassfeld. In: R.-D. Kahlke (Ed.), *Das Pleistozän von Untermaßfeld bei Meiningen (Thüringen)* (pp. 277–374) Mainz: Monographien Römisch-Germanisches Zentralmuseum.

Kahlke, R.-D. (2000). The early Pleistocene (Epivillafranchian) faunal site of Untermassfeld (Thuringia, Central Germany). Synthesis of new results. In D. Lordkipanidze, O. Bar-Yosef, & M. Ottes (Eds.), *Early humans at the gate of Europe* (pp. 123–138). Liège: Etudes et Recherches Archéologiques de l'Université de Liège.

Kaiser, T. M., & Croitor, R. (2004). Ecological interpretations of early Pleistocene deer (Mammalia, Cervidae), from Ceysseguet (Haute-Loire, France). *Geodiversitas, 26*, 661–674.

Kaiser, T. M., & Franz-Odendaal, T. A. (2004). A mixed-feeding *Equus* species from the Middle Pleistocene of South Africa. *Quaternary Research, 62*, 316–323.

Kay, R. F. (1977). Diets of early Miocene African hominoids. *Nature, 268*, 628–630.

King, G., & Bailey, G. (2006). Tectonics and human evolution. *Antiquity, 80*, 265–286.

Kingdon, J. (1974). *East African mammals.* Chicago: Chicago University Press.

Klein, R. G. (1999). *The human career: human biological and cultural origins.* Chicago: University of Chicago Press.

Krebs, C. J. (2001). *Ecology: the experimental analysis of distribution and abundance.* San Francisco: Benjamin Cummings.

Kroll, E. M. (1994). Behavioral implications of Plio-Pleistocene archaeological site structure. *Journal of Human Evolution, 27*, 1–3.

Larick, R., & Ciochon, R. L. (1996). The African emergence and early Asian dispersal of the genus *Homo. American Scientist, 84*, 538–551.

Leakey, M. G. (1993). Evolution of *Theropithecus* in the Turkana Basin. In N. G. Jablonski (Ed.), *Theropithecus. The rise and fall of a primate genus* (pp. 85–24). Cambridge: Cambridge University Press.

Legendre, S. (1986). Analysis of mammalian communities from the late Eocene and Oligocene of southern France. *Paleovertebrata, Montpellier, 16*, 191–212.

Legendre, S. (1987a). Mammalian fauna as paleotemperature indicators: Concordance between oceanic and terrestrial paleontological evidence. *Evolutionary Theory, 8*, 77–86.

Legendre, S. (1987b). Les Communautés de Mammiferes et leur milieu en Europe Occidentale de L'Eocene Superieur et Oligocene. *Revue de Paléobiologie, 6*, 183–188.

Legendre, S., Badgley, C., & Gingrich, P. D. (1991). Cenograms and environmental parameters for mammalian faunas. *Journal of Vertebrate Paléobiologie, 11*, 42.

Legendre, P., & Legendre, L. (1998). *Numerical ecology*. Amsterdam: Elsevier.

Lewis, M.E., & Werdelin, L (2010). Carnivoran dispersal out of Africa during the Early Pleistocene: relevance for hominins? In: J. G. Fleagle, J. J. Shea, F. E. Grine, A. L. Baden, & R. E. Leakey (Eds.), *Out of Africa I: the first hominin colonization of Eurasia* (pp. 13–26). Dordrecht: Springer.

Lorch, Y. (1966). A Pleistocene florule from the Jordan Valley. *Israel Journal of Botany, 15*, 31–34.

Lyman, R. L. (1994). *Vertebrate taphonomy*. Cambridge: Cambridge University Press.

Maddock, L. (1979). The "migration" and grazing succession. In A. R. E. Sinclair & M. Norton-Griffith (Eds.), *Serengeti. Dynamics of an ecosystem* (pp. 104–129). Chicago: The University of Chicago Press.

Martínez-Navarro, B. (2010). Early Pleistocene faunas of Eurasia and hominin dispersals. In J. G. Fleagle, J. J. Shea, F. E. Grine, A. L. Baden, & R. E. Leakey (Eds.), *Out of Africa I: the first hominin colonization of Eurasia* (pp. 207–224). Dordrecht: Springer.

Martínez-Navarro, B. (2004). Hippos, pigs, bovids, saber-toothed tigers, monkeys, and hominids: Dispersals through the Levantine corridor during Late Pliocene and Early Pleistocene times. In N. Goren-Inbar & J. D. Speth (Eds.), *Human paleoecology in the Levantine corridor* (pp. 37–52). Oxford: Oxbow Books.

Martínez-Navarro, B., & Palmqvist, P. (1995). Presence of the African Machairdont *Megantereon whitei* (Broom, 1937) (Felidae, Carnivora, Mammalia) in the Lower Pleistocene site of Venta Micene (Orce, Granada, Spain), with some considerations on the origin, evolution and dispersal of the genus. *Journal of Archaeological Science, 22*, 569–582.

Martínez-Navarro, B., & Palmqvist, P. (1996). Presence of African Saber tooth felid *Megantereon whitei* (Broom, 1937)(Mammalia, Carnivora, Machairodontinae) in Appolonia - 1 (Mygdonia Basin, Macedonia, Greece). *Journal of Archaeological Science, 23*, 869–872.

Martínez-Navarro, B., Rook, L., Segid, A., Yosief, D., Farretti, M.P., Shoshani, J. et al. (2004). The large fossil mammals from Buia (Eritrea): systematics, biochronology and paleoenvironments. *Rivista Italiana di Paleontologia E Stratigrafia, 110* (Suppl.), 61–88.

Martínez-Navarro, B., Belmaker, M., & Bar-Yosef, O. (2009). The large carnivores from 'Ubeidiya (early Pleistocene, Israel): biochronological and biogeographical implications. *Journal of Human Evolution, 56*, 514–524.

Mattiolio, L., Capitani, C., Avanainelli, E., Bertelli, I., Gazzola, A., & Apollonio, M. (2004). Predation by the wolves (*Canis lupus*) on roe deer (*Capreolus capreolus*) in north-eastern Apennine, Italy. *Journal of Zoology, London, 264*, 249–258.

Mazza, P. (1991). Interrelationship between Pleistocene Hippopotami of Europe and Africa. *Bolletino della Società Paleontological Italiana, 30*, 153–186.

Mazza, P. P. A., Bertini, A., & Magi, M. (2004). The Late Pliocene site of Poggio Rosso (central Italy): Taphonomy and paleoenvironment. *Palaios, 19*, 227–248.

Mendelssohn, H., & Yom-Tov, Y. (1999). *Fauna palaestina. Mammals of Israel*. Jerusalem: The Israel Academy of Sciences and Humanities.

Michel, P. (1988). A new Pleistocene Mellivorinae (Carnivora, Mustelidae) from the Bouknadel Quarries (near Rabat, Marocco) *Mellivora carolae* N. Sp. *Comptes Rendue de l'Academie des Science Serie II, 306*, 935–938.

Montuire, S. (1994). *Communautés de Mammiferes et environnements:l'apport des faunes aux reconstitutions des millieux en Europe depuis le Pliocene et l'impact des changments climatiques sur la diversite*. Ph.D. dissertation, Universite Montpellier, France.

Montuire, S. (1995). Evolution climatique et diversité chez les memmifères en Europe Centrale depuis le Pliocene. *Geobios, 8*, 313–327.

Montuire, S. (1998). Paleoecological analysis of a first sequence of mammalian faunas and evolution of environments in Germany during the Pleistocene. *Comptes Rendus de l'Academie des Sciences Serie II A Sciences de la Terre et des Planètes, 326*, 221–226.

Montuire, S., & Desclaux, E. (1997). Paleoecological analysis of mammalian faunas and environmental evolution in the South of France during the Pleistocene. *Boreas Oslo, 26*, 355–365.

Morgan, M., Badgley, C., Gunnell, G., Gingerich, P. D., Kappelman, J. W., & Maas, M. (1995). Comparative paleoecology of Paleogene and Neogene mammalian faunas: body-size structure. *Palaeogeography, Paleoenvironment, Palaeoecology, 115*, 287–317.

Nathan, R. (2001). Dispersal biogeography. In S. A. Levin (Ed.), *Encyclopedia of biodiversity* (pp. 127–152). San Diego: Academic.

Nathan, R., Perry, G., Cronin, G. T., Strand, A. E., & Cain, M. L. (2003). Methods for estimating long-distance dispersal. *Oikos, 103*, 261–273.

Nesbit Evans, E. M., Van Covering, J. A. H., & Andrews, P. (1981). Paleoecology of Miocene sites in Western Kenya. *Journal of Human Evolution, 10*, 99–116.

Palmqvist, P. (2002). On the community structure of the large mammals assemblage from Dmanisi (East Georgia, Caucasus). In M. De Renzi, M. Alonso, M. Belinchòn, E. Peñalver, P. Montoya, & A. Màrquez-Aliaga (Eds.), *Current topics on taphonomy and fossilization* (pp. 151–160). Valencia: Ajuntament de Valencia.

Palmqvist, P., & Arribas, A. (2001). Taphonomic decoding of the paleobiological information locked in a lower Pleistocene assemblage of large mammals. *Paleobiology, 27*, 512–530.

Palmqvist, P., Groöcke, D. R., Arribas, A., & Fariña, R. A. (2003). Paleoecological reconstruction of a lower Pleistocene large mammal community using biogeochemical ($\delta^{13}C$, $\delta^{15}N$, $\delta^{18}O$ Sr:Zn) and ecomorphological approaches. *Paleobiology, 29*, 205–229.

Patel, B. A., Gilbert, C. C., & Ericson, K. E. (2007). Cercopithecoid cervical vertebral morphology and implications for the presence of *Theropithecus* in early Pleistocene Europe. *Journal of Human Evolution, 52*, 113–129.

Petter, G. (1973). Carnivores Pléistocènes du Ravin D'Olduvai (Tanzania). In L. S. B. Leakey, R. J. G. Savage, & S. C. Coryndon (Eds.), *Fossil Vertebrate of Africa* (Vol. 3, pp. 43–100). London: Academic.

Petter, G. (1987). Small carnivores (Viverridae, Mustelidae, Canidae) from Laetoli. In M. D. Leakey & J. M. Harris (Eds.), *Laetoli: A Pliocene Site in Northern Tanzania* (pp. 194–234). Oxford: Clarendon.

Pickford, M. (1993). Climatic change, biogeography, and *Theropithecus*. In N. G. Jablonski (Ed.), Theropithecus: *The rise and fall of a primate genus* (pp. 227–243). Cambridge: Cambridge University Press.

Potts, R. (1998). Environmental hypotheses of hominin evolution. *Yearbook of Physical Anthropology, 41*, 93–136.

Potts, R. (2002). Complexity and adaptability in human evolution. In M. Goodman & A. S. Moffat (Eds.), *Probing human origins* (pp. 33–58). Cambridge, MA: American Academy of Arts and Sciences.

Potts, R., & Teague, R. L. (2010). Behavioral and environmental background to 'Out of Africa I' and the arrival of *Homo erectus* in East Asia. In: Fleagle, J.G., Shea, J.J., Grine, F.E., Baden, A.L., Leakey, R.E. (Eds.), *Out of Africa I: the first hominin colonization of Eurasia* (pp. 67–85). Dordrecht: Springer.

Rabinovich-Vin, A. (1986). *Parent rock, soil and vegetation in Galilee*. Nature Reserves Authority. Tel Aviv: Kibbutz Hameuchad (in Hebrew).

Rabinovich, R. (2002). Man versus carnivores in the Middle-Upper Paleolithic of the southern Levant. In H. Buitenhuis, A.M. Choyke,

M. Mashkour, & A. H. Al-Shiyab (Eds.), Archaeozoology of the Near East (pp. 22–39). Groningen, The Netherlands, Yarmouk University: ARC-Publications 62.

Reed, K.E. (1996). *The paleoecology of Makapansgat and other African plio-pleistocene hominid localities*. Ph.D. dissertation, SUNY, Stony Brook, NY.

Reed, K. E. (1997). Early hominid evolution and ecological change through the African Plio-Pleistocene. *Journal of Human Evolution, 32*, 289–322.

Reed, K. E. (1998). Using large mammals communities to examine ecological and taxonomical structure and predict vegetation in extant and extinct assemblages. *Paleobiology, 24*, 384–408.

Ron, H., & Gvirtzman, G. (2001). *Magenetostratigraphy of Ruhama bad-land Quaternary deposit: a new age of the Lower Paleolithic site* (pp. 95). Elat: Abstracts of the Annual Meeting of the Israel Geological Society, March.

Ron, H., Porat, N., Ronen, A., Tchernov, E., & Horwitz, K. L. (2003). Magnetostratigraphy of the Evron Member – implications for the age of the Middle Acheulian site of Evron Quarry. *Journal of Human Evolution, 44*, 633–639.

Ronen, A. (1991). The Lower Palaeolitic site Evron-Quarry in western Galilee, Israel. Sonderveröffentlichungen Geologisches Institut der Universität zu Köln 82, 187–212.

Ronen, A., Burdukiewicz, J.-M., Laukhin, S.A., Winter, Y., Tsatskin, A., Dayan, T. et al. (1998). The Lower Palaeolithic Site Bizat Ruhama in the Northern Negev, Israel. Preliminary Report, 1996 Excavations. Archäologisches Korrespondenzblatt RGZM 28, 163–173.

Rook, L., Martinez-Navarro, B., & Howell, F. C. (2004). Occurrence of *Theropithecus* sp. in the Late Villafranchian of Southern Italy and implication for Early Pleistocene "out of Africa" dispersals. *Journal of Human Evolution, 47*, 267–277.

Rosenfeld, A., Honigstein, A., & Scarpa, D. (2002). Plio-Pleistocene lakes in Southern Israel - Ostracods and Hominids. *GSI, Current Research, 13*, 86–93.

Rosenzweig, M. L. (1995). *Species diversity in space and time*. Cambridge: Cambridge University Press.

Sagi, A. (2005). *Magnetostratigraphy of 'Ubeidiya Formation, Northern Dead Sea Transform, Israel*. M.Sc. dissertation, The Hebrew University, Israel.

Sagi, A., Belmaker, M., Ron, H., Enzel, Y., Agnon, A., & Bar Yosef, O. (2005). Paleomagnetic dating of 'Ubeidiya formation. In S. Abramovich (Ed.), *Abstracts of the Israel Geological Society Annual Meeting 2005 (pp. 101)*. Israel: Israel Geological Soceity.

Schaller, G. B. (1972). *The Serengeti Lion. A study of predator–prey relationship*. Chicago: The University of Chicago Press.

Schick, K. D., & Zhuan, D. (1993). Early Paleolithic of China and eastern Asia. *Evolutionary Anthropology, 2*, 22–35.

Scholes, R. J., & Archer, S. R. (1997). Tree–grass interactions in savannas. *Annual Review of Ecological Systems, 28*, 517–544.

Spasov, N. (1998). State of the large mammals in the Rila National Park. Report submitted to the GEF Project for Biodiversity Conservation, Sofia.

Steudel, K. L. (1994). Locomotor energetics and hominid evolution. *Evolutionary Anthropology, 3*, 42–48.

Suwa, G., Nakaya, H., Asfaw, B., Saegusa, H., Amzaye, R., Kono, R. T., et al. (2003). Plio-Pleistocene terrestrial mammal assemblage from Konso, Southern Ethiopia. *Journal of Vertebrate Paleontology, 23*, 901–916.

Tchernov, E. (1980). *The Pleistocene birds of 'Ubeidiya, Jordan Valley*. Jerusalem: The Israel Academy of Science and Humanities.

Tchernov, E. (1984). Faunal turnover and extinction rate in the Levant. In P. S. Martin & R. G. Klein (Eds.), *Quaternary extinctions* (pp. 528–552). Tucson: The University of Arizona Press.

Tchernov, E. (1986). *Les Mammifères Du Pléistocene Inférieur De La Vallée Du Jordain a Oubeidiyeh*. Paris: Association Paléorient.

Tchernov, E. (1987). The age of the 'Ubeidiya Formation, an early hominid site in the Jordan Valley, Israel. *Journal of Earth Sciences, 36*, 3–36.

Tchernov, E. (1988a). La biochronologie du site de 'Ubeidiya (Vallée du Jourdain) et les plus anciens hominidés du Levant. *L'Anthropologie (Paris), 92*, 839–861.

Tchernov, E. (1988b). The biogeographical history of the southern Levant. In Y. Yom-Tov & E. Tchernov (Eds.), *The zoogeography of Israel* (pp. 159–250). Dordrecht: Dr. Junk Publishers.

Tchernov, E., & Belmaker, M. (2004). The biogeographic history of the fauna in the Southern Levant and the correlation with climatic change. In A. Dolev & A. Perevolototsky (Eds.), *The red book. Vertebrates of Israel* (pp. 21–25). Jerusalem: The Israel Nature and Park Authority and the Society for the Protection of Nature in Israel.

Tchernov, E., Kolska Horwitz, L., Ronen, A., & Lister, A. (1994). The faunal remains from Evron Quarry in relation to other Paleolithic hominid sites in the Southern Levant. *Quaternary Research, 42*, 328–339.

Thomas, H., Geraads, D., Janjou, D., Vaslet, D., Memesh, A., Billiou, D., et al. (1998). First Pleistocene faunas from the Arabian Peninsula: An Nafud desert, Saudi Arabia. *Comptes Rendue de l'Académie des Science Serie II de la Terre et des Planètes, 326*, 145–152.

Tobias, P. (1966a). *A Member of the Genus* Homo *From 'Ubeidiya*. Jerusalem: Publications of the Israel Academy of Sciences and Humanities.

Tobias, P. V. (1966b). Fossil hominid remains from 'Ubeidiya, Israel. *Nature, 211*, 130–133.

Turner, A. (1984). Hominids and fellow travelers: human migration into high latitudes as part of a large mammal community. In R. A. Foley (Ed.), *Hominid evolution and community ecology* (pp. 193–217). London: Academic.

Turner, A. (1987). *Megantereon cultridens* (Cuvier) (Mammalia, Felidae, Machairodontinae) from Plio – Pleistocene deposits from Africa and Eurasia, with comments on dispersal and the possibility of a new world origin. *Journal of Paleontology, 61*, 1256–1268.

Turner, A. (1992). Large carnivores and earliest European hominids: changing determinants of resource availability during the Lower and Middle Pleistocene. *Journal of Human Evolution, 22*, 109–126.

Turner, A. (1997). *The big cats and their fossil relatives. An illustrated guide to their evolution and natural history*. New York: Columbia University Press.

Turner, A. (1999). Assessing early human settlement of Eurasia: Late Pliocene dispersions from Africa. *Antiquity, 73*, 563–569.

Valverde, J. A. (1967). Remarques sur la structure et l'évolution des communautés de vertébrés terrestres. *Revue d' Ecologie (Terre et Vie), 111*, 121–154.

Vaufrey, R. (1951). Etude Paleontologique, I: Mammiferes. In R. Neuville (Ed.), *Le Palèolithique et le Mesolithique du Dèsert de Judee* (pp. 198–217). Paris: Institut de Paleontologie Humaine.

Vekua, A. (1995). *Die Wirbeltierfauna Des Villafranchium Von Dmanisi Und Ihr Biostratigraphische Bedeutung*. Mainz: Jahrbuch des Roemisch-Germanischen Zentralmuseums.

Voorhies, M. R. (1969). Taphonomy and population dynamics of an early Pliocene vertebrate fauna. Contributions to Geology Special Papers 1 (pp 1–69). Knox County, Nebraska: University of Wyoming.

Vrba, E. S. (1980). The significance of bovid remains as indicators of environment and predation patterns. In A. K. Behrensmeyer & A. P. Hill (Eds.), *Fossils in the making* (pp. 247–271). Chicago: University of Chicago Press.

Vrba, E. S. (1988). Late Pliocene climatic events and hominid evolution. In F. E. Grine (Ed.), *Evolutionary history of the "robust" australopithecines* (pp. 405–426). New York: Aldine de Gruyter.

Vrba, E. S. (1995). The fossil record of African antelopes (Mammalia, Bovidae) in relation to human evolution and paleoclimate. In E. Vrba, G. Denton, T. Partridge, & L. Burckle (Eds.), *Paleoclimate and evolution, with emphasis on human origins* (pp. 385–424). New Haven: Yale University Press.

Walker, A. (1993). Taphonomy. In A. Walker & R. Leakey (Eds.), *The Nariokotome Homo erectus skeleton* (pp. 40–53). Cambridge, MA: Harvard.

Walker, A., & Leakey, R. (Eds.). (1993). *The Nariokotome* Homo erectus *skeleton*. Cambridge: Harvard University Press.

Walker, A., & Shipman, P. (1996). *The wisdom of the bones: in search of human origins*. New York: Knopf, distributed by Random House.

Werdelin, L., & Lewis, M. E. (2005). Plio-Pleistocene Carnivora of eastern Africa: species richness and turnover patterns. *Zoological Journal of the Linnean Society, 144*, 121–144.

Werner, N. Y., & Mokady, O. (2004). Swimming out of Africa: mitochondrial DNA evidence for late Pliocene dispersal of a cichlid from Central Africa to the Levant. *Biological Journal of the Linnean Society, 82*, 103–107.

Wilf, P., Beard, K. C., Davies Vollum, S. K., & Norejko, J. W. (1998). Portrait of a late Paleocene (early Clarkforkian) terrestrial ecosystem, Big multi quarry and associated strata, Washakie Basin, Southwestern Wyoming. *Palaios, 13*, 514–532.

Wynn, J. G. (2004). Influence of Plio-Pleistocene aridification on human evolution: Evidence from paleosols of the Turkana Basin, Kenya. *American Journal of Physical Anthropology, 123*, 106–118.

Yom-Tov, Y., Tchernov, E., & Dumont, H. J. (Eds.). (1988). *The zoogeography of Israel*. Dordrecht: Dr. W. Junk Publishers.

Zaidner, Y. (2003a). The use of raw material at the Lower Palaeolithic site of Bizat Ruhama, Israel. In J. M. Burdukiewicz & A. Ronen (Eds.), *Lower Palaeolithic small tools in Europe and the Levant* (pp. 121–131). Oxford: BAR.

Zaidner, Y. (2003b). The lithic assemblage of Bizat Ruhama: Lower Paleolithic site in southern coastal plain, Israel. M.A. dissertation, Haifa University, Haifa, Israel.

Zaidner, Y., Ronen, A., & Burdukiewicz, J.-M. (2003). L'industrie microlithique de Paléolithique inférior de Bizat Ruhama, Israël. *L'Anthrolopologie, 107*, 203–222.

Zhu, R. X., An, Z., Potts, R., & Hoffman, K. A. (2003). Mangenetostraigraphic dating of early humans in China. *Earth Science Reviews, 61*, 341–359.

Chapter 13
Early Pleistocene Faunas of Eurasia and Hominin Dispersals

Bienvenido Martínez-Navarro

Abstract During Neogene–Quaternary times, all the dispersals of African taxa into the European continent are related to important faunal turnovers in the Eurasian faunas. Only on rare occasions have a few taxa of Ethiopian origin penetrated into the Northern Continents and vice versa. The finding of African species in Eurasia and of Eurasian species in Africa, although rare, is always very significant and provides important climatic, ecologic and geographic information. The Levantine Corridor – situated in the eastern Mediterranean Basin – is accepted as the major route out of Africa into Eurasia and *vice versa.* The fact that the Levant is an extension of the East African Rift and forms an intercontinental bottleneck, and that the climatic and ecological conditions prevailing in East Africa extended north at the Plio-Pleistocene times, makes this region the most important key area for explaining these dispersal phenomena. Although a large number of African origin large mammals are recorded in the Levant, only a few of these species penetrated into the Eurasian middle latitudes during the Plio-Pleistocene transition and Early Pleistocene times; these taxa are *Theropithecus oswaldi*, *Megantereon whitei*, and *Hippopotamus antiquus*. The dispersal of this fauna is associated with the first colonization of the Northern Continent by the genus *Homo,* and it reveals a new paleoecological picture of this event. The development of the social behavior, as has been detected in Dmanisi (Georgia), together with systematic carnivorous behavior by hominins, was necessary for colonizing the middle latitudes of Eurasia and survival in seasonal climates with winters and summers, where vegetable resources were not available throughout the year. One million years later, during the Early-Middle Pleistocene transition, there was another dispersal of African taxa, associated with the arrival of the Acheulean culture into Europe and Asia.

Keywords Early Pleistocene • *Hippopotamus antiquus* • *Homo* • *Megantereon whitei* • *Pachycrocuta brevirostris* • *Theropithecus oswaldi*

B. Martínez-Navarro (✉)
ICREA, IPHES, Universitat Rovira i Virgili, Pl. Imperial Tarraco 1, 43005, Tarragona, Spain
e-mail: bienvenido@icrea.es

Introduction

During the Plio-Pleistocene of Eurasia,[1] the large mammal assemblages are basically composed of a fauna of Palearctic origin, especially bovids and cervids, and some Nearctic groups, such as the one-toed equids of the genus *Equus*. The arrival of a few species of Ethiopian origin into Eurasia is rare, but always significant. The Levantine Corridor is the common and major route of communication between Africa and Eurasia, and the record in this area shows an important mixed fauna of Holarctic and Ethiopian origin during Late Pliocene and Pleistocene times. The species lists of the large mammals from the Late Pliocene site of Bethlehem and the important Early Pleistocene sites of 'Ubeidiya, Evron Quarry or Gesher Benot Ya'aqov (GBY), provide good examples of an admixed fauna. This is the consequence of the geographic position of this region, as a bottleneck between both continents, but also because the Levantine Rift Valley has a "northward extension of the tropical "Sudanese" climate, with tropical biota similar to those of the African hominin savanna homeland" (Por 2004: 5). On rare occasions during the Middle and Late Pliocene, and the Early Pleistocene, Ethiopian fauna penetrated north of the Taurus–Zagros mountain range and dispersed toward the Eurasian interior, and Holarctic fauna also penetrated southward into the African continent (see Fig. 13.1).

Middle and Late Pliocene Large Mammal Assemblages

The Middle and Late Pliocene of Europe are dominated by the Early and Middle Villafranchian faunas, MN 16b and MN 17, respectively. The large mammal assemblages are composed of cervids [*Eucladoceros* (different species), *Pseudodama* (different species), or *Croizetoceros ramosus*], bovids [*Leptobos*

[1]This manuscript was written using the former definition of the Plio-Pleistocene boundary as the top of the Olduvai normal subchron.

J.G. Fleagle et al. (eds.), *Out of Africa I: The First Hominin Colonization of Eurasia*, Vertebrate Paleobiology and Paleoanthropology,
DOI 10.1007/978-90-481-9036-2_13, © Springer Science+Business Media B.V. 2010

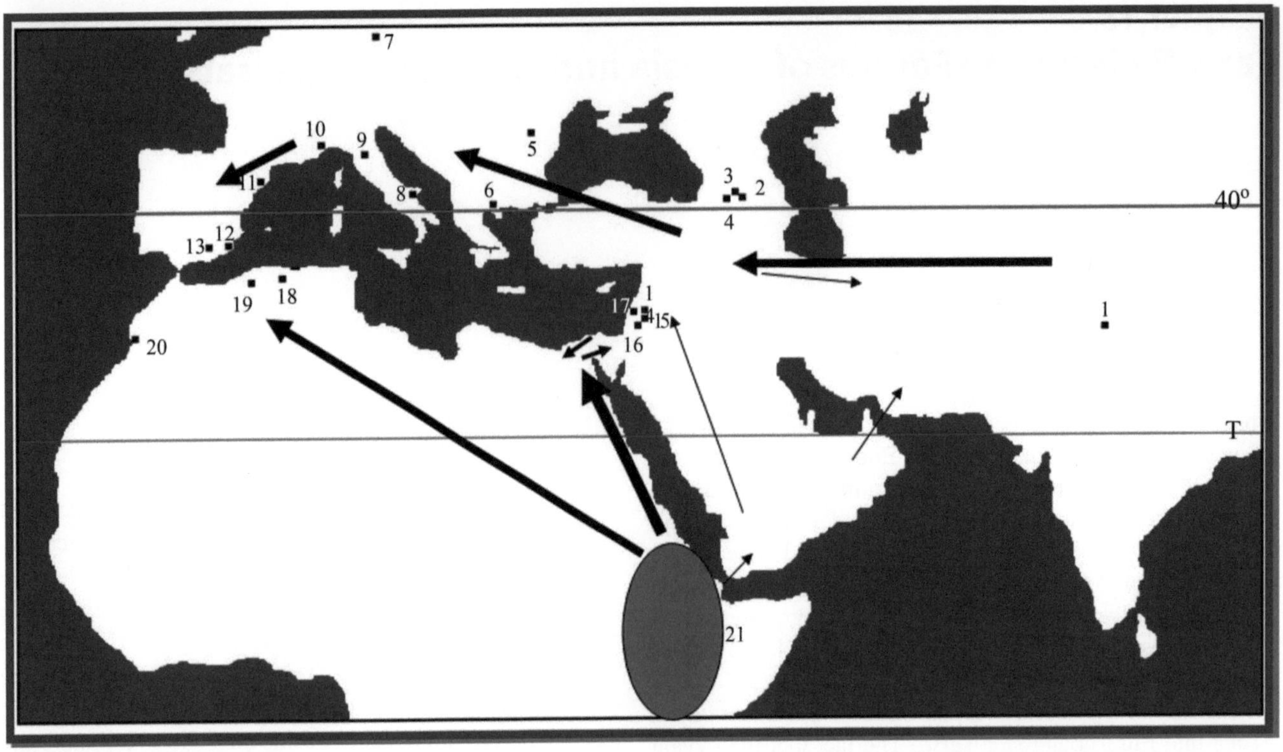

Fig. 13.1 Geographic situation of some of the most important Late Pliocene (LP) and Early Pleistocene (EP) localities of southern and western Asia, Europe and northern Africa: (1) Upper Siwaliks (Tatrot :LP, Pinjor: LP + EP, and Boulder Conglomerate: EP); (2) Kuabebi: LP; (3) Dmanisi: EP; (4) Akhalkalaki: EP; (5) Dacic Basin*: LP + EP; (6) Mygdonia Basin, where together with LP sites is located the EP site of Apollonia-1; (7) Untermassfeld*: EP; (8) Pirro Nord: EP; (9) Upper Valdarno: LP + EP; (10) Vallonnet: EP; (11) Incarcal: EP; (12) Cueva Victoria: EP; (13) Guadix–Baza Basin, where are located the sites of Huélago (LP), Fonelas* (LP), Fuente Nueva-1 (LP), Venta Micena (EP), Barranco León (EP) and Fuente Nueva-3 (EP); (14) Gesher Benot Ya'aqov, GBY (EP); (15) 'Ubeidiya (EP); (16) Bethlehem* (LP); (17) Evron Quarry (EP); (18) Ain Hanech (EP); 19) Ternifine (EP); 20) Ahl al Oughlam* (LP); (21) East African sites (LP + EP). The arrows mark the possible routes and intensity of faunal dispersals. T: Tropic of Cancer (* not directly studied by the author)

(*L. etruscus* and *L. elatus*), *Gazellospira torticornis, Gazella borbonica, Gallogoral meneghinii,* or *Procamptoceras*], a suid (*Sus strozii*), a rhinoceros (*Stephanorhinus etruscus*), the last hipparionines and the arrival of the first one toed horses (*Equus stenonis*) of American origin, and the first true modern elephants corresponding to the genus *Mammuthus* (*M. rumanus* in the MN16, and *M. meridionalis* in the MN17). The carnivores are characterized by the arrival of a middle-large-sized true *Canis* (recorded at the site of Vialette, France, during the Early Villafranchian, ~3.0 Ma (Heintz et al. 1974; Lacombat et al. 2008), the persistence of the raccoon dog *Nyctereutes megamastoides*, the presence of two hyaenas, one a scavenger, *Pliocrocuta perrieri*, and another a cursor and hunter, *Chasmaporthetes lunensis*, the arrival of the machairodonts *Homotherium crenatidens* and *Megantereon cultridens*, and the puma (*Puma pardoides*), or the large cheetah (*Acinonyx pardinensis*).

The boundary between the Early Villafranchian (MN16a) and the Middle Villafranchian (MN16b) was described as the Eurasian "Elephant-*Equus*" event, around 2.5 Ma (Lindsay et al. 1980; Azzaroli et al. 1988). The authors thought that this event marked the appearance in Eurasia of the true elephant (*Mammuthus meridionalis*) and the one-toed horse (*Equus stenonis*). But in fact, the appearance of *Mammuthus*, the modern elephant of African origin (Kalb 1995; Lister and Sher 2001; Lister and van Essen 2003; Lister et al. 2005), in Europe is coincidental with the beginning of the Villafranchian at ~3.2 Ma, based on the record of the primitive form *Mammuthus rumanus,* found in the Dacic Basin (Romania) at Tulucesti and Cernatesti (Radulesco and Samson 2001). The same authors also said that in the Early Pliocene Ruscinian fauna of Malusteni, also in the Dacic Basin, and dated ~4.0 Ma, the presence of a monodactyl horse (*Plesippus* cf. *euxinicus*) was also recorded. At the site of Vialette, at ~3.0 Ma (Bout 1960; Lacombat et al. 2008), a form that could be related to *Equus livenzovensis* (a primitive form of *Equus stenonis*) is also present.

During the Late Pliocene, the record of African fauna in the Levant, at the site of Bethlehem, is marked only by the presence of the genus *Giraffa*, but the finding of African faunal elements in the region of the Caucasus is especially important, specifically at the site of Kuabebi (2.5–2.6 Ma).

Fossil taxa identified from this site include the procaviid *Kuabebihyrax kachethycus* and the giraffid *Giraffa* sp. (Vekua 1972). Other giraffids are also found in Europe, in the Balkans and on the Iberian Peninsula where the short necked form *Mitilanotherium martinii* has been recorded in deposits of the Dacic Basin (Romania), in Wolacks (Greece), and in the Guadix–Baza Basin (Spain) at the sites of Huélago and Fonelas (Radulesco and Samson 1990; Sickenberg 1967; Alberdi et al. 2001; Arribas et al. 2001), although this genus is probably of Eurasian origin. The presence of the ostrich *Struthio transcaucasicus* at Kuabebi does not indicate that this taxon has an African origin, because the ostrich is known in other areas of Eurasia during the Late Miocene and Pliocene. Also *Pachystruthio* sp. (probably a synonym of *Struthio*) has been found in the Pliocene of Romania (Radulesco and Samson 2001).

In Central and Southern Asia, as in Europe, two dispersal events involving African antelopes have been detected during the Late Pliocene, one at around 3.0 Ma and the other at 2.6 Ma (Vrba 1995a, b). The first (at 3.0 Ma) is marked by the presence of the Hippotragini *Sivatragus brevicornis* (Pilgrim 1939) in the Tatrot Formation of the Upper Siwaliks in the Indian Subcontinent. The second (at 2.6 Ma) is marked by the presence – in the Pinjor Formation, also in the Upper Siwaliks – of two Hippotragini, *Sivatragus bohlini* and *Oryx sivalensis*, two Reduncini, *Vishnocobus patulicornis* and *Sivacobus palaeindicus,* and the Alcelaphini *Damalops palaeindicus* (Pilgrim 1939). The last species was also found in Tajikistan (Dmitrieva 1977).

Another taxon that has been suggested to have a possible Ethiopian origin is the Late Pliocene European and Central Asian large terrestrial cercopithecid *Paradolichopithecus* [*P. arvernensis* in Europe at Graunceanu (Romania), Vatera (Lesvos Island, Greece), Senèze (France); La Puebla de Valverde, Cova Bonica and Moreda (Spain); and *P. sushkini* in Central Asia at Kuruk (Tajikistan)]. The postcranial anatomy of this genus resembles that of the extant large baboons and suggests an African origin, as was proposed by Mashchenko (1994) who classified it as *Papio* and not as *Paradolichopithecus*, but the cranial features resemble those of macaques (Szalay and Delson 1979; Van der Geer and Sondaar 2002), especially the presence of a maxillary sinus, which is a significant synapomorphy with *Macaca* (Nishimura et al. 2007). These last data suggest a Holarctic rather than an African origin for *Paradolichopithecus*.

Also in the Late Pliocene, important groups of mammals penetrated into Africa. This includes the genus *Equus*, known from the Lake Turkana Basin around 2.3 Ma (Eisenmann 1983; Harris et al. 1988); later the one-toed horses became a very important element of the African fauna. The route of dispersal may have been the Straits of Bab-el-Mandeb, situated at the southern edge of the Red Sea (Tchernov 1992; Turner 1999).

In connection with this event, it is important to note that in the Late Pliocene deposits of Ahl al Oughlam in Morocco there occurs a typical assemblage of large and small African mammals together with three carnivores of Euro-Asiatic origin: the bear *Ursus* cf. *etruscus,* the hyaena *Pliocrocuta perrieri latidens* and the raccoon dog *Nyctereutes abdeslami* (Geraads 1997). The genera *Ursus, Pliocrocuta* and *Nyctereutes* are well-known from Late Pliocene deposits on the Iberian Peninsula – for example, at La Puebla de Valverde (Kurtén and Crusafont 1977). These genera are also known in Asia and *Nyctereutes* is recorded in Bethlehem (Hooijer 1958). The genus *Nyctereutes* (*N. terblanchei*) is also recorded at the South African deposits (Ewer 1956; Ficcarelli et al. 1984). But, if the interchange of the above-mentioned fauna across the Levantine Corridor between Eurasia and Africa was made via the Straits of Bab-el-Mandeb or by the Peninsula of Sinai, it is difficult to explain why the genera *Ursus* and *Pliocrocuta* are not found in other Northern, Eastern and Southern African sites. It is not so preposterous to consider the possibility of selective species interchanges across the Straits of Gibraltar during Late Pliocene times, although at the moment is not possible to prove this hypothesis. The Eurasian Caprini *Capra primaeva* is also found in the Late Pliocene site of Ain Brimba, Tunis (Arambourg 1979).

In this context, the idea that Late Pliocene hominins could also have dispersed out of Africa in conjunction with these dispersive events continues to be the subject of lively debate. For the moment, it is unquestionable that Pliocene hominins have been found only in Africa, in the region of the Rift Valley, and in some South African caves. Only one Middle Pliocene occurrence is known west of the Rift, viz. *Australopithecus bahrelghazali* from Chad (Brunet et al. 1995) and no unquestionable Pliocene hominins have been found in Eurasia. Only some lithic artifacts have been published at the Late Pliocene sites of Yiron, Israel (Ronen 1991) and at Riwat and Pabbi Hills, Pakistan (Dennell et al. 1988, 2006) that could represent a possible first entrance of hominins into Eurasia. But these few findings are not solid enough at the moment to be considered evidence of a clear Late Pliocene hominin dispersal out of Africa.

The Plio-Pleistocene Transition and the Early Pleistocene Large Mammal Assemblages

The Plio-Pleistocene transition marks a great change. The first arrival of hominins in Eurasia has been detected at the site of Dmanisi, Georgia (Gabunia et al. 2000, Vekua et al. 2002; Lordkipanidze et al. 2005, 2007), and in eastern and southeastern Asia (Swisher et al. 1994; Zhu et al. 2004). This arrival is in conjunction with other endemic African species.

The influence of African forms is especially important in the Levant at the site of 'Ubeidiya (Israel), but some have been found also in Orce (sites of Venta Micena, Fuente Nueva-3 and Barranco León-5), Cueva Victoria or Incarcal (Spain), Sainzelles (France), Pirro Nord (Italy), Untermassfeld (Germany), Apollonia and Ravin de Voulgarakis (Greece), Dmanisi and Akhalkalaki (Georgia), Mirzapur (India), and also probably in China and in Indonesia, and in many other Eurasian assemblages.[2]

In this context, it is important to note that, although most of the Plio-Pleistocene biozones are based on the record of different species of arvicolids, hominins are large mammals and they have to be considered as part of a community of macromammals (see Fig. 13.2).

The Pachycrocuta brevirostris *Event*

In most of Eurasia, the Plio-Pleistocene transition and the Early Pleistocene assemblages are characterized by the presence of the large supercarrion-eating hyaenid *Pachycrocuta brevirostris*. Its first record in Europe is just below the beginning of the Olduvai normal subchron, around 2.0 Ma, in Olivola, Italy (Napoleone et al. 2003). It marks the beginning of the Late Villafranchian (Torre et al. 1996). This arrival is also coincidental with the first record in Eurasia of *Panthera gombaszoegensis*, probably coming from Africa. The species *Pachycrocuta brevirostris* replaced the smaller Pliocene form *Pliocrocuta perrieri*. The Asian or African origin of *Pachycrocuta brevirostris* is still controversial, but the oldest records of *Pachycrocuta* are found in eastern Africa during Middle Pliocene times from ~3.5–2.5 Ma (Werdelin 1999). It is a small form called *Pachycrocuta* sp., and a similar form to *P. brevirostris* persists in South Africa until the base of the Early Pleistocene, *Pachycrocuta bellax* (Randall 1981). The extinction of this form in East Africa is coincidental with the development of lithic artifacts by hominins at 2.5–2.6 Ma (Semaw et al. 2003), and in South Africa, probably with the arrival of Acheulean tools. The colonization of Europe by *P. brevirostris* coincides with the great explosion of the large modern Holarctic canids in the continent, *Lycaon falconeri, Canis etruscus* and *Canis arnensis*. This event is known in the literature as the "the wolf event" (Azzaroli 1983; Azzaroli et al. 1988; also see Sardella and Palombo 2007), but this name is incorrect, because the record of middle-large-sized true dogs in Europe is known from the Early Villafranchian at the French site of Vialette, as has been noted before. As suggested here it is probably more appropriate to replace this event with "the *Pachycrocuta brevirostris* event", because of the extensive record and great impact of this giant hyenid in most of the Early Pleistocene assemblages of Eurasia (Howell and Petter 1980; Werdelin and Solounias 1991; Turner and Antón 1996), from the Iberian Peninsula to China and Indonesia.

Pachycrocuta brevirostris, described for the first time at the Early Pleistocene site of Sainzelles in the French Central Massif by Aymard during the nineteenth century, is a large hyaena, weighing more than 100 kg (and some individuals probably more than 150 kg). It is a large sized non-predator, supercarrion-eating scavenger hyaena (Palmqvist et al., in preparation). In terms of taphonomy, this species is the most important agent of fossil accumulation at Early Pleistocene sites in Europe and Asia, i.e., Venta Micena in Spain (Palmqvist et al. 1996; Arribas and Palmqvist 1998; Martínez-Navarro and Palmqvist 1999; Palmqvist and Arribas 2001). It persists until the end of the Early-Middle Pleistocene transition. Most of the localities where *P. brevirostris* is found in Eurasia are exclusively paleontological sites, but in some instances, it is present in archeological assemblages, where lithic artifacts are recorded, i.e., Fuente Nueva-3 and Barranco León-5, Spain (Martínez-Navarro et al. 2004c; Palmqvist et al. 2005), Vallonnet, France (Moullé 1992), and recently at the site of Pirro Nord, Italy (Arzarello et al. 2006). In all the cases, the lithics are pre-Acheulean or Mode I tools.

The giant and short faced hyena *Pachycrocuta brevirostris* was a super scavenger, with proportionately short and very robust forelimbs not adapted to be a cursorial predator, and a very thick and stout dentition, adapted to be a bone-cracking specialist for eating the marrow content and the brains. With its large size and robustness, its survival strategy was to rob the prey from other large hunting and not bone-cracking carnivores (like *Megantereon whitei, Lycaon lycaonoides*, *Panthera gombaszoegensis,* or *Acinonyx pardinensis*). This species is the most important, and the most direct competitor with hominins for the carrion from the large mammal carcasses during all the Eurasian Early Pleistocene. Human behavior in the middle latitudes was conditioned by this supercarrion eating giant hyaena, which is a more efficient and dangerous scavenger than the African spotted hyaena *Crocuta crocuta*. This last taxon dispersed during the Early Pleistocene into the Levant, and it is recorded at the site of 'Ubeidiya at ~1.4 Ma, but it apparently did not disperse into other Eurasian territories until the Early-Middle Pleistocene transition, around 0.8 Ma (García 2003; Sardella 2004), when *P. brevirostris* disappeared from the Eurasian assemblages.

The Early Pleistocene Faunas

The most important faunistic change in Europe and western Asia lies just after the Plio-Pleistocene boundary, after the Tasso Faunal Unit – associated with Olivola Faunal Unit

[2]The most important taxa of African origin and their influence in the Eurasian assemblages are discussed in the following chapters.

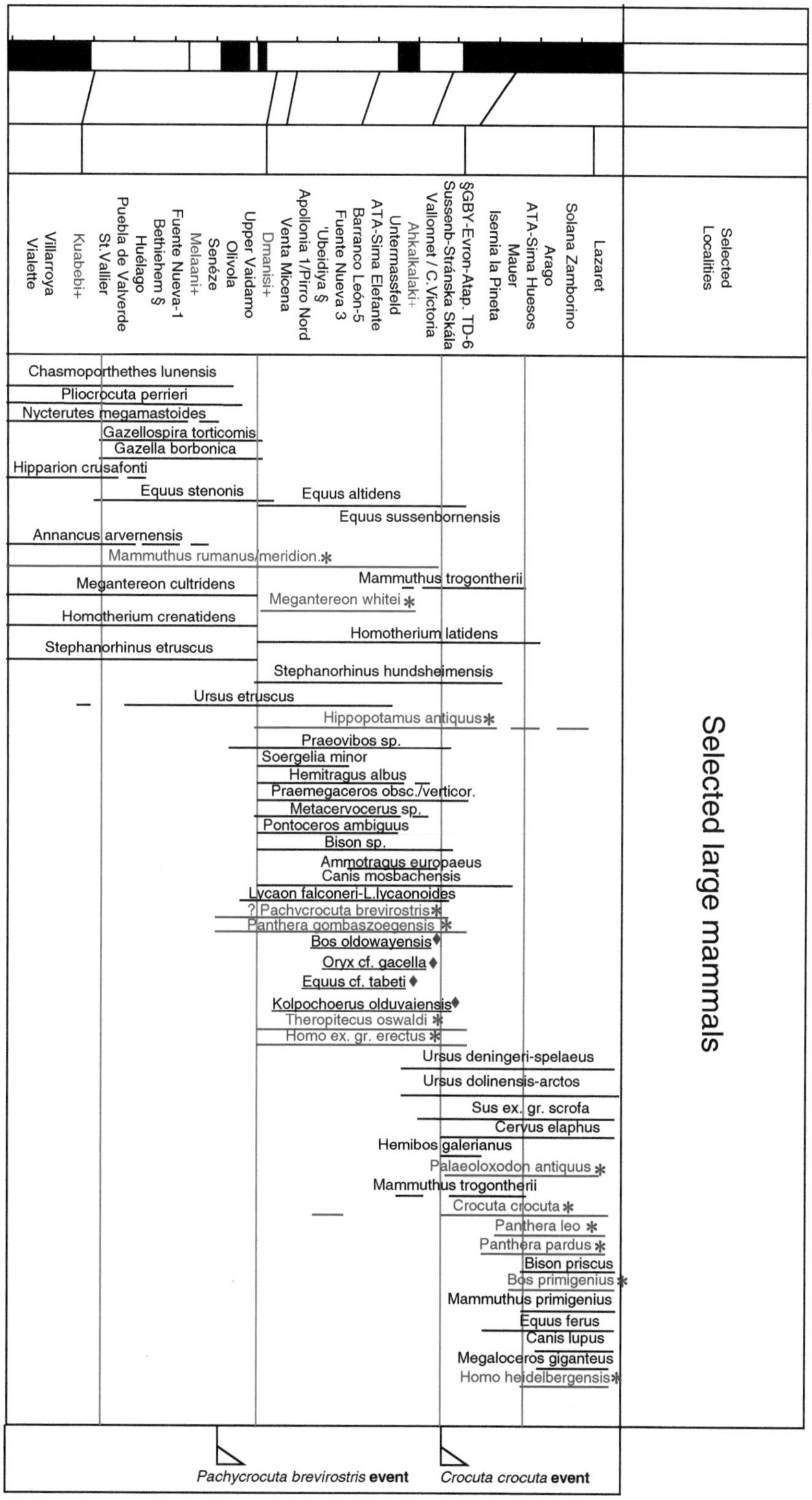

Fig. 13.2 Biostratigraphic chart of selected large mammals from the Middle-Late Pliocene and Pleistocene in different sites of the Levantine Corridor (§), the Caucasian Region (+) and Europe (all others). The African origin mammals only found in the Levant (♦) and the African origin mammals found in several areas of Europe or Asia (*). The figure shows four important faunal turnovers: (1) at 2.6 Ma; (2) at the Plio-Pleistocene boundary (at about 1.8 Ma) where hominins arrive into Eurasia, located at the site of Dmanisi, together with other African origin species; (3) at the Early-Middle Pleistocene transition (at about 0.9 Ma) when late Villafranchian faunas are replaced by Galerian/Cromerian faunas; and (4) at around 0.5 Ma, when the classical Acheulian culture invades Europe. In the Levant, it seems that there is another turnover at 1.4 Ma, the age of 'Ubeidiya, but it could be the result of a gap in the Early Pleistocene record of this area

(Palombo 2004) – if we follow the Italian terminology, which is dated inside the Olduvai normal subchron, where the ungulates are still dominated by the cervids *Pseudodama nestii* or *Eucladoceros dicranios,* or the large bovids by *Leptobos*. After that, an important replacement of ungulates and carnivores took place. The paradigmatic site where this faunistic change has been detected is Dmanisi in Georgia (Vekua 1995), dated 1.81 Ma (de Lumley et al. 2002), but it is also recorded in many other sites at Europe, especially at Venta Micena, Spain, dated ~1.5–1.6 Ma.

In the base and the first half of the Early Pleistocene, the arrival of several new species is detected, with the appearance of two new perissodactyls (*Stephanorhinus hundsheimensis* and the zebra-like *Equus altidens*), new cervids indicated by the presence of a new form, *Metacervoceros* (Abbazzi, in press) different from *Pseudodama nestii*, which is included in the genus *Cervus* by Croitor (2005),[3] and a new form of large deer corresponding to the genus *Praemegaceros*. The bovids in all the Early Pleistocene European sites are dominated by the presence of *Bison*, including Dmanisi, Venta Micena, Apollonia (Greece), Pirro Nord (Italy), Sainzelles (France), Untermassfeld (Germany), and all the other sites; spiral horn-cored antelopes different from *Gazellospira* are present at Dmanisi (Bukhsianidze 2005), and Apollonia (Kostopoulos 1997) and other sites of Eastern Europe and the Middle East; new Caprini (*Capra* or *Hemitragus*) are also recorded, and new forms of Ovibovini are common, such as *Praeovibos* and the rare brachydont form *Soergelia* (*S. minor* in Venta Micena and Dmanisi, and *S. briggitae* in Apollonia), which has a very strange orientation in the horn-cores, going outward, upward and forward. This form of brachydont *Soergelia* is considered by the author an important biostratigraphic marker for the first half of the Early Pleistocene in Europe. The most significant new carnivore that arrived at the Plio-Pleistocene boundary is probably the African form *Megantereon whitei* detected at Dmanisi, Venta Micena, Apollonia, Pirro Nord and other sites (Martínez-Navarro and Palmqvist 1995, 1996; Rook et al. 2004; Palmqvist et al. 2007), described below. But other carnivores, like the canids *Lycaon lycaonoides*, a more evolved form than *L. falconeri* (Martínez-Navarro and Rook 2003), and *Canis mosbachensis*, also arrive at the same time.

The second half of the Early Pleistocene is marked by a small faunal turnover, which is especially recorded at the sites of Fuente Nueva-3 and Barranco León-5 in Orce (Spain), by the extinction of *Soergelia minor* and the arrival of a large hypsodont Caprini, named *Ammotragus europaeus* (Moullé et al. 2004) and the large stenonoid horse *Equus sussenbornensis.* Until now, the oldest record of human presence in Western Europe has been found in these Orce sites (Fuente Nueva-3 and Barranco León-5) in connection with this small faunal turnover at around 1.2–1.3 Ma (Martínez-Navarro et al. 1997, 2004c; Oms et al. 2000) (see Fig. 13.2), but the new findings at Pirro Nord (Arzarello et al. 2006), perhaps indicate that the arrival of hominins into this region was a little bit earlier.

Another important event that marks the upper part of the Early Pleistocene in Europe, is the arrival of the suids which were not known in Europe after the Late Pliocene form *Sus strozzii* became extinct. The oldest records are at the sites of Sima del Elefante (Atapuerca, Spain), dated 1.2 Ma, associated with a human mandible (Carbonell et al. 2008), and at the site of Untermassfeld, 1.0–1.1 Ma, where the species is ascribed to *Sus scrofa priscus* by Guérin and Faure (1997).

Unfortunately, during the Early Pleistocene, the faunal connections between Europe and Eastern Asia are not well known, but current research suggests that they are more important than have been previously detected.

The Early-Middle Pleistocene Faunal Transition

At the end of the Early Pleistocene a very important faunal turnover is recorded in all of Eurasia in connection with the cold climates that announce the Middle Pleistocene. In Europe the new faunal assemblages correspond to the Galerian (if we follow the Italian terminology) or to the Cromerian (if we follow the British terminology). Most of this new fauna is basically of Asian origin, but some new African origin species colonize Eurasia.

This event is coincidental with the extinction of *Pachycrocuta brevirostris* and most of the Late Villafranchian carnivore guild (*Megantereon whitei, Acinonyx pardinensis, Lycaon lycaonoides*, and others), and the record of hominins from the lowermost levels of the Gran Dolina of Atapuerca, Spain (Bermúdez de Castro et al. 1997; Carbonell et al. 2005) and Ceprano, Italy (Manzi et al. 2001), around 0.8 Ma, and with an important change in the large mammal assemblages, with new fauna mostly of Holarctic origin, and new few elements of African origin – mostly carnivores. Of special interest is the arrival of the spotted hyaena *Crocuta crocuta* (García 2003; Sardella 2004), but also the elephant *Palaeoloxodon antiquus* (Lister 2004) and taxa of Indian origin, such the large Bovini *Hemibos galerianus* (Martínez-Navarro and Palombo 2004).

The spotted hyaena *Crocuta crocuta* is a typical African form known on that continent from the Pliocene until present. During the Early Pleistocene, this species dispersed out of Africa; although there are other citations based on problematic and questionable fossil material, it is only clearly recorded at

[3]This author differentiates the presence of three Early Pleistocene genera of small-middle size cervids in Europe, including *Cervus, Dama*, and *Metacervoceros*.

the site of 'Ubeidiya (Israel) at ~1.4 Ma (Ballesio 1986; Martínez-Navarro et al. 2009). The body size of *C. crocuta* is around 45–75 kg and it is a social hunter and scavenging hyena very well documented during the Middle and Late Pleistocene of Europe and Asia. This generalist's social hunting and scavenging behavior is the key to its survival in the changing climatic conditions of the Middle and Late Pleistocene, where a super specialist scavenging giant hyaena like *Pachycrocuta brevirostris* couldn't survive, nor could other super specialist predators carnivores like *Megantereon whitei, Acinonyx pardinensis,* or *Lycaon lycaonoides.* Similar extinction of super specialists carnivores is detected in Africa (Werdelin and Lewis 2005). The oldest European records of *Crocuta crocuta* are around 0.8–0.9 Ma at the lowermost levels of Gran Dolina of Atapuerca, northern Spain (García 2003) and at Ponte Galeria, Central Italy (Sardella 2004), the type locality of the Galerian. The extinction of *Pachycrocuta brevirostris* and the arrival of *Crocuta crocuta* into Europe mark the beginning of the modern faunas on this continent and we can call this faunal turnover as "the *Crocuta crocuta* event".

Palaeoloxodon antiquus, the straight-tusked elephant, is a common form in the Middle Pleistocene of Europe, the Middle East and under variable forms in eastern and southern Asia. This form is an immigrant, derived from the African Pliocene and Early Pleistocene lineage *Palaeoloxodon recki* (Lister 2004).[4] According to Lister (2004), during the late Early Pleistocene in Europe, *Mammuthus meridionalis*, the common Late Pliocene an Early Pleistocene form adapted to forest and semi-open environments, can be found together with the large steppe mammoth *Mammuthus trogontherii*, which is an eastern Asian immigrant at that time. This coexistence is also documented in Italy by Palombo and Ferretti (2005). However, the last appearance of *Mammuthus meridionalis* in Europe is in conjunction with the arrival of the better adapted *P. antiquus,* another generalist forest species that directly competes with *M. meridionalis. Palaeoloxodon antiquus* survives during all the Middle Pleistocene and it is more abundant than *Mammuthus* (*M. trogontherii* and later *M. primigenius*) in the southern latitudes of Eurasia and during the interglacial periods.

Other African immigrants arrive later into Eurasia, like the lion *Panthera leo* (its oldest record in Europe is at the site of Isernia la Pineta, around 0.6 Ma, Sala 1990), or the leopard, *Panthera pardus,*[5] for which the oldest record is probably at Valdemino Cave, Italy, around 0.6 Ma (Nocchi and Sala 1997). Also the large Bovini *Bos primigenius* arrives in a similar time-frame (Martínez-Navarro et al. 2007).

[4]Todd (2005) does not admit this determination and she classifies the forms ascribed to *Paleoloxodon* by Lister as *Elephas*. M.R. Palombo (personal communication, 2006) follows the same classification as Todd.

[5]A form of small panther, *Panthera pardus*, is cited in several sites of the Late Pliocene and the Early Pleistocene of Europe, but this form has been reclassified as *Puma pardoides* (Hemmer 2001; Argant 2004).

There is a debate about the chronology of different sites of Central Europe, especially Stranska Skala, where a very important lithic artifact collection has been found (Valoch 1995), with a faunal assemblage composed of *Castor fiber, Homotherium moravicum* (synonym of *H. latidens*), *Hyaena brevirostris* (synonym of *Pachycrocuta brevirostris*), *Crocuta crocuta, Canis mosbachensis, Xenocyon spelaeoides* (synonym of *Lycaon lycaonoides*, see Martínez-Navarro and Rook 2003), *Vulpes* cf. *praeglacialis, Vulpes* cf. *angustidens, Mustela* sp., *Ursus deningeri, Equus sussenbornensis, Bison* cf. *schoetensacki, Bison priscus, Bos primigenius, Capreolus* cf. *sussenbornensis, "Cervus" elaphoides,* and *Cervus* sp. (Musil 1971, 1995; Kahlke 1995). Although there are some differences based on the identification of problematic elements (i.e., *Bos primigenius, Bison priscus* or *Crocuta crocuta,* which are classical species of the European Middle Pleistocene), this faunal list is consistent with the latest Villafranchian assemblages, like Vallonnet in France (Moullé 1992), dated around 0.9–1.0 Ma. Other localities such as Süssenborn, Voigtstedt, Gombaszög, or Mosbach, also in Central Europe, and dated in the Early-Middle Pleistocene transition or in the base of the Middle Pleistocene, have similar faunal assemblages as Stranska Skala and most of the recorded species are Villafranchian forms.

The Early Pleistocene African Taxa in Eurasia and Their Connection with *Homo* Dispersals

As previously noted, the presence of mixed Ethiopian and Holarctic faunas in the Levant is continuous during Early Pleistocene times. The fauna from 'Ubeidiya is the most diverse example of this blended fauna (Tchernov 1986). The African large mammal species identified at this site are: the equid *Equus tabeti* (Eisenmann 1986), the suid *Kolpochoerus olduvaiensis* (Geraads et al. 1986), the hippopotamid *Hippopotamus gorgops* (Faure 1986), but the main representatives are the ruminants, the giraffid *Giraffa* sp., and the bovids *Oryx* cf. *gazella, Pelorovis oldowayensis* (Geraads 1986), the carnivores *Herpestes* sp., *Crocuta crocuta* (Ballesio 1986) or *Megantereon* cf. *whitei* (Martínez-Navarro et al. 2009), and the cercopithecid primate *Theropithecus* sp. (Belmaker 2002). Most of this fauna, while reaching the Levant, never penetrated north of the Taurus–Zagros Mountain range.

Although the finding in the Levant of a big African bovid assemblage in connection with the extension of African environments (but see Belmaker 2010) is very important because none of these elements has been found in other Early Pleistocene localities of Eurasia. Probably the most significant and best known of these ruminants is the large buffalo *Pelorovis oldowayensis.*

During the Early Pleistocene, only a few of the taxa found in 'Ubeidiya are recorded in other areas of Eurasia, the hippo

Hippopotamus (the species *H. antiquus,* which is a sister form of *H. gorgops*), the sabertoothed tiger *Megantereon whitei,* the monkey *Theropithecus oswaldi*, and maybe the pig *Kolpochoerus.*

The Plio-Pleistocene Transition and Significant Early Pleistocene Taxa

Pelorovis oldowayensis *(Now* Bos oldowayensis *After Martínez-Navarro et al. 2007)*

The large buffalo *Pelorovis sensu stricto*[6] is a common Bovini in the African savannas. This lineage includes the Late Pliocene forms from Africa and the Early Pleistocene forms from Africa and the Levantine Corridor. Gentry (1967) and Vrba (1987) argue that *Pelorovis* evolved from the Early Pliocene African form *Simatherium,* although Gentry (2006) considers that the cranial and horn-core anatomy this hypothesys is not correct. Then, the origin of this genus remains open. In a recent paper, Martínez-Navarro et al. (2007) suggests, after an anatomical study of the skull and the dentition, that the African form *Pelorovis s.s.* is the direct ancestor of the Middle Pleistocene Eurasian species *Bos primigenius*. This interpretation has important consequences for the systematics of this lineage because they propose that both, *Pelorovis s.s.* and *Bos*, should be included in only one genus: *Bos.*

This large Bovini was unknown outside of Africa until the first record in the Levant at the site of 'Ubeidiya, where a big skull classified as *Pelorovis oldowayensis* was found (Geraads 1986). In the most recent field seasons at 'Ubeidiya (directed by E. Tchernov, O. Bar-Yosef, and G. Bosinski), new material of this species has been uncovered (Martínez-Navarro et al. in prep. A). Also a form related to this genus has been identified at the site of Gesher Benot Ya'aqov (Martínez-Navarro et al. 2000). It is represented by a skull from the old collections and several teeth and postcranial elements from the new collections excavated by Naama Goren-Inbar.

The species from Gesher Benot Ya'aqov has similar anatomy in the skull, the dentition and the postcranial elements to that of the late Early Pleistocene of Buia, Eritrea (Martínez-Navarro et al. 2004b). This form is more derived than that from 'Ubeidiya, which is similar to that of the basal Early Pleistocene North African site of Ain Hanech, Algeria (personal observation), described as *Bos bubaloides* (Arambourg 1979), and that from Olduvai, Tanzania (personal observation) described by Gentry and Gentry (1978), which is the type locality for *Pelorovis oldowayensis*. This last form is also more derived than its Late Pliocene ancestor described as *Pelorovis turkanensis* from Koobi Fora and West Turkana in East Africa (Harris 1991), based especially on the anatomy of the cheek teeth, but also in the skull. A form called *Pelorovis* cf. *oldowayensis* is also identified at the Early Pleistocene deposits of the An Fafud desert in the north of Saudi Arabia near the Jordan border (Thomas et al. 1998).

Following this reasoning, Martínez-Navarro et al. (2007) recognize the existence of the following species in the lineage *Pelorovis s.s.*: *Bos turkanensis* for the Late Pliocene forms of Africa, *Bos oldowayensis* for the Early Pleistocene forms of Africa and the Levantine Corridor, and *Bos primigenius* for the Middle Pleistocene-Recent forms of Eurasia. The classification of the Early-Middle Pleistocene transition specimens from Africa and the Levantine Corridor remains open until new and definitive material is found.

Bos oldowayensis has never been found in other regions of Eurasia outside the Levant, but its arrival into this area of western Asia, together with other African ruminants of open environments, suggests the extension of the African savannas into the middle latitudes in connection with gallery forest environments during Early Pleistocene times, as indicated by other species as well. In any case, during the Early Pleistocene this African lineage arrived as far as the Levantine Corridor, but it only colonized other regions of Eurasia later, during the Middle Pleistocene. Martínez-Navarro et al. (2007) also suggest a parallelism between of the dispersal of the Acheulean culture into Eurasia and the dispersal of this lineage of Bovini, because they are found together in 'Ubeidiya at ~1.4 Ma, at GBY at 0.8–0.9 Ma and finally in Europe at 0.5–0.6 Ma.

Kolpochoerus olduvaiensis

A significant finding in the Levant was the presence of African pigs, specifically the genus *Kolpochoerus*, the ancestor of the extant *Hylochoerus*, the forest hog. It is found in 'Ubeidiya, where it is cited as *K. olduvaiensis* (Geraads et al. 1986), and in Evron Quarry, cited as *K. evronensis* (Tchernov et al. 1994). In North Africa, this genus is well known and cited as *K. phacochoeroides* in several sites, including Ain Hanech, Algeria (Sahnouni et al. 2002) and Ahl al Oughlam, Morocco (Geraads 1993, 2004).

The specimens from Evron were ascribed previously to *Metridiochoerus* (Haas 1970), but they clearly correspond to *Kolpochoerus* (Geraads et al. 1986). The genus

[6]The Late Pleistocene giant African buffalo which is included in the name *Pelorovis sensu lato*, named *Bubalus antiquus* (Duvernois 1851), *Homoïoceras antiquus* (Bate 1949) and *Pelorovis antiquus* (Gentry and Gentry 1978), has a different anatomy in the skull and postcranial skeleton from that of *Pelorovis,* and has been included within the genus *Syncerus*, as *S. antiquus* (see Klein 1994; Hadjouis 2002, or Hadjouis and Sahnouni 2006, and references therein).

Metridiochoerus, ancestor of the wart hog *Phacochoerus*, is also cited in northern Africa at the late Early Pleistocene site of Ternifine in Algeria.

The African suids are basically characterized by the development of the zygomatic arch and the development of a complicated talonid in the third molars, with several pairs of cuspids. The enamel is also very thick.

The genus *Kolpochoerus* has been interpreted as an African form evolved from an Asian immigrant during the Middle Pliocene (Harris and White 1979; Harris 1983; Pickford 1994). The oldest recognized species of the genus was *K. afarensis* from Hadar (Cooke 1978), but the study of the Early and Middle Pliocene suid material from Ethiopia and Chad determined the presence of a primitive species which gave rise to the *Kolpochoerus* lineage on the African continent, *K. deheinzelini* (Brunet and White 2000). It evolved from the Late Miocene Asian species *Propotamochoerus hysudricus*. *Kolpochoerus deheinzelini* gave rise to *K. cookei*, a small-sized and hypsodont species which is only found in Hadar, and to *K. afarensis*, which gave rise to *K. majus* and to *K. limnetes-K. olduvaiensis* (Brunet and White 2000). The last representatives of *Kolpochoerus* in Africa are *K. olduvaiensis* in Olduvai Bed IV, Tanzania (0.78 Ma) and *K. majus* in Asbole, Lower Awash (0.6–0.8 Ma), and in Bodo, Middle Awash (0.6 Ma), both in Ethiopia (White 1995; Geraads et al. 2004). In the Early Pleistocene an evolved form of *Kolpochoerus olduvaiensis* dispersed into the Levant, but its record in other regions of Asia is questioned (Martínez-Navarro et al. in prep. B).

Hippopotamus antiquus

Although the species cited in 'Ubeidiya are *Hippopotamus gorgops* and *H. behemoth* (Faure 1986), this last taxon ascribed to an endemic form, all of the specimens published from the site fall within the range of variation of an advanced form of *Hippopotamus gorgops*, similar to that found at the late Early Pleistocene site of Buia, Eritrea (Martínez-Navarro et al. 2004b). *Hippopotamus gorgops* has never been described in areas north of the Taurus–Zagros Mountain range, but this species is the sister taxon of the Early Pleistocene European form *Hippopotamus antiquus*.

At the base of the Early Pleistocene, a large form of the African megaherbivore genus *Hippopotamus*, related to *H. gorgops*, penetrated into western Eurasia. It gave rise to the northern species *H. antiquus*. This species has been described from the latest Pliocene of the Italian Upper Valdarno at the Tasso Faunal Unit (Gliozzi et al. 1997), although a new reinterpretation of this finding suggests that the hippopotamus described from this site by Nesti at the beginning of the nineteenth century was found in younger deposits (Napoleone et al. 2003). If so, the oldest record of this species in Europe is found at the site of Venta Micena in Spain at around 1.5–1.6 Ma (Alberdi and Ruiz-Bustos 1985; Martínez-Navarro et al. 2004a,c). Later, this species is found in most of the European and western Asian Early Pleistocene faunal assemblages until the cold climatic change that begins the Middle Pleistocene. During the interglacials of the Middle Pleistocene a fossil form of the extant *Hippopotamus amphibius* is also recorded in southern Europe.

The presence of hippos in Europe during the Early Pleistocene is very informative; they need abundant water and temperate climates for surviving. These are similar ecological necessities as early hominins probably needed. During the Early Pleistocene, large hippos are confined to Africa, the Levantine Corridor and Europe. The movements of these fossil hippos were different from the other terrestrial large mammals because of their capability of crossing long stretches of water. They are not present in Arabia, Central, Eastern and Southern Asia. Their nonappearance in Central and Eastern Asia is related to the absence of water ways between Europe and these oriental regions, and their absence of any record in the Indian subcontinent probably has to be interpreted as due to the difficulties of crossing the Arabian desert and maybe because large hippos forms couldn't compete with the local small endemic species of *Hexaprotodon*, that survive in this region until the Late Pleistocene (see Dennell 2005).

Hippopotamus antiquus, like *H. gorgops*, is a large species, more than two times bigger than the extant *H. amphibius*. Its size, calculated from the distal transversal diameter of the humerus, following the methodology of Damuth and McFadden (1990) and using the samples from the sites of Untermassfeld, Germany (Kahlke 1997), and Incarcal, Spain (Galobart et al. 2003), falls in the range of 3,500–4,200 kg (Martínez-Navarro et al. 2004a), and the average for the extant *H. amphibius* is around 1,500 kg. Also, in a biogeochemical study of the trophic behavior of the community of large mammals from the site of Venta Micena (Spain), the values of $\delta^{15}N$ for *Hippopotamus antiquus* were higher than those of the other herbivores and also higher than those for the machairodonts *Homotherium latidens* and *Megantereon whitei*. This means that *H. antiquus* was eating only aquatic plants which do not fix the atmospheric N_2 as the terrestrial plants do indicating that the species was exclusively aquatic (Palmqvist et al. 2003; Martínez-Navarro et al. 2004a), and not amphibious as is the extant species of the genus, which goes outside water for foraging during the night.

Hippos are common elements in most of the Late Pliocene and Pleistocene African archeological sites, but they also are common in the Levantine Corridor, Anatolia, Georgia[7]

[7]At the site of Akhalkalaki (Georgia), Vekua (1976) described the form *Hippopotamus georgicus*, which is a junior synonym of *H. antiquus*.

and most of the European Early Pleistocene sites. Because of their ecological and climatological necessities, the presence of hippos in Europe always suggests the possibility of finding evidence of human presence, as it is common in several sites, such as Fuente Nueva-3 and Barranco León-5 in Orce, Spain (Martínez-Navarro et al. 2004c).

Aquatic large megaherbivores that probably never or only sporadically go outside water, such *Hippopotamus antiquus*, are not dangerous for hominins – who live on land – and they are easier to hunt using big stones when they are in swamps close to the riverine margins than are other terrestrial large mammals that can run and also be very aggressive, such as bison, rhinos, elephants, horses, etc. For this reason, hippos could be part of the diet of carnivorous hominins. In Africa, hominins also ate other aquatic animals like crocodiles, which are easy to hunt outside water, as is indicated in the site of Buia, Eritrea (Fiore et al. 2004), and in some other localities. During the Early Pleistocene of Europe, *H. antiquus* is found together with pre-Acheulean tools.

Megantereon whitei

A study of the fauna from Venta Micena, Spain (Martínez-Navarro 1991, 1992a, b) revealed the presence of a sabertoothed tiger of African origin belonging to the genus *Megantereon*, and differing from the Late Pliocene Eurasian form *Megantereon cultridens*. The specimens from Venta Micena were classified as *M. whitei* and this species was also noted at Dmanisi at the other longitudinal limit of the Mediterranean Basin (Martínez-Navarro and Palmqvist 1995). *Megantereon whitei* has also been documented at the Greek site of Apollonia-1 (Martínez-Navarro and Palmqvist 1996), the Italian site of Pirro Nord (Rook et al. 2004), and it is probably present at Untermassfeld (Germany), where it has been classified as *Megantereon cultridens adroveri* (Hemmer 2001), a synonym of *M. whitei* that was first described at the site of Venta Micena (Pons-Moyà 1987; Martínez-Navarro and Palmqvist 1995). A form of *Megantereon* is also recorded at 'Ubeidiya (Ballesio 1986; Martínez-Navarro et al., submitted), as well as in extreme southeast Asia, in Java (Indonesia) (Kurtén 1962; De Vos and Aziz 1987), together with *Pachycrocuta brevirostris* (Geraads 1979).

The origin of the genus *Megantereon* is controversial (Lewis and Werdelin 2010). Berta and Galiano (1983) proposed an Early Pliocene North American origin. Werdelin and Lewis (2000, 2002) described a new species, *Megantereon ekidoit*, dating to 3.5 Ma from the South Turkwell hominin site in the Turkana Basin. We consider that this species is a primitive form of *M. whitei* (Palmqvist 2002; Palmqvist et al. 2007), and differs from the Eurasian Late Pliocene form *Meganteron cultridens*, for which the oldest record in Europe is at the site of Villarroya (Spain) at around 3.0 Ma. At the Plio-Pleistocene boundary, *M. whitei* replaced *M. cultridens* in Eurasia (Martínez-Navarro and Palmqvist 1995).

The principal differences between *M. whitei* and *M. cultridens* are based on the reduction of the length of the palate and the mandible, with smaller premolars, the presence of a diastema between the lower third and fourth premolars, and proportionally longer canines in the African species. In general aspects, *M. whitei* is a more specialized hypercarnivorous predator, with a smaller size and more developed hunting capabilities than *Megantereon cultridens*. Probably, the arrival into Eurasia of this specialized African form provoked the rapid extinction of its competitor *M. cultridens* (Martínez-Navarro and Palmqvist 1995, 1996; Palmqvist et al. 2007).

Together with *Hippopotamus antiquus*, *M. whitei* is the best-known species of African origin from the Eurasian Early Pleistocene. It is found in the eastern and western northern Mediterranean areas together with the earliest evidences of human presence outside of Africa (Bar-Yosef and Goren-Inbar 1993; Martínez-Navarro et al. 1997; Oms et al. 2000; Gabunia et al. 2000; Vekua et al. 2002; Belmaker et al. 2002; Lordkipanidze et al. 2005, 2007). In East and South Africa, *M. whitei,* although it is not abundant in the fossil record, sometimes is also found in association with hominins in Middle-Late Pliocene and Early Pleistocene sites (Leakey 1976; Howell and Petter 1976; Turner 1987; Martínez-Navarro and Palmqvist 1995; Werdelin and Lewis 2000, 2002).

The African and Eurasian record of *M. whitei* shows that this species evolved in Africa during the Late Pliocene and dispersed at the Plio-Pleistocene boundary into Eurasia through the Levantine Corridor. Its last record is around 1.1 Ma in Europe and Africa at the sites of Untermassfeld and Swartkrans, respectively.

Megantereon whitei was a super-predator felid which inhabited mixed habitats, had powerful forelimbs, elongated and non-crenulated upper canines, and a short mandible. It was well-adapted to hunt but its masticatory structure only allowed it to eat the soft parts of its prey, leaving most of the carcasses intact for scavengers (Martínez-Navarro and Palmqvist 1996; Palmqvist et al. 2007), especially for the large hyaena *Pachycrocuta brevirostris*, but probably also for hominins. Dmanisi is the oldest site outside Africa with hominins but also with *Megantereon whitei*.

Although another genus of sabertoothed tiger, *Homotherium*, is also found in the Late Pliocene and Early Pleistocene of Eurasia and Africa, the systematics and connection of the African and Eurasian forms is not clear at the moment. Also there is another form, the false sabertoothed tiger *Dinofelis*, which is very well known in the Pliocene and Early Pleistocene of Africa, and it is also recorded in the Pliocene of Eurasia, but at the moment it has been never recorded in the Eurasian Pleistocene (Werdelin and Lewis 2001).

Theropithecus oswaldi

This species is a giant grazer and granivorous cercopithecid. The first record of this taxon outside Africa, in Asia, was at the late Early Pleistocene site of Mirzapur (India), situated in the Lower Boulder Conglomerate (Gupta and Sahni 1981; Delson 1993; Pickford 1993) dated by paleomagnetism at 1.0 Ma (Azzaroli and Napoleone 1982). At the beginning of the 1990s, fossil teeth of the large-sized African cercopithecid *Theropithecus* cf. *oswaldi* were uncovered in association with *P. brevirostris* at the karstic site of Cueva Victoria in Cartagena, southeast Spain (Gibert et al. 1995) (dated ~1.0 Ma). Recently, another fossil has been ascribed to this species in the same site, a middle phalanx previously misinterpreted as *Homo* (Martínez-Navarro et al. 2005, 2008). Later, the presence of *Theropithecus* was documented in the Early Pleistocene of the Levant based on the finding of a calcaneus at the site of 'Ubeidiya (Belmaker 2002), dated ~1.4 Ma. Finally, based on the finding of three cervical vertebrae, this species has been identified at the site of Pirro Nord in Italy, 1.3–1.6 Ma (Rook et al. 2004), with similar fauna of arvicolids (*Allophaiomys ruffoi,* synonym of *A. pliocaenicus*) and large mammals, as the site of Venta Micena, dated around 1.5–1.6 Ma. Although the attribution of this material from Pirro Nord to *Theropithecus* has recently been questioned by Patel et al. (2007), who argue that the identification is not conclusive, and the vertebrae might belong to *Paradolicopithecus*, a large late Pliocene monkey found in Europe and Central Asia (see above). However, new anatomical, but especially biochronologic arguments (Rook and Martínez-Navarro, in preparation) suggest that it is probably correct. *Theropithecus* is the only large monkey found in the Early Pleistocene of Europe and the Levantine Corridor. *Paradolichopithecus*, the large Late Pliocene monkey found in Europe and Asia is always found with an older faunal assemblage than that of Pirro Nord. The record of species of this Italian locality includes *Megantereon whitei, Pachycrocuta brevirostris, Lycaon lycaonoides, Canis mosbachensis, Bison* and others. This is a typical Early Pleistocene assemblage and is very similar to that of Cueva Victoria, and most of these species are also recorded at 'Ubeidiya.

This large monkey, *Theropithecus oswaldi,* is also found in the late Early Pleistocene North African site of Ternifine (Delson and Hoffstetter 1993). This genus of cercopithecid is well known from the beginning of the Pliocene in the majority of African faunal assemblages associated with hominins. Both *Theropithecus* and hominins have their origin in East Africa, and their evolution and dispersal are parallel until the Middle Pleistocene (Pickford 1993), when *Theropithecus* became restricted to a single species of small size, *T. gelada*, that now only survives in the mountains of Ethiopia. *Theropithecus oswaldi* evolved from *T. darti*, and the members of this lineage are commonly found in African assemblages from the beginning of the Late Pliocene (3.3 Ma) until the Middle Pleistocene (0.5 Ma) (Delson 1993; Leakey 1993; Pickford 1993). In the Late Pliocene, a smaller species of *Theropithecus* monkey (*Theropithecus atlanticus*) is known in North Africa, where it is well-documented at the interesting site of Ahl al Oughlam (Casablanca, Morocco) (Alemseged and Geraads 1998). *T. atlanticus* is an evolved branch of *T. darti*, and it is apparently a North African Late Pliocene isolated species. Hughes et al. (2008), using the Stepping Out cellular automata model to explain the possible dispersal of *Theropithecus oswaldi* outside Africa, suggest that in order to reach Europe this species needed to be tolerant of a wide variety of habitats.

Early Pleistocene Eurasian Species in Africa

A few large mammal species of Eurasian origin penetrated into Africa during the Early Pleistocene, especially into North Africa. An example of considerable interest is the entrance into Africa, during the Plio-Pleistocene transition, of the widely distributed Palearctic carnivore *Canis* (*Xenocyon*) ex gr. *falconeri* (Rook 1994), a synonym of *Lycaon lycaonoides* (Martínez-Navarro and Rook 2003), the ancestor of the extant tetradactyl painted dog *Lycaon pictus*. The fossil form *L. lycaonoides* is known in Ain Hanech, where it has been referred to *Canis atrox* (Arambourg 1979), but also in the East and the South of the continent, including Olduvai Beds I and II (Ewer 1965), and Kromdraai A (Turner 1986).

During all of the Early Pleistocene, the species *Lycaon lycaonoides* also persists throughout all of Eurasia, including some archeological assemblages such 'Ubeidiya in Israel, Fuente Nueva-3 and Barranco León-5 in Spain, or Vallonnet in France.

The extant lycaons are the most social large carnivores in the world and the most efficient predators of Africa. The groups are composed of social family clans that help all the members and permit the survival of pathological individuals until advanced ages. This behavior has also been described in fossil lycaons at the site of Venta Micena (Palmqvist et al. 1999; Martínez-Navarro and Rook 2003). Lycaons eat meat but do not fracture bones, leaving intact the marrow that can be eaten by large bone-cracking scavengers, such *P. brevirostris* and hominins.

Discussion

On the basis of this study, it is possible to see the great confusion that exists with regard to the systematics of the Late Pliocene and Early Pleistocene large mammals from Eurasia, the Near East and North Africa (see also Belmaker 2010; Lewis and Werdelin 2010). Direct study of the collections from many localities (see Fig. 13.1) has revealed a large number of existing synonyms that are now under revision.

Although there are a few Ethiopian species in Dmanisi, Venta Micena, Cueva Victoria and other Early Pleistocene European deposits such as Pirro Nord in Italy (De Giuli et al. 1987; Rook et al. 2004; Arzarello et al. 2006), Apollonia in Greece (Koufos and Kostopoulos 1997; Kostopoulos 1997), and Untermassfeld in Germany (Kahlke 1997, 2001a, b), the dominant fauna is Holarctic; in North Africa, although a few species are Holarctic, the dominant fauna is Ethiopian; and in the Near East, Holarctic and Ethiopian faunas are well represented and mixed. Although the possibility of faunal interchange during Early Pleistocene times through the Gibraltar Straits, between the Iberian Peninsula and Morocco, remains open (see also Lahr 2010), these data suggest that the faunal dispersals between these continents occurred via the Levantine Corridor.

The biochronological chart of selected large mammals in Europe and the Near East (Fig. 13.2) shows the relative age of some of the best known sites of these regions. The figure shows the important replacement of fauna at the Plio-Pleistocene boundary, which is related to the transition from the cold to the warm period detected at around 1.75 Ma. At this moment hominins arrive in Eurasia. The final Pliocene cold climate provoked an impoverishment of the European fauna (Gliozzi et al. 1997). Most of the new Early Pleistocene species in Europe have an Holarctic origin, and some of them arrived from Africa, although most of the latter never crossed the Taurus–Zagros Mountains during the Early Pleistocene and are only recorded in the Levantine area, including *Equus* cf. *tabeti*, *Pelorovis oldowayensis, Oryx* cf. *gazella* and other ruminants.

White (1995) related the evolution of African omnivores to global climatic changes, based on the timing of the first and last appearance data of the different hominin and suid species. The genus *Theropithecus* probably has be included in this group, as it is also found in most of the African Plio-Pleistocene assemblages with hominins and suids (Pickford 1993). Hominins, *Theropithecus* and suids have parallel evolutionary histories in Africa but, in light of the finding of *Theropithecus* in the Middle East, India, Italy and Spain, and given the discovery of *Kolpochoerus* in the Middle East, we must suppose that they also have parallel dispersals during the Early Pleistocene.

The Early Pleistocene dispersal of fauna and hominins out of Africa is related to changes in climate and to modes of food resource exploitation as well. Late Pliocene African hominins are characterized by an increase in the thickness of their enamel which appears after the change of ecological conditions toward greater aridity at 2.5 Ma (Ramirez-Rozzi et al. 1999). The change in enamel thickness probably reflects an adaptation to more fibrous and abrasive plant foods. A similar and parallel process is detected in *Kolpochoerus* – increase in the size of the third molar of *K. limmetes* (Harris and White 1979) and in enamel thickness, 2.7 mm in the M_3 of *Kolpochoerus* from Evron – and in *Theropithecus oswaldi* – 1.5 mm enamel thickness in the molars from East African specimens and only 1.16 mm in the more folivorous Late Pliocene species *T. brumpti* (Benefit 1999). At the Early-Middle Pleistocene Acheulean site of Gesher Benot Ya'aqov (0.7–0.8 Ma BP), the exploitation of acorns, chestnuts, and other nuts by hominins has also been documented (Goren-Inbar et al. 2002). In addition, this is a typical food eaten by other omnivores, such as pigs (i.e., *Kolpochoerus*). Although *Theropithecus oswaldi* is a classical granivorous species maybe it was also possible for this species to eat nuts.

In clear competition with hyaenas, especially the large *Pachycrocuta brevirostris*, another resource for hominins is carrion, most of it left by large carnivores, especially *Megantereon whitei,* a non-cursorial flesh-eater adapted to mixed habitats. This saber-toothed tiger has elongated and non-crenulated upper canines, a short mandible and powerful forelimbs. It is well-adapted to hunt but its masticatory structure only allows it to eat the soft parts of its prey, leaving intact most parts of the carcasses (Martínez-Navarro and Palmqvist 1996; Palmqvist et al. 2007). Other flesh-eaters and non-bone-cracking predators, like *Lycaon lycaonoides* or *Homotherium latidens*, could also be part of this scenario, producing partial carcasses for scavenging. *Kolpochoerus* and possibly *Theropithecus* could also have been opportunistic scavengers, but this is unlikely to have been a systematic behavior in these species as it was in *Pachycrocuta brevirostris* and more probably in *Homo* (Martínez-Navarro 2004).

After the discovery of the latest hominin skull and mandible in Dmanisi, belonging to an old individual lacking teeth, social behavior has been interpreted as crucial for hominin dispersal out of Africa (Lordkipanidze et al. 2005). However, we have to consider that another crucial characteristic for hominin dispersal is carnivorous trophic behavior (see also Lewis and Werdelin 2010; Potts and Teague, 2010). Herbivores, especially ruminants, are obligated to survive in conditioned climatic regions because they are specialized to eat specific vegetation directly related to the climate (determined by latitude), but carnivores eat meat, and it does not matter if it is from an equid, a bovid, a cervid, an elephant, a rhino, etc. Meat is meat and it is present everywhere. When hominins arrived into the middle latitudes of Eurasia, they were obligated to adapt themselves to seasonal climates with summers and winters, where it was not possible to find vegetable resources year round, especially in winter. Although they could store acorns and nuts for some periods, probably the best manner for survival was eating animal resources (Martínez-Navarro et al. 1998, 2004; Martínez-Navarro and Palmqvist 1999). Although the carnivorous behavior of hominins was probably developed during the Late Pliocene in tropical Africa, the dependence on meat was more important during the Early Pleistocene in Eurasia than in Africa. Unfortunately, no direct biogeochemical data are currently available to test this hypothesis.

Direct competition for the carrion between hominins and *Pachycrocuta brevirostris* has been detected at the site of Fuente Nueva-3. At the moment, we do not know how intense this rivalry was. Maybe it was lesser than we have considered, because the extant hyaena *Crocuta crocuta* normally acts during the night when it behaves as hunter, and it is more diurnal (but also nocturnal) when it acts as scavenger. We have to suppose that the Eurasian Early Pleistocene hominins normally were only diurnal having some moments during the day without direct interaction with this large size super scavenging hyaena, *P. brevirostris*.

Vegetarian behavior is enough for longitudinal dispersals, but carnivorous behavior is better for latitudinal dispersals.

Conclusions

At the Plio-Pleistocene transition, hominins arrived into Eurasia following the route of the Levantine Corridor. This Out-of-Africa dispersal of the genus *Homo* is related to general ecological conditions that permitted the dispersal of other large mammals, particularly *Hippopotamus antiquus*, *Megantereon whitei* and probably *Theropithecus oswaldi*, which was favored by the northward expansion of African mixed habitats – savannas and gallery forest – and their particular patterns of foraging (but see Belmaker 2010). These three species probably dispersed into Eurasia at the same moment as *Homo* during the Plio-Pleistocene transition (Rook et al. 2004; Martínez-Navarro 2004).

Hominins were able to disperse into the middle latitudes of Eurasia because of the development of social behavior as has been detected in Dmanisi (Lordkipanidze et al. 2005), but also because of the development of a systematic carnivorous trophic behavior, which was probably more necessary for survival in the seasonal climates with summers and winters around the 40th parallel than in the tropic climates (Martínez-Navarro et al. 1998; Martínez-Navarro and Palmqvist 1999).

Acknowledgments I want to thank R. E. Leakey, J. G. Fleagle, J. J. Shea and F. Grine for inviting me to participate in the second Workshop in Human Evolution at Stony Brook University. I also thank the Leakey Foundation, the CARE Archaeological Foundation, the Duques de Soria Foundation, the Italian CNR, the Museum National d'Histoire Naturelle (Paris), the Spanish Ministry of Science and Education (ref. CGL2006-13808-CO2-01), and the Consejería de Cultura (Junta de Andalucía, Spain), for helping me to study the European, Caucasian, Israeli, Indian, and African collections. I thank the late F. Clark Howell, O. Bar-Yosef, the late E. Tchernov, N. Goren-Inbar, H. de Lumley, E. Carbonell, L. Rook, A. Vekua, D. Lordkipanidze, M. R. Palombo, M. G. Leakey, E. Delson, R. Sardella. C. Petronio, G. Koufos, D. Kostopoulos, A. B. Shabel, E. Hovers, A. Currant, P.-E. Moullé, P. Tassy, R. Rabinovich, L. K. Horwitz, M. Belmaker, E. Cioppi, T. Deng, M.T. Alberdi, A. Galobart, P. Palmqvist, I. Toro, M. P. Espigares, S. Ros-Montoya, F. Burjachs and F. Lacombat, for helping me during the study of the different fossil collections and for critical discussions about the ideas written in this manuscript. Finally, I want to thank the corrections and suggestions made to the first version of the manuscript by J. G. Fleagle, E. Delson and two anonymous referees.

References

Abbazzi, L. (in press). The cervid fauna from Barranco León and Fuente Nueva 3. In: I. Toro, B. Martínez-Navarro. B., & J. Agustí (Eds.), El Pleistoceno inferior de Fuente Nueva-3 y Barranco León. Spain: Monografías Consejería de Cultura, Junta de Andalucía.

Alberdi, M. T., & Ruiz-Bustos, A. (1985). Descripción y significado bioestratigráfico del *Equus* e Hipopótamo en el yacimiento de Venta Micena (Granada). *Estudios Geológicos, 41*, 251–261.

Alberdi, M. T., Alonso, M. A., Azanza, B., Hoyos, M., & Morales, J. (2001). Vertebrate taphonomy in circum-lake environments: three cases in the Guadix-Baza Basin (Granada, Spain). *Palaeogeogrphy, Palaeoclimatology, Palaeoecology, 165*, 1–26.

Alemseged, Z., & Geraads, D. (1998). *Theropithecus atlanticus* (Thomas, 1884) (Primates: Cercopithecidae) from the late Pliocene of Ahl al Oughlam, Casablanca, Morocco. *Journal of Human Evolution, 34*, 609–621.

Argant, A. (2004). Les Carnivores du gisement Pliocène final de Saint-Vallier (Drôme, France). *Geobios, 37*, 133–182.

Arambourg, C. (1979). *Vertebres Villafranchiens d'Afrique du Nord (Artiodactyles, Carnivores, Primates, Reptiles, Oiseaux)*. Paris: Singer-Polignac.

Arribas, A., & Palmqvist, P. (1998). Taphonomy and palaeoecology of an assemblage of large mammals: hyaenid activity in the Lower Pleistocene site at Venta Micena (Orce, Guadix-Baza Basin, Granada, Spain). *Geobios, 31*, 3–47.

Arribas, A., Riquelme, J. A., Palmqvist, P., Garrido, G., Hernández, R., Laplana, C., et al. (2001). Un nuevo yacimiento de grandes mamíferos villafranquienses en la cuenca de Guadix-Baza (Granada): Fonelas P-1, primer registro de una fauna próxima al límite Plio-Pleistoceno en la Península Ibérica. *Boletín Geológico y Minero, 112*, 3–34.

Arzarello, A., Marcolini, F., Pavia, G., Pavia, M., Petronio, C., Petrucci, M., et al. (2006). Evidence of earliest human occupation in Europe:

the 1 site of Pirro Nord (Southern Italy). *Naturwissenschaften, 93*, 107–112.

Azzaroli, A. (1983). Quaternary mammals and the "End-Villafranchian" dispersal event – A turning point in the history of Eurasia. *Palaeogeography, Palaeoclimatology, Palaeoecology, 44*, 117–139.

Azzaroli, A., & Napoleone, G. (1982). Magnetostratigraphic investigation in the Upper Siwaliks near Pinjor, India. *Rivista Italiana di Paleontologia e Stratigrafia, 87*, 739–762.

Azzaroli, A., De Giuli, C., Ficcarelli, G., & Torre, D. (1988). Late Pliocene to early Mid-Pleistocene mammals in Eurasia: Faunal succession and dispersal events. *Palaeogeography, Palaeoclimatology, Palaeoecology, 66*, 77–100.

Ballesio, R. (1986). Les carnivores du Pléistocène d'Oubeidiyeh (Israël). In E. Tchernov (Ed.), *Les Mammifères du Pléistocène inférieur de la Vallée du Jordain a Oubeidiyeh* (Vol. 5, pp. 63–92). Paris: Mémoires et Travaux du Centre de Recherche Français de Jerusalem.

Bar-Yosef, O., & Goren-Inbar, N. (1993). The lithic assemblage of 'Ubeidiya. A Lower Paleolithic site in the Jordan Valley, Vol 34 (pp. 1–266). Monographs of the Institute of Archaeology, The Hebrew University of Jerusalem, Jerusalem: Qedem.

Bate, D. M. A. (1949). A new African fossil long-horned Buffalo. *Annales Magazine of Natural History, London, 2*, 396–398.

Belmaker, M. (2002). First evidence of the presece of *Theropithecus* sp. in the Southern Levant. *Israel Journal of Zoology, 48*, 165.

Belmaker, M. (2010). Early Pleistocene faunal connections between Africa and Eurasia: an ecological perspective. In: J. G. Fleagle, J. J. Shea, F. E. Grine, A. L. Baden, & R. E. Leakey (Eds.), *Out of Africa I: the first hominin colonization of Eurasia* (pp. 183–205). Dordrecht: Springer.

Belmaker, M., Tchernov, E., Condemi, S., & Bar-Yosef, O. (2002). New evidence for hominid presence in the Lower Pleistocene of the Southern Levant. *Journal of Human Evolution, 43*, 43–56.

Benefit, B. R. (1999). Biogeography, dietary specialization, and the diversification of African Plio-Pleistocene monkeys. In T. R. Bromage & F. Schrenk (Eds.), *African biogeography, climate change, and human evolution* (pp. 172–188). New York/Oxford: Oxford University Press.

Bermúdez de Castro, J. M., Arsuaga, J. U., Carbonell, E., Rosas, A., Martinez, I., & Mosquera, M. (1997). A hominid from the Lower Pleistocene of Atapuerca, Spain: possible ancestor to Neanderthals and modern humans. *Science, 276*, 1392–1395.

Berta, A., & Galiano, H. (1983). *Megantereon hesperus* from the late Hemphillian of Florida with remarks on the phylogenetic relationship of machairodonts (Mammalia, Felidae, Machairodontinae). *Journal of Paleontology, 57*, 892–899.

Bout, P. (1960). *Le Villafranchien du Velay et du Bassin hydrographique moyen et supérieur de l'Allier. Corrélations françaises et européennes*. Le Puy: Imprimerie Jeanne d'Arc. 344 p.

Brunet, M., & White, T. D. (2000). Deux nouvelles espèces de Suini (Mammalia, Suidae) du continent Africain (Éthiopie; Tchad). *Comptes Rendus Academie des Sciences, Paris, 332*, 51–57.

Brunet, M., Beauvilain, A., Coppens, Y., Heinz, E., Moutaye, A. H. E., & Pilbeam, D. (1995). The first australopithecine 2,500 kilometres west of the Rift Valley (Chad). *Nature, 378*, 273–275.

Bukhsianidze, M. (2005). *The Fossil Bovidae of Dmanisi*. Ph.D. International Doctorate "Environmental, Humans and Compartmental Dynamics" XVI cycle 2001–2004, University of Ferrara, Italy, 192 pp.

Carbonell, E., Bermúdez de Castro, J. M., Arsuaga, J. L., Allué, E., Bastir, M., Benito, A., et al. (2005). An Early Pleistocene hominin mandible from Atapuerca-TD6, Spain. *Proceedings of the National Academy of Sciences of the United States of America, 102*, 5674–5678.

Carbonell, E. B., de Castro, J. M., Parés, J. M., Pérez-González, A., Cuenca-Besós, G., Ollé, A., et al. (2008). The first hominin of Europe. *Nature, 452*, 465–469.

Cooke, H. B. S. (1978). Pliocene-Pleistocene Suidae from Hadar, Ethiopia. *Kirtlandia, 29*, 1–63.

Croitor, R. (2005). Early Pleistocene small-sized deer from the Mediterranean area. In H. de Lumley, F. Lacombat, & P.-E. Moullè (Eds.), *Abstracts of the colloque "Cadre Biostratigraphique de la fin du Pliocene et du Pleistocene inferieur (3Ma a 780 000 ans)"* (pp. 63–64). Tende, Alpes-Maritimes, France: Musée Dé partemental des Marveilles.

Damuth, J., & MacFadden, B. J. (1990). *Body size in mammalian paleobiology: Estimation and biological implications*. Cambridge: Cambridge University Press.

De Giuli, C., Masini, F., & Torre, D. (1987). The latest Villafranchian faunas in Italy: the Pirro Nord fauna (Apricena, Gargano). *Palaeontographia Italica, 74*, 51–62.

Delson, E. (1993). *Theropithecus* fossils from Africa and India and the taxonomy of the genus. In N. G. Jablonski (Ed.), Theropithecus: *The rise and fall of a primate genus* (pp. 157–189). Cambridge: Cambridge University Press.

Delson, E., & Hoffstetter, R. (1993). *Theropithecus* from Ternifine, Algeria. In N. G. Jablonski (Ed.), Theropithecus: *The rise and fall of a primate genus* (pp. 191–208). Cambridge: Cambridge University Press.

Dennell, R. W. (2005). Early Pleistocene hippopotamid extinctions, monsoonal climates, and river system histories in South and South West Asia: comment on Jablonski (2004) "The hippo's tale: how the anatomy and physiology of Late Neogene *Hexaprotodon* shed light on late Neogene environmental change". *Quaternary International, 117*, 119–123.

Dennell, R. W., Rendell, H. R., & Hailwood, E. (1988). Early toolmaking in Asia: two-million-year-old artifacts in Pakistan. *Antiquity, 62*, 98–106.

Dennell, R., Coard, R., & Turner, A. (2006). The biostratigraphy and magnetic polarity zonation of the Pabbi Hills, northern Pakistan: An Upper Siwalik (Pinjor Stage) Upper Pliocene-Lower Pleistocene fluvial sequence. *Palaeogeogrphy, Palaeoclimatology, Palaeoecology, 234*, 168–185.

De Vos, J., & Aziz, F. (1987). Note on two upper canines of *Megantereon* sp. (Mammalia, Felidae) from the Pleistocene of Java. *Proceedings, B90*, 57–63.

Dmitrieva, E. L. (1977). Tajikistan's and India's fossil Alcelaphinae. *Journal of the Palaeontological Society of India, 20*, 97–101.

Duvernois, G. L. (1851). Note sur une espèce fossile de Buffle (*Bubalus* (*Arni*) *antiquus*) découverte en Algérie. *Compte Rendu Hebdomadaire des Séances de l'Académie des Sciences de Paris, 33*, 595–597.

Eisenmann, V. (1983). Family Equidae. In J. M. Harris (Ed.), *Koobi Fora Research Project 2: The fossil ungulates: Proboscidea, Perissodactyla and Suidae* (pp. 156–214). Oxford: Clarendon.

Eisenmann, V. (1986). Les Equidés du Pléistocène d'Oubeidiyeh. In E. Tchernov (Ed.), *Les Mammifères du Pléistocène inférieur de la Vallée du Jordain a Oubeidiyeh* (Vol. 5, pp. 191–212). Paris: Mémoires et Travaux du Centre de Recherche Français de Jerusalem.

Ewer, R. F. (1956). The fossil carnivores of the Transvaal caves: Canidae. *Proceedings. Zoological Society of London, 126*, 97–119.

Ewer, R. F. (1965). Large carnivora. In L. S. B. Leakey (Ed.), *Olduvai Gorge, 1959–1961* (Vol. 1, pp. 109–123). Cambridge: Cambridge University Press.

Faure, M. (1986). Les Hippopotamidés du Pléistocène ancien d'Oubeidyeh (Israël). In E. Tchernov (Ed.), *Les Mammifères du Pléistocène inférieur de la Vallée du Jordain a Oubeidiyeh* (Vol. 5, pp. 107–142). Paris: Mémoires et Travaux du Centre de Recherche Français de Jerusalem.

Ficcarelli, G., Torre, D., & Turner, A. (1984). First evidence for a species of racoon dog, *Nyctereutes* TEMMINCK, 1838, in South-African Plio-Pleistocene deposits. *Bolletino della Societá Paleontologica Italiana, 23*, 125–130.

Fiore, I., Bondioli, L., Coppa, A., Macchiarelli, R., Russom, R., Kashay, H., et al. (2004). Taphonomic analysis of the late Early Pleistocene bone remains from Buia (Dandiero Basin, Danakil Depression,

Eritrea): evidence for large mammal and reptile butchering. *Rivista Italiana Paleontologia e Stratigrafia, 110* (Suppl.), 89–97.

Gabunia, L., Vekua, A., Lordkipanidze, D., Swisher, C., III, Ferrign, R., Justus, A., et al. (2000). Earliest Pleistocene hominid cranial remains from Dmanisi, Republic of Georgia: taxonomy, geological setting, and age. *Science, 288*, 1019–1025.

Galobart, A., Ros, X., Maroto, Jy, & Vila, B. (2003). Descripción del material de hipopótamo (*Hippopotamus antiquus* Desmarest, 1822) de los yacimientos del Pleistoceno inferior de Incarcal (Girona, NE de la Península Ibérica). *Paleontología i Evolució, 34*, 153–173.

García, N. (2003). Osos y otros carnívoros de la Sierra de Atapuerca (560 p). Asturias: Fundación Oso de Asturias.

Gentry, A. W. (1967). *Pelorovis oldowayensis* Reck, an extinct bovid from East Africa. Bulletin of the British Museum (Natural History). *Geology, 14*, 245–299.

Gentry, A. W. (2006). A new bovine (Bovidae, Artiodactyla) from the Hadar Formation, Ethiopia. *Transactions of the Royal Society of South Africa, 61*(2), 41–50.

Gentry, A. W., & Gentry, A. (1978). Fossil Bovidae (Mammalia) of Olduvai Gorge, Tanzania. Part II. Bulletin of the British Museum (Natural History). *Geology, 30*, 1–83.

Geraads, D. (1979). Nouvelles donnés sur *Hyena brevirostris bathygnata* DUBOIS (Carnivora, Mammalia) du Pléistocène de Java. *Comptes Rendus Somm. Societé. Géologique France, 2*, 80–82.

Geraads, D. (1986). Les ruminants du Pléistocène d'Oubeidiyeh (Israël). In E. Tchernov (Ed.), *Les Mammifères du Pléistocène inférieur de la Vallée du Jordain a Oubeidiyeh* (Vol. 5, pp. 143–181). Paris: Mémoires et Travaux du Centre de Recherche Français de Jerusalem.

Geraads, D. (1993). *Kolpochoerus phacochoeroides* (Thomas, 1884) (Suidae, Mammalia) du Pliocène supérieur de Ahl al Oughlam (Casablanca, Maroc). *Geobios, 26*, 731–743.

Geraads, D. (1997). Carnivores du Pliocéne terminal de Ahl al Oughlam (Casablanca, Maroc). *Geobios, 30*, 127–164.

Geraads, D. (2004). New skulls of *Kolpochoerus phacochoeroides* (Suidae: Mammalia) from the late Pliocene of Ahl al Oughlam, Morocco. *Palaeontologica Africana, 40*, 64–83.

Geraads, D., & Tchernov, E. (1983). Fémurs humains du Pléistocène moyen de Gesher Benot Ya'acov (Israel). *L'Anthropologie, 87*, 138–141.

Geraads, D., Guérin, C., & Faure, M. (1986). Les suidés du Pléistocéne ancien d'Oubeidiyeh (Israël). In E. Tchernov (Ed.), *Les Mammifères du Pléistocène inférieur de la Vallée du Jordain a Oubeidiyeh* (Vol. 5, pp. 93–105). Paris: Mémoires et Travaux du Centre de Recherche Français de Jerusalem.

Geraads, D., Alemseged, Z., Reed, D., Wynn, J., & Roman, D. C. (2004). The Pleistocene fauna (other than Primates) from Asbole, lower Awash Valley, Ethiopia, and its environmental and biochronological implications. *Geobios, 37*, 697–718.

Gibert, J., Ribot, F., Gibert, L., Leakey, M. G., Arribas, A., & Martínez-Navarro, B. (1995). Presence of the cercopithecid genus *Theropithecus* in Cueva Victoria (Murcia, Spain). *Journal of Human Evolution, 28*, 487–493.

Gliozzi, E., Abbazzi, L., Argenti, P., Azzaroli, A., Caloi, L., Capasso Barbato, L., et al. (1997). Biochronology of selected mammals, molluscs and ostracods from the Middle Pliocene to the Late Pleistocene in Italy. The state of the art. *Rivista Italiana di Paleontologia e Stratigrafia, 103*, 369–388.

Goren-Inbar, N., Sharon, G., Melamed, Y., & Kislev, M. (2002). Nuts, nut cracking, and pitted stones at Gesher Benot Ya'aqov. *Proceedings of the National Academy of Sciences of the United States of America, 99*, 2455–2460.

Gupta, V. J., & Sahni, A. (1981). *Theropithecus delsoni*, a new cercopithecine species from the Upper Siwaliks of India. *Bulletin of the Indian Geological Association, 14*, 69–71.

Guérin, C., & Faure, M. (1997). The Wild Boar (*Sus scrofa priscus*) from the Post-Villafranchian Lower Pleistocene of Untermassfeld. In: Kahlke, R.D. (Ed.), Das Pleistozän von Untermassfeld bei Meiningen (Thüringen). *Römisch-Germanisches Zentralmuseum, 1*, 375–384.

Haas, G. (1970). *Metridichoerus evronensis* n. sp., a new Middle Pleistocene phacochoeroid from Israel. *Israel Journal of Zoology, 19*, 179–181.

Hadjouis, D. (2002). A new Bovini in the Upper Pleistocene fauna of Algeria. *L'Anthropologie, 106*, 377–386.

Hadjouis, D., & Sahnouni, M. (2006). *Pelorovis howelli* nov. sp. (Mammalia, Artiodactyla): a new bovine from the Lower Pleistocene site of Aïn Hanech (El-Kherba locus), Northeastern Algeria. *Geobios, 39*, 673–678.

Harris, J.M. (1983). Family Suidae. In, Harris, J.M. (Ed.), *Koobi Fora Research Project, Vol 2, The Fossil Ungulates: Proboscidea, Perissodactyla, and Suidae*. Chapter 6 (pp. 215–302). Oxford: Clarendon.

Harris, J. M. (1991). Family Bovidae. In J. M. Harris (Ed.), *Koobi Fora Research Project, vol 3, The fossil ungulates: Geology, fossil Artiodactyls and paleoenvinronments* (pp. 139–320). Oxford: Clarendon.

Harris, J. M., & White, T. D. (1979). Evolution of the Plio-Pleistocene African Suidae. *Transactions of the American Philosophical Society, 69*, 1–128.

Harris, J. M., Brown, F. H., & Leakey, M. G. (1988). Stratigraphy and paleontology of Pliocene and Pleistocene localities West of Lake Turkana, Kenya. *Contributions in Science, Natural History Museum of Los Angeles County, 399*, 1–128.

Heintz H., Guérin C., Martin R., Prat F., 1974. Principaux gisements villafranchiens de France : liste faunique et biostratigraphie. *Mémoire du B.R.G.M.* (Vème congrès du Néogène méditerranéen, Lyon, 1971, 1, Paris), 78, 169–182.

Hemmer, H. (2001). Die Feliden aus dem Epivillafranchium von Untermassfeld. In R. D. Kahlke (Ed.), *Das Pleistozän von Untermassfeld bei Meiningen (Thüringen)* (Vol. 3, pp. 699–782). Bonn: Dr. Rudolf Habelt GMBH.

Hooijer, D.A. (1958). An Early Pleistocene mammalian fauna from Bethlehem. *Bulletin of the British Museum (Natural History) Geology, 3*, 267.

Howell, F. C., & Petter, G. (1976). Carnivora from the Omo Group Formations, Southern Ethiopia. In Y. Coppens, F. C. Howell, G. L. Isaac, & R. C. Leakey (Eds.), *Earliest man and environments in the Lake Rudolf Basin* (pp. 314–331). Chicago: University of Chicago Press.

Howell, F. C., & Petter, G. (1980). The *Pachycrocuta* and *Hyaena* lineages (Plio-Pleistocene and extant species of the Hyaenidae); their relationships with Miocene ictitheres: *Palhyaena* and *Hyaenictitherium*. *Geobios, 13*, 579–623.

Hughes, J. K., Elton, S., & O'Regan, H. J. (2008). *Theropithecus* and "Out of Africa" dispersal in the Plio-Pleistocene. *Journal of Human Evolution, 54*, 43–77.

Kalb, J. E. (1995). Fossils elephantoids, Awash paleolake basins, and the Afar triple junction. *Palaeogeogfraphy, Palaelclimatology, Palaeoecology, 114*, 357–68.

Kahlke, H.-D. (1995). Neue Cerviden-Reste der Stránská Skála bei Brno. In R. Musil (Ed.), *Stránská Skála Hill. Excavation of open-air sediments 1964–1972* (pp. 85–91). Brno Anthropos Series 26. Brno: Moravian Museum.

Kahlke, R. D. (Ed.). (1997). *Das Pleistozän von Untermassfeld bei Meiningen (Thüringen) (Vol 1, pp. 1–418)*. Mainz, Germany: Römisch-Germanisches Zentralmuseum.

Kahlke, R. D. (Ed.). (2001a). *Das Pleistozän von Untermassfeld bei Meiningen (Thüringen) (Vol. 2, pp. 419–698)*. Mainz, Germany: Römisch-Germanisches Zentralmuseum.

Kahlke, R. D. (Ed.). (2001b). *Das Pleistozän von Untermassfeld bei Meiningen (Thüringen) (Vol. 3, pp. 699–1030)*. Mainz, Germany: Römisch-Germanisches Zentralmuseum.

Klein, R. (1994). The long-horned African buffalo (*Pelorovis antiquus*) is an extinct species. *Journal of Archaeological Science, 21*, 725–733.

Kostopoulos, D. (1997). The Plio-Pleistocene artiodactyls (Vertebrata, Mammalia) of Macedonia; 1, The fossileferous site "Apollonia-1", Mygdonia Basin of Greece. *Geodiversitas, 19*, 845–875.

Koufos, G., & Kostopoulos, D. (1997). New carnivore material from the Plio-Pleistocene of Macedonia (Greece) with the description of a new canid. *Münchner Geowiss. Abh. (A), 34*, 33–63.

Kurtén, B. (1962). The sabre-toothed cat *Megantereon* from the Pleistocene of Java. *Zoologische Mededelingen, Uitgegeven Door Het Rijksmuseum van Natuurlijke Historie the Leiden Deel XXXVIII, 6*, 101–104.

Kurtén, B., & Crusafont, M. (1977). Villafranchian carnivores (Mammalia) from La Puebla de Valverde (Teruel, Spain). *Commentationes Biologicae, 85*, 1–39.

Lacombat, F., Abbazzi, L., Ferretti, M. P., Martínez-Navarro, B., Moulle, P.-E., Palombo, M. R., et al. (2008). New data on the Early Villafranchian fauna from Vialette (Haute-Loire, France) based on the collection of the Crozatier Museum (Le Puy-en-Velay, Haute-Loire, France). *Quaternary International, 179*, 64–71.

Lahr, M.M. (2010). Saharan corridors and their role in the evolutionary geography of 'Out of Africa I'. In J. G. Fleagle, J. J. Shea, F. E. Grine, A. L. Baden, & R. E. Leakey (Eds.), *Out of Africa I: the first hominin colonization of Eurasia* (pp. 27–46). Dordrecht: Springer.

Leakey, M. G. (1976). Carnivora of the Lake Rudolf succession. In Y. Coppens, F. C. Howell, G. L. Isaac, & R. C. Leakey (Eds.), *Earliest man and environments in the Lake Rudolf Basin* (pp. 302–313). Chicago: University of Chicago Press.

Leakey, M. G. (1993). Evolution of the *Theropithecus* in the Turkana Basin. In N. G. Jablonski (Ed.), Theropithecus. *The rise and fall of a primate genus* (pp. 85–123). Cambridge: Cambridge University Press.

Lewis, M.E., & Werdelin, L. (2010). Carnivoran dispersal out of Africa during the Early Pleistocene: relevance for hominins? In: J. G. Fleagle, J. J. Shea, F. E. Grine, A. L. Baden, & R. E. Leakey (Eds.), *Out of Africa I: the first hominin colonization of Eurasia* (pp. 13–26). Dordrecht: Springer.

Lindsay, E. H., Opdyke, N. D., & Johnson, N. M. (1980). Pliocene dispersal of the horse *Equus* and Late Cenozoic mammalian dispersal events. *Nature, 287*, 135–138.

Lister, A. M. (2004). Ecological interactions of Elephantids in Pleistocene Eurasia: *Palaeoloxodon* and *Mammuthus*. In N. Goren-Inbar & J. D. Speth (Eds.), *Human Paleoecology in the Levantine Corridor* (pp. 53–60). Oxford: Oxbow Books.

Lister, A. M., & Sher, A. V. (2001). The origin of the Wooly Mammoth. *Science, 294*, 1094–1097.

Lister, A.M., & van Essen, H. (2003). *Mammuthus rumanus* (Stephanescu), the earliest mammoth in Europe. In: A. Petculescu, & E. Stiuca (Eds.), *Advances in vertebrate paleontology "hen to panta"* (pp. 47–52). Bucharest: Romanian Academy Institute of Speleology "Emil Racovita".

Lister, A. M., Sher, A. V., van Essen, H., & Wei, G. (2005). The pattern and process of mammoth evolution in Eurasia. *Quaternary International, 126–128*, 49–64.

Lordkipanidze, D., Vekua, A., Ferring, R., Rightmire, G. P., Agusti, J., Kiladze, G., et al. (2005). The earliest toothless hominin skull. *Nature, 434*, 717–718.

Lordkipanidze, D., Jashashvili, T., Vekua, A., Ponce de León, M. S., Zollikofer, C. P. E., Rightmire, G. P., et al. (2007). Postcranial evidence from early *Homo* from Dmanisi, Georgia. *Nature, 449*, 305–310.

de Lumley, H., Lordkipanidze, D., Féraud, G., Garcia, T., Perrenoud, C., Falguères, C., et al. (2002). Datation par la méthode 40Ar/39Ar de la couche de cendres volcaniques (coucheVI) de Dmanissi (Géorgie) qui a livré des restes d'hominidés fossiles de 1, 81 Ma. *Comptes Rendus Palevol Academie Sciences, Paris, 1*, 181–189.

Manzi, G., Mallegni, F., & Ascenzi, A. (2001). A cranium for the earliest Europeans: Phylogenetic position of the hominid form Ceprano, Italy. *Proceedings of the National Academy of Sciences of the United States of America, 98*, 10011–10016.

Martínez-Navarro, B. (1991). *Revisión sistemática y estudio cuantitativo de la fauna de macromamíferos del yacimiento de Venta Micena (Orce, Granada)*. Ph.D. thesis, Universidad Autónoma of Barcelona.

Martínez-Navarro, B. (1992a). Revisión sistemática de la fauna de macromamíferos del yacimiento de Venta Micena (Orce, Granada, España). In J. Gibert (Ed.), *Proyecto Orce-Cueva Victoria (1988–1992): Presencia humana en el Pleistoceno inferior de Granada y Murcia* (pp. 21–85). Granada, Spain: Ayuntamiento de Orce.

Martínez-Navarro, B. (1992b). *Megantereon* sp. (Carnivora, Mammalia) de Venta Micena (Orce, Granada, España). Revista Española de Paleontología. *Extra*, 113–117.

Martínez-Navarro, B. (2004). Hippos, pigs, bovids, sabertoothed tigers, monkeys and hominids: dispersals during Late Pliocene and Early Pleistocene times through the Levantine Corridor. In N. Goren-Inbar & J. D. Speth (Eds.), *Human Paleoecology in the Levantine Corridor* (pp. 37–51). Oxford: Oxbow Books.

Martínez-Navarro, B., & Palombo, M. R. (2004). Occurrence of the Indian genus *Hemibos* (Bovini, Bovidae, Mammalia) at the Early-Middle Pleistocene transition in Italy. *Quaternary Research, 61*, 314–317.

Martínez-Navarro, B., & Palmqvist, P. (1995). Presence of the African Machairodont *Megantereon whitei* (Broom, 1937) (Felidae, Carnivora, Mammalia) in the Lower Pleistocene site of Venta Micena (Orce, Granada, Spain), with some considerations on the origin, evolution and dispersal of the genus. *Journal of Archaeological Science, 22*, 569–582.

Martínez-Navarro, B., & Palmqvist, P. (1996). Presence of the African saber-toothed felid *Megantereon whitei* (Broom, 1937) (Mammalia, Carnivora, Machairodontinae) in Apollonia-1 (Mygdonia Basin, Macedonia, Greece). *Journal of Archaeological Science, 23*, 869–872.

Martínez-Navarro, B., & Palmqvist, P. (1999). Venta Micena (Orce, Granada, Spain): Human activity in a hyena den during the Lower Pleistocene. In: S. Gaudzinski & E. Turner (Eds.), Proceedings of the workshop "The role of early humans in the accumulation of European Lower and Middle Palaeolithic bone assemblages", Monrepos, 1995. *Monographien des Römisch-Germanischen Zentralmuseums, 42*, 57–71.

Martínez-Navarro, B., & Rook, L. (2003). Gradual evolution in the African hunting dog lineage. Systematic implications. *Comptes Rendus Palevol, 2*, 695–702.

Martínez-Navarro, B., Turq, A., Agustí, J., & Oms, O. (1997). Fuente Nueva-3 (Orce, Granada, Spain) and the first human occupation of Europe. *Journal of Human Evolution, 33*, 611–620.

Martínez-Navarro, B., Palmqvist, P., Arribas, A., & Turq, A. (1998). La adaptación a una dieta carnívora: clave de la primera dispersión humana fuera de África en el Pleistoceno inferior. Senderos de la Evolución Humana, Estudios en homonaje a Phillip V. Tobias. *Ludus Vitalis, 1*, 179–204.

Martínez-Navarro, B., Rabinovich, R., & Goren-Inbar, N. (2000). Preliminary study of the fossil Bovidae assemblage from the late Lower Pleistocene archaeological site of Gesher Benot Ya'aqov (northern Israel). Abstracts of 2000 Meeting of the INQUA-SEQS "The Plio-Pleistocene boundary and the Lower/Middle Pleistocene transition: type areas and sections". Bari, Italy.

Martínez-Navarro, B., Palmqvist, P., Madurell, J., & Pérez-Claros, J. A. (2004a). *El registro de Hippopotamus antiquus en el Pleistoceno inferior de Europa: implicaciones paleoambientales e inferencias paleobiológicas* (pp. 121–122). Alcalá de Henares (Madrid): Abstracts of the XX Jornadas de Paleontología.

Martínez-Navarro, B., Rook, L., Segid, A., Yosief, D., Ferretti, M. P., Shoshani, J., et al. (2004b). The large fossil mammals from Buia (Eritrea): Systematics, biochronology and paleoenvironments. *Rivista Italiana di Paleontologia e Stratigrafia, 110*, 61–88.

Martínez-Navarro, B., Toro, I., & Agustí, J. (2004c). Las asociaciones de grandes mamíferos de Fuente Nueva-3 y Barranco León-5 (Orce, Granada, España): Resultados preliminares. In E. Baquedano & S.

Rubio (Eds.), *Miscelánea Homenaje a Emiliano Aguirre Paleontología* (pp. 292–305). Alcalá de Henares, Madrid: Museo Arqueológico Regional.

Martínez-Navarro, B., Claret, A., Shabel, A. B., Pérez-Claros, J. A., Lorenzo, C., & Palmqvist, P. (2005). Early Pleistocene "hominid remains" from southern Spain and the taxonomic assignment of the Cueva Victoria phalanx. *Journal of Human Evolution, 48*, 517–523.

Martínez-Navarro, B., Pérez-Claros, J. A., Palombo, M. R., Rook, L., & Palmqvist, P. (2007). The Olduvai buffalo *Pelorovis* and the origin of *Bos*. *Quaternary Research, 68*, 220–226.

Martínez-Navarro, B., Palmqvist, P., Shabel, A. B., Pérez-Claros, J. A., Lorenzo, C., Claret, A., et al. (2008). Reply to Gibert (2008) on the supposed human phalanx from Cueva Victoria (Cartagena, Spain). *Journal of Human Evolution, 54*, 157–161.

Martínez-Navarro, B., Belmaker, M., & Bar-Yosef, O. (2009). The large carnivores from 'Ubeidiya (early Pleistocene, Israel): Biochronological and biogeographical implications. *Journal of Human Evolution, 56*, 514–524.

Martínez-Navarro, B., Belmaker, M., & Bar-Yosef, O. (in prep A). Bovids from 'Ubeidiya.

Martínez-Navarro, B., Shabel, A.B., et al. (in prep. B). Revision of the record of *Kolpochoerus* (Suidae, Mammalia) in Asia.

Maschenko, E. N. (1994). Comparative morphological analysis of the skull and lower jaw of the Late Pliocene baboon *Papio suschkini*. In: Tatarinov, L. P. (Ed.), Paleonteriologia. Nauka, Moscow, pp. 15–57 (in Russian).

Moullé, P.E. (1992). Les grands mammifères du Pléistocène inférieur de la grotte du Vallonnet (Roquebrune-Cap-Martin, Alpes Maritimes). *Etude paleontologique des carnivores, Equidé, Suidé et Bovidés*. Ph.D. thesis, Muséum National d'Histoire Naturelle, Paris.

Moullé, P.-E., Echassoux, A., & Martínez-Navarro, B. (2004). *Ammotragus europaeus*: une nouvelle espèce de Caprini (Bovidae, Mammalia) du Pléistocène inférieur à la grotte de Vallonnet (France). *Comptes Rendus Palevol, 3*, 663–673.

Musil, R. (Ed.) (1971). *Stránská Skála I 1910–1945. Anthropos Studie z oboru anthropologie, Paleoethologie, Paleontologie et Kuartérni Geologie c. 20 (N.S. 12) Studia Musei Moraviae*. Czech Republic: Vydává Moravské museum – Ústav Anthropos.

Musil, R. (1995). Large fauna of talus cones at the Stránská Skála Hill. In Musil, R. (Ed.), *Stránská Skála Hill Excavation of open-air sediments 1964–1972* (pp. 65–83). Anthropos series 26. Brno: Moravian Museum.

Napoleone, G., Albanielli, A., Azzaroli, A., Bertini, A., Magi, M., & Mazzini, M. (2003). Calibration of the Upper Valdarno Basin to the Plio-Pleistocene for correlating the Apennine continental sequences. Il Quaternario. *Italian Journal of Quaternary Sciences, 16(Bis)*, 131–166.

Nishimura, T. D., Takai, M., & Maschenko, E. N. (2007). The maxillary sinus of *Paradolichopithecus sushkini* (late Pliocene, southern Tajikistan) and its phyletic implications. *Journal of Human Evolution, 52*, 637–646.

Nocchi, G., & Sala, B. (1997). The fossil rabbit from Valdemino cave (Borgio Verezzi, Savona) in the context of western Europe Oryctolagini of Quaternary. *Paleovertebrata, 26*, 167–187.

Oms, O., Parés, J. M., Martínez-Navarro, B., Agustí, J., Toro, I., Martínez-Fernández, G., et al. (2000). Early human occupation of Western Europe: Paleomagnetic dates for two paleolithic sites in Spain. *Proceedings of the National Academy of Sciences of the United States of America, 97*, 10666–10670.

Palmqvist, P. (2002). On the presence of *Megantereon whitei* at the South Turkwell hominid site, Northern Kenya. *Journal of Paleontology, 76*, 920–930.

Palmqvist, P., & Arribas, A. (2001). Taphonomic decoding of the paleobiological information locked in a Lower Pleistocene assemblage of large mammals. *Paleobiology, 27*, 512–530.

Palmqvist, P., Martínez Navarro, B., & Arribas, A. (1996). Prey selection by terrestrial carnivores in a Lower Pleistocene paleocommunity. *Paleobiology, 22*, 514–534.

Palmqvist, P., Arribas, A., & Martínez-Navarro, B. (1999). Ecomorphological study of large canids from the lower Pleistocene of southeastern Spain. *Lethaia, 32*, 75–88.

Palmqvist, P., Gröcke, D. R., Arribas, A., & Fariña, R. (2003). Paleoecological reconstruction of a lower Pleistocene large mammals community using biogeochemical ($\delta^{13}C$, $\delta^{15}N$, $\delta^{18}O$, Sr:Zn) and ecomorphological approaches. *Paleobiology, 29*, 205–229.

Palmqvist, P., Martínez-Navarro, B., Toro, I., Espigares, M. P., Ros-Montoya, S., Torregrosa, V., et al. (2005). A re-evaluation of the evidence of human presence during Early Pleistocene times in southeastern Spain. *L´Anthropologie, 109*, 411–450.

Palmqvist, P., Torregrosa, V., Pérez-Claros, J. A., Martínez-Navarro, B., & Turner, A. (2007). A re-evaluation of the diversity of *Megantereon* (Mammalia, Carnivora, Machairodontinae) and the problem of species identification in extinct carnivores. *Journal of Vertebrate Paleontology, 27*, 160–175.

Palmqvist, P., Pérez-Claros, J.A., Martínez-Navarro, B., Torregrosa, V., & De Renzi, M. (in prep.). The giant hyena *Pachycrocuta brevirostris*: modelling the bone-cracking behavior of an extinct carnivore.

Palombo, M. R. (2004). Biochronology of Plio-Pleistocene mammalian faunas on the Italian Peninsula: knowledge, problems and perspectives. *Il Quaternario, 17*, 565–582.

Palombo, M. R., & Ferretti, M. P. (2005). Elephant fossil record from Italy: knowledge, problems, and perspectives. *Quaternary International, 126–128*, 107–136.

Patel, B. A., Gilbert, C. C., & Ericson, K. E. (2007). Cercopithecoid cervical vertebral morphology and implications for the presence of *Theropithecus* in early Pleistocene Europe. *Journal of Human Evolution, 52*, 113–129.

Pickford, M. (1993). Climatic change, biogeography, and *Theropithecus*. In N. G. Jablonski (Ed.), Theropithecus: *The rise and fall of a primate genus* (pp. 227–243). Cambridge: Cambridge University Press.

Pickford, M. (1994). Fossil Suidae of the Albertine Rift, Uganda-Zaire. In B. Senut & M. Pickford (Eds.), *Geology and Paleobiology of the Albertine Rift Valley, Uganda-Zaire, Paleobiology* (Vol. II, pp. 339–373). Orléans: CIFEG.

Pilgrim, G.E. (1939). The fossil Bovidae of India. Memoirs of the Geological Survey of India Palaeontologica Indica, Calcutta (new series) 26(1), 1–356, 8 plt.

Pons-Moyà, J. (1987). Los carnívoros (Mammalia) de Venta Micena (Granada, España). Paleontologia i Evolució. *Memòria Especial, 1*, 109–127.

Por, D. F. (2004). The Levantine waterway, Riparian archaeology, paleolimnology, and conservation. In N. Goren-Inbar & J. D. Speth (Eds.), *Human paleoecology in the Levantine Corridor* (pp. 5–20). Oxford: Oxbow Books.

Potts, R., & Teague, R. (2010). Behavioral and environmental background to 'Out-of-Africa I' and the arrival of *Homo erectus* in East Asia. In: J. G. Fleagle, J. J. Shea, F. E. Grine, A. L. Baden, & R. E. Leakey (Eds.), *Out of Africa I: the first hominin colonization of Eurasia* (pp. 67–85). Dordrecht: Springer.

Radulesco, C., & Samson, P. (1990). The Plio-Pleistocene mammalian succession of the Oltet Valley, Dacic Basin, Romania. *Quatärpaläontologie, 8*, 225–232.

Radulesco, C., & Samson, P. (2001). Biostratigraphy and evolution of the Early Pliocene to Early Pleistocene mammalian faunas of Romania. *Bolletino della Societá Paleontologica Italiana, 40*, 285–291.

Ramirez-Rozzi, F., Walker, C., & Bromage, T. (1999). Early hominid dental development and climate change. In T. G. Bromage & F. Schrenk (Eds.), *African biogeography, climate change, and human evolution* (pp. 349–363). New York/Oxford: Oxford University Press.

Randall, R. M. (1981). Fossil Hyaenidae from Makapansgat limeworks deposit, South Africa. *Palaeontologia Africana, 24*, 75–85.

Ronen, A. (1991). The Yiron-gravel lithic assemblage artifacts older than 2.4 My in Israel. *Archäologisches Korrespondenzblatt, 21*, 159–164.

Rook, L. (1994). The Plio-Pleistocene Old World *Canis* (*Xenocyon*) ex. gr. *falconeri*. *Bollettino della Societá Paleontologica Italiana, 33*, 71–82.

Rook, L., Martínez-Navarro, B., & Howell, F. C. (2004). Occurrence of *Theropithecus* sp. in the Late Villafranchian of southern Italy and implication for Early Pleistocene "out of Africa" dispersals. *Journal of Human Evolution, 47*, 267–277.

Rook, L., & Martínez-Navarro, B. (in prep.). *Theropithecus* in the Early Pleistocene of Europe and its importance as faunal marker for fossil hominins dispersal outside Africa.

Sahnouni, M., Hadjouis, D., Made, J. V. D., Abd-el-Kader, D., Canals, A., Medig, M., et al. (2002). Further research at the Oldowan site of Ain Hanech, North-eastern Algeria. *Journal of Human Evolution, 43*, 925–937.

Sala, B. (1990). Panthera leo fossilis (v. Reich., 1906) (Felidae) de Isernia la Pineta (Pléistocène moyen inférieur d'Italie. *Geobios, 2*, 189–194.

Sardella, R. (2004). Late Villafranchian and Galerian carnivores (Mammalia) from the Tyrrenian coastal area of Central Italy. Abstracts of the118° International Seckenberg Conference 2004, Weimar, Germany, p. 228.

Sardella, R., & Palombo, M. R. (2007). The Pliocene-Pleistocene boundary: Which significance for the so called "wolf event"? Evidences from western Europe. *Quaternaire, 18*, 65–71.

Semaw, S., Rogers, M. J., Quade, J., Renne, P. R., Butler, R. F., Dominguez-Rodrigo, M., et al. (2003). 2.6-Million-year-old stone tools and associated bones from OGS-6 and OGS-7, Gona, Afar, Ethiopia. *Journal of Human Evolution, 45*, 169–177.

Sickenberg, O. (1967). Die unterpleistozäne Fauna von Wolaks (Griech.-Mazedonien). I. Eine neue Giraffe (*Macedonitherium martinii* nov. gen. nov. spec.) aus dem unteren Pleistozän von Griecheland. *Annales Géologiques des Pays Helléniques Athen, 18*, 34–54.

Swisher, C. C., Curtis, G. H., Jacob, T., Getty, A. G., Suprijo, A., & Widiasmoro. (1994). Age of the earliest known hominids in Java, Indonesia. *Science, 263*, 1118–1121.

Szalay, F. S., & Delson, E. (1979). *Evolutionary history of the primates*. New York: Academic.

Tchernov, E. (Ed.). (1986). *The Lower Pleistocene mammals of 'Ubeidiya (Jordan Valley), Vol. 5, 405 p.* Paris: Memoires et Travaux du Centre de Recherche Français de Jerusalem, Association Paléorient.

Tchernov, E. (1992). The Afro-Arabian component in the Levantine mammalian fauna – a short bio-geographical review. *Israel Journal of Zoology, 38*, 155–192.

Tchernov, E., Horwitz, L. K., Ronen, A., & Lister, A. (1994). The faunal remains from Evron Quarry in relation to other Lower Paleolithic hominid sites in the Southern Levant. *Quaternary Research, 42*, 328–339.

Thomas, H., Geraads, D., Janjou, D., Vaslet, D., Memesh, A., Billou, D., et al. (1998). First Pleistocene faunas from the Arabian Peninsula: An Nafud desert, Saudi Arabia. *Comptes Rendus Academie Sciences, Paris, 326*, 145–152.

Todd, N. E. (2005). Reanalysis of African *Elephas recki*: implications for time, space and taxonomy. *Quaternary International, 126–128*, 65–72.

Torre, D., Albanielli, A., Bertini, A., Ficarrelli, G., Massini, F., & Napoleone, G. (1996). Paleomagnetic calibration of Plio-Pleistocene mammal localities in central Italy. *Acta Zoologica Cracovensia, 39*, 559–570.

Turner, A. (1986). Miscellaneous carnivore remains from Plio-Pleistocene deposits in the Sterkfontein Valley (Mammalia, Carnivora). *Annals of the Transvaal Museum, 34*, 203–236.

Turner, A. (1987). *Megantereon cultridens* (Cuvier) (Mammalia, Felidae, Machairodontinae) from Plio-Pleistocene deposits in Africa and Eurasia, with comments on dispersal and the possibility of a new world origin. *Journal of Palaeontology, 61*, 1256–1268.

Turner, A. (1999). Assessing earliest human settlement of Eurasia: Late Pliocene dispersions from Africa. *Antiquity, 73*, 563–570.

Turner, A., & Antón, M. (1996). The giant hyena, *Pachycrocuta brevirostris* (Mammalia, Carnivora, Hyaenidae). *Geobios, 29*, 455–468.

Valoch, K. (1995). Early human activities at Stránská Skála Hill. In: R. Musil (Ed.), *Stránská Skála Hill. Excavation of open-air sediments 1964–1972* (pp. 159–167). Brno Anthropos series 26. Brno: Moravian Museum.

Van der Geer, A. A. E., & Sondaar, P. Y. (2002). The postcranial elements of *Paradolichopithecus arvernensis* (Primates, Cercopithecidae, Papionini) from Lesvos, Greece. *Annales Géologiques des Pays Helléniques, 39*, 71–86.

Vekua, A. (1972). *Kvabebskaja fauna akcagylskich pozvonocnych*. (in Russian). Nakua, Moskow. 351 pp.

Vekua, A. (1976). The Lower Pleistocene mammalian fauna of Akhalkalaki (southern Georgia, USSR). *Palaeontographia Italica, 74*, 63–96.

Vekua, A. (1995). Die Wirbeltierfauna des Villafranchian von Dmanisi und ihre biostratigraphische Bedeutung. Jarhb. Ger.- Röm. *Zentralmuseum Mainz, 42*, 77–180.

Vekua, A., Lordkipanidze, D., Rightmire, P., Agusti, J., Ferring, R., Maisuradze, G., et al. (2002). A new skull of early *Homo* from Dmanisi, Georgia. *Science, 297*, 85–89.

Vrba, E. S. (1987). A revision of the Bovini (Bovidae) from a preliminary revised checklist of Bovidae from Makapansgat. *Palaeontologia Africana, 26*, 33–46.

Vrba, E. S. (1995a). On the Connections between paleoclimate and evolution. In E. S. Vrba, G. H. Denton, T. C. Partridge, & L. H. Burckle (Eds.), *Paleoclimate and evolution with emphasis on hominid origins* (pp. 22–45). New Haven: Yale University Press.

Vrba, E. S. (1995b). The fossil record of African antelopes (Mammalia, Bovidae) in relation to human evolution and paleoclimate. In E. S. Vrba, G. H. Denton, T. C. Partridge, & L. H. Burckle (Eds.), *Paleoclimate and evolution with emphasis on hominid origins* (pp. 385–424). New Haven: Yale University Press.

Werdelin, L. (1999). *Pachycrocuta* (hyaenids) from the Pliocene of East Africa. *Paläontologische Zeitschrift, 73*, 157–165.

Werdelin, L., & Lewis, M. E. (2000). Carnivora from the South Turkwell hominid site, northern Kenya. *Journal of Paleontology, 74*, 1173–1180.

Werdelin, L., & Lewis, M. E. (2001). A revision of the genus *Dinofelis* (Mammalia, Felidae). *Zoological Journal of the Linnean Society, 132*, 147–258.

Werdelin, L., & Lewis, M. E. (2002). Species identification in *Megantereon*: a reply to Palmqvist. *Journal of Paleontology, 76*, 931–933.

Werdelin, L., & Lewis, M. E. (2005). Plio-Pleistocene Carnivora of eastern Africa: species richness and turnover patterns. *Zoological Journal of the Linnean Society, 144*, 121–144.

Werdelin, L., & Solounias, N. (1991). The Hyenidae: taxonomy, systematics and evolution. *Fossils and Strata, 30*, 1–104.

White, T. D. (1995). African omnivores: global climatic change and Plio-Pleistocene hominids and suids. In E. S. Vrba, G. H. Denton, T. C. Partridge, & L. H. Burckle (Eds.), *Paleoclimate and evolution with emphasis on hominid origins* (pp. 369–384). New Haven: Yale University Press.

Zhu, R. X., Potts, R., Xie, F., Hoffman, K. A., Deng, C. L., Shi, C. D., et al. (2004). New evidence on the earliest human presence at high northern latitudes in northeast Asia. *Nature, 431*, 559–562.

Chapter 14
Fossil Skulls from Dmanisi: A Paleodeme Representing Earliest *Homo* in Eurasia

G. Philip Rightmire and David Lordkipanidze

Abstract The Plio-Pleistocene site of Dmanisi has yielded much evidence bearing on the morphology and behavior of the earliest hominins from western Eurasia. Human remains, animal bones and stone artifacts were deposited at Dmanisi during a brief interval following the close of the Olduvai Subchron (1.77 million years ago). The hominin fossils now include an adult braincase (D2280), the partial skull of a young adult (D2282/D211), a large mandible (D2600), a small subadult skull (D2700/D2735), an edentulous cranium with lower jaw (D3444/D3900), and postcranial bones attributed to several individuals. The crania have capacities ranging from 600 to 775 cm^3. Supraorbital tori and other vault superstructures are only moderately developed. Although there is variation related to ontogenetic age and sex dimorphism, it is appropriate to group the Dmanisi individuals together. Most probably, all of the hominins are sampled from one paleodeme. This population resembles *Homo habilis* in brain volume and some aspects of craniofacial morphology, but many of these features can be interpreted as symplesiomorphies. Other discrete characters and measurements suggest that the Dmanisi skulls are best placed with *H. erectus*. There are numerous similarities to individuals from the Turkana Basin in Kenya, but a few features link Dmanisi to Sangiran in Java. This evidence can be read to support an early dispersal of *H. erectus* from Africa to Eurasia. However, an alternative hypothesis must also be considered. Morphological comparisons of the fossils and findings from geochronology are consistent with the view that *H. erectus* evolved in Asia. Only later did representatives of this species disperse from western Asia to Africa and eastward toward Java and China.

G.P. Rightmire (✉)
Department of Anthropology, Peabody Museum, Harvard University, Cambridge, MA 02138, USA
and
Department of Anthropology, Binghamton University, NY 13902, USA
e-mail: gprightm@fas.harvard.edu

D. Lordkipanidze
Georgian National Museum, 3, Rustaveli Avenue, 0105 Tbilisi, Georgia
e-mail: dlordkipanidze@museum.ge

Keywords Cranial anatomy • Character analysis • Systematics • Human evolution • Species • *Homo habilis* • *Homo erectus*

Introduction

Hominins are known to have originated in Africa and were restricted to that continent for approximately two-thirds of their evolutionary history. Only about 2 million years ago is there evidence to suggest movement of *Homo* into other regions. Such traces consist of flaked stones along with broken animal bones, but the assemblages are not securely dated, and interpreting them has been problematical. The record improves with the onset of the Pleistocene. One important datum is 'Ubeidiya in the southern Levant, where undoubted artifacts occur with rich faunas in several different levels dating from 1.4 million years or earlier (Tchernov 1987; Sagi et al. 2005). To the west, the Orce localities of Barranco León and Fuente Nueva in Spain have produced credible evidence for human occupation at ca. 1.3–1.2 million years (Oms et al. 2000). To the north at Dmanisi in the Georgian Caucasus, stone tools, cut-marked animal bones, and numerous well preserved hominin fossils are close to 1.8 million years old, while sites in Java and in China demonstrate that *Homo* had reached the Far East 1.7 million years ago (Swisher et al. 1998; Larick et al. 2001; Zhu et al. 2004, 2008). These signals suggest dispersal(s) across southern Eurasia, very early in the Pleistocene. Identifying the populations involved, working out the routes by which this colonization occurred, and documenting the ways in which the hominins adapted to new environments are important tasks, addressed by contributors to this volume.

We discuss the evidence bearing on these questions as it is accumulating from Dmanisi, Georgia. Excavations at this site have yielded hundreds of flakes, cores, and stone manuports attributed to an Oldowan-like industry, along with a diverse fauna and many hominin bones in excellent condition (Lordkipanidze et al. 2007). The D2280 partial cranium, the D2282/D211 and D2700/D2735 skulls, and the D3444/D3900 edentulous individual have been described as *H. erectus*

J.G. Fleagle et al. (eds.), *Out of Africa I: The First Hominin Colonization of Eurasia*, Vertebrate Paleobiology and Paleoanthropology,
DOI 10.1007/978-90-481-9036-2_14,

by Gabunia et al. (2000), Vekua et al. (2002), Lordkipanidze et al. (2005, 2006), and Rightmire et al. (2006). However, the large D2600 mandible has been attributed to a (new) species *H. georgicus* by Gabunia et al. (2002). Also, three of the crania have been referred to *H. georgicus* by de Lumley et al. (2006). Here we present updated descriptions and further comparisons of the Dmanisi skulls with other specimens representing *H. habilis*, *H. rudolfensis*, African *H. erectus*, and *H. erectus* from the Far East. In some aspects of their morphology, the Dmanisi fossils do seem to differ from *H. erectus*. But new discoveries from Africa emphasize that the range of variation within this species was probably greater than has been appreciated, perhaps reflecting a high level of sex dimorphism (Potts et al. 2004; Spoor et al. 2007). It is not clear that either the high corpus of D2600 or the low capacities measured for several of the crania provide an appropriate basis for claiming the Dmanisi paleodeme as a new taxon (Rightmire et al. 2008).

The Site

Dmanisi is situated in the southeastern region of Georgia, on a promontory at the confluence of the Masavera and Pinezaouri Rivers. The fauna (including *Homo*) and accompanying stone tools are derived from sediments overlying an extensive lava flow. The ca. 80 m thick Masavera Basalt has been dated radiometrically to 1.85 million years (Gabunia et al. 2000). Lordkipanidze et al. (2007) now recognize two major stratigraphic units above this marker. The lower unit (A) is separated from the upper unit (B) by a minor erosional disconformity. Layer A1 consists of black tuffaceous sand filling low areas on the irregular surface of the basalt. As noted by Gabunia and Vekua (1995), the basalt itself, where it is exposed in the excavations, appears fresh and essentially unweathered. The A1 ash is now known to be close in radiometric age to the lava (de Lumley et al. 2002). Like the basalt and the lowest ash deposits, several additional A layers are of normal geomagnetic polarity. In the vicinity of the A/B contact, the sands contain a ca. 35 cm thick zone of calcite cement. This is the carbonate “Kerki” of other workers. However, this indurated zone is now understood to have formed as a result of groundwater action, and it occurs at varying levels within the excavated area. These subsurface groundwater calcretes envelop buried geological contacts and occur even on higher parts of the Masavera Basalt. The “Kerki” is not a reliable guide to stratigraphy, but it does form a hard protective crust, and this may be one reason why the relatively fragile bones in and below it are so well preserved. Overlying B deposits are composed of horizontally extensive ashfalls, capped by about 50 cm of calcareous soil. Samples from all of the B sediments consistently give reversed geomagnetic polarity (Gabunia et al. 2000; Lordkipanidze et al. 2007). This demonstrates that the A/B erosional surface is coincident with a paleomagnetic boundary.

An important finding is that there are numerous irregular sedimentary structures below the A/B interface, which are best described as infills. Pipes apparently formed during a period of erosion of the A sediments. Surface breaching and collapse of such pipes resulted in short cycles of gully construction, and gullies were filled by material that is penecontemporary with the lowest B horizons. This material includes the hominin mandibles and crania, and many of the animal bones, stone flakes and manuports. Samples from these cavities give reversed geomagnetic polarity, showing that they postdate the A/B contact. The data support correlation of this contact with the upper Olduvai-Matuyama boundary, and indeed several lines of evidence now seem to constrain the age of the fossils and the artifacts. Many of the skeletal elements recovered in the excavations are fragile but relatively unweathered and complete. Probably they could not have been transported very far before burial (Tappen et al. 2002). Stratigraphic and sedimentological arguments, dating, and composition of the vertebrate assemblages all point to the conclusion that material from the infills must have been deposited during a brief interval following the close of the Olduvai Subchron (1.77 million years ago).

The D2280 Cranium

As described by Gabunia et al. (2000) and Rightmire et al. (2006), the D2280 braincase is exceptionally complete (Fig. 14.1). With a capacity of 775 cm^3, it is small in comparison to most other African or Asian *H. erectus*. Nevertheless, it would be premature to characterize D2280 as female. Determination of sex from the cranium alone will always be difficult, particularly when the facial parts are missing, but the D2280 supraorbital torus is thick and projecting, for a vault of this size. Other features including the strong angular torus and deeply incised nuchal lines are in keeping with identification as a male.

The glabellar region and supraorbital structures are intact. The frontal sinus is developed as a series of pockets extending beyond the midline and into the orbital roof particularly on the right side. Glabella itself is only slightly projecting relative to the nasal root. On each side, the supraorbital torus is inclined upward but then straightens to become a horizontal bar. Thickness measured at the center of the orbit is about 11 mm (Table 14.1). Behind the orbits, the frontal narrows sharply. In the midline, there is faint (palpable) keeling anterior to bregma. The surface surrounding this landmark is slightly elevated, and bone thickness is 8 mm. Along the length of the sagittal suture, there is further keeling, especially posteriorly.

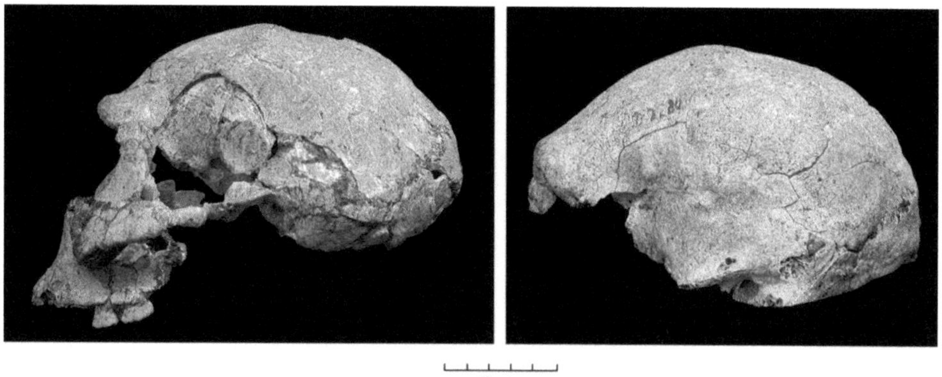

Fig. 14.1 The D2280 (*right*) and D2282 (*left*) crania in lateral view. The D2280 adult consists of almost the entire braincase. D2282 is a young adult, and the facial parts and base of the cranium have been damaged

Near lambda, this midline ridge divides and becomes double. The two limbs dissipate before reaching the lambdoid suture, and in between them there are irregular depressions in the cortex. Thickness at lambda is 7 mm.

The temporal lines are raised anteriorly. On the left, the superior and inferior lines curve downward to pass just forward of asterion, where they produce a strong torus that fills the lateral angle of the parietal bone. The (inferior) line then contributes to the supramastoid crest. This structure is incomplete but could have been moderately prominent. Both of the squamous temporals are damaged. On the more complete left side, vertical elevation of the temporal arch above the zygomatic root is close to 42 mm. Posteriorly, the parietotemporal suture follows a relatively straight course downward toward asterion, as is usual for African and Asian representatives of *H. erectus*.

The upper scale of the D2280 occiput is intact, and its length as measured from lambda to the center of the linear tubercle is 46 mm. The inion-opisthion chord is 47 mm. This approximate equivalence of the occipital and nuchal planes constitutes a resemblance to early African populations. The biasterionic breadth is only 104 mm, so the D2280 occipital bone is narrow in comparison to that of the Koobi Fora, Sangiran and Zhoukoudian individuals.

In side view, the occipital is clearly flexed, and the upper scale must slope gently forward. This part of the squama is limited below by a transverse torus. The torus resembles a low ridge. There is little supratoral hollowing, and no retromastoid processes are present. Centrally, a linear tubercle is formed at the confluence of the (superior) nuchal lines. As in other *H. erectus*, the linear tubercle and muscle markings immediately adjacent to it are elevated relative to the nuchal area below. However, even the most projecting portion of the torus is quite lightly built. Here the Dmanisi occiput resembles that of OH 12. Such morphology is in keeping with the small size and overall gracility of both D2280 and the hominin from Olduvai Bed IV.

The mastoid process is laterally projecting and flattened on its posterior aspect, so as to be coplanar with the adjacent nuchal area of the occipital. It is here or at the supramastoid crest that the cranium reaches its maximum width. Muscle markings are faint, but probably the superior nuchal line can be traced a few mm below asterion, as it passes toward the mastoid crest. This crest is not strongly expressed, and above it the supramastoid sulcus is present only as a shallow groove. Where the mastoid apex has been broken away, a honeycomb-like pattern of small pockets extends anteriorly to the tympanic plate and medially toward the occipitomastoid suture. The inner aspect of this damaged but roughly horizontal section is triangular in form and oriented toward the midline, as described for Chinese *H. erectus* by Weidenreich (1943). In its original state, the mastoid process would likely have been robust, with some medial inclination of its long axis. The juxtamastoid eminence could not have been prominent. Posteriorly, this ridge sends a spur across the occipitomastoid suture, where it is continuous with the lateral margin of the attachment for the superior oblique muscle. Thus, in D2280, the juxtamastoid process may also be termed an occipitomastoid crest.

The mandibular fossa is partially preserved on each side. On the right, width of the cavity measured from the entoglenoid apex to the ectoglenoid process is ca. 31 mm. The entoglenoid is blunt and composed mainly of squamous temporal bone. The sphenoid may produce a small inferior projection, but this structure is applied to the medial side of the entoglenoid pyramid and does not contribute to the wall of the cavity itself. On the left, the sphenotemporal suture can be followed as it crosses the preglenoid planum toward the foramina ovale and spinosum. All of the foramen ovale and much of the foramen spinosum are still incorporated within the sphenoid boundary, and there is no (sphenoid) spine.

No bar-like articular tubercle is developed in the Dmanisi hominin. Instead, the articular surface is depressed and curves forward to form an arc between the entoglenoid

Table 14.1 Cranial Measurements (mm) for the Dmanisi hominins and selected representatives of earlier *Homo*

	Dmanisi				Early *Homo*		*H. erectus* (Africa)			*H. erectus* (Asia)		
	D2700	D2280	D2282	D3444	ER1813	ER1470	ER3733	ER3883	WT15000	Sangiran 2	Sangiran 4	Sangiran 17
Whole vault												
Cranial length	155	177	–	163	145	168	182	182	–	–	–	207
Basion-nasion length	92	–	–	–	82?	–	107	102	–	–	–	115
Basion-prosthion length	100?	–	–	–	94?	–	118	–	–	–	–	129?
Basion-bregma height	101	–	–	–	98?	–	111?	102	106?	–	–	114?
Max. cranial breadth	126	136[a]	–	132	113	>138	142	140	131	141	147	161
Biauricular breadth	119	132[a]	–	120	112	135?	132	129	–	126	132	140
Frontal bone												
Supraorbital torus thickness												
Central	8	11	10.5	10	9	8	8	11	–	12	–	17
Lateral	6	9	5.5	9	6.5	6.5	9	7	–	8	–	13
Min. frontal breadth	67	75	66	67.5	65	71	83	80	73	82	–	95
Max. frontal breadth	85?	105	87?	91?	–	92	110	105	–	102	–	119
Biorbital chord	90	105?	96?	98	91	109	109	110	96	–	–	115
Postorbital constriction index[b]	74.4	71.4	68.7	68.8	71.4	65.1	76.1	72.7	76.0	–	–	82.6
Frontal sag. chord	89	101	–	93	80	93	104	101	–	–	–	118?
Frontal sag. arc	95	108	>95	101	90	104	119	118	–	–	–	–
Frontal angle	150	149	–	148	139	140	139	140	–	–	–	–
Parietotemporal region												
Max. biparietal breadth	117	119	116?	122	100?	120	131	134	128?	137	140	142
Parietal sag. chord	87	91	82	98	74?	84	82	90	93	98?	–	108?
Parietal sag. arc	91	96	85	105	77?	89	85	95	107	103?	–	–
Lambda-asterion chord	65	70	68	71	64	80	81	74	63	82?	80?	74
Lambda-asterion arc	70	75	72	74	69	88	88	79	76	92?	87?	–
Occipital bone												
Biasterionic breadth	105	104	103?	104	93?	108?	119	115	106	122	126?	124
Occipital sag. chord	70?	76?	–	79	78?	86?	88	75?	69	>71	82	81?
Occipital sag. arc	87?	97?	–	95	96?	105?	118	101?	93	–	108	–
Occipital angle	115.6	108?	–	117	114?	–	103	101	–	–	105	100
Lambda-inion chord	45?	46?	46?	50	55?	60?	57	48	38	45?	47	52
Inion-opisthion chord	39?	47?	–	42.5	40?	45?	53	51	50	45?	56	57
Occipital scale index[c]	86.6	102.1	–	85.0	72.7?	75.0?	92.9	106.2	131.5	100.0?	119.1	109.6
Foramen magnum length	30	–	–	–	–	–	37	33	36	–	40?	39
Foramen magnum breadth	28?	–	–	28	–	–	32?	26?	27	–	31	29?
Foramen magnum area[d]	660?	–	–	–	–	–	930?	674?	763	–	974?	888?

[a]Obtained by doubling the measurement to the midline.

[b]Calculated as the ratio of min. frontal breadth to the biorbital chord.

[c]Calculated as the ratio of the inion-opisthion chord to the lambda-inion chord.

[d]Calculated as π (1/2 length) (1/2 breadth).

process and the zygomatic root. Anteriorly, this surface merges smoothly into the flattened expanse of the preglenoid planum. On the right, where the lateral aspect of the joint is more complete, it is apparent that the cavity extends well out onto the underside of the zygomatic process. As in other *H. erectus*, the inner section of the mandibular fossa lies below the braincase, while the outer part is lateral to the cranial wall above.

The roof of the fossa is moderately high ("deep"). This surface is bounded posteriorly by the squamotympanic fissure, which reaches almost to the margin of the auditory meatus.

Laterally, the postglenoid process extends downward as a rounded lip, filling the space between the anterior wall of the meatus and the outer border of the zygomatic root. This postglenoid tubercle is comparable to that of the Turkana specimens (e.g., KNM-ER 3883). The mandibular cavity is slightly longer in its antero-posterior dimension than that of the East African hominins, but medially it seems less constricted. There is little development of a recess between the entoglenoid process and the tympanic plate.

The tympanic plate is approximately vertical. Where it contributes to the border of the auditory porus, the bone is only moderately thickened. The plate is less heavily constructed than in the Far Eastern specimens. Inferiorly also, the plate is relatively delicate, and its posterior part is fused directly to the mastoid. Slightly more than midway along its length, the tympanic is drawn downward to produce a spine. This petrosal spine is prominent, and its posterior aspect carries a vertical groove. The groove ascends into a cavity, but there is no sign of any styloid process. Also, there is no supratubarius process (termed the infratubarius process by Delson et al. 2001) of the form seen in KNM-ER 3733 from Koobi Fora and the Sangiran hominins.

On the left, the petrous temporal remains in its original position, but its apex is badly broken. On the right, more of the apex has survived, but it has been pulled away from the sphenoid greater wing. This bony tip is cemented by matrix to the remaining piece of basioccipital. It is not possible to determine whether the petrous surface is smooth and relatively compact, as would be expected in *H. erectus*, or roughened as in recent humans. On its medial side, the pyramid is very closely applied to the occipital. As a result, the foramen lacerum is reduced almost to a groove.

The D2282/D211 Skull

In this individual, parts of the face including the zygomatic and maxillary bones are present. However, all of the interorbital region is missing, the cheeks are cracked on both sides, and the lower aspect of the facial skeleton has been pulled forward. The mastoid portions of the temporal bones and the occiput are partly crushed, and the remaining sections of the cranial base have been forced upward. The frontal and parietals are intact and in relatively good condition (Figs. 14.1 and 14.2). A preliminary CT reconstruction of the cranium (by C. Zollikofer and M. Ponce de Leon) suggests a capacity of 650–660 cm^3.

D2282 is smaller than D2280 in its principal dimensions. The torus is perfectly preserved over the left orbit, where it reaches a maximum thickness of 10.5 mm. Behind the brows, the frontal profile rises relatively steeply. The temporal lines are slightly crested and reach medially almost to the center of the orbital margin before turning to the rear. Postorbital narrowing is very marked. Vault thickness at bregma is 7 mm. Surrounding this landmark, there is an area of cortex that is raised. This elevation extends for a short distance along the coronal and the sagittal sutures. Inside these borders, the parietal surfaces are flat or even slightly depressed. Posteriorly, there is some additional midline keeling, but this sagittal ridge is low and does not bifurcate as it approaches lambda. Lambdoid thickness is 6 mm.

The occiput has sustained heavy damage, as a result of the taphonomic process that crushed the base and thrust portions of it upward into the endocranial cavity. The junction of the upper and lower occipital scales has been displaced anteriorly. The superior nuchal line is visible on the right side and can

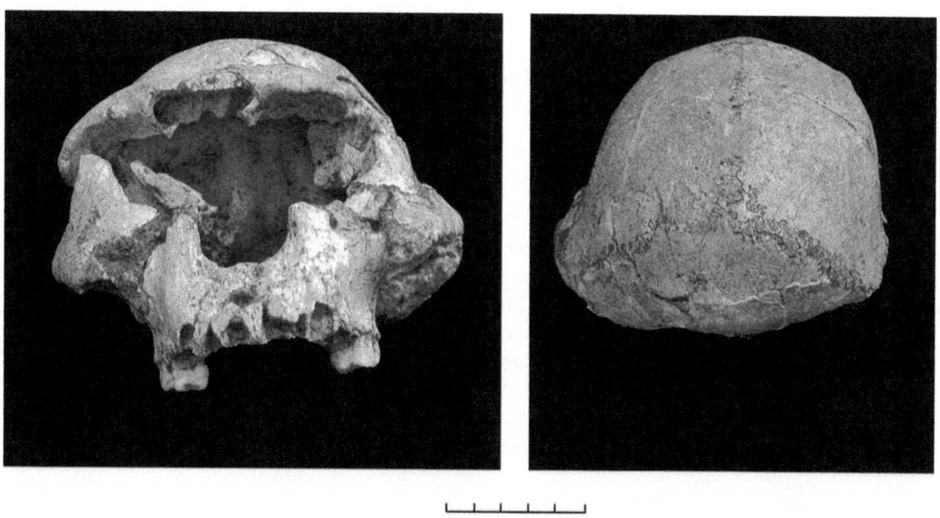

Fig. 14.2 Facial and occipital views of the D2282 cranium. The zygomatic and maxillary bones are present, but the lower part of the facial skeleton has been pulled forward away from its original position. The mastoid processes and the occiput are partly crushed, and the remaining sections of the cranial base have been forced upward

be followed for a short distance on the left as well. The linear tubercle could not have been prominent. A rough approximation of length for the upper scale is 46 mm. To either side of the midline, the surface is almost smooth, and neither a transverse torus nor any retromastoid processes can be discerned.

On the left, the occipitomastoid suture can be tracked anteriorly toward a bulge that must represent the juxtamastoid eminence. The adjacent digastric fossa is obliterated, and the mastoid process has been broken and compacted upward. Nevertheless, it appears that the posterolateral face of the mastoid is flattened as in D2280, and the supramastoid region is quite laterally projecting. On the right, some of the margin of the foramen magnum is preserved. This affords a guide to the location of opisthion (which is missing). No accurate measurement can be made, but it is clear that the opisthion-inion distance must be shorter (by 8–10 mm) than the inion-lambda chord.

The mandibular fossa is slightly narrower and shallower than that of D2280. The ectoglenoid process is blunt and prominent, but the adjacent articular surface is concave from side to side, and there is no raised (articular) tubercle. On the left, much of the tympanic plate is intact. Laterally, the plate is thickened where it forms the anterior wall and rim of the auditory meatus. The petrosal crest is slightly damaged, but at about its midpoint there are the remnants of a stout, knob-like spine. Medially, the crest is more delicately constructed. At its terminus there is an area where the bone has been eroded, and if a (small) supratubarius process was expressed, it has been lost. The petrous apex is more complete than in D2280. Its surface is quite smooth, rather than irregular or pitted as in recent humans. It is not possible to judge the condition of the foramen lacerum.

Measurements of the face are provided in Table 14.2. The least damaged facial bones are the maxillae. The frontal processes and associated nasal walls are broken out, but many other structures are preserved. The lower part of the paranasal sinus is present on both sides. On the left, the roots of M^1 and M^2 protrude into this space, which is partially clogged with matrix. It is evident that the sinus is set well below the level of the nasal floor, as in the Zhoukoudian fossils (Weidenreich 1943). The cavity also extends relatively far anteriorly, almost to the canine alveolus.

The nasal sill is slightly weathered. The surviving bony contours suggest that at most there was a small tubercle marking the anterior attachment of the nasal septum. A faint spinal crest can be followed laterally toward the nasal margin. Insofar as it is preserved, the border of the aperture is rounded. The nasal floor is stepped in the terminology of McCollum et al. (1993) but continuous-discrete in the revised scoring of McCollum (2000). The subnasal clivus is flattened from side to side and slopes steeply downward. The canine juga are prominent and take the form of rounded pillars reaching upward to thicken the walls of the nose. These pillars are bounded by deep furrows. On the left, this maxillary

Table 14.2 Facial Measurements (mm) for the Dmanisi hominins and selected representatives of earlier *Homo*

	Dmanisi			Early *Homo*		*Homo erectus*			
	D2700	D2282	D3444	ER1813	ER1470	ER3733	ER3883	WT15000	Sangiran 17
Nasion-prosthion length	69?	–	–	64	90?	81	–	77?	>75
Biorbital chord	90	96?	98	91?	109	109	110	103	114?
Nasion angle	136	–	142	153	151	155	151	138	141?
Nasal bridge width	18?	–	21	–	–	22	22	32	24?
Nasal bridge height	9.0	–	9.0	–	–	8.0	9.0?	9.5	9.0?
Nasal bridge index[a]	50.0	–	42.8	–	–	36.3	40.9	29.6	37.5
Nasal bridge angle	90	–	98	–	–	108	101	119	106
Orbit breadth	35	–	38	34	41?	44?	45	39?	44
Orbit height	31	–	32	30	36?	35	36	42?	40
Midorbital chord	55	–	51	60	64?	73	–	70	66?
Naso-orbital angle	129	–	130	–	136?	135	–	123	123
Nasal breadth	28	27	28	24	27	36?	–	36	29
Nasal height	50	–	50	44	58?	53	–	57	52?
Clivus length	>20	28	–	24?	36?	30?	–	22	25?
Bimaxillary chord	97?	91[b]	93	86?	98?	101	–	100	116?
Subspinale angle	143?	154?	140	144?	161?	143	–	133	125?
Prosthion angle	107?	107?	–	108?	112?	102	–	103	–
Cheek height	28	30?	25?	27	40?	34	–	30	37
Max. malar height	39	43.5?	40	–	–	53?	>58	53?	57?
Palate breadth	37?	39	–	35?	–	–	–	40	–
Palate length	55?	54	–	54?	–	–	–	–	–

[a] Calculated as the ratio of nasal bridge height to nasal bridge width.

[b] Obtained by doubling the measurement to the midline.

sulcus can be followed vertically for 11 or 12 mm, where it meets the infraorbital foramen. This is the condition described by Weidenreich (1943) for *H. erectus* at Zhoukoudian. The zygomaticoalveolar crest takes its origin above M^1. On the left, a deep incisure is present, and the pillar is oriented almost horizontally. There is no prominent malar tubercle.

The D2282 palate is shallow rather than highly arched. As is true for other *H. erectus*, the incisive canal is situated well behind the alveolar margin. At least 12 mm separate this opening from orale, and the canal itself must pass posteriorly and upward toward the nasal floor. The palatal surface is marked by many small pores, and its rugosity is similar to that observed in recent humans. Toward the rear, there is a low palatine torus, restricted to the midline.

It is evident that the D2282 face fits well with the D211 mandible. The two specimens were found close together and probably represent one individual. Relative amounts of wear and orientation of the occlusal facets on the cheek teeth are consistent with this conclusion, but D2282 lacks third molars. Some of the socket/crypt is present on the left, but there is no trace of M^3 on the right. In the D211 mandible, the M_3s are (newly) erupted and exhibit small wear facets on the mesiobuccal cusps (protoconids). These observations do not preclude linking the jaw with the cranium, and if this is the correct inference, then the skull is that of a young adult.

The D2700/D2735 Subadult Skull

This skull is in remarkably fine condition (Figs. 14.3 and 14.4). Four teeth (right M^1 and M^2, left P^4 and M^2) were present in the D2700 maxilla when it was recovered. In addition, the right C (D2732), P^3 (D3672), P^4 (D2719) and M^3 (D2711) and the left I^2 (D2677), M^1 (D2710) and M^3 (D2720) were found as isolated specimens. An isolated (right?) I^1 (D2736)

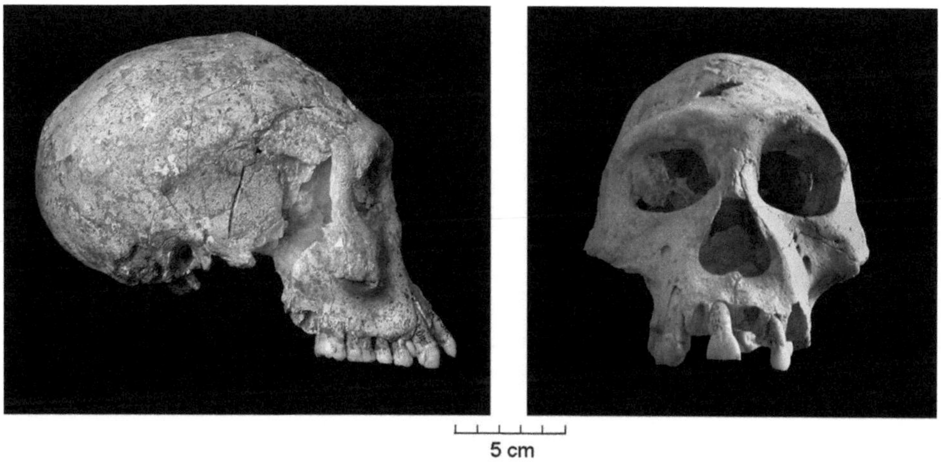

Fig. 14.3 Lateral and facial views of the D2700 cranium. This subadult is exceptional in its state of preservation. The maxillae are slightly damaged anteriorly and the zygomatic arches are broken. In other respects, the face and braincase including the base are largely complete and undistorted

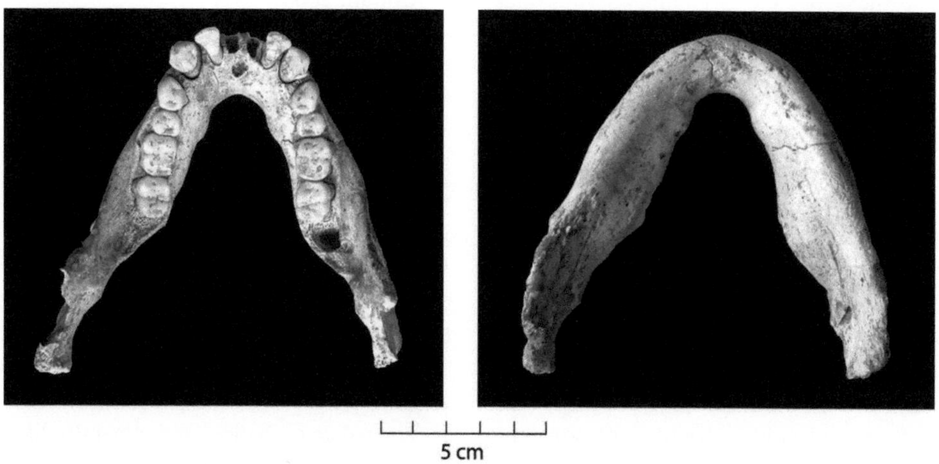

Fig. 14.4 Occlusal and basal views of the D2735 mandible and dentition

is most probably associated with this same individual. At the time of its discovery, the D2735 mandible contained eight teeth (right P_3 to M_2 and left P_3 to M_2). The isolated right (D2854) and left (D3698) I_2s can be replaced in the jaw, along with the right C (D2678) and left C (D2732).

The M^3s are partially erupted, and the Dmanisi individual thus has a dental age greater than that of the Nariokotome *H. erectus* juvenile (in which the M3s are unerupted). The D2700 maxillary dentition more closely resembles that of OH 13, in which the M^3s are just emerging. By modern standards, Tobias (1991) estimates the age of the Olduvai hominin as 14–16 years, while noting that an age of 13–15 years may be more appropriate, if dental development was accelerated in *H. habilis* relative to recent populations. There is now little doubt that earlier species of *Homo* grew up more rapidly than *H. sapiens* (Dean 2006). In D2700, the synchondrosis between the basioccipital and the sphenoid remains unfused, and this also underlines the immature status of the specimen.

Cranial capacity is ca. 600 cm^3 (Vekua et al. 2002). In most of its vault dimensions, D2700 is smaller than D2280 and closer in size to D2282 (Table 14.1). Facial dimensions are also reduced (Table 14.2). The face is surmounted by thin but well defined supraorbital tori that curve gently upward from an inflated glabellar prominence. The frontal profile slopes upward gradually and shows only a trace of keeling. The parietals are long sagittally, and they exhibit a midline ridge that becomes increasingly prominent near lambda. The parasagittal surfaces of the parietals are depressed in relation to both the frontal and the occiput. Together with the inward sloping of the cranial walls above the supramastoid crests, this gives the rear of the D2700 braincase a low and transversely flattened appearance, particularly characteristic of Asian *H. erectus*.

Viewed from the side, the occiput is flexed, but the upper scale seems more vertical in orientation than in D2280, and the occipital angle suggests that curvature is less pronounced. The occipital scale index of 86.6 is comparable to values obtained for early *Homo* from East Africa. The linear tubercle is present in very low relief. From it, the superior nuchal lines can be followed laterally, but there is no true transverse torus. Such cranial superstructures would likely have become more pronounced if growth had continued in this subadult (Rightmire et al. 2006). The external occipital crest is stronger than in D2280 and passes from the linear tubercle through the confluence of the inferior nuchal lines toward the rim of the foramen magnum.

The mandibular fossa is shallower than that of D2280. It is relatively open anteriorly, and there is no tubercle separating this surface from the flattened preglenoid planum. As with D2280, the entoglenoid pyramid is of squamous temporal origin. Posteriorly, the entoglenoid process is closely applied to the tympanic plate, and there is little development of any recess. The postglenoid tubercle is very strong and laterally extensive. Although there is damage to the mastoid region(s), it appears that the tympanic plate is fused posteriorly with the mastoid process. Where it forms the lateral rim of the auditory porus, the tympanic is less thickened than that of D2282. The petrosal crest is also relatively delicate. Medially, the crest is quite sharp. Here the tympanic anatomy is revealed clearly, and the crest ends in a small tubercle, tucked against the posterior-most aspect of the sphenoid wing. This represents the supratubarius process of Weidenreich (1943), but it is less robust than the processes seen in most other *H. erectus*.

On the left, the petrous temporal is intact. As in D2280 and D2282, the long axis of the pyramid is angled so as to lie more nearly in the sagittal plane, relative to the transverse orientation of the tympanic plate. On its surface, the apex carries a thin crest but is otherwise smooth and unpitted. At this stage of growth, the petrous bone has not quite filled the space between the basioccipital and the sphenoid, and the foramen lacerum is still open.

The face of D2700 is diminutive in comparison to that of both adult (KNM-ER 3733) and juvenile (KNM-WT 15000) specimens of *H. erectus* from East Africa. It is also reduced in size relative to the long face and massive cheek bones that characterize some early *Homo* (e.g., KNM-ER 1470) from Koobi Fora. Glabella is rounded but does not project much beyond the nasal root. The nasal bones are slightly angled to one another ("peaked"), but there is no keel. In side view, D2700 strongly resembles certain smaller individuals referred to *H. habilis*, particularly KNM-ER 1813. The lateral margin of the nasal aperture slopes forward, to merge inferiorly with the nasal floor at a point well in front of rhinion. This is the pattern documented for ancient *Homo*, whereas in recent humans the wall of the aperture is likely to be vertical, so that its floor does not extend beyond the overhanging roof. The nasal sill is smooth as in D2282, and the topography of this region is stepped (continuous-discrete following McCollum 2000).

The clivus is flattened from side to side and slopes forward. The canine juga are expanded. These pillars contribute to the massive build of the maxillary walls, although they do not thicken the margins of the piriform opening itself. Clivus length must be close to that of D2282. This part of the face would almost certainly have grown a little more, as the individual matured. Patterns of growth modeled for early African *H. erectus* by Richtsmeier and Walker (1993) suggest that there would be increases during adolescence especially in lateral and midline facial heights. Also, Antón (2003) notes that the morphology of D2700 would have been altered with advancing age.

The palate is shallow and generally like that of KNM-ER 1813 in its length/width proportions. The incisive canal is poorly preserved, but it is clear that the canal opens well behind

Table 14.3 Measurements (mm) for the Dmanisi mandibles

	D2735	D211	D2600
Symphysis height	34	31.0[a]	49
Symphysis thickness	16	17.2[a]	21
Corpus height at P_3	26.5	26.5[a]	44
Corpus breadth at P_3	18.5	18.0[a]	22
Corpus height at M_1	22.5	–	41
Corpus breadth at M_1	19	17.8[a]	(21)
Robusticity index (M_1)	84.4	–	(51)
Cross-sectional area (M_1)[b]	335.7	–	(676)

Where damage is appreciable ( ) indicates that only an estimate is possible.

[a] Dimensions from Van Arsdale (2006).

[b] Calculated as π (1/2 height) (1/2 breadth).

the sockets for the central incisor teeth, as in *H. erectus*. The bone surface presents fewer pores and irregularities than are documented for D2282, and in this respect D2700 is (marginally) more similar to some Asian specimens (cf. Sangiran 17) than to African *H. erectus*.

The D2735 mandible is complete but for the condyles and a little of the angle on each side. Minimum height at the symphysis is 34 mm, while the corresponding height for D211 is 31 mm. Corpus breadths are similar in these two Dmanisi jaws, but D2735 is slightly more robust (Table 14.3). This is probably a reflection of its subadult status. D2735 is smaller in size but comparable in its proportions to the mandible of KNM-WT 15000. Both D2735 and D211 differ from D2600. The latter specimen has a large but less robust corpus and is substantially higher at the symphysis.

The D2735 symphyseal face is flattened, and marginal tubercles are absent. The components of a bony chin are thus poorly expressed in comparison to the condition in D211, where both a central eminence and lateral tubercles are present. Laterally, there is slight bulging of the canine juga. Otherwise, the wall of the corpus shows little relief. A lateral prominence flows anteriorly from the root of the ascending ramus but subsides below M_1. On the right, a single upward-opening mental foramen is situated 13.5 mm below the P_3/P_4 septum. On the left, the foramen is again single.

The alveolar planum slopes downward and posteriorly without producing a hollowed shelf. The superior transverse torus is less clearly defined than in D211 and gives way inferiorly to a shallow depression, within which there is a faint vertical ridge. This ridge joins the inferior torus, which is constituted from little more than the rounded posterior margins of the digastric fossae. The internal alveolar prominence is swollen. Especially anteriorly, at the level of the P_4, there is an irregular bulge that may be called a mandibular torus. Posteriorly, the shelf is crested as it passes the position of the M_3 and swings upward to join the triangular torus of the ramus. The shape of the relatively broad and low ramus is like that of KNM-WT 15000.

The D3444/D3900 Skull

This fourth skull resembles the other specimens but also extends the range of variation observed within the Dmanisi paleodeme. The vault is almost complete and exhibits minimal distortion. There is damage to the cranial base, and on the left, the petrous temporal is broken out. Some of the facial skeleton is in good condition, but there is damage to the lower margin of the right cheek. The maxillary alveolar processes are heavily resorbed, and no teeth are present (Fig. 14.5). As judged from sinus development and the state of suture closure, this individual is fully adult.

The face of D3444/D3900 is striking in appearance. The advanced state of alveolar bone atrophy indicates substantial tooth loss several years before death. We are aware of no other fossil hominins that display comparable masticatory impairment. It is evident that D3444/D3900 survived for a significant period without consuming foods that required heavy chewing. Inferences regarding subsistence must be drawn with caution. However, it may be hypothesized either that this individual was able to survive without help by utilizing softer plant foods and extracting animal brain and marrow with stone tools and manuports, or that the hominins could offer assistance to one another beyond the level observed in non-human primates (Lordkipanidze et al. 2005, 2006). In any case, it is apparent that the Dmanisi population was capable of exploiting a wide spectrum of resources, on an individual basis and/or as a group. Animal carcass

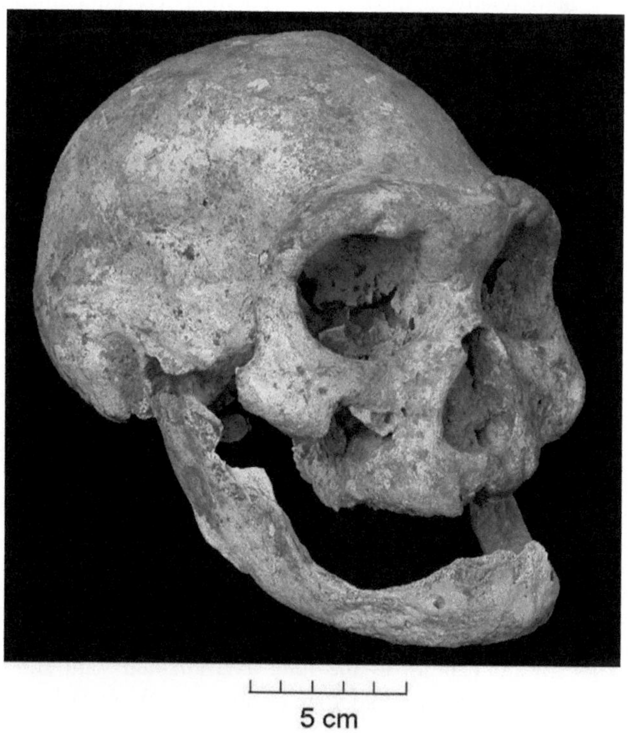

Fig. 14.5 Oblique view of the D3444/D3900 skull

processing is well documented at the site (Lordkipanidze et al. 2007), and both meat and softer tissues may have been incorporated in the diet. It is reasonable to assume that the Dmanisi hominins survived the relatively harsh Caucasus winters by increasing their consumption of animal remains.

In its vault dimensions, D3444 is generally a little larger than D2700 but smaller than D2280. Cranial capacity as measured from a CT reconstruction of the endocast is close to 625 cm^3 (Lordkipanidze et al. 2006). The glabellar prominence is broad and projecting. Maximum thickening of the brow occurs medially, and there is little definition of a supratoral sulcus. Least frontal width measured at the temporal lines is 67.5 mm, and it is apparent that marked frontal narrowing is a characteristic common to all the Dmanisi individuals. Behind the brows, the frontal profile is quite flat. On the parietal vault, a blunt sagittal keel becomes more prominent posteriorly, where the bone is heaped up on both sides of the sagittal suture but seems to subside centrally. Here the pattern of (double) keeling resembles that in D2280.

At their closest approach to the midline, the temporal lines are ca. 70 mm apart. On each side, the inferior line produces a strong lateral torus as it curves toward the supramastoid crest. The crest itself is much more prominent in D3444 than in D2700. Below the supramastoid crest, there is a shallow sulcus, which extends posteriorly onto the angle of the parietal bone. In the floor of the sulcus, there is a clear incisure, where a tongue of the parietal is inserted between the squamous and mastoid portions of the temporal bone. The squamous temporal is highest anteriorly. Its upper border is straight relative to the (arched) condition in recent humans and slopes posteroinferiorly to meet the parietal incisure.

The occiput is flexed, and its upper scale is oriented vertically. Height of this plane is 50 mm, while the chord from inion to opisthion is 42.5 mm. The index of scale lengths is very close to the figure obtained for D2700. The transverse torus is low and mound-like, and more similar to that in KNM-ER 3733 than to the sharply sculpted torus of D2280. The nuchal plane shows only moderate relief. If allowance is made for the fact that surface bone has been lost in some areas, the morphology is comparable to that observed for D2700. No external occipital crest is preserved in the interval between the transverse torus and the inferior nuchal lines, but a thin crest passes from this confluence to the rim of the foramen magnum.

D3444 is the only Dmanisi cranium to retain a complete mastoid process, and this structure is short and cylindrical in form. The processes are very inturned, so that the distance between their tips is only 102 mm. By contrast, the breadth measured higher at the (strong) supramastoid crests is 132 mm. This inward canting of the long axis coupled with short length allows the mastoid apex to protrude downward only slightly. As a result, the digastric incisure is very shallow. This impression is bounded medially by a low ridge. Still more medially, there is a channel, situated fully on the temporal but bounded by bone that is heaped up along the occipitomastoid suture. This channel trends anteriorly toward the stylomastoid foramen, and in its floor, there is a narrow groove, possibly for the occipital artery. The adjacent occipitomastoid junction is raised to form an eminence. This eminence follows the scar left by the superior oblique muscle, and it may be termed an occipitomastoid crest.

In these anatomical details, D3444 resembles the condition noted for several of the crania from Sambungmacan and Ngandong in Java. Although the mastoid structures are quite small in the Dmanisi specimen, compared for example to the much more rugose Ngandong 12 individual, the low ridge medial to the digastric fossa and the stronger occipitomastoid crest correspond to the components of the juxtamastoid eminence described originally by Weidenreich (1951). In the Ngandong crania, there is a "paramastoid" crest, separated by the groove for the occipital artery from the more medial occipitomastoid crest. Both structures contribute to what may generally be termed a juxtamastoid complex (e.g., Rightmire 1990).

The glenoid fossa is close in width and overall proportions to that of D2280. As in the latter specimen, there is no clearly defined articular tubercle. The D3444 postglenoid process is lip-like with a flattened anterior face. It is decidedly less massive than that in D2700. The rear of the mandibular cavity is composed of the tympanic plate, oriented almost vertically. The plate curves slightly as it reaches inward. Here there is a crevice-like constriction, where the entoglenoid pyramid meets the tympanic plate posteriorly. In this feature, D3444 more clearly approaches the condition described for (some of) the Zhoukoudian individuals (Weidenreich 1943) and most African *H. erectus*.

The face is nearly as complete as that of D2700 and shows none of the distortion that has affected D2282. At the frontonasal suture, the nasal bones together are 13 mm wide. These elements narrow slightly in their middle parts and broaden again below. They are broken inferiorly. It is apparent that the nasal saddle is tented but lacks any sharp midline keel. If the nasal bones were complete, the saddle would be elevated, relative to the orbital rims. On the floor of the piriform aperture, the remnants of an incisive crest extend forward to produce a blunt tubercle, which projects slightly beyond the nasal sill. The sill itself is flattened, and it slopes inward more steeply than in D2700. At its posterior pole, there is a distinct step downward onto the surface of the palate (as in the other Dmanisi specimens).

Enough of the clivus has survived to show that there is little swelling laterally near the nasal margins. Canine juga are expressed in D2700 and are more prominent in D2282, so their absence in D3444 is unexpected. It appears that the canines and also the incisors were lost before death and that the surface of the clivus was remodeled as the sockets were resorbed. The entire alveolar process has been altered as a consequence of tooth loss (Lordkipanidze et al. 2005, 2006). Shape of the dental arcade cannot be determined with any confidence, but in its proportions, the palate must be similar to that of other Dmanisi individuals.

The D3444 midface is robust in its construction and (even after allowance is made for bone loss due to remodeling) somewhat orthognathic in comparison to that of D2700. In neither D3444 nor D2700 is the maxillary sulcus expressed as clearly as it is in D2282 and some African and Far Eastern *H. erectus*. Indeed the infraorbital region of D3444 is flattened. There is minor hollowing both at the base of the cheek, where the maxillary zygomatic process merges with the wall of the nasal aperture, and also laterally along the zygomaxillary suture. However, no true "canine fossa" is developed. Where it arises from the body of the maxilla, the zygomaticoalveolar pillar is quite massive, with an antero-posteriorly thickened root. This morphology is characteristic also of D2700 and D2282. The pillar turns laterally to become almost horizontal, and the form of the (relatively open) incisure is like that in D2700.

The D3900 mandible is relatively well preserved. All of the teeth are missing, and it is clear that in most cases the sockets have been fully resorbed. Much bone mass has been lost, and the body of the mandible is now very attenuated, as in an individual of advanced age. What remains of the symphysis is low, and there is a distinct midline eminence. However, this blunt projection must be an artifact of the remodeling process. On the left, all of the angle and the ramus have been broken away. On the right, the mandibular angle is present, but the ascending portion is damaged. Overall, D3900 has been reduced in size, and the specimen is now slightly smaller than D2735.

The Dmanisi Paleodeme

In light of the many metric and anatomical similarities of the crania and mandibles described above, it is reasonable to group them together. Along with the postcranial bones, this material constitutes one of the most significant assemblages known from any Plio-Pleistocene locality. In contrast to the situation at Olduvai Gorge, Koobi Fora, or Sangiran, where the fossils are scattered through a substantial thickness of sediments representing many millennia of deposition, the Dmanisi hominins may be regarded as essentially contemporary. Howell (1999) recommends application of the term paleodeme to such a set of specimens.

Where there are crania and jaws without securely associated postcranial bones to work with, one must be cautious in assigning sex to individuals. On the basis of its relatively massive supraorbital torus, and rugose occiput, D2280 is perhaps better characterized as a male than as a female. D2282 is less well preserved, but characters such as the thinner brow, small (?) mastoid process, and smooth occipital are in keeping with an assessment as female. The D211 mandible is thus also likely to be female, while the D2600 mandible almost certainly represents a larger male. The D2700 subadult possesses expanded jugal pillars, and the cheek is comparatively deep. It is likely that the brow ridges, mastoid and supramastoid crests, and occipital torus would have developed further as the individual reached maturity. These considerations suggest that D2700/D2735 may be a young male. Several immature vertebrae, ribs, a clavicle, and broken right and left humeri presumed to be associated with this skull (Lordkipanidze et al. 2007) do not help to settle the question of sex determination. D3444/D3900 is plausibly a male, as indicated by cresting in the mastoid region and the relatively prominent occipital torus. The face is also robust, with comparatively heavy brows, large orbits and deep cheeks. Edentulousness and bone atrophy may indicate advanced age.

Although it is difficult to identify the sex of individual specimens, it is quite possible to assess overall variation within the assemblage and to infer from this the level of sex dimorphism characteristic of the Dmanisi population. Skinner et al. (2006) have raised this question with their treatment of mandibular dimensions. They emphasize the degree of size difference present in the Dmanisi mandibles, remarking on the contrast between D2600 and the smaller D211 jaw. Skinner et al. (2006) find that it is particularly corpus height that sets these two mandibles apart. The D2600/D211 height differences at M_1 and at the symphysis are large enough to occur only rarely in a series of resampling experiments carried out for reference populations of extant apes and humans. Ratios of corpus breadth at least as large as the Dmanisi values are encountered more frequently in all of the living hominoids. On the strength of these findings, Skinner et al. (2006) conclude that either the Dmanisi population was so sexually dimorphic as to raise doubts about its status within *Homo*, or the D2600 mandible should be grouped separately from the other specimens. However, the corpus height measurements utilized by these authors are uncorrected for damage that is clearly present at the base, particularly in the case of D211 (Rightmire et al. 2008). When the jaws are measured so as to avoid the damaged portions, and a larger suite of characters is employed, resampling results show that the Dmanisi hominins differ no more than would be expected for individuals within a relatively dimorphic ape population (Van Arsdale 2006). Given this outcome, there are no compelling anatomical grounds for sorting D2600 to a hypodigm different from that containing the other Dmanisi fossils. This species apparently possessed a level of sex dimorphism greater than is characteristic of recent *Homo*. However, it is probably not surprising to encounter novel patterns of craniofacial or dental variation in Plio-Pleistocene populations. Ancient taxa such as *H. erectus* may differ from *H. sapiens* and also fail to match expectations based on extant apes, as noted by Plavcan (2002).

While there is variation within the paleodeme, the Dmanisi hominins share a common bauplan. The crania fall near or below the limits for size observed in early *H. erectus*. All have capacities smaller than KNM-ER 3733, KNM-ER 3883, or Sangiran 2, and two of the adults have volumes less than the 691 cm^3 reported for KNM-ER 42700 by Spoor et al. (2007). The subadult but nearly full size brain of D2700 lies close to the mean (610 cm^3) calculated for *H. habilis* (*sensu stricto*) by Wood (1992). Brow ridges are only moderately thickened. Supratoral hollowing is minimal, but there is very marked postorbital constriction. The frontal profile is flattened. There is commonly an eminence at bregma, coupled with midline keeling on the parietals. The surfaces of the parietals themselves tend to be depressed, relative to the sagittal midline and the coronal/lambdoid margins. This gives the vault a poorly filled appearance. Cresting in the mastoid region is variable, but the cranium is much broader at the supramastoid crests than higher on the parietals. The occiput is flexed (especially D2280 and D3444), but a transverse torus is not uniformly present. The upper scale is greater than or about equal to the lower (nuchal) plane in length, as in African specimens of *H. erectus*. The glenoid fossa extends well out onto the underside of the zygomatic root. This cavity exhibits a number of features that appear to be primitive, in comparison to later Pleistocene species of *Homo*. Orientation of the tympanic bone relative to the petrous pyramid is like that in *H. erectus*, but the tympanic plate is delicate in its morphology. As currently known, the facial skeleton is small in comparison to that of most other Plio-Pleistocene hominins. The midface seems relatively broad, but in profile there is a resemblance to *H. habilis*. The nasal and subnasal prognathism that contributes to this similarity is more pronounced in D2700 than in D3444. The border of the nasal aperture slopes forward, and there is little relief on the surface of the nasal sill. The massive zygomaticoalveolar pillar is flexed to produce a clear incisure. The palate is shallow, and the incisive canal is situated posteriorly. The D211 and D2735 mandibles are robust (but smaller in their symphyseal and corpus dimensions than D2600). They exhibit a thickened vertical symphysis with variable expression of a mental eminence, a sloping alveolar planum, and internal transverse tori.

Comparisons with Early *Homo* from Eastern Africa

It is appropriate first to compare the Dmanisi fossils with representatives of ancient *Homo*. D2280 (the largest of the four Dmanisi crania) is similar in gross size to KNM-ER 1470, often referred to *H. rudolfensis*. However, detailed inspection shows that there are important differences. The D2280 braincase is longer but relatively narrow (see Table 14.1). Breadths taken of the mid-vault (but not the frontal) and the cranial base are less than in the Koobi Fora individual. The parietal itself is long sagittally, but the lambda-asterion arc is 13 mm less than in KNM-ER 1470. For D2280, the parietal arc (96 mm) and the occipital arc (97 mm) are about the same. In KNM-ER 1470, the parietal is short (89 mm), and the long (105 mm) occipital arc contributes a proportionately greater amount to the total (sagittal) curvature measured for the braincase. The occiput itself is shaped differently in the two specimens. For D2280, the upper scale and the nuchal plane are subequal in size (the index is 102.1), but for *H. rudolfensis*, this ratio drops to 75.0.

Apart from these differences in vault proportions, D2280 has a bar-like supraorbital torus, parietal keeling, moderate development of a supramastoid crest, and a deep mandibular fossa. KNM-ER 1470 displays neither a bregmatic eminence nor any sign of a sagittal keel, but the mastoid region is more heavily pneumatized, and there is a massive supramastoid crest. The glenoid cavity is shallow. D2280 lacks facial parts, but the other Dmanisi individuals differ from KNM-ER 1470 in nasal proportions and length of the nasoalveolar clivus. In D2700 and D2282, palate breadth is far less than any reasonable estimate for the Koobi Fora specimen. Also, the Dmanisi faces are moderately projecting, as is apparent from the (low) nasion angle and the angles measured at subspinale or at prosthion. As has been recognized before, the *H. rudolfensis* nasal aperture and clivus are lengthened, and the palate is enlarged. The cheek bones are deep and very prominent. The face is comparatively flat, especially below the nose, and the morphology of this region resembles that observed in species of *Paranthropus*.

While it is improbable that D2280 or the other Dmanisi hominins can be linked to the hypodigm that includes KNM-ER 1470, there are resemblances to the smaller East African crania referred to *H. habilis*. D2700 has been likened to KNM-ER 1813 by Vekua et al. (2002), and these individuals are comparable in size. Although it is subadult, the Dmanisi vault is a little larger in its principal dimensions. Maximum length, biparietal and biauricular breadths, and basibregmatic height all exceed those for KNM-ER 1813, which has an endocranial volume of only 509 cm^3 (Holloway 1983). However, the form of the supraorbital torus is similar in the two specimens, as is the degree of postorbital constriction. D2700 has a somewhat flatter frontal, and there is midline keeling on the parietals. Surprisingly, the Dmanisi cranium also has a shorter occipital arc, although rounding of the posterior braincase as registered by the occipital angle is about like that in KNM-ER 1813. The D2700 occiput is smooth, while the Koobi Fora hominin possesses at most a rudimentary transverse torus. Even if the supraorbital and

Table 14.4 Traits shared by the Dmanisi assemblage with earlier *Homo* and/or *Australopithecus* sp.

- Average cranial capacity close to 665 cm^3
- Marked postorbital constriction
- Mastoid region inflated and laterally projecting (D3444, probably D2282), as in earlier *Homo*
- Occipital transverse torus variably expressed or absent
- Mandibular fossa, hollowed articular "tubercle" and postglenoid process as in earlier *Homo*
- Nasocanine (midfacial) contour projecting (D2700)
- Canine jugum prominent, bounded by furrow-like maxillary sulcus
- Zygomaticoalveolar incisure present
- Palatal opening to incisive canal situated relatively far posteriorly
- Hard palate shallow rather than deeply arched
- Mandibles show thickened vertical symphysis with little/no mental eminence, sloping alveolar planum, and internal transverse tori

occipital tori of D2700 were to become slightly more prominent as this individual grew into an adult, there would be resemblances to *H. habilis*.

Both individuals have faces that are diminutive in comparison to those of other *Homo*. In D2700, nasion is set further forward relative to the orbital margins, but orbital proportions, shape of the nasal bones, the midfacial profile (the "nasocanine contour" of Kimbel et al. 1984) and clivus length are similar to (those of) KNM-ER 1813. The clivus of D2700 is relatively flattened, and the canine juga are more prominent. But cheek height is about equal to that in the small African individual, and both faces exhibit some localized excavation of the infraorbital surfaces.

These comparisons demonstrate similarities with *H. habilis*, but conclusions must be drawn with caution. Some of the characters considered seem to be primitive, in the sense that they are common not only to earlier *Homo* but also species of *Australopithecus* and/or extant apes (Table 14.4). The more conspicuous resemblances of D2700 to KNM-ER 1813 may be of little help in identifying the taxon to which D2700 belongs. The two individuals have endocranial capacities that are among the smallest recorded for any representatives of *Homo*, but this observation serves only to distinguish them (marginally) from *Australopithecus* and (more convincingly) from Late Pleistocene humans. Both the prominence of the D2700 "nasocanine contour" and its set relative to the prognathic clivus are characteristic not only of KNM-ER 1813 but also *A. afarensis* and African apes (Kimbel et al. 1984). Flexion of the zygomaticoalveolar pillar occurs in KNM-ER 1813, probably OH 24, and *H. erectus*. The trait is present also in individuals referred to *A. afarensis* (Kimbel et al. 1994, 2004). All in all, it is difficult to identify traits shared by the Dmanisi skulls and KNM-ER 1813 that are not either size-related or anatomically primitive. By themselves, these similarities neither support (nor rule out) a close link to *H. habilis*.

Resemblances to *Homo erectus*

In many aspects of their anatomy, the Dmanisi skulls parallel the condition documented for *H. erectus*. This is apparent in the low cranial profile, flattened frontal, sagittal keeling, reduced width of the parietal vault in relation to the cranial base, cresting at the parietal angle, shape of the temporal squama, morphology of the mastoid process and its associated crests, angled occiput (D2280 and D3444), disposition of the digastric groove and juxtamastoid eminence, depth and architecture of the mandibular fossa, and orientation of the petrous axis. In the elevation of the nasal saddle, forward slope of the border of the nasal aperture, lack of surface relief on the nasal sill, flexion of the malar pillar, and posterior location of the palatal incisive canal, the facial skeleton is also like that of *H. erectus*. Additional similarities include the shape of the mandibular corpus and the occurrence of multiple mental foramina (in D211 and D2600). Most measurements of tooth size place the Dmanisi hominins within the range observed for *H. erectus*.

The larger Dmanisi crania particularly resemble specimens from the Turkana Basin. This is apparent when D2280 is compared to KNM-ER 3733 (Fig. 14.6). A link between the Caucasus and East Africa is favored by Gabunia et al. (2000), who point to several features including vault bone thickness, "angular" sagittal profiles, and proportions of the facial skeleton that are plausibly diagnostic for *H. erectus* (= *ergaster*). Rosas and Bermúdez de Castro (1998) also describe traits of the D211 mandible and teeth that seem to underscore the primitive morphology of these specimens, while generally aligning them with KNM-ER 992 from Koobi Fora and KNM-WT 15000 from Nariokotome.

In order to make a convincing case for linking the Dmanisi fossils broadly with *H. erectus* (as opposed to earlier *Homo*), or more narrowly with African *H. erectus*, it is necessary to weigh the anatomical evidence carefully. Although studies of the Caucasus material reveal numerous resemblances to *H. erectus*, not all such traits are equally useful. Some are distributed widely in other taxa and thus have descriptive value without being diagnostic. Also, polarities are not easily determined. This difficulty arises (partly) from the rather poor condition of fossils belonging to the most appropriate reference groups. Crania recovered from Olduvai Gorge, the Turkana Basin localities, and from sites in southeast Asia are frequently damaged, and fragile structures of the face and base are often missing. Despite the patchy nature of the record, it can be ascertained that several of the features linking the Dmanisi population with *H. erectus* are probably primitive retentions. Examples include marked postorbital constriction, proportions of the mandibular fossa and hollowing of the articular "tubercle," a prominent canine jugum associated with a maxillary sulcus, presence of a zygomaticoalveolar incisure, a relatively posterior placement in the

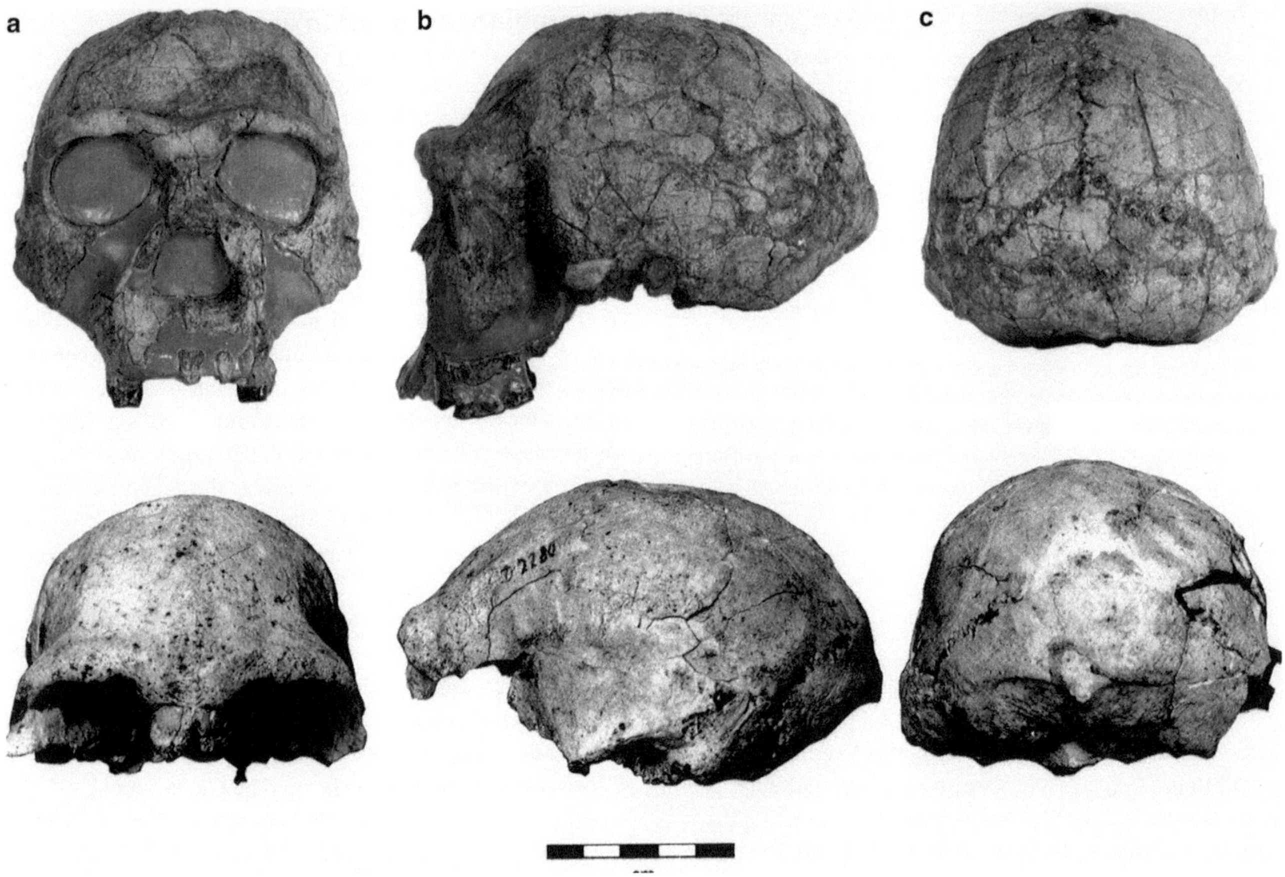

Fig. 14.6 Facial (**a**), *left* lateral (**b**) and occipital (**c**) views of KNM-ER 3733 (*above*) and D2280 (*below*). The two crania are similar in size. They share many characters of vault form and cranial base morphology that are likely to be plesiomorphic for *Homo*, along with other features that are more clearly diagnostic for *H. erectus*

palate of the opening to the incisive canal, and aspects of mandibular morphology (see Table 14.4).

Other traits documented for the Dmanisi crania are variable within assemblages of *H. erectus* and earlier *Homo*. Such characters are difficult to analyze and must generally be accorded low utility for sorting specimens at the species level. An example is the development of a supratubarius process at the medial terminus of the tympanic plate. This process is poorly expressed in the Dmanisi individuals, while it is larger in some (not all) African and Asian *H. erectus*. A small process is present in OH 24. Facial morphology also varies. Vertical thickness of the supraorbital tori, glabellar projection, and supratoral hollowing differ even within groups of Far Eastern *H. erectus*, probably as a consequence of size-related sex dimorphism. Forward protrusion of the incisive (intermaxillary) crest may or may not produce an anterior nasal spine, and transverse cresting on the nasal sill is absent, slight or occasionally marked. D2282 and D2700 have sills that lack strong relief, as is the norm for *H. erectus*. Topography of the nasal floor where it meets the surface of the palate can be scored as smooth or stepped (McCollum et al. 1993). In the Dmanisi hominins, the back of the sill drops in elevation, but both this (stepped) condition and a smooth state occur in African *H. erectus* (Rightmire 1998). McCollum (2000) notes that the contour in this region may better be described as continuous for (all) earlier *Homo* specimens, even if there is some depression of the palatine process of the maxilla.

Some of the morphology described for the Dmanisi hominins is more clearly diagnostic for *H. erectus* (Table 14.5). This systematic link is strengthened if it is agreed that several characters that are variable within groups may still carry important information. The bar-like and projecting supraorbital torus (D2280 and D3444) differs from that of *H. habilis*, and this frontal morphology is best matched in populations of *H. erectus*. A petrous pyramid that is smooth and dense (rather than eroded) in appearance, and constriction of the foramen lacerum, will probably be confirmed in the cranial bases of all the Dmanisi adults. This complex of traits is not presently documented for early *Homo*, but it is routinely expressed in *H. erectus*.

As has been noted, the Dmanisi crania are broadly similar to those referred to African *H. erectus*. There are also good indications that the skulls share at least a few (specialized?)

Table 14.5 Derived characters of the Dmanisi assemblage likely to be diagnostic for African and/or Far Eastern *H. erectus*

- Bar-like and projecting supraorbital torus (D2280 and D3444)
- Eminence at bregma accompanied by parietal sagittal keeling
- Depression of parietal surfaces giving rear of vault a low and transversely flattened appearance
- Angular torus present (D2280, D3444)
- Temporal squama low with straight upper border passing downward toward asterion
- Supramastoid crest coupled with a mastoid process that is inturned at its tip and flattened posteriorly
- Occipital bones (D2280, D3444) flexed as in African and Asian *H. erectus*
- Expression of a juxtamastoid complex (D3444)
- Tympanic plate oriented coronally relative to petrous axis, giving a petrotympanic angle of 140°–150°
- Inferior surface of petrous pyramid smooth ("dense"), foramen lacerum restricted
- Nasal saddle prominent in relation to orbital rims, as in later hominins

characters with populations from the Far East. An example is provided by parasagittal flattening of the posterior vault. Flattening or even depression of the parietal surfaces is pronounced at Dmanisi, and it is common in the *H. erectus* crania from Sangiran in Java. A related character may be the angular torus, well expressed in D2280 and D3444. Here the temporal line produces a rounded bulge filling the mastoid angle of the parietal bone, as in the Sangiran, Ngandong, and Zhoukoudian specimens. Also, while expression of a juxtamastoid complex in D3444 is unlikely to link the Dmanisi population directly with the much later Ngandong group, it suggests that an evolutionary relationship between the Caucasus and Java is certainly plausible. This evidence from the mastoid region emphasizes the scope of variation to be expected within both western and Far Eastern *H. erectus*.

Dmanisi and Human Evolution

Anatomical assessments place the Dmanisi population with *H. erectus*. Small cranial volume and other plesiomorphic retentions do not offer a basis for excluding the Dmanisi crania and associated jaws from a hypodigm encompassing individuals from the Turkana Basin, other African localities, Sangiran, and additional sites in the Far East. Indeed the D2280 and D3444 adults exhibit many of the attributes that are listed in definitions and/or diagnoses available for *H. erectus* (e.g., Howell 1978, 1980; Groves 1989; Rightmire 1990; Antón 2003; Antón et al. 2007). The D2282/D211 young adult is less well preserved but also shows affinities with this group, while some of the traits that seem to set the D2700/D2735 subadult apart from *H. erectus* are likely to reflect incomplete growth. Although the D2600 mandible has been distinguished from D211 on the basis of corpus size (Skinner et al. 2006) and various other features (Gabunia et al. 2002), it can be argued that the two jaws differ no more than would be expected for individuals drawn from a single, sexually dimorphic population (Rightmire et al. 2008).

If *H. erectus* is regarded as a widespread, polytypic, and perhaps highly dimorphic taxon, there remains the question of how the Dmanisi paleodeme is related to African and Asian populations of this species. Certainly there is an overall resemblance to the skulls from Koobi Fora, but there are also differences. The latter are partly size-related but include aspects of vault shape and facial morphology. Similarities to specimens from Sangiran are less numerous, but in key features such as parasagittal flattening and the possession of both paramastoid and occipitomastoid crests, the Dmanisi crania are like those from the Far East. Also, it is important to point out that the Caucasus population presents other characters that seem to be unique. In D2280 and D3444, there is clearly a double sagittal keel on the parietal vault. Keeling is common in *H. erectus*, but the bone is heaped up to form a single ridge along the sagittal suture. Other possible autapomorphies at Dmanisi include the relatively thin and delicate tympanic plate, and absence of the supratubarius process (although a trace occurs in D2700).

As further comparative analyses are carried out, several phylogenetic hypotheses may be considered. One is that *H. erectus* evolved in eastern Africa. According to this widely accepted view, a population of early *Homo* (*H. habilis* or *H. rudolfensis*, documented at Olduvai and in the Turkana Basin) gave rise to *H. erectus*. The first representatives of the latter species are found at Koobi Fora, and there are numerous additional fossils from Olduvai Gorge, Nariokotome, Konso, and Bouri (Daka). Groups of *H. erectus* then ventured out of Africa, leaving abundant traces of their passing in the Jordan Valley, and to the north, in the Georgian Caucasus. From sites such as Dmanisi, the hominins could presumably have spread westward into Europe and also across southern Asia to the Far East (Fig. 14.7). This scenario implies that differences between African *H. erectus* and the Dmanisi fossils reflect geographic distance, adaptation to new environments in western Asia, or drift in small isolates.

Problems with such an "African origins" hypothesis have been noted, and it is increasingly clear that alternatives must be explored (Asfaw et al. 2002; see Dennell and Roebroeks 2005 for a review). One is suggested by the distribution of craniofacial characters within these ancient populations. As has been emphasized, the Georgian crania are small, show postorbital constriction, and lack strong crests or tori. In D2700, where the face is well preserved and not altered by remodeling associated with tooth loss, the prominent midfacial profile and forward sloping clivus appear to be shared with *H. habilis*, some australopiths and African apes. Such traits are primitive, but in the Dmanisi assemblage they are

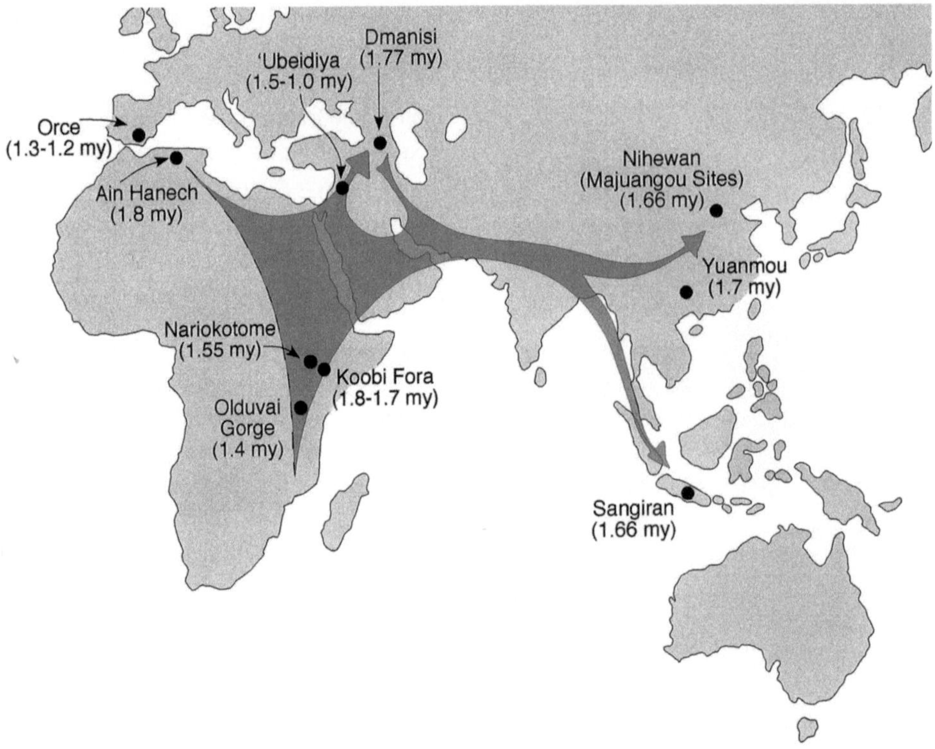

Fig. 14.7 Map illustrating an African origin for *H. erectus*. Early representative of this species are documented at Koobi Fora, Nariokotome, and Olduvai Gorge, and ancient stone artifacts are present at Ain Hanech in North Africa. From Africa, populations of *H. erectus* disperse to western Asia and also eastward toward the southeast Asian tropics. An early entrance by hominins into China is indicated by artifacts from the Nihewan Basin and other sites

coupled with other morphology that is clearly derived in the direction of *H. erectus*. This combination of numerous plesiomorphic characters, some synapomorphies, and a few unique traits, sets the Dmanisi paleodeme apart from both African and East Asian populations. A reasonable explanation for the occurrence of such a pattern is that Dmanisi documents a group evolving (very) early in the history of *H. erectus*. This proposition is supported by stratigraphic and paleomagnetic analyses demonstrating that the site was occupied close to 1.8 million years ago. The fossils are at least as ancient as those from Koobi Fora. KNM-ER 3733, recognized as one of the oldest crania with anatomy diagnostic for *H. erectus*, is dated to ca. 1.75 million years (Feibel et al. 1989). The KNM-ER 2598 occipital is often cited as evidence for a first appearance of *H. erectus* at ca. 1.9 million years, but if this specimen was collected from a lag surface, it may well have come originally from a higher level in the deposits, as noted by Suwa et al. (2007).

If the Dmanisi hominins are both anatomically primitive and geologically ancient, then an Asian origin for *H. erectus* must be considered. It may be argued that this species did not evolve directly from *H. habilis* or *H. rudolfensis* in eastern Africa. Indeed, it is increasingly apparent that *H. habilis* overlapped with *H. erectus* in the Turkana Basin, perhaps for a period of several hundred thousand years. Spoor et al. (2007) suggest that this lengthy episode of sympatry rules out any anagenetic relationship between the two species, and such a claim may be reasonable. However, there are no firm grounds for excluding *H. habilis* as the ancestor to later humans in another region. Although the number of lineages represented by the earliest *Homo* fossils is disputed, there is general agreement that crania (e.g., KNM-ER 1813) included within the hypodigm of *H. habilis* (*sensu stricto*) constitute possible structural antecedents to *H. erectus* (Lieberman et al. 1996; Strait et al. 1997; Kimbel et al. 2004). A population composed of such small-brained and lightly built individuals may well have been ancestral to the Dmanisi paleodeme.

According to this scenario, early (pre-*erectus*) *Homo* dispersed from Africa into western Asia, sometime prior to 1.8 million years ago (Fig. 14.8). These hominins were efficient bipeds of relatively small body mass. They were able to make simple chopping tools and utilize flakes and cobbles to obtain meat and marrow from animal carcasses. Such skills would have been important to survival at higher latitudes, especially during the cold winters, when plant foods were probably scarce (Lordkipanidze et al. 2005). There is little hard evidence to support such claims, as hominin fossils and stone artifacts of the requisite age have not yet been documented unequivocally in the Levant or in Arabia.

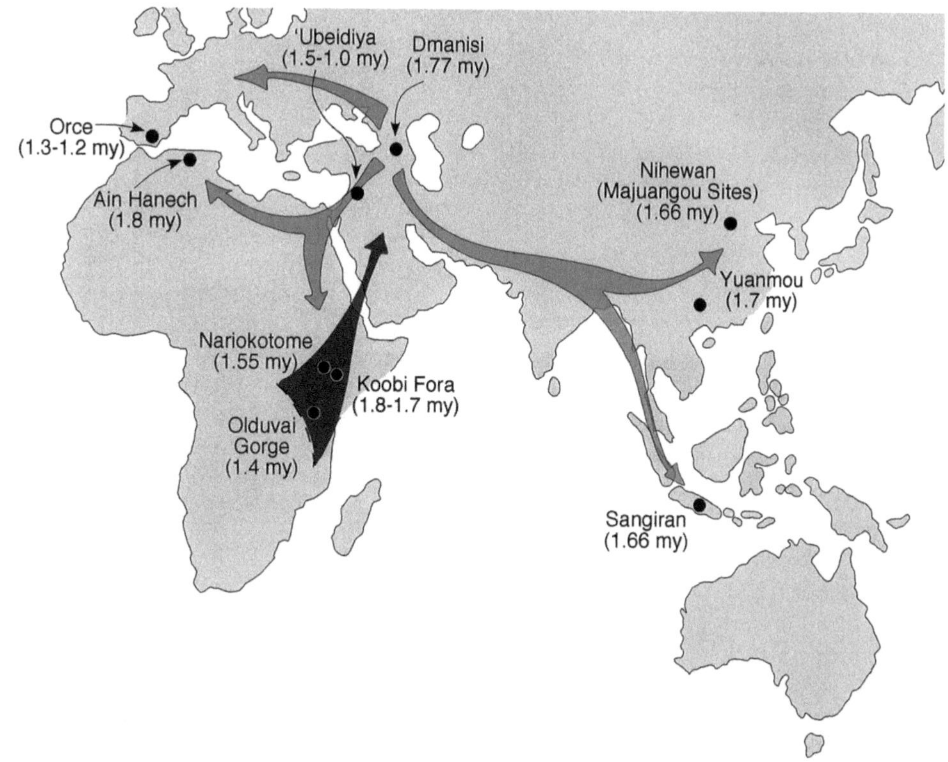

Fig. 14.8 Map illustrating an alternative "Asian origins" hypothesis. Here early *Homo* (cf. *H. habilis*) disperses from East Africa into southwestern Asia prior to 1.8 Ma (as suggested by the *dark arrow*). These early hominins evolve into *H. erectus* as documented by the paleodeme at Dmanisi. Later, the new species spreads from this Asian locus back to Africa, eastward toward Java and China, and eventually into Europe (*lighter arrows*)

Nevertheless, such a founding population could later have evolved the anatomical bauplan seen at Dmanisi. As a number of the characters displayed by the Dmanisi skulls are interpreted to be apomorphic for *H. erectus*, this population can in turn be viewed as closely related to other paleodemes in East Africa. This "Asian origins" hypothesis can accommodate the observation that the Koobi Fora crania (KNM-ER 3733, KNM-ER 3883) have larger capacities than those at Dmanisi, heavier superstructures, and bases that are more *erectus*-like. The greater stature and modern limb proportions exhibited by the Nariokotome skeleton may also be regarded as derived traits.

Given either an "African origins" or an "Asian origins" view, it is reasonable to suppose that the Dmanisi paleodeme is related not only to the hominins of Africa but also to populations of the Far East. Dates emerging from field work at Sangiran in Java (Swisher et al. 1998; Larick et al. 2001), the Nihewan Basin in northern China (Zhu et al. 2001, 2003, 2004), and Yuanmou to the southwest (Zhu et al. 2008) suggest that these areas were inhabited at least 1.7 million years ago, by people who must have moved through the southern parts of Asia. Here caution is appropriate, however. While there is clear evidence for a ca. 1.77 million year old human presence at Dmanisi, there is no certainty that hominins managed to colonize this region on a long-term basis. Indeed, it seems likely that many of the earliest dispersals eastward into Asia resulted in occupations that were ephemeral, and the Early Pleistocene record does not document any continuity of populations through southern Asia to the Far East (Dennell 2003). The routes taken by the first colonists to cross this landscape, the dynamics of such populations in respect to environmental change, and the extent of gene exchange among parapatric groups, are entirely unknown. There is no doubt, however, that *H. erectus* had settled both the south Asian tropics and more temperate parts of China, very early in the Pleistocene.

Summary

Dmanisi opens a rare and revealing window on the most ancient hominins documented in western Eurasia. The skulls described in this report demonstrate that the Caucasus people were small, with brain volumes close to those of African *H. habilis* or *H. rudolfensis*. Nevertheless, craniofacial morphology suggests that the Dmanisi paleodeme can be placed in the species *H. erectus*. These hominins were able to make stone tools, and simple choppers, flakes and numerous manuports occur even in the earliest levels at the site.

Along with the bones of animals (some bearing cutmarks or percussion damage), the stone artifacts attest to the ability of the hominins to obtain meat and marrow. Whether the Dmanisi population was able to establish a long-term presence in the Caucasus region is not clear, but it is evident that hominins were dispersing from Africa close to 1.8 million years ago, if not before.

One possibility is that the earliest travelers from Africa were representatives of *H. erectus*. This hypothesis implies an African origin for the species, presumably from earlier *Homo*. However, the relatively primitive morphology of the Dmanisi fossils suggests that an alternative hypothesis should be considered. It can be argued that the first hominins to leave Africa were similar to *H. habilis*, and that such a founding population gave rise later to *H. erectus* as documented at Dmanisi. This "Asian origins" view holds that *H. erectus* evolved in Eurasia and then returned to Africa. At about the same time, other populations of *H. erectus* must have dispersed from Georgia eastward toward Java and China. Such an interpretation is consistent with the distribution of anatomical characters in the ancient fossil assemblages, and it can be accommodated within the current geochronological framework established for the Early Pleistocene.

Acknowledgements We deeply appreciate the efforts of all of the students and professionals who have joined in the excavations and worked in the laboratory at Dmanisi, and we are grateful for the continuing support of the Georgian Academy of Sciences. For assistance with preparation of the hominin fossils at the Georgian National Museum and aspects of anatomical analysis, we are indebted to M. Bukhsianidze, T. Jashashvili, G. Kiladze, M. Ponce de Leon, A. Vekua and C. Zollikofer. R. Ferring and M. Tappen offered valuable advice. S. Kauffman and A. Hull helped with editing the photographs and preparing figures. G.P.R. thanks the American Philosophical Society, the Eckler Fund of Binghamton University, the Leakey Foundation and the National Science Foundation. D.L. acknowledges support from the Leakey Foundation and the National Geographic Society.

References

Antón, S. C. (2003). Natural history of *Homo erectus*. *Yearbook of physical anthropology, 46*, 126–170.

Antón, S. C., Spoor, F., Fellmann, C. D., & Swisher, C. C., III. (2007). Defining *Homo erectus*: Size considered. In W. Henke & I. Tattersall (Eds.), *Handbook of paleoanthropology* (Vol. 3, pp. 1655–1693). Berlin: Springer.

Asfaw, B., Gilbert, W. H., Beyene, Y., Hart, W. K., Renne, P. R., WoldeGabriel, G., et al. (2002). Remains of *Homo erectus* from Bouri, Middle Awash, Ethiopia. *Nature, 416*, 317–320.

de Lumley, H., Lordkipanidze, D., Feraud, G., Garcia, T., Perrenoud, C., Falgueres, C., et al. (2002). Datation par la méthode 40Ar/39Ar de la couche de cendres volcaniques (couche VI) de Dmanissi (Géorgie) qui a livré des restes d'hominidés fossils de 1, 81 Ma. *Comptes Rendus Palevol, 1*, 181–189.

de Lumley, M. A., Gabunia, L., Vekua, A., & Lordkipanidze, D. (2006). Les restes humains du Pliocene final et du debut du Pleistocene inferieur de Dmanissi, Georgie (1991–2000). *L'Anthropologie, 110*, 1–110.

Dean, M. C. (2006). Tooth microstructure tracks the pace of human life-history evolution. *Proceedings of the Royal Society B, 273*, 2799–2808.

Delson, E., Harvati, K., Reddy, D., Marcus, L. F., Mowbray, K., Sawyer, G. J., et al. (2001). The Sambungmacan 3 *Homo erectus* calvaria: A comparative morphometric and morphological analysis. *The Anatomical Record Part A, 262*, 380–397.

Dennell, R. (2003). Dispersal and colonization, long and short chronologies: How continuous is the Early Pleistocene record for hominids outside East Africa? *Journal of Human Evolution, 45*, 421–440.

Dennell, R., & Roebroeks, W. (2005). An Asian perspective on early human dispersal from Africa. *Nature, 438*, 1099–1104.

Feibel, C. S., Brown, F. H., & McDougall, I. (1989). Stratigraphic context of fossil hominids from the Omo Group deposits: Northern Turkana basin, Kenya and Ethiopia. *American Journal of Physical Anthropology, 78*, 595–622.

Gabunia, L., & Vekua, A. (1995). A Plio-Pleistocene hominid from Dmanisi, East Georgia, Caucasus. *Nature, 373*, 509–512.

Gabunia, L., Vekua, A., Lordkipanidze, D., Swisher, C. C., III, Ferring, R., Justus, A., et al. (2000). Earliest Pleistocene hominid cranial remains from Dmanisi, Republic of Georgia: taxonomy, geological setting, and age. *Science, 288*, 1019–1025.

Gabunia, L., de Lumley, M. A., Vekua, A., Lordkipanidze, D., & de Lumley, H. (2002). Découverte d'un nouvel hominidé a Dmanissi (Transcaucasie, Géorgie). *Comptes Rendus Palevol, 1*, 242–253.

Groves, C. P. (1989). *A theory of human and primate evolution*. Oxford: Oxford University Press.

Holloway, R. L. (1983). Human brain evolution: A search for units, models and synthesis. *Canadian Journal of Anthropology, 3*, 215–230.

Howell, F. C. (1978). Hominidae. In V. J. Maglio & H. B. S. Cooke (Eds.), *Evolution of African mammals* (pp. 154–248). Cambridge: Harvard University Press.

Howell, F. C. (1999). Paleo-demes, species clades, and extinctions in the Pleistocene hominin record. *Journal of Anthropological Research, 55*, 191–243.

Howells, W. W. (1980). *Homo erectus* – who, when and where: A survey. *Yearbook of Physical Anthropology, 23*, 1–23.

Kimbel, W. H., White, T. D., & Johanson, D. C. (1984). Cranial morphology of *Australopithecus afarensis*: A comparative study based on a composite reconstruction of the adult skull. *American Journal of Physical Anthropology, 64*, 337–388.

Kimbel, W. H., Johanson, D. C., & Rak, Y. (1994). The first skull and other new discoveries of *Australopithecus afarensis* at Hadar, Ethiopia. *Nature, 368*, 449–451.

Kimbel, W. H., Rak, Y., & Johanson, D. C. (2004). *The skull of Australopithecus afarensis*. Oxford: Oxford University Press.

Larick, R., Ciochon, R. L., Zaim, Y., Sudijono, S., Rizal, Y., Aziz, F., et al. (2001). Early Pleistocene $^{40}Ar/^{39}Ar$ ages for Bapang Formation hominins, Central Jawa, Indonesia. *Proceedings of the National Academy of Sciences of the United States of America, 98*, 4866–4871.

Lieberman, D. E., Wood, B. A., & Pilbeam, D. R. (1996). Homoplasy and early *Homo*: An analysis of the evolutionary relationships of *H. habilis* and *H. rudolfensis*. *Journal of Human Evolution, 30*, 97–120.

Lordkipanidze, D., Vekua, A., Ferring, R., Rightmire, G. P., Agusti, J., Kiladze, G., et al. (2005). The earliest toothless hominin skull. *Nature, 434*, 717–718.

Lordkipanidze, D., Vekua, A., Ferring, R., Rightmire, G. P., Zollikofer, C. P. E., Ponce de Leon, M. S., et al. (2006). A fourth hominin skull from Dmanisi, Georgia. *The Anatomical Record Part A, 288A*, 1146–1157.

Lordkipanidze, D., Jashashvili, T., Vekua, A., Ponce de Leon, M. S., Zollikofer, C. P. E., Rightmire, G. P., et al. (2007). Postcranial evidence from early *Homo* from Dmanisi, Georgia. *Nature, 449*, 305–310.

McCollum, M. A. (2000). Subnasal morphological variation in fossil hominids: A reassessment based on new observations and recent developmental findings. *American Journal of Physical Anthropology, 112*, 275–283.

McCollum, M. A., Grine, F. E., Ward, S. C., & Kimbel, W. H. (1993). Subnasal morphological variation in extant hominoids and fossil hominids. *Journal of Human Evolution, 24*, 87–111.

Oms, O., Pares, J. M., Martinez-Navarro, B., Agusti, J., Toro, I., Martinez-Fernandez, G., et al. (2000). Early human occupation of western Europe: Paleomagnetic dates for two Paleolithic sites in Spain. *Proceedings of the National Academy of Sciences of the United States of America, 97*, 10666–10670.

Plavcan, J. M. (2002). Taxonomic variation in the patterns of craniofacial dimorphism in primates. *Journal of Human Evolution, 42*, 579–608.

Potts, R., Behrensmeyer, A. K., Deino, A., Ditchfield, P., & Clark, J. (2004). Small mid-Pleistocene hominin associated with East African Acheulean technology. *Science, 305*, 75–78.

Richtsmeier, J. T., & Walker, A. (1993). A morphometric study of facial growth. In A. Walker & R. Leakey (Eds.), *The Nariokotome Homo erectus skeleton* (pp. 391–410). Cambridge: Harvard University Press.

Rightmire, G. P. (1990). *The evolution of Homo erectus. Comparative anatomical studies of an extinct human species*. Cambridge: Cambridge University Press.

Rightmire, G. P. (1998). Evidence from facial morphology for similarity of Asian and African representatives of *Homo erectus*. *American Journal of Physical Anthropology, 106*, 61–85.

Rightmire, G. P., Lordkipanidze, D., & Vekua, A. (2006). Anatomical descriptions, comparative studies and evolutionary significance of the hominin skulls from Dmanisi, Republic of Georgia. *Journal of Human Evolution, 50*, 115–141.

Rightmire, G. P., Van Arsdale, A. P., & Lordkipanidze, D. (2008). Variation in the mandibles from Dmanisi, Georgia. *Journal of Human Evolution, 54*, 904–908.

Rosas, A., & Bermúdez de Castro, H. M. (1998). On the taxonomic affinities of the Dmanisi mandible (Georgia). *American Journal of Physical Anthropology, 107*, 145–162.

Sagi, A., Belmaker, M., Ron, H., Enzel, Y., Agnon, A., R. & Bar-Yosef, O. (2005). *Paleomagnetic dating of 'Ubeidiya formation*. Israel Geological Society Annual Meeting 2005, Jerusalem.

Skinner, M. M., Gordon, A. D., & Collard, N. J. (2006). Mandibular size and shape variation in the hominins at Dmanisi, Republic of Georgia. *Journal of Human Evolution, 51*, 36–49.

Spoor, F., Leakey, M. G., Gathogo, P. N., Brown, F. H., Antón, S. C., McDougall, I., et al. (2007). Implications of new early *Homo* fossils from Ileret, east of Lake Turkana, Kenya. *Nature, 448*, 688–891.

Strait, D. S., Grine, F. E., & Moniz, M. A. (1997). A reappraisal of early hominid phylogeny. *Journal of Human Evolution, 32*, 17–63.

Suwa, G., Asfaw, B., Haile-Selassie, Y., White, T., Katoh, S., WoldeGabriel, G., et al. (2007). Early Pleistocene *Homo erectus* fossils from Konso, southern Ethiopia. *Anthropological Science, 115*, 133–151.

Swisher, C. C., III, Scott, G. R., Curtis, G. H., Butterworth, J., Antón, S. C., Jacob, T., et al. (1998). The antiquity of *Homo erectus* in Java: $^{40}Ar/^{39}Ar$ dating and paleomagnetic study of the Sangiran area. Dual Congress 1998 (abstract), Sun City.

Tappen, M., Ferring, R., Lordkipanidze, D., Vekua, A., & Kiladze, G. (2002). Preliminary observations on the vertebrate taphonomy of the Dmanisi locality in the Republic of Georgia. In M. de Renzi, M. V. P. Alonso, M. Belichón, E. Pañalver, P. Montoya, & A. Márquez-Aliga (Eds.), *Current topics on taphonomy and fossilization* (pp. 161–170). Valencia: Ajuntament de Valencia.

Tchernov, E. (1987). The age of the 'Ubeidiyu formation, an early hominid site in the Jordan Valley. *Israel Journal of Earth Science, 36*, 3–36.

Tobias, P. V. (1991). *Olduvai Gorge. Vol. 4. The Skulls, Endocasts and Teeth of Homo habilis*. Cambridge: Cambridge University Press.

Van Arsdale, A. P. (2006). *Mandibular variation in early Homo from Dmanisi, Georgia*. PhD dissertation, University of Michigan, Ann Arbor, MI.

Vekua, A., Lordkipanidze, D., Rightmire, G. P., Agusti, J., Ferring, R., Maisuradze, G., et al. (2002). A new skull of early *Homo* from Dmanisi, Georgia. *Science, 297*, 85–89.

Weidenreich, F. (1943). The skull of *Sinanthropus pekinensis*: A comparative study of a primitive hominid skull. *Palaeontologica Sinica (New Series D), 10*, 1–484.

Weidenreich, F. (1951). Morphology of solo man. *Anthropological Papers of the American Museum of Natural History, 43*, 205–290.

Wood, B. (1992). Origin and evolution of the genus *Homo*. *Nature, 355*, 783–790.

Zhu, R., Hoffman, K. A., Potts, R., Deng, C. L., Pan, Y. X., Guo, B., et al. (2001). Earliest presence of humans in northeast Asia. *Nature, 413*, 413–417.

Zhu, R., An, Z., Potts, R., & Hoffman, K. A. (2003). Magnetostratigraphic dating of early humans in China. *Earth Science Reviews, 61*, 341–359.

Zhu, R., Potts, R., Xie, F., Hoffman, K. A., Deng, C. L., Shi, C. D., et al. (2004). New evidence on the earliest human presence at high northern latitudes in northeast China. *Nature, 431*, 559–562.

Zhu, R. X., Potts, R., Pan, Y. X., Yao, H. T., Lü, L. Q., Zhao, X., et al. (2008). Early evidence of the genus *Homo* in East Asia. *Journal of Human Evolution, 55*, 1075–1085.

Part V
Summary, Synthesis and Future Directions

Introduction

In the two chapters of this section, the authors attempt to summarize the status of our current understanding of the earliest dispersal(s) of hominins from Africa to Eurasia, with a focus on unresolved issues and priorities for future research. In "Out of Africa I: current problems and future prospects", Robin Dennell offers a detailed summary of the most recent research on early hominin dispersal and how it has called into question earlier views of a Middle Pleistocene dispersal of *Homo erectus*. In particular he focuses on problems with our understanding of (1) the systematics, adaptations, and biogeography of *Homo erectus*; (2) climate and paleoecology of Asia in the Late Pliocene; and (3) the many gaps in our knowledge of Asian paleontology and opportunities for new research in unexplored areas.

In "Summary and Prospectus", John Fleagle and John Shea try to identify the common themes in the preceding chapters of the volume and use this summary to identify priorities for future research on early hominin dispersal. Among the common themes and debated topics are the nature of the stimulus for the initial hominin dispersal from Africa; the number, timing, and likely routes of dispersal; the identity of the dispersing taxa, both hominins and potential "fellow travelers"; our ability to identify hominin presence through either fossils or archeological remains; and the need for accurate dating.

The Editors

Chapter 15
"Out of Africa I": Current Problems and Future Prospects

Robin Dennell

Abstract This paper examines current weaknesses in the Out of Africa I model, and indicates why it is less robust than often proposed. It examines the main gaps in our understanding of hominin dispersals and settlement in Asia before 0.5 Ma, and highlights those shortcomings that have a realistic chance of being rectified, and which most impede fruitful discussion of early Asian prehistory. It also proposes that any discussion of early hominin dispersals out of (and possibly into) Africa before 0.5 Ma needs to take much fuller account than has often been the case of recent data on climatic change across southern Asia.

Keywords Asia • Colonization • Dispersal • *Homo erectus* • Pleistocene • Pliocene

Introduction

For the last 20 years, 'Out of Africa I' – the proposition that *H. erectus*[1] originated in East Africa and colonized Asia in the Early Pleistocene – has been a useful way of making sense of a large set of Late Pliocene and Early Pleistocene fossil hominin and archeological data from East Africa, and a very small amount of Early Pleistocene evidence from Asia. According to this model, the genus *Homo* and the species *H. erectus* both originated in Africa, as did tool-making and many of the other skills needed to survive in the drier and more strongly seasonal conditions that became more widespread in the Late Pliocene (see e.g., Ruddiman et al. 1989; Bonnefille 1995; Vrba 1995; Bobe and Behrensmeyer 2004; Fernández and Vrba 2006). At some point in the Late Pliocene or Early Pleistocene, some groups of *H. erectus* supposedly moved out of Africa and colonized southern Asia. Two pieces of information often highlighted in this model are the Nariokotome *H. erectus* skeleton WT15000, dated to 1.53 Ma (Walker and Leakey 1993), and the re-dating by Swisher et al. (1994) of the earliest hominins at Mojokerto and Sangiran in Java to 1.81 ± 0.04[2] and 1.66 ± 0.04 Ma respectively. The Nariokotome skeleton helped to crystallize the awareness that the transformation from *Australopithecus* to *Homo* involved "not only an expansion of the brain and a reduction of the cheek teeth, but a change in walking and climbing behaviour" (McHenry and Coffing 2000: 125). Its body size and proportions as well the size of its brain seemed to provide an explanation of how and why Asia was colonized. "Here at last", suggested Tattersall (1997: 47), "we have early hominids who were at home in the open savanna"; with its tall, thin body and long limbs, it was far better adapted for life in the grasslands that were expanding in East Africa at that time than its contemporaries and immediate predecessors. Additionally, its larger brain would have made it a more skillful competitor with other predators, but its energetically expensive brain would also have required a larger proportion of meat in its diet (see e.g., Aiello and Wheeler 1995). The unexpectedly early dates obtained by Swisher's team for the earliest Javan hominins implied that *H. erectus* had already left Africa before the development of the Acheulean handaxe and cleaver ca. 1.5 Ma, and did not need these items during its dispersal across southern Asia: indeed, the new dates neatly explained why handaxes and cleavers are virtually absent east of the "Movius Line". The implication that *Homo erectus* had expanded very rapidly across the whole of southern Asia shortly after it first appeared in East Africa focused attention on its distinctive anatomy (particularly obligate bipedalism [see e.g., Bramble and Lieberman 2004], modern body proportions [Ruff 2002], a larger brain, and probable thermoregulatory efficiency in

[1] Because of continuing debate over whether early East African specimens attributed by some to *H. ergaster* should be included in the taxon *H. erectus* (see below), I use the term *H. erectus* to include both the early East African and the early East Asian specimens that some classify as *H. erectus* sensu stricto (s. s.).

R. Dennell (✉)
Department of Archaeology, University of Sheffield, S1 4 ET, UK
e-mail: r.dennell@sheffield.ac.uk

[2] As discussed below, it now appears that the Mojokerto specimen is considerably younger than 1.8 Ma.

J.G. Fleagle et al. (eds.), *Out of Africa I: The First Hominin Colonization of Eurasia*, Vertebrate Paleobiology and Paleoanthropology,
DOI 10.1007/978-90-481-9036-2_15, © Springer Science+Business Media B.V. 2010

sweating and cooling [Wheeler 1991, 1992]) that would have enabled it to cope with the hot and dry savannah grasslands of both Africa and Asia, in which there was fierce competition for meat resources that were often plentiful but usually dispersed.

The addition of European evidence to the Out of Africa I model indicates that hominins had entered southern Europe by perhaps as early as 1.2–1.4 Ma (Azarello et al. 2006; Carbonell et al. 2008) and northern Europe by ca. 0.7 Ma (Parfitt et al. 2005) as part of the same process of colonization from a source that was directly or indirectly African. As such, the Out of Africa I model gives a satisfying coherence to Old World prehistory before 1.0 Ma in both Asia and Europe, and confirms the primacy of Africa in human evolution. At the close of the twentieth century, the Out of Africa I model found widespread acceptance amongst paleoanthropologists as a way of envisaging early human evolution in the Old World.

Part I: Problems with the "Out of Africa I" Model

Despite its apparent coherence, at the beginning of the twenty-first century, there are a number of reasons why this "grand narrative" should be questioned (see e.g., Dennell 1998; Dennell and Roebroeks 2005). One is that early *Homo erectus* was not as distinctive from other forms of early *Homo* (see below), and another is that savannah grasslands did not become prevalent in East Africa until long after 1.8 Ma, and perhaps not until the Middle Pleistocene (Cerling 1992). The scenario whereby *Homo erectus* was the first hominin "at home in the savannah" probably overcooks the evidence, and stands in danger of losing both its main protagonist and its setting. A more open-minded approach may be preferable, whereby early *H. erectus* was simply one of several hominins that might have had the ability to disperse out of Africa at a time when environments were becoming more open.

There are also several long-standing problems that need to be clarified before we can be confident as to when and which hominins first dispersed out of Africa. Four that are particularly important are: (I) the shortcomings of the East African Late Pliocene fossil hominin record (II) doubts about where *H. erectus* originated; (III) doubts over the taxonomic unity of early *H. erectus* in Africa and Asia; and (IV) the inadequacy of the Late Pliocene fossil vertebrate record of Southwest Asia prior to 1.8 Ma. All four reasons are thrown into sharp focus by interpretations of the continuing discoveries of hominin remains and stone artefacts at the 1.75 Ma-old site of Dmanisi, Georgia. Each point can be taken in turn.

The Enigmatic Origin of H. erectus in Africa

"The origin of the genus *Homo* remains poorly understood. Definitions of *Homo* have changed through time, but usually have emphasized large cranial capacity, relatively small dentition, obligate bipedality, and the ability to make tools" (Bobe and Behrensmeyer 2004: 414). One reason why the origin of *Homo* remains unclear is that the East African fossil hominin record is poor between 2.0 and 2.4 Ma (Kimbel 1995: 435), and thus there is "little consensus on which species of *Australopithecus* is the closest to *Homo*" (McHenry and Coffing 2000: 126). At present, the earliest specimens of our genus date from ca. 2.4 Ma. The key evidence comprises a temporal bone from the Chemeron Formation, Kenya (Hill et al. 1992; Deino and Hill 2002; Sherwood et al. 2002) and several teeth from the Omo sequence (Suwa et al. 1996). None of these is identified beyond generic level. The dental evidence from Omo is particularly difficult to classify because of the considerable overlap in morphology between early *Homo* and late, non-robust australopithecines (Suwa et al. 1996: 247). There is also a mandible from Uraha, Malawi (Schrenk et al. 1993), that is attributed to *H. rudolfensis* (see below), and also dated to ca. 2.4 Ma, although Deino and Hill (2002) suggest that it could be much younger. Another specimen, this time a maxilla (AL666-1) from the Hadar Formation, Ethiopia that is identified as *Homo* with affinities to *H. habilis*, is robustly dated by fauna and argon–argon dating to 2.33 ± 0.07 Ma (Kimbel et al. 1997). There is also a recent find in the same age-range of a juvenile lower molar from West Turkana, ca. 2.3–2.4 Ma and attributed to *Homo* (Prat et al. 2005). Shortly after 2.0 Ma, there are several specimens from Olduvai, and Koobi Fora. Those from Bed I and lower Bed II at Olduvai are ca. 1.9–1.5 Ma, and usually classified as *Homo habilis;* its anatomy is now reasonably well documented from cranial, dental and post-cranial evidence, including its hand and feet, and a post-cranial skeleton (OH 62). Specimens attributed to *H. habilis* from Koobi Fora have been dated to 1.9–1.5 Ma, but recently the age of some important specimens has been revised from 1.88–1.90 to 1.65 Ma (Gathogo and Brown 2006). The type specimen of *H. ergaster* (KNM ER 992, a mandible) is ca. 1.49 Ma; cranium KNM ER 1470 that is often classified as *H. rudolfensis* is ca. 1.88–1.90 Ma (Schwartz and Tattersall 2003: 133).

The origin of *H. erectus* in East Africa remains unclear. One reason is there are only three sets of finds (KNM-ER 803, 1808 and WT 150000) where post-cranial material has been found in direct association with cranio-dental specimens of early East African *H. erectus*; another is that one of its contemporaries, *H. rudolfensis,* is known only from cranial material, and thus isolated post-cranial specimens could belong to either. The earliest well-dated specimens often attributed to *Homo erectus* are a pelvic specimen, KNM-ER

3228, dated at ca. 1.9 Ma, and femora and other leg bones (KNM-ER 1472 and 1481) dated at ca. 1.89 Ma, although McHenry and Coffing (2000: 128) suggest that these might have belonged to *H. rudolfensis*, the remains of which were found nearby. There is also an occipital specimen (KNM ER 2598) from Ileret, dated at 1.89 Ma that may have belonged to *H. erectus*. An important cranial specimen, KNM-ER 3733, dated to either 1.78 Ma or 1.65 Ma, according to Gathogo and Brown (2006) may also belong to *H. erectus*, although some researchers (e.g., Wood 1984; Schwartz 2000; Schwartz and Tattersall 2003: 591–593) feel that it does not fit well with the cranium from the 1.53 Ma skeleton WT 15000 or the isolated 1.57 Ma cranium, KNM-ER 3833. On current evidence, the least ambiguous specimens are dated to ca. 1.5–1.7 Ma. Evidence for a local, East African origin of *H. erectus* remains limited and ambiguous. As example, Wood (1984: 99) noted that "the evidence for incipient *H. erectus* traits in *H. habilis* is equivocal". Likewise, Walker and Shipman (1996: 109) saw the differences between *H. habilis* and *H. erectus* as too great to have developed anagenetically in the time available, and regard *H. erectus* as "without a clear ancestor, without a past". We are thus in the curious position of assuming that *H. erectus* was the first hominin to migrate from East Africa into Asia, whilst lacking conclusive evidence that it was in fact indigenous to East Africa.

The distinctiveness of early *H. erectus* also appears contentious. For example, Wood and Collard (1999) have claimed that there were fundamental differences between *H. ergaster* (i.e., early African *H. erectus*) and earlier types of hominins. However, its dental maturation does not appear to have been any faster than its contemporaries (Dean et al. 2001), and its limb proportions may also have been similar to *H. habilis* (see Haeusler and McHenry 2004). If the latter are correct in suggesting that *H. habilis* was also relatively long-legged, it becomes harder to explain why *H. erectus* was the only hominin that ever dispersed out of Africa.

The Implications of Dmanisi

The problems of establishing the origins of *H. erectus* have been compounded by the difficulties in assimilating the 1.75 Ma-old hominins from Dmanisi, Georgia, within a simple model of African origin and Asian dispersal by an early East African population of *H. erectus* (see Rightmire et al. 2010). The first discovery, of mandible D211, could be safely classed as *H. erectus* (Gabunia and Vekua 1995), although Bräuer and Schultz (1996) suggested that it most resembled late examples of that taxon. Following the discoveries of crania D2880 and D2882, with cranial capacities of only 775 and 650 cm^3, respectively, Gabunia et al. (2000a: 1025) suggested that "these hominids may represent the species that initially dispersed from Africa and from which the Asian branch of *H. erectus* was derived". It was, in other words, a very early version of early East African *H. erectus*, as well as a putative ancestor of the East Asian *H. erectus* s. s. The discovery of the third cranium (D2700), with a cranial capacity of only 600 cm^3, mandible D2735, and 10 isolated teeth led to a different conclusion, this time that they were most like *H. habilis*: "The Dmanisi specimens are the most primitive and small-brained fossils to be grouped with this species or any taxon linked unequivocally with genus *Homo* and also the ones most similar to the presumed *habilis*-like stem" (Vekua et al. 2002: 85). They also suggested that "this population is closely related to *Homo habilis* (sensu stricto)", and that "the ancestors of the Dmanisi population dispersed from Africa before the emergence of humans identified broadly with the *H. erectus* grade". Subsequently, Gabunia et al. (2002) have proposed that mandible D2600 should be assigned to a new taxon, *Homo georgicus,* which they consider to be a very early form of the genus *Homo* that dispersed from Africa to Eurasia between 2 and 1.8 Ma. This suggestion is broadly consistent with the assessment of Schwartz and Tattersall (2003: 490) that "none of the Dmanisi fossils can be regarded as belonging to either Asian *Homo erectus* or to the species containing its supposed African relatives".

In the most recent development, Rightmire et al. (2006; also 2010) propose that the Dmanisi crania are most appropriately classified as *H. erectus*, but are also the most primitive types yet found. (They were unable to agree over the taxonomic status of mandible D2600, the type specimen of *H. georgicus*). The most provocative part of their analysis is their suggestion that the Dmanisi population may be ancestral to early *H. erectus* in East Africa; as they state (2006: 140) "Dating does not presently rule out the possibility that *H. erectus* originated in Eurasia and that some groups then returned to Africa, where they evolved towards *H. erectus ergaster*".

Debate will doubtless continue over whether the Dmanisi hominins should be classified as an extremely primitive form of *H. erectus*, as proposed by Rightmire et al. (2006, 2010) and Antón (2002) and/or under the new taxon *H. georgicus* (Gabunia et al. 2002); or regarded as a composite sample of two populations, neither of which is *H. erectus* (Schwartz 2000). Further cranio-dental discoveries will doubtless cause further changes of view, as will the associated (and long awaited) post-cranial specimens (Lordkipanidze et al. 2007). Current indications are that the Dmanisi individuals were short, as befitting the small size of the crania. Estimates made from the length of the third metatarsal specimen suggest a stature of only 1.48 m (SE 65.4 mm) (Gabunia et al. 2000c), considerably less than that of WT 15000, and at the small end of the female range for early *H. erectus* in Africa (Antón 2002: 151).[3]

To conclude, there is now a rival hypothesis to the canonical one that *H. erectus* originated in Africa and was the first hominin to disperse into Asia. The alternative proposition is that *H. erectus* originated in Asia from a population of *Homo* that dispersed out of Africa before 1.8 Ma; some of this Asian population of early *H. erectus* then dispersed back into East Africa, and others dispersed eastwards to Java. The trinomena *H. erectus georgicus, H. erectus ergaster* and *H. erectus erectus* are available to differentiate these populations (Rightmire et al. 2006).

Rightmire is not the first to propose that *H. erectus* originated in Asia, as Swisher et al.'s (1994, p. 1118) suggestion that "*Homo erectus* may have evolved outside Africa" stemmed from the observation that the oldest specimens of *H. erectus* in Java might antedate the earliest examples in East Africa. White (1995: 383) also suggested on the basis of general morphological criteria that "It seems more likely that *Homo erectus* is an immigrant from Asia to the eastern African area than an anagenetic, *in situ* derivative from *Homo habilis*". Likewise, Asfaw et al. (2002: 319) stated that: "Uncertainties surrounding the taxon's appearance in Eurasia and southeast Asia make it impossible to establish accurately the time or place of origin of *H. erectus*. Available evidence is insufficient to detect the direction of its geographic dispersal. Given new perspectives afforded by the discoveries at Dmanisi in Eurasia, the assumption that the earliest *H. erectus* populations migrated from Africa to Eurasia, rather than invading Africa from Eurasia, is premature". However, Rightmire et al.'s (2006, 2010) arguments for an Asian origin of *H. erectus* are undoubtedly the most convincing as they rest upon detailed, first-hand examination of the Dmanisi material, most of which had not been discovered at the time of the assessments by either Swisher or Asfaw's team, or White.

The Taxonomic Unity of H. erectus*: Coherent Taxon or a Dust-Bin Category?*

The discussions over how to classify the Dmanisi hominins highlight the long-running debate between "lumpers" and "splitters" over whether *H. erectus* was the first hominin to inhabit both Africa and Asia, or whether its coherence is more apparent than real. For the lumpers, the *Bauplan* of *H. erectus* is shared by all Asian hominin specimens prior to the late Middle Pleistocene as well as those classified as *H. erectus* from East Africa (see e.g., Antón 2002; Antón and Swisher 2004; Asfaw et al. 2002; Rightmire et al. 2006); for the splitters, far too much morphic diversity has been herded into one category (see e.g., Schwartz and Tattersall 2000, 2003: 591–593). We have already noted how these debates have surfaced in recent discussions of the Dmanisi material. Regarding the larger and longer-investigated hominin sample from Java, the same type of discussions continue over whether they should be assigned to *H. erectus* (and thus essentially part of the same population as represented by WT15000 in Kenya, or the Dmanisi specimens), *H. erectus* s. s. (and probably derived from African populations of *Homo*), or even seen as a composite sample of *H. erectus* s. s., a different ("*Meganthropus*") type of hominin, and even perhaps some non-hominin taxa (e.g., Tyler 1992; Krantz 1994; Schwartz and Tattersall 2003: 591–593).

To conclude, there is little prospect of any consensus being reached amongst specialists over the taxonomic unity of *H. erectus* during the Early Pleistocene. For that reason alone, it seems advisable to be cautious about assuming too readily that *H. erectus* was the only type of hominin in Asia during the Late Pliocene and Early Pleistocene, or that it originated in Africa.

The Asian Late Pliocene Fossil Record: Absence of Evidence Versus Evidence of Absence

A topic that has rarely been addressed in discussions of when hominins first dispersed into Asia is the poor quality of the Asian vertebrate fossil record in the Late Pliocene and Early Pleistocene, particularly when considered against the size of the continent. As example, the key Early Pleistocene Asian fossil hominin localities are Dmanisi and Sangiran, 5,300 air-miles apart – a distance comparable to that between Boxgrove, UK, and Sterkfontein, South Africa. Dmanisi is also ca. 2,300 miles from the nearest relevant fossil locality in East Africa. The area of Asia south of latitude 40° N. (i.e., south of a line from Dmanisi to the Nihewan Basin) is ca. 10 million square miles, or approximately the same area as Africa. Two sets of fossil hominin specimens from an area this vast are grossly insufficient to indicate that hominins were absent before 1.8 Ma, or that *H. erectus* was the first and only African hominin to enter Asia before modern humans. (For example, Argue et al. 2006: 373 suggest that the ancestor of *H. floresiensis* "was in the process of evolving from *Australopithecus* to *Homo* when it diffused from Africa … before the appearance of the fully derived *Homo* morphology, that is, prior to about 2 Ma").

[3] Antón (2002) also points out that the Dmanisi specimen differs from the KNM ER-803 specimen not only in size but also in the morphology of the lateral facets. These differences further accentuate the differences between the earliest-known Asian hominins and contemporaneous populations of *H. erectus* in East Africa.

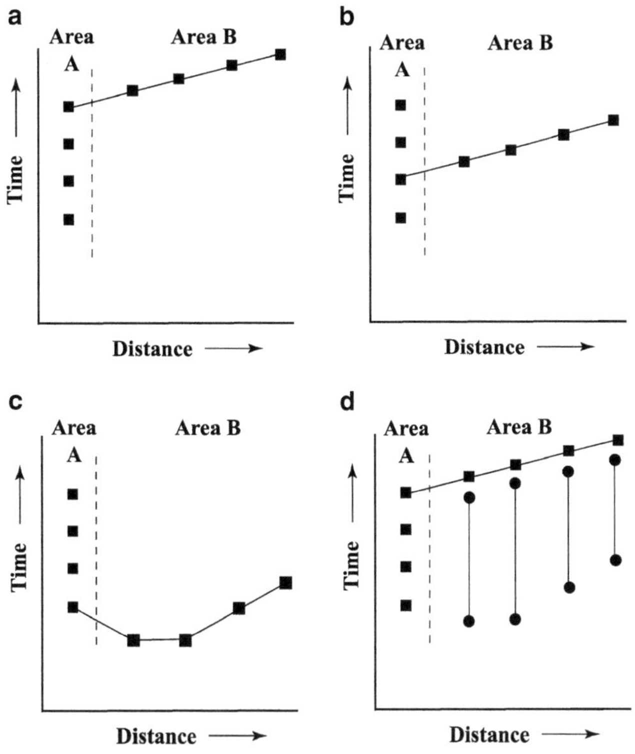

Fig. 15.1 Differing assessments of when and where a taxon originated

The dangers of relying upon a small number of observations that purportedly show the first appearance of hominins in various parts of Asia were discussed in Dennell and Roebroeks (2005) and are summarized in Fig. 15.1a–d. In the first example (Fig. 15.1a), an initial set of observations suggests that a taxon evolved in one area, and then migrated into an adjacent territory some time after its first appearance – in much the same way that the timing of the dispersal of *Homo erectus* was envisaged in the 1980s. In the second example (Fig. 15.1b), the timing of this dispersal appears to have happened at a much earlier date, shortly after the taxon first appeared in its core area: this of course is how Out of Africa I is now commonly envisaged as a result of the re-dating of the earliest hominins in Java, the discovery of those at Dmanisi, and most recently, the dating of the earliest sites in northern China (see below). However, for all we know, further discoveries might show a fundamentally different pattern (shown in Fig. 15.1c), in which the taxon actually originated in the area in which it was thought to have colonized, and then migrated into the area in which it was thought to have evolved.[4] As seen above, Rightmire et al.'s (2006) assessment of the Dmanisi evidence suggests that this might be the case with *H. erectus*. In order to be confident that a set of observations about the first appearance of a taxon outside its apparent area of origin will not turn out to be incorrect about both the timing of its dispersal, and also its direction, first appearance dates need to be matched as closely as possible by dates showing the last probable absence (LPA). Although the absence of a taxon is almost impossible to demonstrate, we can be more confident if the fossil data sets prior to its first appearance are comparable in quality to subsequent ones.

At present, it is simply not possible to provide accurate dates of when hominins were last absent across southern Asia, and their absence cannot therefore be demonstrated across most of Asia during the Late Pliocene and Early Pleistocene. Southwest Asia is a particularly crucial gap in current knowledge.

Southwest Asia

Southwest Asia covers 2.4 million square miles and is considerably larger than the combined areas of the EU (1.69 million square miles for all 27 states), or Kenya, Ethiopia, Somalia and Tanzania (1.27 million square miles (see Dennell 1998, Table 9.3). Its Late Pliocene vertebrate fossil record prior to Dmanisi is almost wholly unknown. There is a mid-Pliocene assemblage from Çalta, in western Turkey, which has a small number of taxa that include *Giraffa, Hipparion, Chasmaportetes* and *Nyctereutes;* isotopic analyses indicate an open steppic environment (Bocherens and Sen 1998). The best-known assemblage is that from Bethlehem (Gardner and Bate 1937; Hooijer 1958). As it contained the three-toed horse *Hipparion* that was replaced by the one-toed *Equus* after ca. 2.5 Ma in southern Asia (Lindsay et al. 1980), its age is probably ≥2.3 Ma. The assemblage is very small, with only 11 taxa,[5] dominated by animals with an adult body weight of >60 kg and thus larger than adult Late Pliocene hominins. These fossils were found in very coarse gravel (with clasts up to 0.5 m long) in a clay matrix, in which "small animals would have been unable to withstand the conditions of the bone-bearing deposit" (Hooijer 1958: 289). Most of the identifiable specimens were poorly-preserved teeth, and the sample had undergone extreme attrition. At Kvabebi, Georgia, dated to >2.6 Ma (i.e., earlier than the earliest stone tools in Africa (Semaw et al. 2003), there are 21 mammalian taxa indicative of a riverine and marshy environment (Vekua 1995; Hemmer et al. 2004). Two other Georgian localities, Kochachuri and Calka are slightly earlier than Dmanisi and

[4]For example, it was commonly believed in the 1920s and 1930s that hominins evolved in Central Asia and later migrated into Africa (see e.g., Black 1925); not until the 1960s could it be clearly shown that the reverse pattern was more likely; see e.g., Dennell (2001).

[5]*Nyctereutes megamastoides*, *Homotherium* sp., *Archidiskodon* cf. *planifrons*, *Hipparion* sp., *Dicerorhinus etruscus*, *Sus* cf. *strozzii*, *Giraffa* cf. *camelopardalis*, *Leptobos* sp., and *Gazellospira torticornis* (Hooijer 1958:289). Gardner and Bate (1937) also recognized *Testudo* (two types), *Hippopotamus* and *Stegodon*, but the last two of these were not confirmed by Hooijer (1958).

yielded small assemblages dominated by large taxa (Vekua 1995).

The Bethlehem assemblage in particular cannot be regarded as a reliable indication that hominins were absent in the Levant in the Late Pliocene. Small assemblages, with only a few taxa that are mainly from large animals, are most unlikely to contain the remains of hominins that were probably rare, and whose skeletons are fragile (compared to horse or elephant, for example). By way of comparison, the three sites in Southwest Asia that have produced hominin remains >0.7 Ma all have large fossil assemblages with several taxa: 21 taxa at Dmanisi, 33 (excluding microfauna) at 'Ubeidiya (Tchernov 1987), and 18 at Gesher Benot Ya'aqov (see Martínez-Navarro 2004). Even so, very few hominin specimens were found at 'Ubeidiya[6] and Gesher Benot Ya'aqov (GBY): one incisor from 'Ubeidiya (Belmaker et al. 2002), and two femora of uncertain provenance from GBY that were found in a box of cervid long-bones (Geraads and Tchernov 1983). The same is true of large primates – just one bone attributed to *Theropithecus* from 'Ubeidiya (Belmaker 2002).

To conclude, the fossil vertebrate record of Asia in the Late Pliocene is too poor to demonstrate that hominins were absent before 1.8 Ma. Put more provocatively, we cannot be certain that hominins did not disperse out of Africa shortly after or even before stone tool-making became routine ca. 2.6 Ma.

Rather than simply criticizing the current problems concerning the Out of Africa I model, and especially the lamentably poor amount of relevant information from Late Pliocene and Early Pleistocene Asia, this paper takes a positive approach by proposing how the fossil hominin and archeological record for the Early Pleistocene could be substantially improved by even modest investments of carefully-targeted fieldwork. In the next section, we consider how the paleoclimatic data now available from Asia could be used to model the likely nature of early hominin settlement in Asia.

Part II: Climatic Change and Hominin Settlement in Asia During the Late Pliocene and Early Pleistocene

The last 10–15 years have seen an enormous output of papers on Asian paleoclimatology that provides the basis for modelling hominin settlement in Asia during the Late Pliocene and Early Pleistocene. Examples are studies of the loess and paleosol sequences of the Chinese Loess Plateau (e.g., Liu et al. 1999; Ji et al. 2001; Lu et al. 1999) and Tajikistan, Central Asia (e.g., Dodonov 2002; Dodonov and Baiguzina 1995; Yang and Ding 2006); sediment cores from Lake Baikal (e.g., Williams et al. 1997), stable isotope studies in South Asia (e.g., Quade et al. 1989, 1993) and North China (e.g., Ding and Yang (2000), as well as marine records from the North Pacific (e.g., Rea et al. 1998; Prueher and Rea 2001), South China Sea (e.g., Tian et al. 2004; Baoqi Huang et al. 2003; Liu et al. 2003), Indian Ocean (e.g., Clemens et al. 1996), and the East Mediterranean (e.g., Kroon et al. 1998). There is also the PRISM reconstruction of Asian vegetation and climate in the Late Pliocene ca. 3.2 Ma that is based largely on inferences from marine paleoclimatic data (Dowsett et al. 1994) but which provides a useful reconstruction of Asian vegetation at this time (see Fig. 15.2). In order to use all this information, we first need an overview of how Asian climate operates south of latitude 40–45° N., i.e., within the areas settled by hominins in the Early Pleistocene.

The climate of Asia south of latitudes 40–45° N is dominated by two weather systems. In Southwest and Central Asia, most precipitation occurs in winter and spring, and is transported by westerly winds blowing inland from the Mediterranean and also the Black and Caspian Seas. As one proceeds inland from the coast of northern Israel, Lebanon and western Turkey (where rainfall can reach 1,000 mm), the rainfall totals decrease dramatically, and almost all of the Arabian Peninsula, much of Syria, Iraq, Iran (particularly on the Iranian Plateau) and Central Asia is semi-desert or desert, with rainfall as low as <50 mm p.a. Exceptions where rainfall is higher are the Caucasian, Tauros and Zagros Mountains, and the southern coasts of the Black and Caspian Seas. In South, Southeast and East Asia, most rainfall occurs in summer and is delivered by the Indian and East Asian summer monsoons that originate in the Indian Ocean, South China Sea and western Pacific. In south, east and north India, Bangladesh, mainland Southeast Asia and South China, rainfall totals are often considerably in excess of 1,000 mm p.a. Areas on the margins of the summer monsoon, such as North China and Northwest India, are semi-arid, and highly vulnerable to the consequences of decreased rainfall.

The paleoclimatic information now available from Asia shows two important features. The first is that its climatic record for the Early Pleistocene exhibits the same high-frequency, low amplitude climatic changes (with an average frequency of 41 ka) that are indicated in the marine records of the North Atlantic and West Pacific. The best terrestrial records of these changes are the loess and paleosol sequences of the Chinese Loess Plateau and Tajikistan, and the sediment records from Lake Baikal. Relevant off-shore marine records are from the North Pacific, South China Sea, Indian Ocean and East Mediterranean. Overall, Asian climatic

[6] Two teeth and four cranial fragments were found in the early excavations at 'Ubeidiya that were assigned to *Homo erectus* (Tobias 1966). Tchernov (1987) recorded the teeth as in situ and included them in his analysis of the 'Ubeidiya fauna. However, fluorine tests suggested that the cranial fragments were recent, and probably intrusive (Molleson and Oakley 1966).

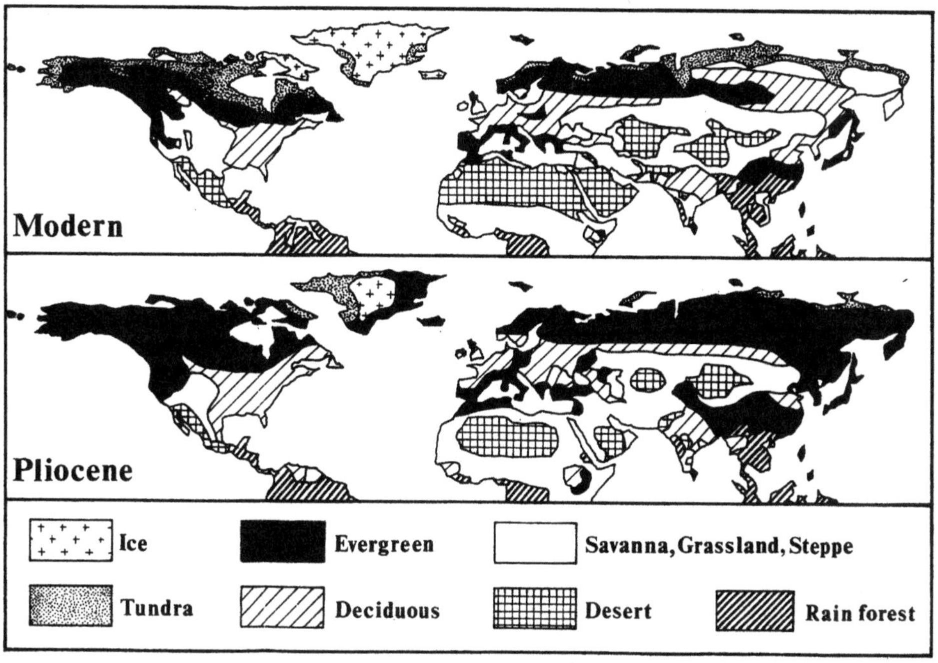

Fig. 15.2 The PRISM reconstruction of Late Pliocene vegetation in Eurasia ca. 3.2 Ma. The most striking contrasts between the Pliocene ca. 3 Ma and today are in the extent of grasslands and the deserts. As shown, in the Late Pliocene, savannah or steppe grasslands extended from northern China to West Africa, and the present-day desert barrier between the Sahara and Arabia did not exist. Conditions for hominin dispersals out of (and perhaps into) Africa were thus more favourable then than in recent times (Dowsett et al. 1994, Fig. 11; reproduced with permission from Elsevier)

change was broadly synchronous with that in the North Atlantic and western Pacific after 2.5 Ma except in North China, where it this synchronicity was delayed until 1.67 Ma because of the regional effects of uplift of the Tibetan Plateau (Tungsheng Liu et al. 1999).

The second feature is that in the Late Pliocene and the Pleistocene, aridity would have been the main determinant affecting mammalian populations (including hominins) in those areas now receiving <500 mm p.a because both the Mediterranean and monsoon rainfall systems weakened during the cool periods that corresponded to high-latitude glaciations. Dodonov and Baiguzina (1995) provided a simple but useful model of how these fluctuated during the Middle Pleistocene (see Fig. 15.3) that can be taken as representative for earlier periods. In cold periods, the Mediterranean system was weakened because evaporation rates were lower, and winds blowing inland across Southwest and Central Asia were blocked by northerly winds originating in areas of high pressure over Scandinavia. These strengthened considerably after ice-sheets developed over Scandinavia during Marine Isotope Stage (MIS) 16, ca. 650–620 ka (Head and Gibbard 2005: 13), so inland Southwest and Central Asia were probably less arid in cold periods in the Early Pleistocene than during the Middle Pleistocene, when most of Asia's deserts developed (see Dennell 2009). The summer monsoon weakened in cold periods because evaporation rates over the Indian Ocean and South China Sea were lower, and its onset was delayed by a strengthened winter monsoon that would have driven cold, dry air southwards from Siberia and Mongolia towards the Indian Ocean and South China Sea. In East Asia, a strengthened winter monsoon also reduced the ability of the summer monsoon to penetrate North China (hence the alternation of loesses and paleosols in the Chinese Loess Plateau sequences). There would therefore have been significant vegetational shifts between glaciations and interglacials across Southwest and Central Asia and North China between desert and semi-desert, semi-desert and grassland, grassland and open woodland, and so on, even if these shifts were not as severe in the Middle Pleistocene.

As a consequence of these periodic fluctuations in the Early Pleistocene, most of Arabia, much of inland Southwest and Central Asia and North China would have been too arid for hominin settlement in cool, dry periods. Core areas of continuous settlement are thus likely to have been those where rainfall was always sufficient to sustain viable populations of hominins and the resources upon which they depended. These would most likely have been the Levant (northern Israel, Lebanon and western Syria), western Turkey, the Caucasus and southern Caspian; peninsular India; Southeast Asia and South China. Marginal areas of discontinuous settlement would have been in inland Southwest Asia (including the Arabian and Sinai Peninsulas), Central Asia and North China. (Interestingly, both "gateways" into Asia, across the Sinai Peninsula or the Bab-el

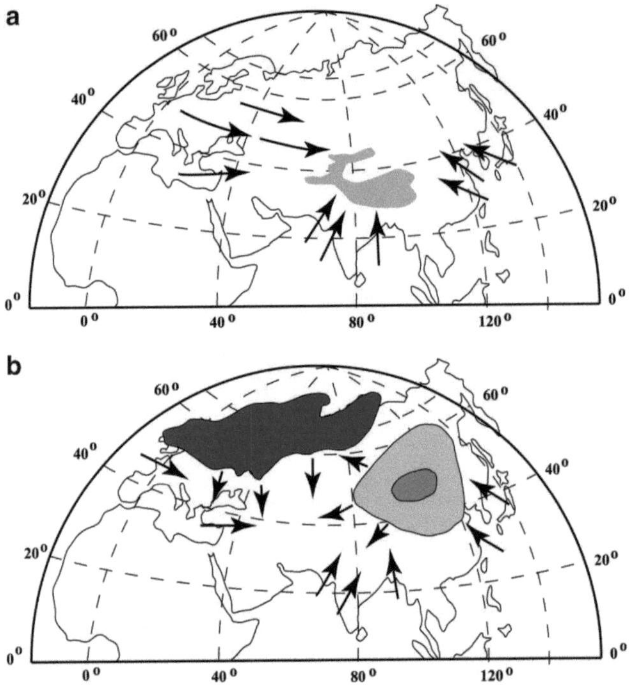

Fig. 15.3 A simple model of Asian climate in glacial and interglacial periods. This figure shows the dynamics of air-mass dynamics in Eurasia during interglacial (**a**) and glacial (**b**) periods. The *black area* shows ice-sheets (particularly extensive in the Middle Pleistocene). The *dark and light grey areas* show respectively the central and peripheral parts of the Siberian-mongolian high-pressure system. In (**a**), easterly winds from the Mediterranean and across the Black Sea provide most of the precipitation during spring and winter; the summer monsoon is also able to penetrate into northern India and China. In (**b**), strong northerly winds from the European ice-sheets block easterly winds from the Mediterranean, and thus Southwest and Central Asia become more arid. Because the high-pressure zone over Siberia and Mongolia is strengthened, the summer monsoon over India, Southeast Asia and China is weakened, and thus rainfall is reduced (Dodonov and Baiguzina 1995; redrawn with permission from Elsevier)

Mandab Strait, would have also been in areas where settlement was probably marginal). These periodic fluctuations in rainfall make it likely that the Asian Paleolithic record prior to 1.0 Ma is spatially and chronologically discontinuous (see Fig. 15.4), and heavily biased towards the warmer and wetter periods that correspond to high latitude interglacials. The hominin record across southern Asia before 1.0 Ma is thus likely to have been one of "repeated, short-lived and modest dispersal events, rather than continuous residence" (Dennell 2003: 434). We therefore need to be able to determine whether hominins were "visitors" or long-term residents (i.e., during cool and arid periods and well as warm and moist ones). A second consequence is that inferred long-term continuities in lithic traditions should not be mistaken for long-term residential continuity in hominin settlement as these continuities could indicate repeated episodic periods of settlement during moist periods by groups whose technology remained unchanged in their core areas. Overall, early Asian prehistory is likely to comprise the same type of histories of colonization and abandonment, expansion and contraction, as recorded for Europe in the Middle Pleistocene (see e.g., Gamble 1999; Roebroeks 2001). To offer a European perspective on the possible magnitude of these changes at the northern limits of hominin settlement, areas such as Britain at 50 N. were probably uninhabited for 80% of the last 500,000 years (see Stringer 2006).

To date, there is no equivalent of the PRISM reconstruction of Late Pliocene climate and vegetation in Asia (ca. 3.2 Ma) for the Early Pleistocene at ca. 1.8 Ma. The most important changes that occurred in the intervening period were the onset of northern hemisphere glaciation ca. 2.4 Ma (as indicated by ice-rafted debris in the North Atlantic [Shackleton et al. 1984]) and rapid uplift of the northern Tibetan Plateau that had major consequences on the strength of the East Asian winter monsoon (Li et al. 1997a; Hongbo Zheng et al. 2000; Tapponnier et al. 2001). I have suggested (Dennell 2004a, 2009, in press; Dennell and Roebroeks 2005) that most of the Asian grasslands of the Late Pliocene were still in place in the Early Pleistocene. Recently, I have attempted to synthesise current paleoclimatic data to provide an outline of Asian average rainfall totals during moist, warm episodes (i.e., interglacials) in the Early Pleistocene (see Fig. 15.5). As way of health warnings, there is an urgent need for paleoclimatic data from Southwest and mainland Southeast Asia and India, and my reconstruction also indicates four current areas of uncertainty: (1) the timing and extent of the Akchagyklian Transgression that linked the Black and Caspian Seas (see Mitchell and Westaway 1999); (2) changes to the size of the Aral Sea and its possible connections to the Caspian (see Boomer et al. 2000); (3) the Early Pleistocene elevation of the Tien Shan (Abdrakhmatov et al. 1996; Zhencheng Sun et al. 1999); and (4) the height of the northern Tibetan Plateau in the Early Pleistocene (see above).

These problems notwithstanding, the main features of this reconstruction are that in the Late Pliocene and Early Pleistocene, there were no ice-sheets over northern Europe, and therefore no strong northerly winds blowing southwards and blocking the westerly winds from the East Mediterranean that brought rainfall eastwards across Southwest and much of Central Asia. Apart from small areas of Arabia and North China, few areas of Asia south of latitude 40° N. are likely to have received <100 mm. Loess was deposited in cold periods in both Central Asia and North China, but on a much smaller scale than during the Middle Pleistocene (see Dodonov 2002; Sun and Liu 2000; Yang and Ding 2006). The summer monsoon penetrated further inland than during much of the Middle Pleistocene, and thus the northern limit of the semi-arid zone (i.e., <600 mm) in North China lay further north. Additionally, the desert barrier between the Sahara and Arabia was greatly reduced, and faunal movements (including hominins) would have been possible between Africa and

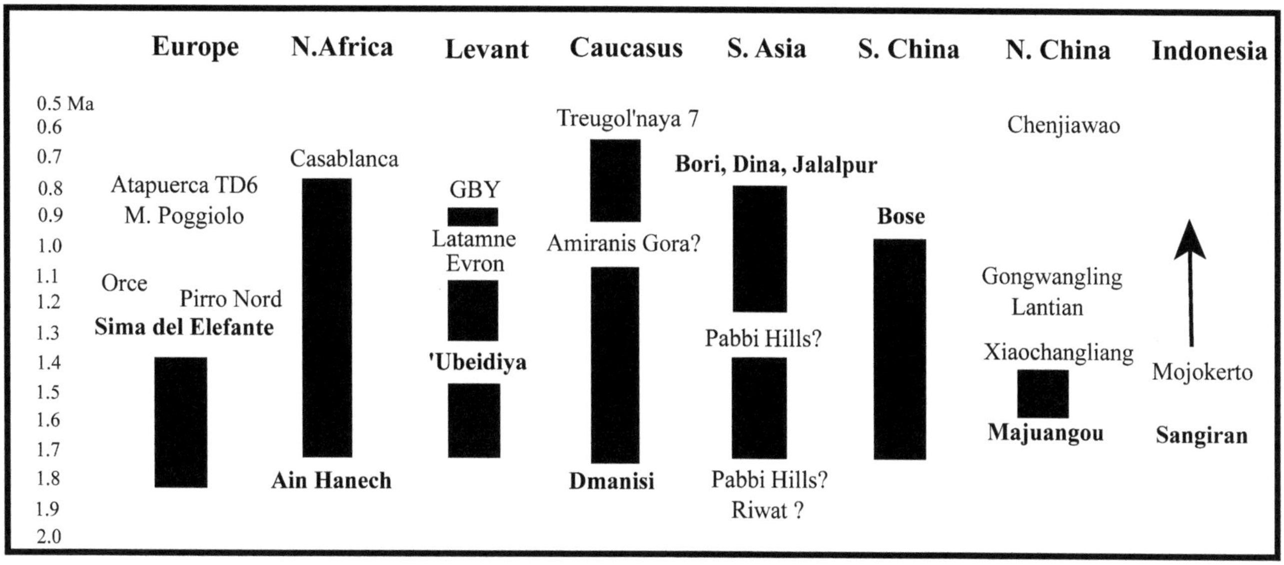

Fig. 15.4 Continuities and discontinuities in hominin settlement in the Early and Middle Pleistocene. Sites in *bold* refer to those where there is widespread acceptance that the dating, context and identification of skeletal remains and/or flaked stone are hominin and/or artefacts respectively. *Shaded areas* denote periods for which there is no definite evidence that hominins were present. Controversial dates or finds are indicated by a "?". The following sites are considered too unreliable to include: Erq el Ahmar and Yiron (Israel), and Longgupo and Renzidong (South China). Yuanmou is reluctantly included given doubts over the age and precise stratigraphic context of the hominin remains and artefacts. Mojokerto (Java) is shown as having a maximum age of 1.49 Ma (see Huffman et al. 2006 and Morwood et al. 2003). Sites in *large bold font* are the earliest known in their region

Asia (and vice versa), at least in moist periods. The primary constraints to movement across most of continental Asia in moist periods would probably have been topographic (notably the mountain ranges of the Himalayas, Karakorum, northern Tibet, and the Tien Shan, etc.) rather than climatic.

Part III: A Research Agenda for Improving Our Understanding of Our Early Hominin Settlement in Asia

It is already obvious that the fossil and archeological evidence for hominins across Asia before 1 Ma is extremely sparse, especially when considered against its size. It is equally obvious that there is no Asian equivalent of the African Rift Valley, and no quick and easy solution for drastically increasing the coverage of early Pleistocene archeological and fossil hominin sites across Asia. Nevertheless, we can take a positive approach, and suggest how the limited funds available might best be spent in improving our current evidence of well-dated finds of artifacts and hopefully hominin remains, preferably in association with climatic and environmental (particularly faunal and botanical) evidence. The following section thus proposes how a 10–15 year research programme of carefully targeted, modestly-funded, 3–5 year field projects could greatly improve our understanding of the earliest hominin settlement of Asia.

The primary need is to find evidence that can be securely dated. There are five types of contexts (or "taphonomic traps") in which there are reasonably good chances of finding Late Pliocene and Early Pleistocene hominin fossil and/or archeological remains that could be dated, and placed in a climatic context. These are the margins of extinct lakes; floodplains and river channels; caves; loess and paleosol sequences; and in a rare number of cases, deposits sealed by volcanic outflows of lava, basalt or ash. Of these, lake margins, flood-plain and river banks are obvious sources of water and food, and sometimes near stone sources, as well as ideal preservational environments. Caves are obvious taphonomic traps and were often used by early hominins in the Middle Pleistocene, although Early Pleistocene examples are rare in Eurasia. Predators such as *Pachycrocuta brevirostris* that often preyed upon hominins also sometimes left their remains in caves (as at locality 1, Zhoukoudian [Boaz et al. 2004]). Loess and paleosol sequences can provide stratified and dated occurrences of artifacts, but rarely expose major archeological sites or useful faunal (and fossil hominin) evidence. Prior to the Upper Pleistocene, the only archeological evidence from loess and paleosol sequences in Asia comes from paleosols (as at Kuldara, Tajikistan; see below). Typically, small numbers of artifacts are found in restricted exposures within a steep loess section, and large-scale

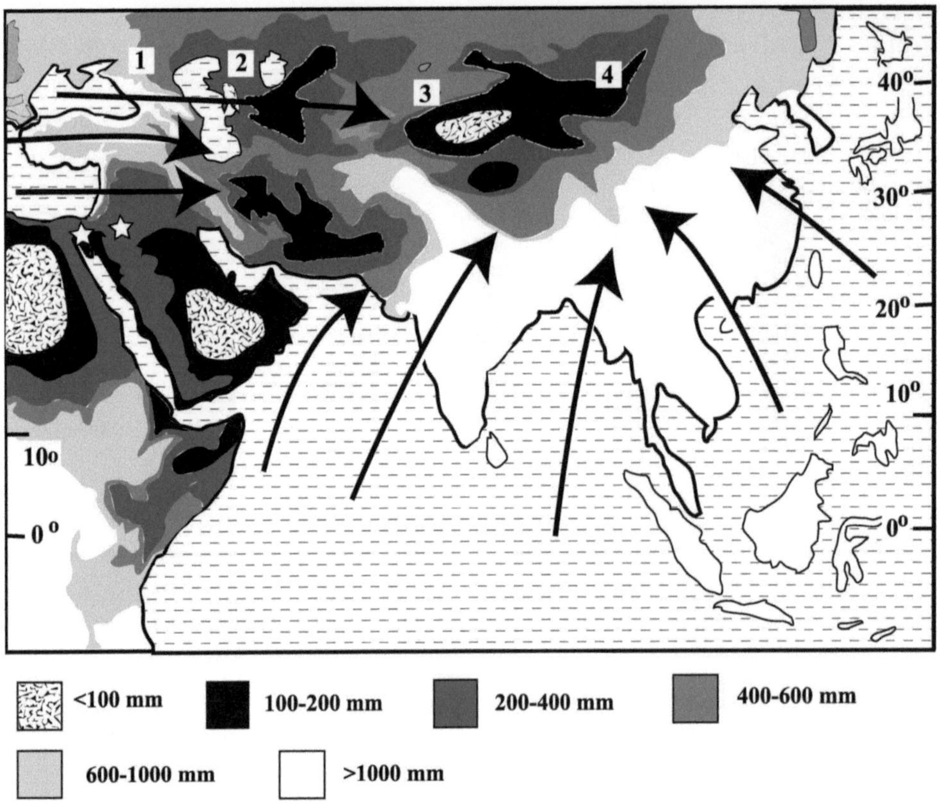

Fig. 15.5 Estimated rainfall in moist periods of the Late Pliocene or Early Pleistocene. This map attempts to indicate the probable level of precipitation across southern Asis during moist (i.e., interglacial) parts of the Early Pleistocene. *Arrows* show the main rain-bearing winds: westerly ones in winter and spring from the East Mediterranean and Black, and southwest and southeast ones in summer from the Indian and East Asian monsoon. The *asterisks* denote Nahal Zihor (Israel; Chapter 4) and Ain Nefud (Saudi Arabia), where there were substantial lake systems during the Early Pleistocene in areas now receiving <100 mm p.a. Ain Nefud also indicates the local presence of hippopotamus, a metre-long fish, elephant, horse, *Pelorovis* and other large bovids as well as large carnivores, none of which could survive in the area today. In the Late Pliocene and Early Pleistocene, there were no ice-sheets over northern Europe, and therefore no northerly winds blowing southwards and blocking westerly winds from the East Mediterranean bringing rainfall eastwards across Southwest and much of Central Asia. Apart from small areas of Arabia and North China, few areas of Asia are likely to have received <100 mm. Loess was deposited in cold periods in both Central Asia and North China, but on a much smaller scale than during the Middle Pleistocene. The summer monsoon penetrated further inland than during much of the Middle Pleistocene, and thus the northern limit of the semi-arid zone (i.e., <600 mm) lay further north. On this reconstruction, the desert barrier between the Sahara and Arabia was greatly reduced, and faunal movements (including hominins) would have been possible between Africa and Asia (and vice versa)

excavations are impossible. Nevertheless, they can be dated, and are thus an important source of evidence on when, and how often, hominins were in a region. Volcanic outflows are useful because they are amenable to radiometric dating techniques. Bailey et al. (2000) also point out that the disruptive effects of volcanic outflows would also have provided early hominins with a variety of opportunities, such as vantage points, refuge areas, natural traps for catching animals and of course a source of stone, and thus might have formed important components of early hominin landscapes.

These different types of taphonomic traps are unevenly distributed across Asia, and there are large areas where it is most unlikely that survey or excavation could produce useful – i.e., dateable – evidence. There are two additional constraints. One is that Late Pliocene and Early Pleistocene examples of paleolakes and fluvial sequences must also be in areas that have experienced subsequent uplift and re-exposure (such as the Sangiran Basin), down cutting and erosion (such as the Nihewan Basin), or deflation (as in the Arabian Peninsula). The second is that there must originally have been sources of flakable stone that were accessible within a 10 km radius. Whilst the ability to flake stone was undoubtedly one of the most important developments in hominin evolution by opening up a wide range of possibilities for obtaining resources and making other tools more easily (particularly from wood), it was also a restrictive development in that it tied hominins to those areas where stone was abundant. All available data (e.g., Blumenschine et al. 2007; Féblot-Augustins 1997; Roebroeks et al. 1988) indicates that Early (and most Middle) Pleistocene hominins had very limited abilities to transport stone, and almost all was used and discarded within 10 km of its source. Consequently, areas lacking stone but rich in plant and animal resources would probably have been beyond the abilities of early hominins to exploit. Unfortunately for hominins, such situations were probably common in large parts of Asia (such as some loess landscapes, and in the flood plains of many major river systems; see below for South Asia). The search for evidence for early Pleistocene hominins in Asia needs therefore also to focus not only on those locations where stratified deposits can be found but which would also have been near sources of flakeable stone.

Asia south of latitudes 40–45° N. can be divided into four regions on the basis of the types of stratified contexts most likely to produce evidence of Late Pliocene and Early Pleistocene hominins. Each region can be taken in turn; Dennell (2009) provides summaries and discussions of current archeological and environmental evidence.

Southwest Asia: From the Mediterranean and Red Sea to the Indus

As indicated in the previous section, the Levant and western Turkey would probably have been "core" areas of settlement because of their proximity to the Mediterranean that provides most of the annual rainfall. Arabia and inland Southwest Asia (including the Iranian Plateau and Baluchistan) that are now largely desert would always have been more marginal for hominins. Almost all the current (and well-dated) evidence for early hominins in this region comes from the two "flagship" sites of 'Ubeidiya (Israel) and Dmanisi (Georgia), both of which had locally available sources of stone and were probably also in core areas of early hominin settlement (see Fig. 15.6). 'Ubeidiya has at least 65 archeological sites that date to ca. 1.4 Ma (Bar-Yosef 1994, 1998; Bar-Yosef and Goren-Inbar 1993) and are from a variety of lakeside environments (Mallol 2006). It also has a rich record of vertebrate fossils (Tchernov 1987; Gaudzinski-Windheuser 2005), one hominin incisor (Belmaker et al. 2002) and is the oldest Acheulean site outside Africa. The hominin evidence from Dmanisi came from volcanic deposits formed when a basalt outflow created a small lake (Gabunia et al. 2000b; Gabunia and Vekua 1995), and is thus taphonomically an instance of both a lacustrine and volcanic preservation. The hominin remains (Rightmire and Lordkipanidze 2010) are the oldest examples of *H. erectus* in Asia, and the associated lithic industry is a very primitive, and classified as pre-Oldowan on technological grounds (Lumley et al. 2005). Faunal and botanical evidence indicates a variety of local environments and a Mediterranean type of climate (Gabunia et al. 2000a, b).

The best prospects for finding Late Pliocene and/or Early Pleistocene hominin remains and archeological assemblages in Southwest Asia will be from former lakes such as 'Ubeidiya. Given the intensity of research in Israel before and since independence, it is most unlikely that a second

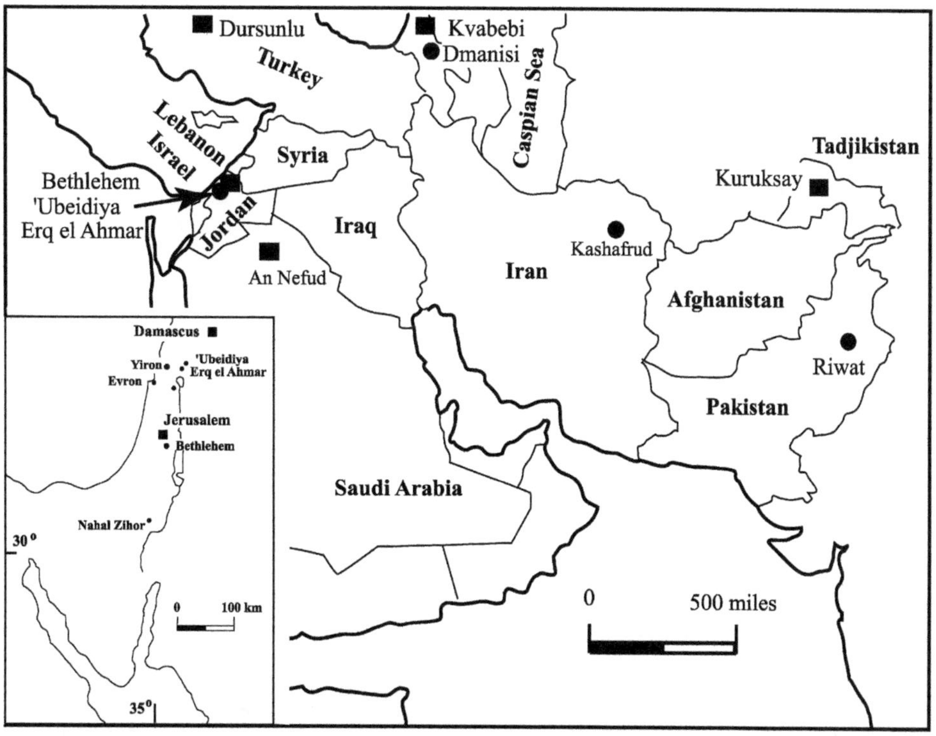

Fig. 15.6 Location of sites in Southwest and Central Asia mentioned in the text

Early Pleistocene one will be found in the Jordan Rift Valley, although Gesher Benot Ya'aqub (Goren-Inbar et al. 2000) provides an outstanding early Middle Pleistocene example. There may be some prospects for further research at Erq el Ahmar, also in the Jordan Rift Valley, where two "pebble tools" were reported (Verosub and Tchernov 1991). Although palemagnetic work has implied that these are Early Pleistocene in age (Braun et al. 1990; Ron and Levi 2001), the paleomagnetic sections could not be correlated to where the claimed artifacts were found, and the artifacts (if such they are) remain undated.

However, at least four other extinct lake systems are known in Southwest Asia, and doubtless there are others that await discovery. Any found in the Sinai Peninsula or southern Arabia would probably contain important information on dispersals near the two gateways into (and out of) Asia. The first and most westerly paleolake is at Dursunlu in western Turkey which was exposed by local mining activities. Here, a core >50 m long taken near the quarry showed reddish alluvial mud in the basal 5 m; then an upward-shallowing lake sequence with intermittent fluctuations, and marsh deposits at the top. This site had a faunal assemblage that is consistent with an Early Pleistocene age, and included a Proboscidean (cf. *Mammuthus*), an indeterminate hippopotamus, pigs, three types each of deer and bovids (Güleç et al. 1999). The bird remains indicated lacustrine conditions, open, steppic vegetation and a Mediterranean climate not unlike the present (Louchart et al. 1998). A flaked stone assemblage (N = 127) was recovered from silt sediments and from tip heaps in the quarry, and probably originated from above the fauna. Sixty-seven pieces of stone were regarded as most probably the result of hominin flaking. Most artifacts were plain flakes and flake fragments (some retouched), but there were also a few cores, a polyhedron and a chopper; almost all (95%) were quartz. Paleomagnetic and microfaunal evidence indicates an age for this assemblage of ca. 0.85–0.90 Ma (Güleç et al. 2009). Dursunlu is thus the first indication of stratified, Early Pleistocene archeological material from western Turkey, and other similar quarries and open-cast mines should be explored.

Three paleolakes are known from the arid interior of Southwest Asia. The first is the Nahal Zivor in the Negev Desert of southern Israel (Ginat et al. 2003) which now receives <50 mm p.a. This lake system lies within the Zeheiha Formation that is correlated with the Erq el Ahmar Formation, and is thus probably older than 'Ubeidiya. It had three sedimentary cycles. Each lake is thought to have been 3–5 m deep; i.e., deep enough to maintain a viable fish population, and also extensive enough to prevent large stones being transported into the lake centre. On the marginal facies, there were coarse alluvial and colluvial sediments which indicate small streams flowing into the lake. Some pedogenic features were present which indicated that the lake periodically dried out. The paleosols that developed when the lake was drying out indicate a semi-arid climate, with an annual rainfall of ca. 150–200 mm. Ginat et al. (2003) estimate that the lake system at Nahal Zihor probably lasted between 45,000 and 150,000 years under a semi-arid climate, and suggest that the lake was maintained by rainfall as well as some ground water seepage. Each freshwater phase may have lasted between 3,000 and 10,000 years, and the paleosols may have taken 10–20,000 years to develop. Three concentrations of stone artifacts were found that might be linked to the ancient shoreline. Two were at the contact between the lake marginal sediments and the outcropping limestone on the southern shoreline of the lake, and the third was found on exposures of weathered green limestone. A few flint flakes were also found in the lacustrine sediments of the marginal facies. The investigators suggest that the other artifacts were weathered out of the lake sediments, and were thus contemporary with them. These concentrations included handaxes, picks, and some chopping tools that were described as similar to those from 'Ubeidiya. Although there are good circumstantial grounds for linking the archeological sites to the paleolakes at Nahal Zihor, this association needs to be demonstrated by the discovery of artefacts *in* as well as *on* the lakeshore sediments.

The second example is a series of paleolakes in the An Nefud desert of northern Saudi Arabia (Thomas et al. 1998). Here, the lake sediments (see Fig. 15.7) filled concave depressions on the top of white aeolian sands of an earlier dune system and were commonly covered laterally by the present-day dunes. The investigators suggested that originally there were several isolated lakes that were probably not synchronous. Three vertebrate fossil localities were reported (fossil collecting was on a small scale, and the survey was paleontological rather than archeological). At one, fossils were collected from the surface of the lacustrine deposits from which they have probably been recently eroded, and at the others, the fossils came from a thin siltstone between the basal aeolian sand and a 10 cm-thick layer of lacustrine carbonates. The fossils were all consistent with an Early Pleistocene age, and many showed affinities to finds from East Africa and 'Ubeidiya. They included a maxilla of a large fish, estimated to have been >1 m long, so the lake was clearly quite large, and a carapace fragment of *Geochelone sulcata*, the largest African land tortoise, and now found only in the Sahel. Carnivores were represented by specimens of *Crocuta crocuta*, *Panthera gombaszoegensis* (found also at 'Ubeidiya), and the fox, *Vulpes vulpes.* Herbivores included *Elephas*, possibly *E. recki*, *Pelorovis oldowayensis, and Equus* (the last being compared to those from Olduvai Upper Bed II). The pygmy hippopotamus *Hexaprotodon* was also represented; this is found today only in West Africa, but in the Early Pleistocene was also present in Northern Pakistan and Java; its presence here confirms that the lake was large, as they prefer standing water 2–5 m deep (Jablonski 2004).

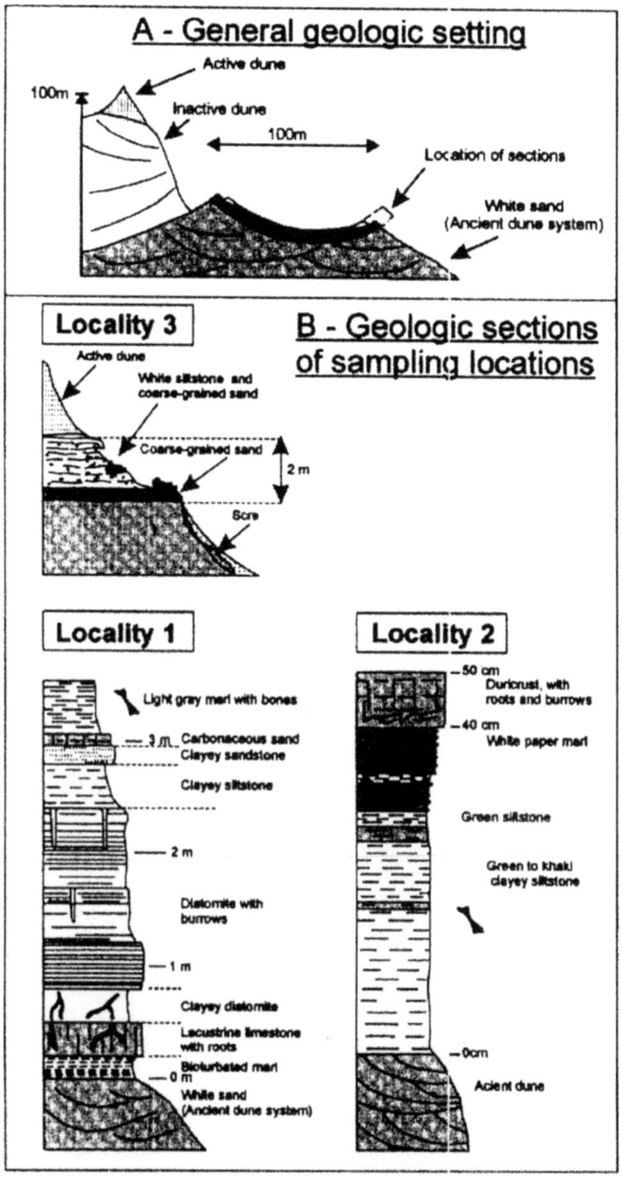

Fig. 15.7 The stratigraphic sequence of the An Nefud lakes, Saudi Arabia. Although it is unlikely that sections of palaeo-lakes as thick and extensive as those at 'Ubeidiya will be found elsewhere in inland Southwest and in Central Asia, shallow lake sequences like those at An Nefud offer considerable opportunities for finding archeological and faunal material in Late Pliocene and Early Pleistocene contexts that can be dated, particularly if the Olduvai Sub-chron can be detected palaeomagnetically (Thomas et al. 1998, Fig. 15.2)

A camel, oryx and some kind of alcelaphine were also present. $\partial^{13}C$ measurements taken from tooth fragments of *Pelorovis, Elephas* and the alcelaphine indicated a C_4 grassland environment.

This study is particularly interesting in showing the presence of substantial lakes in Arabia during the Early Pleistocene which were able to support populations of large fish and hippopotamus, as well as good quality grassland in an area that is now desert. This area should therefore have been highly attractive to hominins at this time, and it would be worthwhile making a thorough investigation of the paleoshoreline for stone artifacts and nearby stone sources. In addition to containing well-preserved fossil remains from a wide range of vertebrate taxa (that might even include hominins), a major attraction of investigating this type of extinct lake system is that it could be dated by paleomagnetism. At present, the most likely age on faunal grounds is ca. 1.2–1.4 Ma, i.e., comparable to that of 'Ubeidiya and Olduvai Upper Bed II. Because the Early Paleolithic evidence from the Arabian Peninsula consists almost entirely of surface (and undateable) occurrences of Oldowan- and Acheulean-type artifacts (Petraglia 2003), extinct lakes such as those at An Nefud provide the best prospects of dating the Early Paleolithic from the important region near the gateways to (and from) Africa. Extinct lakes in Arabia (and elsewhere in Southwest Asia; see below) should be an obvious target for early paleolithic fieldwork.

The third is Kashafrud in northeast Iran 40–80 km east of Meshad. Here, Arai and Thibault (1975) studied a series of sections along the edge of what appears to have been a large, but probably shallow lake. Interestingly, they found what were described as "pebble tools" on the surface but also in a stony alluvial layer underlying a sand unit. A further dozen were found in the same layer at another section, at Baghbaghu. At present, the age of these deposits and the stone tools (assuming that they are not geofacts) cannot at present be determined. However, the suggested Early Pleistocene date is reasonable, given other similar lakes of this age in Saudi Arabia and Israel. It would be well worth re-examining these deposits to verify the identification of the flaked stone as artefacts, and if they are artifacts, to confirm their association with the lake and obtain a clearer idea of their age. Because the current evidence for the Lower Paleolithic in Iran remains pitifully small and undated (see Smith 1986; Biglari et al. 2000), paleolakes such as that at Kashrufrud are the best opportunities for finding dateable evidence.

The second best prospects for finding datable material are in areas of former volcanic activity; Dmanisi of course provides an excellent example. Yiron in Israel offered some prospects as artifacts were reported from a gravel that allegedly underlay a basalt dated at 2.4 Ma (Ronen 1991), but these claims have not found widespread acceptance. A more recent and better example is from Kalatepe Deresi 3 in Central Anatolia (Slimak et al. 2004), which lies on a volcanic plateau at 1,600 m a.s.l. Excavations into the side of a ravine revealed seven occupation horizons (with 1,124 artifacts to date), of which the uppermost five had Middle Paleolithic artifacts and evidence of Levallois and discoid debitage. Six tephra layers, the oldest dated to ca. 160 ka, are interstratified between archeological levels I and II. The presence of an equid mandible in level II holds out promise of some faunal data in future excavations. The assemblages from the lowest two horizons are described as Lower Paleolithic, and include a biface from the fifth and lowest level which may be late Early Pleistocene or early Middle Pleistocene in age (Tyron et al. 2009). Exploration of earlier volcanic landscapes in Anatolia and the Caucasus may prove fruitful. Another intriguing example of what might be obtained from volcanic areas is a report of hominin footprints (undated but probably Middle Pleistocene) from a quarry in Anatolia (Ozansoy 1969). As there are several extinct (and dormant) volcanoes in Anatolia, the Caucasus and Elburz Mountains, other volcanic deposits in Southwest Asia might also be worth prospecting for archeological material.

It is most unlikely that caves will provide any useful information on Early Pleistocene hominins in Southwest Asia. Although numerous Middle Pleistocene caves are known from this region, such as Yarimburgaz (Kuhn et al. 1996) and Karain (Otte et al. 1995, 1998) in western Turkey, Kudaro I and III, Azokh and Tcona in the Caucasus (Ljubin and Bosinski 1995) and Tabun, Qesem, Zuttiyeh, Umm Qatafa (see Bar-Yosef 1998) in Israel, and Jabrud (Rust 1950) in Syria, no caves so far excavated contain Early Pleistocene deposits with the possible exception of the basal (and archeologically sterile) layers of Azokh in Azerbaijan (Ljubin and Bosinski 1995). It is thus likely that Early Pleistocene caves (and their contents) in these regions have been destroyed by erosion. Given the heavy investment of fieldwork in Israel, it is also most unlikely that there are any major cave systems that await discovery, although the discovery of remnant cave deposits at Misliya (Weinstein-Evron et al. 2003) shows that there is some scope for new discoveries.

Central Asia and North China

This enormous area covers ca. 3.0 million square miles and encompasses the "Five Stans" of Kazakhstan, Kyrgyzstan, Tajikistan, Turkmenistan and Uzbekhistan, Xinjiang Province of Northeast China, and China north of the Qinling Mountains. Much of this region is now desert, particularly the Karakum and Kizyllkum Deserts of Turkmenistan and Uzbekistan, and the Taklamakam, Badan Jarain, Tengger, Gobi and Ordos Deserts in the "Tibetan corridor" between Central Asia and North China. Most of these deserts developed in the Middle Pleistocene (Dennell 2009). Both Central Asia and North China have harsh winters and hot summers. They differ in that Central Asia receives most of its rainfall in winter and spring, whereas North China lies on the northern margin of the East Asian summer monsoon. Reductions in rainfall in both Central Asia and North China would cause much of both regions to revert from grassland to semi-desert, and it is likely that they were marginal regions of early hominin settlement, and abandoned during the driest parts of periods equivalent to high-latitude glaciations. Although data are currently lacking on the level of precipitation during the Early Pleistocene, various studies of the Chinese Loess Plateau show that rainfall might have decreased by 25–30% (Florindo et al. 1999; Maher et al. 1994; Maher and Thompson 1995) or even up to 75% during the last glaciation (Xiuming Liu et al. 1995). Core populations in Central Asia are likely to have been along the southern coast of the Caspian or the Caucasus, and/or perhaps further south in North Pakistan. Those in China would probably have been south of the Qinling Mountains.

There are two obvious depositional contexts that should be further investigated for evidence of early Pleistocene hominins. The first are the loess and paleosol sequences that are extensive across both regions, and extend back to ca. 2.5 Ma (Dodonov 2002; Liu and Ding 1998). These have the advantage of being amenable to paleomagnetic dating, and are also excellent climatic indicators. To date, the best example of the potential of these sequences for documenting early hominin occupation is from Tajikistan (Fig. 15.6), where Ranov and others (e.g., Dodonov et al. 1992; Ranov 1995) have demonstrated a long sequence of Late, Middle and Early Pleistocene Paleolithic assemblages. Unsurprisingly for a region that is likely to have been marginal to hominin settlement, these assemblages have always been found in paleosols that denote interglacial periods when rainfall was higher (see Fig. 15.8). At present, the earliest evidence for hominins in Central Asia is from Kuldara in Tajikistan (Ranov 1995), which lies at the base of a deep ravine in pedocomplex 11/12, dated to ca. 880–955 ka, and just below the Brunhes-Matuyama boundary. This site contained a small assemblage (96 items of which 40 were indisputable artifacts) of small flakes struck from pebbles. Stone appears to have been a scarce resource in these landscapes, as the stone used for tool-making at Kuldara (and later sites) tends to be a variety of often low-grade small pebbles: as example, 50% of the tools flaked at Kuldara were <40 mm in length, and 70% of flake removals were < 50 mm.

Because hominins were at the same latitude (40–42° N) at Dmanisi to the west, ca. 1.75 Ma, and at Majuangou in northern China to the east at 1.66 Ma (see below), hominins were almost certainly present in Central Asia 600,000 years earlier. The challenge now is to fill that Central Asian gap between 0.9 Ma and 1.7 Ma by further survey of sequences that lie between pedocomplex 11/12 and its base at ca. 2.5 Ma.

The loess deposits of North China are much more extensive, and those on the Chinese Loess Plateau cover ca. 440,000 km^2, or an area twice the size of the U.K. Stone appears to have been absent. However, Early Pleistocene loess is found in thinner sequences across throughout North China, and sometimes accumulated in or over layers containing (late) Early Pleistocene artifacts. (Examples are Lantian [Gongwangling] and Xihoudou; see Hyodo et al. 2002; Zhu

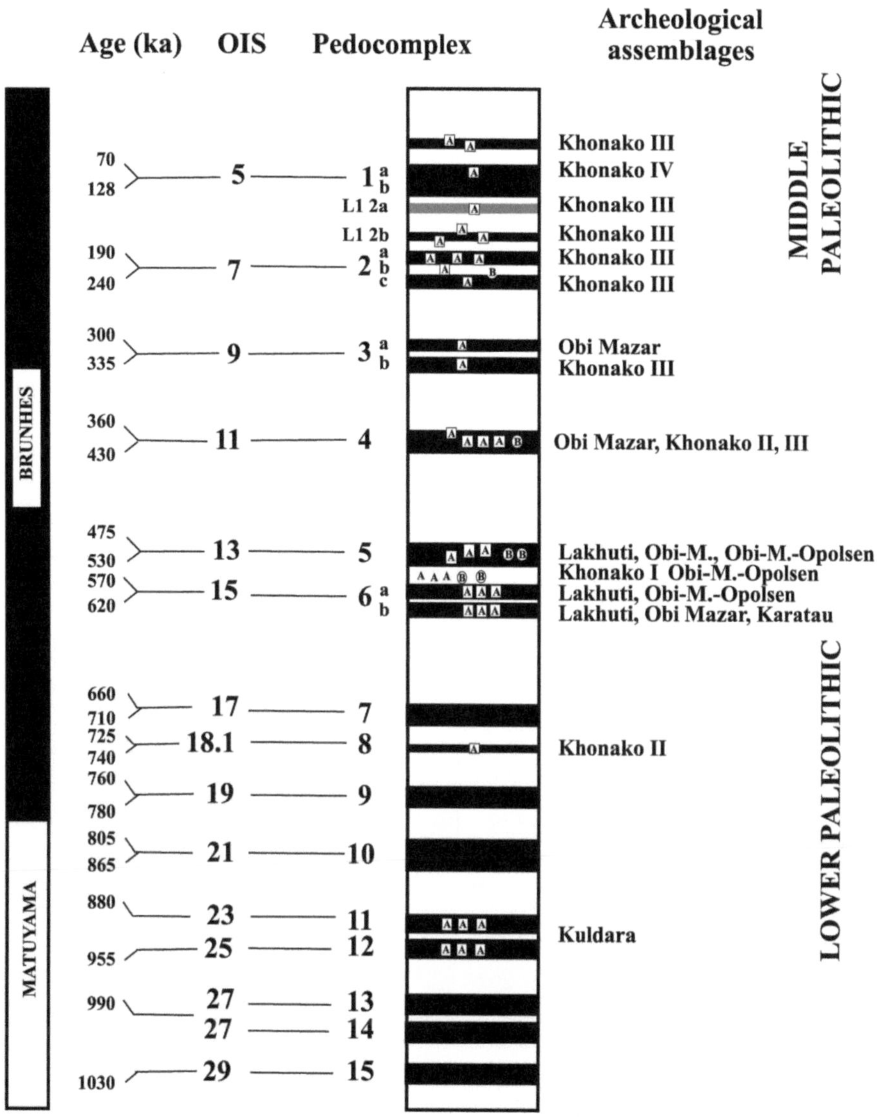

Fig. 15.8 The loess-paleosol sequence of hominin occupation in Pleistocene Tajikistan. This is a model example of how patient and methodical investigations of loess and paleosol sequences can result in a well-dated sequence of archeological finds that document when a loessic environment was inhabited. As hominins were both west and east of Tajikistan ca. 1.66 Ma, Kuldara is unlikely to be the oldest site in Central Asia, and there is considerable scope for further work. *Black bars* denote interglacial pedocomplexes (paleosols); *intervening white parts* denote glacial loess (Dodonov 2002, Tables 9 and 14, and Ranov and Dodonov 2003, Fig. 15.10; redrawn by the author)

et al. 2004). Further survey of Early Pleistocene loess-paleosol sequences adjacent to the Chinese Loess Plateau should be carried out.

The second type of deposits that would repay further exploration are lacustrine. The best example to date is the Nihewan Basin in North China (Fig. 15.9), which lies within the Shaanxi Graben that developed in the Late Pliocene (Youli Li et al. 1998) and contains ca. 700 m of lake and lake margin deposits that are capped by loess from the last glaciation. Low-grade stone (small pebbles and local outcrops of poor chert) was readily available for hominins. Nihewan currently contains evidence of the oldest sites east of Dmanisi (Hou and Zhao 2010). For several years, the oldest site was thought to be Donggutuo, dated to ca. 1.0 Ma (Lanpo and Qi 1987), but recently Xiaochangliang has been dated to 1.36 Ma (Zhu et al. 2001), and the earliest at Majuangou to 1.66 Ma (Zhu et al. 2003, 2004). These sites contain stone artifacts and very poorly-preserved fossil vertebrate remains. Hominin remains have not yet been found, but because vertebrate fossils are poorly preserved and very fragmented, the prospects for finding taxonomically-diagnostic hominin specimens is probably slight. Even older sites than Majuangou might be expected, as there are 17 m of deposits underlying Majuangou III that have yet to be thoroughly investigated (Huang Weiwen, personal communication). Additionally, there may be other lake basins in the Shaanxi Graben that

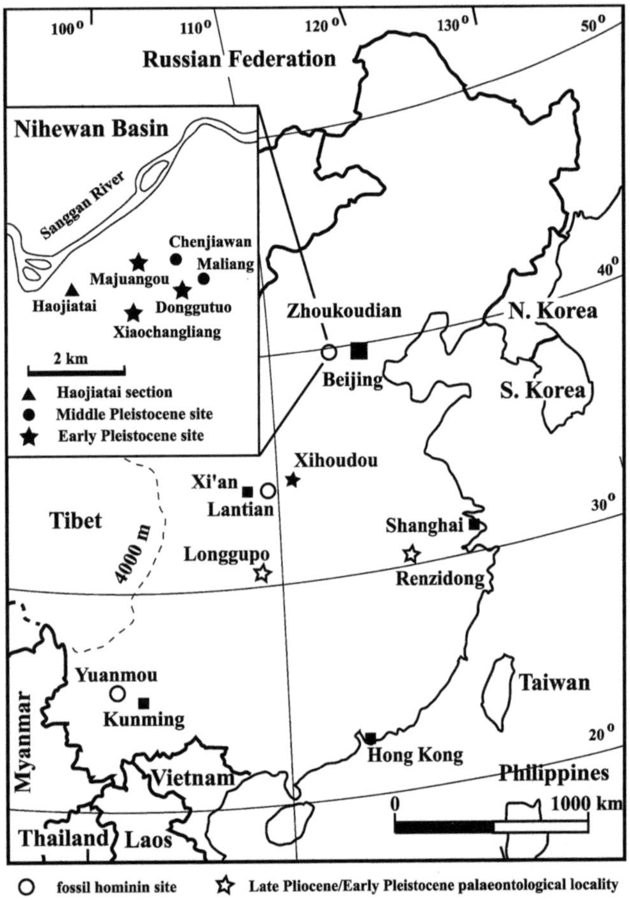

Fig. 15.9 Location of sites in China mentioned in text (Dennell 2009)

have yet to be investigated. (Central Asia and Xijiang Province also contain paleolakes but most of these appear to be Middle or Upper Pleistocene, and often saline; see Yongqiu Wu et al. 2001; Sun et al. 1999).

Two other types of deposits might repay further investigation in Central Asia. The first are Late Pliocene and Early Pleistocene fossil localities, some of which contain the remains of several taxa, including carnivores and primates. Kuruksay, for example, contained evidence of a large canid *Canis etruscus,* several other carnivores (*Ursus cf. etruscus, Pliocrocuta, Lynx, Acinonyx, Megantereon, Homotherium*), and herbivores such as *Archidiskodon, Dicerorhinus, Equus stenonis, Paracamelus, Sivatherium,* and *Damalops palaeindicus* (Sotnikova et al. 1997). The last two of these, *Sivatherium* and *Damalops* are also recorded in both Pakistan (Dennell 2004b) and Africa at this time. There is also a primate that was identified as *Papio sushkini* (Maschenko 1994) but is probably *Paradolichopithecus sushkini* (Nishimura et al. 2007; Takai et al. 2008), a Eurasian type of macaque.

The second are loess and alluvial/palaeosol sequences. For example, there may be loess and paleosol exposures older than Kuldara (i.e., >0.9 Ma) in Tajikistan, and 30–45 m of alluvium and loess are reported in the Batpak Valley of Kazakhstan. These earliest of these deposits are grouped in the Lower Aktasy Formation, which probably lasted from the Olduvai Event (ca. 1.77 Ma) to the Brunhes-Matuyama boundary at 0.78 Ma. Artefacts reported from these deposits are now considered to be geofacts, but as this is one of the few areas with Early Pleistocene deposits (Moloney et al. 2001), further investigation would be useful.

There are few other prospects for finding Early Pleistocene evidence of hominins in Central Asia or North China. Although Early Paleolithic assemblages (including ones with bifaces) have been found at several places in Central Asia, these are all surface finds (Vishnyatsky 1999), and thus undateable. As there are no volcanic deposits, there are no prospects of finding material stratified between lava flows as in Turkey or the Caucasus. The earliest cave sequence from this region is Sel'ungir in Krygyzstan (Islamov 1990) that dates from the late Middle Pleistocene. The only major rivers in Central Asia are the Amur Darya and Su Darya, both of which flow into the Aral Sea and have low gradients, and hence short geological sequences. Both have also changed course frequently throughout the Pleistocene, as has the extent of the saline Aral Sea (Boomer et al. 2000). In contrast, the Yellow River (Huang He) of northern China is ancient and has a long sequence of exposed terraces (Li et al. 1997b). Its weakness is that its floodplain is overwhelmingly composed of silt and clay, and hominins would have found it difficult to cope in a stone-free environment (see below).

South Asia (India, Pakistan and Nepal)

There are four reasons why the absence (or at least, the extreme scarcity) of evidence for Early Pleistocene hominins in South Asia is at first sight surprising and anomalous. First, paleontological and Paleolithic investigations in British India began over 150 years ago, and since then a great deal of research has taken place; secondly, the Upper Siwaliks of northern India, Pakistan and Nepal contain an excellent Late Pliocene to early Middle Pleistocene mammalian fossil record (see Nanda 2002; Dennell et al. 2006; Patnaik et al. 2009); thirdly, peninsular India has a rich archeological record for the Middle, but not for the Early, Pleistocene; (see e.g., Petraglia 1998, in press) and fourthly, hominins were present at Dmanisi ca. 1.7 Ma to the west, and at Sangiran ca. 1.6 Ma to the east, so should also have been recorded by now in South Asia. However, the only indications of Late Pliocene and Early Pleistocene hominins in South Asia are the small and contested artifact assemblage from Riwat, Pakistan (Dennell et al. 1988; Rendell et al. 1989), and the (typologically) Early Paleolithic artefacts found on eroding Late Pliocene and Early Pleistocene deposits in the Pabbi Hills, Pakistan (Hurcombe 2004), that were not found in context but were nevertheless thought to have been derived from the underlying deposits. Despite these reasons, however, there are several reasons why so little evidence for Early Pleistocene hominins has been found in South Asia.

To take the Late Pliocene to early Middle Pleistocene Upper Siwalik sequences first, there are three main reasons why the long-standing absence of clear evidence for hominins is probably not definitive evidence of their absence.

The Upper Siwaliks of North India, Pakistan and Nepal

a) The incompleteness and taphonomic biases of the Upper Siwalik (Pinjor Stage)[7] fossil vertebrate record. Although investigations of the fossil record of the Upper Siwaliks began over a century ago in the British period (e.g., Pilgrim 1913) in what are now India, Nepal and Pakistan, its fossil record for the Late Pliocene and Early Pleistocene is less complete than might be supposed. As example, anthracoceres, *Megantereon*, the large canid *Canis cautleyi*, and the snow leopard *Panthera uncio* are recorded in the Pakistan but not in India, whereas *Camelus*, *Dorcatherium*, *Theropithecus* and small primates are recorded in India, but not in Pakistan (see Dennell 2004b; Dennell et al. 2006). Additionally, the large felid *Homotherium* is not recorded in either country, but was present at Bethlehem (Hooijer 1958) and Dmanisi to the west (Gabunia et al. 2000b; Vekua 1995), Kuruksay to the north (Sotnikova et al. 1997), and Longgupo to the east (Huang Wanpo et al. 1995), so its current absence from the Upper Siwaliks probably reflects the difficulties of tracking large and elusive felids in the fossil record (see Dennell et al. 2008). Secondly, the Upper Siwalik fossil mammalian record appears to be biased towards the preservation of mammals that were >60 kg in adult body weight, and thus larger than *H. erectus*. As example, intensive survey in the Pabbi Hills produced ca. 20,000 fossil specimens from one particularly fossil-rich formation (Sandstone 12), dated to 1.2–1.4 Ma. Fossil preservation was often outstanding, and this unit contained two rich fossil localities that were excavated and which probably resulted from hyaenid activity (Dennell et al. 2005a, b). Despite this, hominin remains were not found, even though it is likely that they were present in South Asia at this time.

b) Large flood plains and the absence of stone. One probable reason why no major paleolithic site has ever been discovered within the Upper Siwaliks is that stone was absent or extremely rare in the large flood plain systems that formed these sequences. These sequences, whether in Pakistan, India or Nepal are dominated by sands, silts and clays. When stone is found in the Late Pliocene and Early Pleistocene parts of Upper Siwalik sequences, it typically comprises thin (<0.5 m) stringers (as at Riwat), interspersed by long periods when the dominant sediments were sands, silts and clays. As these stringers and other conglomerates would have been in the active, year-round channel, stone may have been a seasonal resource that was accessible only in the winter dry season, but not during the summer monsoon when river levels rose; or alternatively, accessible year-round but only in those abandoned channels that contained stone. The scarcity of stone in Upper Siwalik landscapes may well have been a major drawback of these extensive flood plains, even though water, animal and plant resources were plentiful. I have suggested elsewhere (Dennell 2007) that hominin settlement in these floodplains during the Early Pleistocene may have intermittent and discontinuous, and that hominins would have had to have been opportunistic in targeting those areas where and when stone was available. (This situation changed in the Middle Pleistocene, when large units of stone often referred to as "Boulder Conglomerate" were deposited following regional uplift.) One example of hominins targeting areas of stone in river channels is Anangpur, where Acheulean handaxes were found exposed in a quarry in a gravel deposit at the base of a former (post-Siwalik, Middle Pleistocene) paleochannel of the Yammuna, which is one of the southern tributaries of the Ganges (Sharma 1993). Riwat may be another, as here, the stone tool assemblage came from the only source of flakable stone in the entire 70 m of sands and silts.

c) The need for intensive survey. An additional reason why current evidence for hominins in the Upper Siwaliks is so poor is that there has been little intensive, detailed survey. Most paleontological surveys of the Upper Siwaliks have concentrated on finding taxonomically diagnostic specimens (typically dentitions and crania), and the surveys in the Pabbi Hills are probably the only ones in that were systematically collecting all fossil vertebrate specimens, recording their distribution across the landscape, and simultaneously attempting to find stone artifacts and map where stone was occurring (see Dennell 2004b). Coarse grained surveys aimed solely at finding dental and cranial specimens for biostratigraphic purposes are insufficient for finding rare taxa such as hominins (or primates and felids) and for identifying those comparatively rare episodes when stone was available and used.

What is required are further detailed surveys of the kind attempted in the Pabbi Hills that map and target conglomerates and stringers in fluvial and flood plain sequences, and also investigate their lateral facies to see if hominin activities and even hominin remains could be found in situ. It would also be advisable to concentrate on smaller stream systems nearer stone sources (i.e., adjacent to conglomerates as at Riwat, or Dina and Jalalpur [Rendell and Dennell 1985]

[7]The Pinjor Stage of the Upper Siwaliks (as defined in India) extends from ca. 2.5 to 0.6 Ma, when it is succeeded by the fossil-poor Boulder Conglomerate that denotes large influxes of coarse bed load resulting from regional uplift and an increase in river gradients. Its timing however varies between river basins and is not synchronous (see e.g., Opdyke et al. 1979: 30).

but with more extensive lateral exposures). Such systems might be found nearer the southern flanks of the Karakorum Mountains or the Himalayas. Surveys of this kind are low-cost but high-risk in the sense that they need not involve large numbers of people or lavish funding, but may have to cover a great deal of deeply dissected terrain without obtaining clear results.

Peninsular India

Peninsular India, with its rich Middle Pleistocene Acheulean record, presents a completely different set of problems from the Upper Siwaliks. Here, the over-riding problem is the absence of any Early Pleistocene deposits in which Paleolithic material could be preserved. Evidence to date for the Early Paleolithic comes from two main sources. The first are fluvial channels, as at Chirki on the Pravara (Corvinus 1983), Anagwadi in the Ghataprabha Valley (Pappu 1974), the Son and Belan Valleys (Williams and Clarke 1995), the Hiran Valley, Saurashtra, (Marathe 1981), and the Wagan and Kadmali Valleys in the Berach basin, southern Rajasthan (Misra 1967: 203) as well as along the Narmada (see e.g., Patnaik et al. 2009). None of these contain accessible Early Pleistocene deposits, and in most, the only ones preserved are Middle Pleistocene. The second set of data is from areas away from major rivers. Caves such as Bhimbetka III F-23 that have a long (but undated) Acheulean to Holcene record (Misra 1985) are Middle Pleistocene in origin. Open-air sites such as Singi Talav (Gaillard et al. 1985), and those at Paisra (Pant and Jayaswal 1991), Raisen (Jacobson 1985), the Kortallyar Basin (Pappu 2001) and in the Hunsgi-Baichbal Valleys (Paddayya 2001) are in or eroding from Middle Pleistocene deposits. So far, no extensive Early Pleistocene deposits have been found in peninsular India that could be investigated for hominin remains or stone artifacts. A second problem is that fossil material is hardly ever preserved: examples are the small number of poorly-preserved specimens from Chirki or the Hunsgi-Baichbal valleys.

There are two possible exceptions to date that require further investigation. The first is at Attirampakkam, Tamil Nadu, where Pappu (Pappu 2001: 240–241; Pappu et al. 2003) found Acheulean artifacts in a 7 m-deep test pit dug into a deposit of laminated clays that had previously been mapped as Cretaceous. Paleomagnetic dating of these deposits has so far proved inconclusive, but the presence of three fossil teeth, of *Bubalus* or *Bos*, *Equ*us, and a caprine or *Boselaphus* raise the prospect that they may be dated by ESR or Th-U. Even if these sediments are found to be Middle Pleistocene in age, they raise the possibility that other "Cretaceous" clays may be Early Pleistocene in age. The second is the Acheulean workshop site of Isampur in the Hunsgi Valley, for which an estimated age of 1.27 ± 0.17 Ma (Paddayya et al. 2002) was obtained by ESR (the first from a South Asian archeological site). This estimate was an average of 10 determinations on two teeth, assuming a linear uptake (LU) model. (An EU [early uptake] model set a minimum age of 730 ± 100 ka, and recent uptake [RU] indicated a maximum age of 3.112 ± 0.4 Ma.) As all other age estimates from the Hunsgi-Baichbal valleys indicate ages of <500 ka for Acheulean material, further work is needed to validate the Isampur ESR age estimates.

As in northern India, there is no quick solution in peninsular India for finding Early Pleistocene hominin remains. Korisettar (2007) has made an important contribution that may improve the chances of finding such evidence by proposing that the core areas of early paleolithic settlement in peninsular India would have seven "Purana" basins (see Fig. 15.10) where there would have been permanent water, abundant food, and flakeable stone. These seven basins immediately narrow down the areas that need to be surveyed. One necessary objective would be to see if any contain Early Pleistocene deposits, and then intensively survey them.

South China and Mainland Southeast Asia

The Pleistocene faunas of South China and mainland Southeast Asia were unusual in two respects. The first was that this region appears to have been a refuge for several large extinct primates, such as *Gigantopithecus*, *Langsonia* and *Lufengpithecus* as well as *Pongo*, which is now found only on Sumatra and Borneo; the second is it contained a distinctive non-primate fauna that included *Stegodon* and the panda *Ailuripoda* but which excluded western taxa such as giraffids, equids and camelids, that are found in the Upper Siwaliks. The key issue that remains unresolved is whether *Homo* was part of the *Stegodon-Ailuripoda* fauna or part of the indigenous primate fauna before the late Middle Pleistocene (see Ciochon 2009). Evidence that it might have been comes from three sites: the Middle Pleistocene cave of Tham Khuyen in Vietnam, the South Chinese cave of Longgupo, and the open-air faunal locality of Yuanmou. Each is deeply problematic. At Tham Khuyen, nine teeth were identified as *Homo erectus* (Ciochon et al. 1996), but only one is "unequivocally hominind" (Schwartz et al. 1995: 3). Of the three teeth identified as *Gigantopithecus,* only one is unequivocally so (ibid.). The remainder are probably derived from either *Pongo pygmaeus weidenreichi* (Harrison et al. 2002) or a new genus *Langsonia liquidens* (Schwartz et al. 1994, 1995). At Longgupo, the claimed hominin incisors and mandible have not withstood detailed scrutiny, as the incisors are probably hominoid (Schwartz and Tattersall 1996), and the mandible probably belonged to the ape *Lufengpithecus* (Xinxhi Wu 2000), although Harrison et al. (ibid., p. 220)

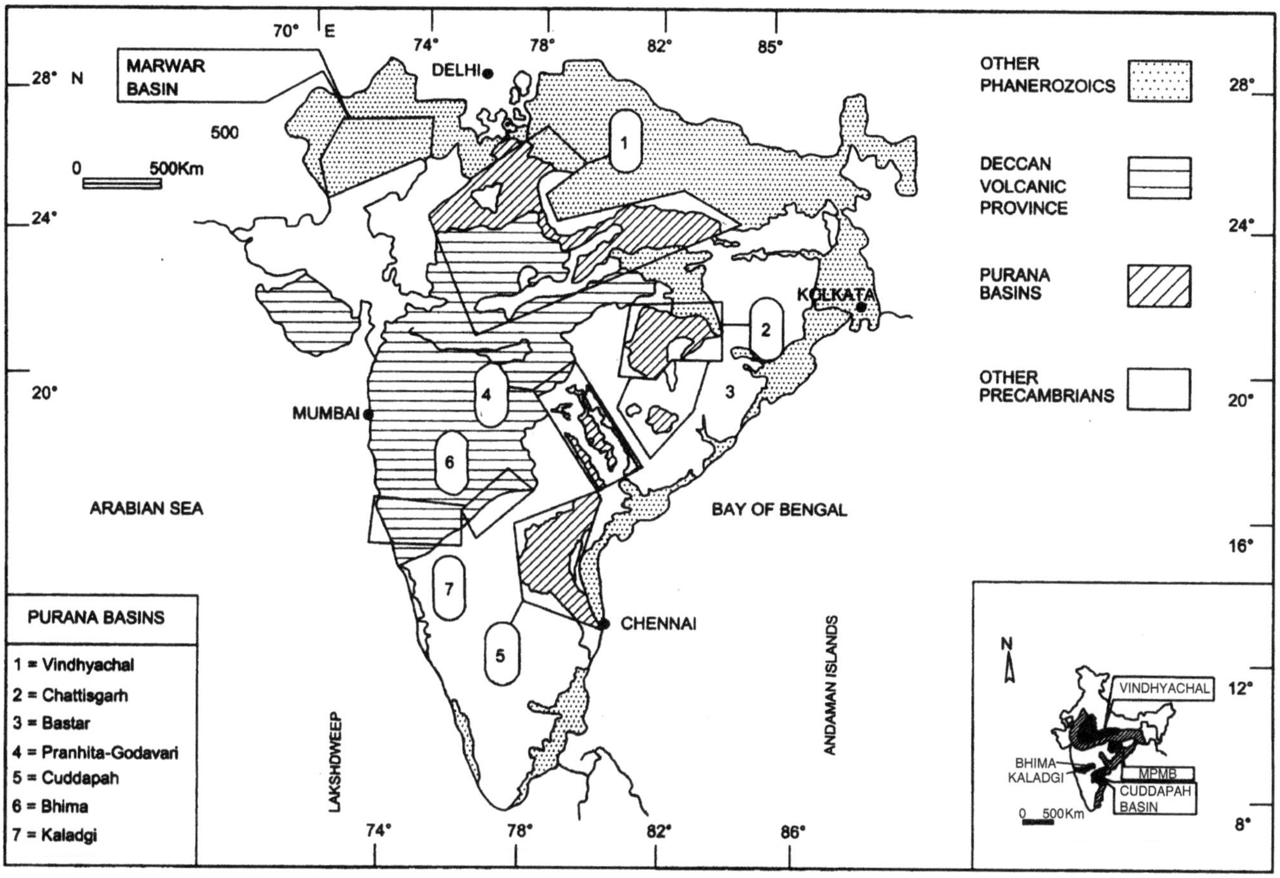

Fig. 15.10 The "Purana Basins" of peninsular India. The figure shows the seven major Purana Basins of peninsular India. These basins are scattered across the peninsula, and some are partly covered by the Deccan Volcanic Province. The smaller Gondwana Basins occur in contiguity with these basins. Together they constitute the core area of Paleolithic occupation in the Indian sub-continent, and are obvious targets for further investigation (Korisettar 2007, Fig. 15.3)

dispute that identification whilst agreeing that it is non-hominin. The flaked artifacts are also unconvincing (personal observation). At Yuanmou, two hominin incisors were found in 1965, and some allegedly associated artifacts a few years later (see Xinzhi Wu and Poirier 1995: 12–16). Estimates of the age of the deposits in which they were (reportedly) found have alternated between the early part of the Early Pleistocene and early Middle Pleistocene (i.e., ca. 1.8 or ca. 0.9 Ma), and the most recent detailed published assessments are that the latter dating is more probable (Urabe et al. 2001; Hyodo et al. 2002). On-going work is said to contradict this view, and maintains that they are Early Pleistocene in age (Potts and Teague 2010).[8] The identification of the incisors as belonging to *Homo* sp. requires detailed justification given that the identification of the "hominin" incisor and mandible at Longgupo, and all but one of the "hominid" teeth from Tham Khuyen has been disproven. The fact that the Yuanmou incisors differ morphologically from those of all *extant* apes (see Potts and Teague 2010) still leaves open the possibility that they derived from an *extinct* ape, of which there are several in this region. Although the Yuanmou incisors differ metrically from *Lufengpithecus* (see Potts and Teague, ibid.), it still needs to be demonstrated that they differ from those of other extinct Southeast Asian apes. As argued by Ciochon (2009, 2010), there is an urgent need to re-examine all Pleistocene dental specimens assigned to *Homo* from this region in view of the uncertainties that *Homo* was part of the *Stegodon-Ailuripoda* fauna in the Early and even Middle Pleistocene. If so, the first indication that it may have been is from Panxian Dadong, South China, in the late Middle Pleistocene (Bekken et al. 2004), apart from what appears to have been a very brief period of occupation ca. 800 ka in the Bose Basin (Hou et al. 2000).

South China and mainland Southeast Asia are thus one of the few regions in southern Asia where a case can be made for arguing that the absence of evidence for hominins in the

[8]Even if the deposits in which the incisors and artefacts were allegedly found are Early Pleistocene in age, the precise stratigraphic context of the incisors and artefacts cannot be established after so many years, and the lack of detailed documentation over their precise stratigraphic provenance remains deeply worrying.

Early Pleistocene is genuine. The best prospects for testing this hypothesis lie in further excavations of caves, of which there are thousands in South China. There may also be prospects in Thailand, although to date the oldest caves appear to be Middle Pleistocene, such as Mae Tha South, Thailand (Pope and Keates 1994: 536), which produced some artifacts that may be ca. 800 ka. Others that have recently been investigated for their cave faunas (Tougard 2001; Tougard and Montuire 2006) include hominin remains (Tougard et al. 1998), albeit limited at present to one tooth from a late Middle Pleistocene context at Thum Wiman Nakin Cave.[9] There are also reasonable prospects for applying uranium-series dating to some of the teeth and associated stalagmites (Esposito et al. 2002). Substantial progress could be expected in Thailand, especially if caves with Early Pleistocene deposits (and even better, hominin remains) are found.

Flood Plains: Yuanmou Revisited?

Member 4 of the Yuanmou Formation represents a continuously aggrading flood plain (Potts and Teague 2010), and is thus not unlike fluvial sequences of the Upper Siwaliks such as the Pabbi Hills. Despite (or possibly because) of the attention focused on establishing the age of the two incisors found in 1965, it is strange that no further evidence of hominins has been found, even though the sections are >80 m deep, laterally extensive, include silts and clays from braided river channels and conglomerates (Urabe et al. 2001: 1678), span much of the Pleistocene, and contain abundant faunal remains (see Pan and Zong 1991). As with Upper Siwalik sequences in India, the Late Pliocene to Middle Pleistocene deposits at Yuanmou may be worth intensive reinspection. There are no detailed accounts of the numbers of fossil vertebrate specimens and skeletal elements of each taxon, or their spatial distribution across the landscape, and it seems probable from published accounts that fossil collecting has often been (as with similar sequences in India) unsystematic and biased towards cranial and dental specimens. As noted above, this type of collecting strategy is likely to miss rare taxa such as hominins. It may also be useful to systematically map and check conglomerate layers for stone tools, particularly if the associated flood plain could be traced laterally. Other Early Pleistocene fluvial sequences in South China would also repay similar investigation. Further fieldwork at Yuanmou would also provide an additional opportunity for testing the hypothesis that *Homo* was (or was not) part of the *Stegodon-Ailuripoda* fauna.

[9] There are also two hominin molars and a cranial fragment of similar age from the cave of Ma U'Oi in northern Vietnam (see Demeter et al. 2005).

Indonesia

A peculiar aspect of the last 20 years paleoanthropological research in Indonesia (Zaim 2010) is that it has focused on a small number of localities that were discovered in the 1930s, such as Sangiran (see e.g., Larick et al. 2001; Ciochon 2009 and references), Mojokerto (see e.g., Huffman et al. 2006; Morwood et al. 2003) or Ngandong (Morwood, personal communication, 2005) or even in the 1890s, as at Trinil (see e.g., Vos et al. 1994; van den Bergh et al. 2001). There is no disputing the value of such reinvestigations, particularly as Sangiran in particular provides the best environmental and chronometric sequence of hominins in Southeast Asia, and the recent re-discovery of the precise provenance of the Mojokerto cranium that is now dated to a maximum of 1.49 Ma (Morwood et al. 2003) clarifies long-standing uncertainties over the age of this important specimen. Nevertheless, these localities are finite resources, and it is to be hoped that in the twenty-first century the number of Early Pleistocene localities is increased. In this respect, the discovery of a hominin fossil at a new Early Pleistocene locality in West Java (Kramer et al. 2005) opens up prospects of a new source of evidence.

Summary

Our knowledge of hominin dispersals across and settlement in Asia before 1.0 Ma is not unlike European maps of Asia in the sixteenth century (see Fig. 15.11) that were based on a few major landmarks, embodied an enormous amount of ignorance and uncertainty, and often placed an unhealthy reliance upon speculation. Our present understanding of when (and which) hominins first left Africa, and of the nature of colonization and settlement in Late Pliocene and Early Pleistocene Asia remains severely limited for three main reasons: the origins of *Homo* and *H. erectus* in East Africa are still unclear; the number of accurate observations of hominins from Late Pliocene and Early Pleistocene Asia is pathetically small; and paleoclimatic data has not been used to its full extent to model the likely nature of early hominin colonization and settlement in Asia. The magnitude of current uncertainties over when and which hominins first left Africa is clearly shown by the rival hypothesis based on the evidence from Dmanisi that *Homo* dispersed out of Africa before 2 Ma and that *H. erectus* originated in Southwest Asia.

In order to move forward in Asia into the twenty-first century, we need to utilize better the paleoclimatic evidence to model the likely nature of early hominin settlement in Asia, and to identify those regions where hominin settlement was probably continuous, and those where it was probably

Fig. 15.11 Robert Thorne's 1527 map of Asia. Robert Thorne's map provides a suitable analogy of our current knowledge of hominin settlement in Asia in the Late Pliocene and Early Pleistocene: a few landmarks, considerable uncertainty, and an often unhealthy reliance upon speculation. The addition of even a dozen Early Pleistocene archeological landmarks in Asia would lead to a considerable improvement in current knowledge

Table 15.1 The most likely opportunities for improving current evidence for hominins in Asia before 1 Ma

		Climatic prediction	
Depositional context	Dating	Continuous occupation	Discontinuous occupation
Lakes	Paleomagnetism, fauna	Levant, western Turkey,	Inland SW Asia, Central Asia, North China, India outside Purana basins
Rivers	Paleomagnetism, fauna	India: Purana basins	Indus/Ganges/Huang He, etc. if/when stone available
Caves	Paleomagnetism, fauna	South China, SE Asia	
Paleosols/loess	Paleomagnetism		Central Asia, North China
Volcanic sequences	Ar/Ar, K/Ar	Indonesia	Anatolia, Elburz, Caucasus

episodic. Above all, we need to increase the number of well-dated and informative sites in Late Pliocene and Early Pleistocene Asia. Although paleoanthropology is an unpredictable science and we will always be surprised by chance discoveries such as Dmanisi, or *H. floresiensis* on Flores (Brown et al. 2004), most progress is achieved by sustained, methodical (and often tedious and uncomfortable) survey and excavation. What is most needed now are small, modestly-funded 3–5 year collaborative research projects that involve archeologists, paleontologists and geologists and which target those areas most likely to produce results: paleolakes in Southwest and Central Asia, and North China; loess sections in Central Asia and North China; fluvial sequences in South Asia (i.e., Upper Siwalik ones in North India, Pakistan and Nepal, and those in the Purana Basins of peninsular India) and perhaps Southeast Asia; caves in South China and Southeast China; and perhaps areas of volcanic activity in parts of Southwest Asia and Indonesia (see Table 15.1). Discoveries in any one of these regions could radically change our understanding of early hominin settlement outside Africa. In most of Asia, the selection and investigation of areas to survey could be greatly facilitated by the type of satellite imagery now freely available from Google Earth and other agencies (see Conroy et al. 2008 for an example). In densely vegetated regions such as South China and Southeast Asia, landforms are more difficult to map from satellite imagery than in semi-arid or arid regions, although new developments in ground-penetrating radar offer some prospects of overcoming these difficulties. Despite the size of Asia, the areas that are worth sustained investigation for information on hominins before 1 Ma are fairly limited, and often straightforward to investigate (leaving aside political issues). Providing the necessary commitment of funding and personnel are in place, the next 10–15 years could prove to be the most exciting yet in investigations of the earliest hominin settlement of Asia.

Acknowledgements I am extremely grateful to the organizers of the "Out of Africa I" workshop at Stony Brook in September 2005 for their invitation to the author to what proved to be an extremely hospitable and stimulating meeting. The British Academy is also thanked for granting the author a Research Professorship for the period 2003–2006 so that he could concentrate on researching early Asian prehistory. The comments and criticisms of two anonymous referees are also gratefully acknowledged.

References

Abdrakhmatov, K. Ye, Makarov, V. I., Molnar, P., Panasyuk, S. V., Prilepin, M. T., Reilinger, R. E., et al. (1996). Relatively recent construction of the Tien Shan inferred from GPS measurements of present-day crustal deformation rates. *Nature, 384*, 450–453.

Aiello, L. C., & Wheeler, P. (1995). The expensive tissue hypothesis. *Current Anthropology, 36*, 199–222.

Antón, S. C. (2002). Evolutionary significance of cranial variation in Asian *Homo erectus*. *American Journal of Physical Anthropology, 118*, 301–323.

Antón, S. C., & Swisher, C. (2004). Early dispersals of *Homo* from Africa. *Annual Review of Anthropology, 33*, 271–296.

Arai, A., & Thibault, C. (1975/1977). Nouvelles précisions à propos de l'outillage paléolithique ancien sur galets du Khorassan (Iran). *Paléorient 3*, 101–108.

Argue, D., Donlon, D., Groves, C., & Wright, R. (2006). *Homo floresiensis*: microcephalic, pygmoid, *Australopithecus*, or *Homo*? *Journal of Human Evolution, 51*, 360–374.

Asfaw, B., Gilbert, W. H., Beyenne, Y., Hart, W. K., Renne, P. R., WoldeGabriel, W., et al. (2002). Remains of *Homo erectus* from Bouri, Middle Awash, Ethiopia. *Nature, 416*, 317–319.

Azarello, M., Marcolini, F., Pavia, G., Pavia, M., Petronio, C., Petrucci, M., et al. (2006). Evidence of earliest human occurrence in Europe: the site of Pirro Nord (Southern Italy). *Naturwissenschaften, 94*, 107–112.

Bailey, G., King, G., & Manighetti, I. (2000). Tectonics, volcanism, landscape structure and human evolution in the African Rift. In G. N. Bailey, R. Charles, & N. Winder (Eds.), *Human ecodynamics* (pp. 31–46). Oxford: Oxbow.

Bar-Yosef, O. (1994). The Lower Palaeolithic of the Near East. *Journal of World Prehistory, 8*, 211–265.

Bar-Yosef, O. (1998). Early colonizations and cultural continuities in the Lower Palaeolithic of western Eurasia. In M. Petraglia & R. Korisettar (Eds.), *Early human behavior in global context: The rise and diversity of the Lower Palaeolithic record* (pp. 221–279). London: Routledge.

Bar-Yosef, O., & Goren-Inbar, N. (1993). *The lithic assemblages of 'Ubeidiya: A Lower Palaeolithic site in the Jordan Valley*. Jerusalem: The Hebrew University of Jerusalem.

Bekken, D., Schepartz, L., Miller-Antonio, S., Yamei, H., & Weiwen, H. (2004). Taxonomic abundance at Panxian Dadong, a Middle Pleistocene cave in south China. *Asian Perspectives, 43*, 333–359.

Belmaker, M. (2002). First evidence of the presence of *Theropithecus* sp. in the southern Levant. *Israel Journal of Zoology, 48*, 165.

Belmaker, M., Tchernov, E., Condemi, S., & Bar-Yosef, O. (2002). New evidence for hominid presence in the Lower Pleistocene of the Southern Levant. *Journal of Human Evolution, 43*, 43–56.

Biglari, F., Nokandeh, G., & Heydari, S. (2000). A recent find of a possible Lower Palaeolithic assemblage from the foothills of the Zagros Mountains. *Antiquity, 74*, 749–750.

Black, D. (1925). Asia and the dispersal of primates. *Bulletin of the Geological Society of China, 4*, 133–183.

Blumenschine, R. J., Masao, F. T., Tactikos, J. C., & Ebert, J. I. (2007). Effects of distance from stone source on landscape-scale variation in Oldowan artifact assemblages in the Palaeo-Olduvai Basin, Tanzania. *Journal of Archeological Science, XX*, 1–11.

Boaz, N. T., Ciochon, R. L., Qinqi, Xu, & Liu, J. (2004). Mapping and taphonomic analysis of the *Homo erectus* loci at Locality 1 Zhoukoudian, China. *Journal of Human Evolution, 46*, 516–549.

Bobé, R., & Behrensmeyer, A. K. (2004). The expansion of grassland ecosystems in Africa in relation to mammalian evolution and the origin of the genus *Homo*. *Palaeogeography, Palaeoclimatology, Palaeoecology, 207*, 399–420.

Bocherens, H., & Sen, S. (1998). Pliocene vertebrate locality of Çalta, Ankara, Turkey. 11. Isotopic composition. *Geodiversitas, 20*, 487–495.

Bonnefille, R. (1995). A re-assessment of the Plio-Pleistocene pollen record of East Africa. In E. S. Vrba, G. H. Denton, T. C. Partridge, & L. H. Burckle (Eds.), *Paleoclimate and human evolution* (pp. 299–310). New Haven/London: Yale University Press.

Boomer, I., Aladin, N., Plotnikov, I., & Whatley, R. (2000). The palaeolimnology of the Aral Sea: a review. *Quaternary Science Reviews, 19*, 1259–1278.

Bramble, D. M., & Lieberman, D. (2004). Endurance running and the evolution of *Homo*. *Nature, 432*, 345–352.

Bräuer, G., & Schultz, M. (1996). The morphological affinities of the Plio-Pleistocene mandible from Dmanisi, Georgia. *Journal of Human Evolution, 30*, 445–481.

Braun, D., Ron, H., & Marco, S. (1990). Magnetostratigraphy of the hominind tool-bearing Erk el Ahmar Formation in the northern Dead Sea Rift. *Israeli Journal of Earth Sciences, 40*, 191–197.

Brown, P., Sutkina, T., Morwood, M.J., Soejono, R.P., Jatniko, & Saptomo, E.W. (2004). A new small-bodied hominin from the Late Pleistocene of Flores, Indonesia. *Nature 431*, 1055–1068.

Carbonell, E., Bermúdez de Castro, J. M., Parés, J. M., Pérez-González, A., Cuenca-Bescós, G., Ollé, A., et al. (2008). The first hominin of Europe. *Nature, 452*, 465–469.

Cerling, T. E. (1992). Development of grasslands and savannahs in East Africa during the Neogene. *Palaeogeography, Palaeoclimatology and Palaeoecology, 97*, 241–247.

Ciochon, R. (2009). The mystery ape of Pleistocene Asia. *Nature, 459*, 910–911.

Ciochon, R. (2010). Divorcing hominins from the *Stegodon-Ailuropoda* Fauna: New views on the antiquity of hominins in Asia. In: J.G. Fleagle, J. J. Shea, F. E. Grine, A. L. Baden, R. E. Leakey (Eds.), *Out of Africa I: the first hominin colonization of Eurasia* (pp. 111–126). Dordrecht: Springer.

Ciochon, R., Long, Vu The, Larick, R., Gonzalez, L., Grün, R., de Vos, J., et al. (1996). Dated co-occurrence of *Homo erectus* and *Gigantopithecus* from Tham Khuyen Cave, Vietnam. *Proceedings of the National Academy of Sciences of the United States of America, 93*, 3016–3020.

Clemens, S. C., Murray, D. W., & Prell, W. L. (1996). Nonstationary phase of the Plio-Pleistocene Asian monsoon. *Science, 274*, 943–948.

Conroy, G. C., Anemonec, R. L., van Regenmorterc, J., & Addisond, A. (2008). Google Earth, GIS, and the Great Divide: a new and simple method for sharing paleontological data. *Journal of Human Evolution, 55*, 751–755.

Corvinus, G. (1983). A survey of the Pravara river system in Western Maharashtra, India. Vol. 2: the excavations of the Acheulean Site of Chirki-on-Pravara, India. In *Tübinger Monographien zur Urgeschichte* (Vol. 7, pp. 1–466). Tübingen: Institut für Urgeschichte.

Dean, C., Leakey, M. G., Reid, D., Schrenk, F., Schwartz, G. T., Stringer, C., et al. (2001). Growth processes in teeth distinguish modern humans from *Homo erectus* and earlier hominins. *Nature, 414*, 628–631.

Deino, A. L., & Hill, A. (2002). $^{40}Ar/^{39}Ar$ dating of Chemeron Formation strata encompassing the site of hominid KNM-BC 1, Tugen Hills, Kenya. *Journal of Human Evolution, 42*, 141–151.

Demeter, F., Bacon, A.-M., Kim Thuy Nguyen, Vu, Long, T., Duringer, P., Rousse, S., et al. (2005). Discovery of a second human molar and cranium fragment in the late Middle to Late Pleistocene cave of Ma U'Oi (Northern Vietnam). *Journal of Human Evolution, 48*, 393–402.

Dennell, R. W. (1998). Grasslands, tool-making and the earliest colonization of south Asia: a reconsideration. In M. Petraglia & R. Korisettar (Eds.), *Early human behavior in global context: The rise and diversity of the Lower Palaeolithic Record* (pp. 280–303). London: Routledge.

Dennell, R. W. (2001). From Sangiran to Olduvai, 1937–1960: the quest for "centres" of hominid origins in Asia and Africa. In R. Corbey & W. Roebroeks (Eds.), *Studying human origins: Disciplinary history and epistemology* (pp. 45–66). Amersterdam: Amsterdam University Press.

Dennell, R. W. (2003). Dispersal and colonisation, long and short chronologies: how continuous is the Early Pleistocene record for hominids outside East Africa? *Journal of Human Evolution, 45*, 421–440.

Dennell, R. W. (2004a). Hominid dispersals and Asian biogeography during the Lower and Early Middle Pleistocene, ca. 2.0–0.5 Mya. *Asian Perspectives, 43*, 205–226.

Dennell, R. W. (2004b). Early hominin landscapes in northern Pakistan: investigations in the Pabbi Hills (with contributions from M. Anwar, M. Beech, R. Coard, L. Hurcombe, H. Rendell, and A. Turner). *British Archeological Reports International Series, 1265*, 1–454.

Dennell, R. W. (2007). "Resource-rich, stone-poor": early hominin land use in large river systems of northern India and Pakistan. In M. D. Petraglia & B. Allchin (Eds.), *The evolution and history of human populations in South Asia: Inter-disciplinary studies in archaeology, biological anthropology, linguistics and genetics* (pp. 41–68). Dordrecht: Springer.

Dennell, R. W. (2009). *The Palaeolithic Settlement of Asia*. Cambridge: Cambridge University Press.

Dennell, R.W. (in press). The colonization of "Savannahstan": issues of timing(s) and patterns of dispersal across Asia in the Late Pliocene and Early Pleistocene. In C.J. Norton, & D.R. Braun (Eds.), *Asian Palaeoanthropology: from Africa to China and beyond.* Springer: Vertebrate Paleobiology and Palaeoanthropology Series.

Dennell, R. W., & Roebroeks, W. (2005). An Asian perspective on early human dispersal from Africa. *Nature, 438*, 1099–1104.

Dennell, R. W., Rendell, H. R., & Hailwood, E. (1988). Early toolmaking in Asia: two-million-year-old artefacts in Pakistan. *Antiquity, 62*, 98–106.

Dennell, R. W., Coard, R., Beech, M., Anwar, M., & Turner, A. (2005a). Locality 642, an Upper Siwalik (Pinjor Stage) fossil accumulation in the Pabbi Hills, Pakistan. *Journal of the Palaeontological Society of India, 50*, 83–92.

Dennell, R. W., Coard, R., Beech, M., Anwar, M., & Turner, A. (2005b). Two Upper Siwalik (Pinjor Stage) fossil accumulations from localities 73 and 362 in the Pabbi Hills, Pakistan. *Journal of the Palaeontological Society of India, 50*, 101–111.

Dennell, R. W., Coard, R., & Turner, A. (2006). The biostratigraphy and magnetic polarity zonation of the Pabbi Hills, northern Pakistan: an Upper Siwalik (Pinjor Stage) Upper Pliocene – Lower Pleistocene fluvial sequence. *Palaeogeography, Palaeoclimatology and Palaeoecology, 234*, 168–185.

Dennell, R. W., Coard, R., & Turner, A. (2008). Predators and scavengers in Early Pleistocene southern Asia. *Quaternary International, 192*, 78–88.

Ding, Z. L., & Yang, S. L. (2000). C3/C4 vegetation evolution over the last 7.0 Myr in the Chinese Loess Plateau: evidence from pedogenic carbonate $d^{13}C$. *Palaeogeography, Palaeoclimatology and Palaeoecology, 160*, 291–299.

Dodonov, A. E. (2002). *Quaternary of Middle Asia: Stratigraphy, correlation and paleogeography*. Moscow: Geos (in Russian).

Dodonov, A. E., & Baiguzina, L. L. (1995). Loess stratigraphy of Central Asia: palaeoclimatic and palaeoenvionmental aspects. *Quaternary Science Reviews, 14*, 707–720.

Dodonov, A. E., Ranow, V. A., & Schäfer, J. (1992). Das Lösspalaolithikum am Obi-Mazar (Tadshikistan). *Jahrbuch des Romisch-Germanischen Zentralmueums Mainz, 39*, 209–243.

Dowsett, H., Thompson, J., Barron, J., Cronin, T., Fleming, F., Ishman, S., et al. (1994). Joint investigations of the Middle Pliocene climate I: PRISM palaeoenvironmental reconstructions. *Global and Planetary Change, 9*, 169–195.

Esposito, M., Reyss, J.-L., Chainmanee, Y., & Jaeger, J.-J. (2002). U-series dating of fossil teeth and carbonates from Snake Cave, Thailand. *Journal of Archeological Science, 29*, 341–349.

Féblot-Augustins, J. (1997). *La Circulation des Matières Premières au Paléolithique*. Liège: Etudes et Recherches Archéologiques de l'Université de Liège.

Fernández, M. H., & Vrba, E. S. (2006). Plio-Pleistocene climatic change in the Turkana Basin (East Africa): Evidence from large mammal faunas. *Journal of Human Evolution, 50*, 595–626.

Florindo, F., Zhu, R., & Guo, B. (1999). Low-field susceptibility and paleorainfall estimates. New data along a N-S transect of the Chinese Loess Plateau. *Physics and Chemistry of the Earth (A), 24*, 817–821.

Gabunia, L., & Vekua, A. (1995). A Plio-Pleistocene hominid from Dmanisi, East Georgia, Caucasus. *Nature, 375*, 509–512.

Gabunia, L., Vekua, A., Lordkipanidze, D., Swisher, C. C., Ferring, R., Justus, A., et al. (2000a). Earliest Pleistocene hominid cranial remains from Dmanisi, Republic of Georgia: taxonomy, geological setting, and age. *Science, 288*, 1019–1025.

Gabunia, L., Vekua, A., & Lordkipanidze, D. (2000b). The environmental contexts of early human occupation of Georgia (Transcaucasia). *Journal of Human Evolution, 38*, 785–802.

Gabounia, L., de Lumley, M.-A., & Berillon, G. (2000c). Morphologie et fonction du troisième métatarsien de Dmanissi, Géorgie Orientale. In D. Lordkipanidze, O. Bar-Yosef, & M. Otte (Eds.), *Early humans at the gates of Europe: Les Premiers Hommes aux Portes de l'Europe* (pp. 29–42). Liège: Etudes et Recherches Archéologiques de l'Université de Liège.

Gabunia, L., de Lumley, M. A., Vekua, A., Lordkipandize, D., & de Lumley, H. (2002). Découverte d'un nouvel hominide à Dmanissi (Transcauacasie, Georgie). *Comptes Rendus Palevol, 1*, 243–253.

Gaillard, C., Misra, V. N., Rajaguru, S. N., Raju, D. R., & Raghavan, H. (1985). Acheulean occupation at Singi Talav in the Thar Desert: a preliminary report on 1981 excavation. *Bulletin of the Deccan College Research Institute, 44*, 141–152.

Gamble, C. (1999). *The Palaeolithic societies of Europe*. Cambridge: Cambridge University Press.

Gardner, E. W., & Bate, D. M. A. (1937). The bone-bearing beds of Bethlehem: their fauna and industry. *Nature, 140*, 431–433.

Gathogo, P. N., & Brown, F. H. (2006). Revised stratigraphy of Area 123, Koobi Fora, Kenya, and new age estimates of its fossil mammals, including hominins. *Journal of Human Evolution, 51*, 471–479.

Gaudzinski-Windheuser, S. (2005). Subsistenzstratagien Frühpleistozäner Hominiden in Eurasien: Taphonomische Faunenbetrachtungen der Fundstellen der'Ubeidiya Formation (Israel). *Römisch-Germanischen Zentralmuseums, 61*, 1–310.

Geraads, D., & Tchernov, E. (1983). Femurs humains de Pleistocene moyen de Gesher Benot Ya'acov (Israel). *L'Anthropologie, 87*, 138–141.

Ginat, H., Zilberman, E., & Saragusti, I. (2003). Early Pleistocene lake deposits and Lower Palaeolithic finds in Nahal (wadi) Zihor, Southern Negev desert, Israel. *Quaternary Research, 59*, 445–458.

Goren-Inbar, N., Feibel, C. S., Versoub, K. L., Melamed, Y., Kislev, M. E., Tchernov, E., et al. (2000). Pleistocene milestones on the Out-of-Africa corridor at Gesher Ya'aqov, Israel. *Science, 289*, 944–947.

Güleç, E., Howell, F.C., & White, T.D. (1999). Dursunlu – a new Lower Pleistocene faunal and artifact-bearing locality in southern Anatolia. *Hominid Evolution: Lifestyles and Survival Strategies, Edition Archaea, Gelsenkirchen*, 349–364.

Güleç, E., White, T., Kuhn, S., Özer, I., Sagir, M., Yilmaz, H., et al. (2009). The Lower Pleistocene lithic assemblage from Dursunlu (Konya), central Anatolia, Turkey. *Antiquity, 83*(319), 11–22.

Haeusler, M., & McHenry, H. M. (2004). Body proportions of *Homo habilis* reviewed. *Journal of Human Evolution, 46*, 433–465.

Harrison, T., Ji, Xueping, & Su, D. (2002). On the systematic status of the Late Neogene hominoids from Yunnan Province, China. *Journal of Human Evolution, 43*, 207–227.

Head, M.J., & Gibbard, P.L. (2005). Early-Middle Pleistocene transitions: an overview and recommendation for the defining boundary. 1–18. In: M.J. Head & P.L. Gibbard (Eds.), *Early-Middle Pleistocene transitions: the land-ocean evidence*. Geological Society of London, Special Publication 247.

Hemmer, H., Kahlke, R.-D., & Vekua, A. K. (2004). The Old World puma – *Puma pardoides* (Owen, 1846) (Carnivora: Felidae) – in the Lower Villafranchian (Upper Pliocene) of Kvabebi (East Georgia, Transcaucasia) and its evolutionary and biogeographical significance. *Neues Jahrbuch für Geologie und Paläontologie, 233*, 197–231.

Hill, A., Ward, S., Deino, A., Curtis, G., & Drake, R. (1992). Earliest *Homo*. *Nature, 355*, 719–722.

Hooijer, D. A. (1958). An early Pleistocene mammalian fauna from Bethlehem. *Bulletin of the British Museum (Natural History): Geology, 3*, 267–292.

Hou, Y.M., & Zhao, L.X. (2010). New archeological evidence for the earliest hominid presence in China. In J.G. Fleagle, J.J. Shea, F.E. Grine, A.L. Baden, & R.E. Leakey (Eds.), *Out of Africa I: The First Hominin Colonization of Eurasia* (pp. 87–95). Dordrecht: Springer.

Hou, Y., Potts, R., Baoyin, Y., Zhengtang, G., Deino, A., & Wei, W. (2000). Mid-Pleistocene Acheulean-like stone technology of the Bose Basin, South China. *Science, 287*, 1622–1626.

Huang, B., Cheng, X., Jian, Z., & Wang, P. (2003). Response of upper ocean structure to the initiation of the North Hemisphere glaciation in the South China Sea. *Palaeogeography, Palaeoclimatology, Palaeoecology, 196*, 305–318.

Huffman, O.F., Zaim, Y., Kappelman, J., Ruez, D.R. jnr., de Vos, J., Rizal, Y., et al. (2006). Relocation of the 1936 Mojokerto skull discovery site near Perning, East Java. *Journal of Human Evolution 48*, 321–363.

Hurcombe, L.M. (2004). The stone artefacts from the Pabbi Hills. In R.W. Dennell (Ed.), *Early hominin landscapes in Northern Pakistan: investigations in the Pabbi Hills. British Archeological Reports (International Series) 1265*, 222–292.

Hyodo, M., Nakaya, H., Urabe, A., Saegusa, H., Shunrong, X., Jiyun, Y., et al. (2002). Paleomagnetic dates of hominid remains from Yuanmou, China, and other Asian sites. *Journal of Human Evolution, 43*, 27–41.

Islamov, Y. I. (1990). Sel'oungour, un nouveau site du paléolithique inférieur en Asie centrale. *L'Anthropologie, 94*, 675–688.

Jablonski, N. (2004). The hippo's tale: how the anatomy and physiology of Late Neogene *Hexaprotodon* shed light on Late Neogene environmental change. *Quaternary International, 117*, 119–123.

Jacobson, J. (1985). Acheulean surface sites in Central India. In V. N. Misra & P. Bellwood (Eds.), *Recent advances in Indo-Pacific prehistory* (pp. 49–58). New Delhi: Oxford and IBH Publishing.

Ji, J., Balsam, W., & Chen, J. (2001). Mineralogic and climatic interpretations of the Luochuan loess section (China) based on diffuse reflectance spectrophotometry. *Quaternary Research, 56*, 23–30.

Kimbel, W. H. (1995). Hominid speciation and Pliocene climatic change. In E. S. Vrba, G. H. Denton, T. C. Partridge, & L. H. Burckle (Eds.), *Paleoclimate and evolution* (pp. 425–437). New Haven/London: Yale University Press.

Kimbel, W. H., Johanson, D. C., & Rak, Y. (1997). Systematic assessment of a maxilla of *Homo* from Hadar, Ethiopia. *American Journal of Physical Anthropology, 103*, 235–262.

Korisettar, R. (2007). Toward developing a basin model for Palaeolithic settlement of the Indian subcontinent: Geodynamics, monsoon dynamics, habitat diversity and dispersal routes. In M. D. Petraglia & B. Allchin (Eds.), *The evolution and history of human populations in South Asia: Inter-disciplinary studies in archaeology, biological anthropology, linguistics and genetics* (pp. 69–96). Dordrecht: Springer.

Kramer, A., Djubiantono, T., Aziz, F., Bogard, J.S., Weeks, R.A., Weinand, D.C., et al. (2005). The first hominid fossil recovered from West Java, Indonesia. *Journal of Human Evolution 48*, 661–667.

Krantz, G. S. (1994). The palate of skull Sangiran 4 from Java. *Courier Forschungsinstitut Senckenberg, 171*, 69–74.

Kroon, D., Alexander, I., Little, M., Lourens, L. J., Matthewson, A., Robertson, A. H. F., et al. (1998). Oxygen isotope and sapropel stratigraphy in the eastern Mediterranean during the last 3.2 million years. Proceedings of the Ocean Drilling Program. *Scientific Results, 160*, 181–189.

Kuhn, S. L., Arsebük, G., & Howell, F. C. (1996). The Middle Pleistocene lithic assemblage from Yarimburgaz Cave, Turkey. *Paléorient, 22*, 31–49.

Lanpo, J., & Qi, W. (1987). Artefacts lithiques provenant du site pléistocène ancien de Donggutuo près de Nihewan (Nihowan), province d'Hebei, Chine. *L'Anthropologie, 91*, 727–732.

Larick, R., Ciochon, R. L., Zaim, Y., Sudijono, S., Rizal, Y., Aziz, F., et al. (2001). Early Pleistocene $^{40}Ar/^{39}Ar$ ages for Bapang Formation hominins, Central Jawa, Indonesia. *Proceedings of the National Academy of Sciences of the United States of America, 98*, 4866–4871.

Li, Ji-Jun, Fang, X.-M., Van der Voo, R., Zhu, J.-J., Niocaill, C. M., Cao, Ji-Xiu, et al. (1997a). Late Cenozoic magnetostratigraphy (11–0 Ma) of the Donghanding and Wangjiashan sections in the Longzhong Basin, western China. *Geologie en Mijnbouw, 76*, 121–134.

Li, Ji-Jun, Fang, X.-M., Van der Voo, R., Zhu, J.-J., Niocaill, C. M., Ono, Y., et al. (1997b). Magnetostratigraphic dating of river terraces: rapid and intermittent incision by the Yellow River of the northeastern margin of the Tibetan Plateau during the Quaternary. *Journal of Geophysical Research, 102*, 10121–10132.

Li, Y., Yang, J., Xia, Z., & Mo, D. (1998). Tectonic geomorphology in the Shanxi Graben system, northern China. *Geomorphology, 23*, 77–89.

Lindsay, E. H., Opdyke, N. D., & Johnson, N. M. (1980). Pliocene dispersal of the horse *Equus* and late Cenozoic mammalian dispersal events. *Nature, 287*, 135–138.

Liu, T., & Ding, Z. (1998). Chinese loess and the paleomonsoon. *Annual Review of Earth and Planetary Sciences, 26*, 111–145.

Liu, X., Rolph, T., Bloemendal, J., Shaw, J., & Liu, T. (1995). Quantitative estimates of palaeoprecipitation at Xifeng, in the Loess

Plateau of China. *Palaeogeography, Palaeoclimatology and Palaeoecology, 113*, 243–248.

Liu, T., Ding, Z., & Rutter, N. (1999). Comparison of Milankovitch periods between continental loess and deep sea records over the last 2.5 Ma. *Quaternary Science Reviews, 18*, 1205–1212.

Liu, Z., Trentesaux, A., Clemens, S., Colin, C., Wang, P., Huang, B., et al. (2003). Clay mineral assemblages in the northern South China Sea: implications for East Asian monsoon evolution over the past 2 million years. *Marine Geology, 201*, 133–146.

Ljubin, V. P., & Bosinski, G. (1995). The earliest occupation of the Caucasus region. In W. Roebroeks & T. van Kolfschoten (Eds.), *The earliest occupation of Europe* (pp. 207–253). Leiden: University of Leiden Press.

Lordkipanidze, D., Jashashvili, T., Vekua, A., de León, M. S. P., Zollikofer, C. P. E., Rightmire, G. P., et al. (2007). Postcranial evidence from early *Homo* from Dmanisi, Georgia. *Nature, 449*, 305–309.

Louchart, A., Mourer-Chauviré, C., Güleç, E., Howell, F.C., White, T.D. (1998). The avifauna of Dursunlu, Turkey, Lower Pleistocene: climate, environment and biogeography. *Comptes Rendus de l'Academie des Sciences Series II (Earth and Planetary Science) 327*, 341–346.

Lu, H., Liu, X., Zjhang, F., An, Z., & Dodson, J. (1999). Astronomical calibration of loess-paleosol deposits at Luochuan, central Chinese Loess Plateau. *Palaeogeography, Palaeoclimatology, Palaeoecology, 154*, 237–246.

Lumley, H. de, Nioradzé, M., Barsky, D., Cauche, D., Celiberti, V., Nioradzé, G., et al. (2005). Les industries lithiques préoldowayennes du début du Pléistocène inférieur du site de Dmanissi en Géorgie. *L'Anthropologie 109*, 1–182.

Maher, B., & Thompson, R. (1995). Paleorainfall reconstructions from pedogenic magnetic susceptibility variations in the Chinese loess and paleosols. *Quaternary Research, 44*, 383–391.

Maher, B. A., Thompson, R., & Zhou, L. P. (1994). Spatial and temporal reconstructions of changes in the Asian palaeomonsoon: A new mineral magnetic approach. *Earth and Planetary Science Letters, 125*, 461–471.

Mallol, C. (2006). What's in a beach? Soil micromorphology of sediments from the Lower Palaeolithic site of 'Ubeidiya, Israel. *Journal of Human Evolution, 51*, 185–206.

Marathe, A. R. (1981). *Geoarchaeology of the Hiran Valley, Saurashta, India*. Pune: Deccan College Post-Graduate and Research Institute.

Martínez-Navarro, B. (2004). Hippos, pigs, bovids, saber-toothed tigers, moneys, and hominids: dispersals through the Levantine Corridor during Late Pliocene and Early Pleistocene times. In N. Goren-Inbar & J. D. Speth (Eds.), *Human paleoecology in the Levantine Corridor* (pp. 37–52). Oxford: Oxbow.

Maschenko, E.N. (1994). *Papio (Paradolichopithecus) sushkini*: a revision of systematics, morphofunctional peculiarities of the skull and mandible (in Russian). *Paleoteriologiya*, 15–57.

McHenry, H. M., & Coffing, K. (2000). *Australopithecus* to *Homo*: transformations in body and mind. *Annual Review of Anthropology, 29*, 125–146.

Misra, V. N. (1967). *Pre- and proto-history of the Berach Basin, South Rajasthan*. Pune: Deccan College Post-Graduate and Research Institute.

Misra, V. N. (1985). The Acheulean succession at Bhimbetka, Central India. In V. N. Misra & P. Bellwood (Eds.), *Recent advances in Indo-Pacific prehistory* (pp. 35–47). New Delhi: Oxford and IBH Publishing.

Mitchell, J., & Westaway, R. (1999). Chronology of Neogene and Quaternary uplift and magmatism in the Caucasus: constraints from K-Ar dating of volcanism in Armenia. *Tectonophysics, 304*, 157–186.

Molleson, T., & Oakley, K. P. (1966). Relative antiquity of the Ubeidiya hominid. *Nature, 209*, 1268.

Moloney, N., Olsen, S. L., & Voloshin, V. (2001). Lower and Middle Palaeolithic occupation in Central Kazakhstan: the Batpak Valley and environs. In S. Milliken & J. Cook (Eds.), *A very remote period indeed: Papers on the Palaeolithic presented to Derek Roe* (pp. 138–143). Oxford: Oxbow Books.

Morwood, M., Sulluivan, O. P., Susanto, E. E., & Aziz, F. (2003). Revised ages for Mojokerto 1, an early *H. erectus* cranium from East Java. *Australian Archaeology, 57*, 1–4.

Nanda, A. C. (2002). Upper Siwalik mammalian faunas of India and associated events. *Journal of Asian Earth Sciences, 21*, 47–58.

Nishimura, T. D., Takai, M., & Maschenko, E. N. (2007). The maxillary sinus of *Paradolichopithecus sushkini* (late Pliocene, southern Tajikistan) and its phylectic implications. *Journal of Human Evolution, 52*, 637–646.

Opdyke, N. M., Lindsay, E., Johnson, G. D., Johnson, N., Tahirkheli, R. A. K., & Mirza, M. A. I. (1979). Magnetic polarity stratigraphy and vertebrate palaeontology of the Upper Siwalik subgroup of northern Pakistan. *Palaeogeography, Palaeoclimatology, Palaeoecology, 27*, 1–34.

Otte, N., Yalçinkaya, I., Taskiran, H., Kozlowski, J. K., Bar-Yosef, O., & Noiret, P. (1995). The Anatolian Middle Paleolithic: new research at Karain Cave. *Journal of Anthropological Research, 51*, 287–299.

Otte, M., Yalçinkaya, Kozlowski, J., Bar-Yosef, O., Lopez Bayon, I., & Taskiran, H. (1998). Long-term technical evolution and human remains in the Anatolian Palaeolithic. *Journal of Human Evolution 34*, 413–431.

Ozansoy, F. (1969). Pleistocene fossil human footprints in Turkey. *Bulletin of the Mineral Research and Exploration Institute of Turkey, 72*, 146–150.

Paddayya, K. (2001). The Acheulean Culture Project of the Hunsgi and Baichbal Valleys, peninsular India. In L. Barham & K. Robson-Brown (Eds.), *Human roots: Africa and Asia in the Middle Pleistocene* (pp. 235–258). Bristol: Western Academic and Specialist Press Ltd.

Paddayya, K., Blackwell, B. A. B., Jhaldiyal, R., Petaglia, M. D., Fevrier, S., Chaderton, D. A., et al. (2002). Recent findings on the Acheulean of the Hunsgi and Baichbal valleys, Karnataka, with special reference to the Isampur excavation and its dating. *Current Science, 83*, 641–647.

Pan, Y., & Zong, G. (1991). Mammalian faunas. In F. Qian & G. Zhou (Eds.), *Quaternary geology and palaeoanthropology of Yuanmou, Yunnan, China* (pp. 94–102). Science Press: Beijing.

Pant, P. C., & Jayaswal, V. (1991). *Paisra: The Stone Age Settlement of Bihar*. Delhi: Agam Kala Prakashan.

Pappu, R. S. (1974). *Pleistocene studies in the Upper Krishna Basin*. Poona: Deccan College Postgraduate and Research Institute.

Pappu, S. (2001). A re-examination of the palaeolithic archeological record of Northern Tamil Nadu, South India. *British Archeological Reports (International Series), 1003*, 1–246.

Pappu, S., Gunnell, Y., Taieb, M., Brugal, J.-P., & Touchard, Y. (2003). Excavations at the Palaeolithic site of Attirampakkam, South India: preliminary findings. *Current Anthropology, 44*, 591–598.

Parfitt, S. A., Barendregt, R. W., Breda, M., Candy, I., Collins, M. J., Coope, G. R., et al. (2005). The earliest record of human activity in northern Europe. *Nature, 438*, 1008–1012.

Patnaik, R., Chauhan, P. R., Rao, M. R., Blackwell, B. A. B., Skinner, A. R., Sahni, A., et al. (2009). New geochronological, palaeoclimatological and Palaeolithic data from the Narmada Valley hominin locality, central India. *Journal of Human Evolution, 56*, 114–133.

Petraglia, M. D. (1998). The Lower Palaeolithic of India and its bearing on the Asian record. In M. Petraglia & R. Korisettar (Eds.), *Early human behavior in global context: The rise and diversity of the lower palaeolithic record* (pp. 343–390). London: Routledge.

Petraglia, M. D. (2003). The Lower Palaeolithic of the Arabian Peninsula: occupations, adaptations, and dispersals. *Journal of World Prehistory, 17*, 141–179.

Pilgrim, G. E. (1913). The correlation of the Siwaliks with mammal horizons of Europe. *Geological Survey Records of India, 43*, 264–326.

Pope, G. C., & Keates, S. G. (1994). The evolution of human cognition and cultural capacity. In R. S. Corruccini & R. Ciochon (Eds.), *Integrative paths to the past: Paleoanthropological advances in honor of F.Clark Howell* (pp. 531–567). Prentice Hall, NJ: Englewood Cliffs.

Potts, R., & Teague, R. (2010). Behavioral and environmental background to 'Out-of-Africa I' and the arrival of *Homo erectus* in East Asia. In: J.G. Fleagle, J.J. Shea, F.E. Grine, A.L. Baden, & R.E. Leakey (Eds.), *Out of Africa I: the first hominin colonization of Eurasia* (pp. 67–85). Dordrecht: Springer.

Prat, S., Brugal, J.-P., Tiercelin, J-J., Barrat, J.A., Bohn, M., Delagnes, A., Harmand, S., Kimeu, K., Kibunjia, M., Texier, P.-J., Roche, H. (2005). First occurrence of early *Homo* in the Nachukui Formation (West Turkana, Kenya) at 2.3–2.4 Myr. *Journal of Human Evolution 49*, 230–240.

Prueher, L. M., & Rea, D. K. (2001). Volcanic triggering of late Pliocene glaciation: evidence from the flux of volcanic glass and ice-rafted debris to the North Pacific Ocean. *Palaeogeography, Palaeoclimatology, Palaeoecology, 173*, 215–230.

Quade, J., Cerling, T. E., & Bowman, J. R. (1989). Development of Asian monsoon revealed by marked ecological shift during the latest Miocene in northern Pakistan. *Nature, 342*, 163–166.

Quade, J., Cerling, T. E., Bowman, J. R., & Jah, A. (1993). Paleoecologic reconstruction of floodplain environments using palaeosols from Upper Siwalik Group sediments, northern Pakistan. In J. F. Schroder (Ed.), *Himalaya to the sea: Geology, geomorphology and the quaternary* (pp. 213–226). London/New York: Routledge.

Ranov, V. (1995). The 'Loessic Palaeolithic' in South Tadjikistan, Central Asia: its industries, chronology and correlation. *Quaternary Science Reviews, 14*, 731–745.

Ranov, V.A., & Dodonov, A.E. (2003). Small instruments of the Lower Palaeolithic site Kuldara and the geoarcheological meaning. In M. Burdukiewicz, A. Ronen (Eds.), Lower Palaeolithic Small Tools in Europe and the Levant. Oxford, British Archeological Reports (International Series) 1115, pp.133–148.

Rea, D. K., Snoeckx, H., & Joseph, L. H. (1998). Late Cenozoic eolian deposition in the North Pacific: Asian drying, Tibetan uplift, and cooling of the northern hemisphere. *Paleooceanography, 13*, 215–224.

Rendell, H. M., & Dennell, R. W. (1985). Dated lower palaeolithic artefacts from northern Pakistan. *Current Anthropology, 26*, 393.

Rendell, H. M., Dennell, R. W., & Halim, M. (1989). Pleistocene and Palaeolithic Investigations in the Soan Valley, Northern Pakistan. *British Archeological Reports International Series, 544*, 1–346.

Rightmire, G.P., Lordkipanidze (2010). Fossil skulls from Dmanisi: a paleodeme representing earliest *Homo* in Eurasia. In J.G. Fleagle, J.J. Shea, F.E. Grine, A.L. Baden, & R.E. Leakey (Eds.), *Out of Africa I: the first hominin colonization of Eurasia* (pp. 225–243). Dordrecht: Springer.

Rightmire, G. P., Lordkipanidze, D., & Vekua, A. (2006). Anatomical descriptions, comparative studies and evolutionary significance of the hominin skulls from Dmanisi, Republic of Georgia. *Journal of Human Evolution, 50*, 115–141.

Roebroeks, W. (2001). Hominid behaviour and the earliest occupation of Europe: an exploration. *Journal of Human Evolution, 41*, 437–461.

Roebroeks, W., Kolen, J., & Rensink, E. (1988). Planning depth, anticipation and the organisation of Middle Palaeolithic technology: the "archaic natives" meet Eve's descendants. *Helinium, 28*, 17–34.

Ron, H., & Levi, S. (2001). When did hominids first leave Africa? New high-resolution paleomagnetic evidence from the Erk-El-Ahmar formation, Israel. *Geology, 29*, 887–890.

Ronen, A. (1991). The Yiron-gravel lithic assemblage: artefacts older than 2.4 My in Israel. *Archaologisches Korrespondenzblatt 21*, 159–164.

Ruddiman, W. F., Sarnthein, M., Backman, J., Baldauf, J. G., Curry, W., Dupont, L. M., et al. (1989). Late Miocene to Pleistocene evolution of climate in Africa and the low-latitude Atlantic: overview to Leg 108 results. *Proceedings of the Ocean Drilling Program, Scientific Results, 108*, 463–484.

Ruff, C. (2002). Variation in human body size and shape. *Annual Review of Anthropology, 31*, 211–232.

Rust, A. (1950). *Die Höhlenfunde von Jabrud (Syrien)*. Karl Wachholtz Verlag, Neumünster.

Schrenk, F., Bromage, T. G., Bernier, C. G., Ring, U., & Juwayeyi, Y. M. (1993). Oldest *Homo* and Pliocene biogeography of the Malawi Rift. *Nature, 365*, 833–36.

Schwartz, J. H. (2000). Taxonomy of the Dmanisi cranium. *Science, 289*, 5776.

Schwartz, J. H., & Tattersall, I. (1996). Whose teeth? *Nature, 381*, 201–202.

Schwartz, J. H., & Tattersall, I. (2000). What constitutes *Homo erectus*? *Acta Anthropologica Sinica Supplement, 19*, 18–22.

Schwartz, J. H., & Tattersall, I. (2003). *The Human Fossil Record. Volume two: Craniodental morphology of Genus* Homo *(Africa and Asia)*. New York: Wiley-Liss.

Schwartz, J. H., Long, V. T., Cuong, N., Kha, L. T., & Tattersall, I. (1994). A diverse hominoid fauna from the late Middle Pleistocene breccia cave of Tham Khuyen, Socialist Republic of Vietnam. *Anthropological Papers of the American Museum of Natural History, 73*, 1–11.

Schwartz, J. H., Long, V. T., Cuong, N. L., Kha, L. T., & Tattersall, I. (1995). A review of the Pleistocene hominoid fauna of the Socialist Republic of Vietnam (excluding Hylobatidae). *Anthropological Papers of the American Museum of Natural History, 76*, 1–24.

Semaw, S., Rogers, M. J., Quade, J., Renne, P. R., Butler, R. F., Dominguez-Rodrigo, M., et al. (2003). 2.6-million-year-old stone tools and associated bones from OGS-6 and OGS-7, Gona, Afar, Ethiopia. *Journal of Human Evolution, 45*, 169–177.

Shackleton, N. J., Backman, J., Zimmerman, H., Kent, D. V., Hall, M. A., Robert, D. G., et al. (1984). Oxygen isotope calibration of the onset of ice-rafting and history of glaciation in the North Atlantic region. *Nature, 307*, 620–623.

Sharma, A. K. (1993). *Prehistoric Delhi and its neighbourhood*. New Delhi: Aryan Books International.

Sherwood, R. J., Ward, S. C., & Hill, A. (2002). The taxonomic status of the Chemeron temporal (KNM-BC-1). *Journal of Human Evolution, 42*, 153–184.

Slimak, L., Roche, H., Mouralis, D., Buitenhuis, H., Balkan-Atlı, N., Binder, D., et al. (2004). Kaletepe Deresi 3 (Turquie), aspects archéologiques, chronologiques et paléontologiques d'une sequence pleistocène en Anatole centrale. *Comptes Rendus Palevol 3*, 411–420.

Smith, P. E. L. (1986). *Palaeolithic archaeology in Iran*. Philadelphia: University Museum.

Sotnikova, M. V., Dodonov, A. E., & Pen'kov, A. V. (1997). Upper Cenozoic bio-magnetic stratigraphy of Central Asian mammalian localities. *Palaeogeography, Palaeoclimatology, Palaeoecology, 133*, 243–258.

Stringer, C. (2006). *Homo Britannicus: The incredible story of human life in Britain*. London: Allen Lane.

Sun, J., & Liu, T. (2000). Stratigraphic evidence for the uplift of the Tibetan Plateau between 1.1 and 0.9 myr ago. *Quaternary Research, 54*, 309–320.

Sun, Z., Feng, X., Li, D., Yang, F., Yonghong, Qu, & Wang, H. (1999). Cenozoic ostracoda and palaeoenvironments of the northeastern Tarim Basin, western China. *Palaeogeography, Palaeoclimatology, Palaeoecology, 148*, 37–50.

Suwa, G., White, T. D., & Howell, F. C. (1996). Mandibular postcanine dentition from the Shungura Formation, Ethiopia: crown morphology, taxonomic allocations, and Plio-Pleistocene hominid evolution. *American Journal of Physical Anthropology, 101*, 247–282.

Swisher III, C.C., Curtis, G.H., Jacob, T., Getty, A.G., & Suprijo, A., & Widiasmoro (1994). Age of the earliest known hominids in Java Indonesia. *Science 263*, 1118–1121.

Takai, M., Maschenko, E. N., Nishimura, T. D., Anezaki, T., & Suzuki, T. (2008). Phylogenetic relationships and biogeographic history of *Paradolichopithecus sushkini* Trofimov 1977, a large-bodied cercopithecine monkey from the Pliocene of Eurasia. *Quaternary International, 179*, 108–119.

Tapponnier, P., Zhiqin, Xu, Roger, F., Meyer, B., Arnaud, N., Wittlinger, G., et al. (2001). Oblique stepwise rise and growth of the Tibet Plateau. *Science, 294*, 1671–1677.

Tattersall, I. (1997). Out of Africa again... and again? *Scientific American, 276*, 46–53.

Tchernov, E. (1987). The age of the Ubeidiya Formation, an Early Pleistocene hominid site in the Jordan Valley, Israel. *Israel Journal of Earth Sciences, 36*, 3–30.

Thomas, H., Geraads, D., Janjou, D., Vaslet, D., Memseh, A., Billiou, D., et al. (1998). Découverte des premieres faunes pleistocène de la peninsule Arabique dans le desert du Nafoud (Arabie Saoudite). *Compte Rendu de l'Academie des Sciences, Paris 326*, 145–152.

Tian, J., Wang, P., & Cheng, X. (2004). Development of the East Asian monsoon and Northern Hemisphere glaciation: oxygen isotope records from the South China Sea. *Quaternary Science Reviews, 23*, 2007–2016.

Tobias, P. V. (1966). Fossil hominid remains from Ubeidiya, Israel. *Nature, 211*, 130–133.

Tougard, C. (2001). Biogeography and migration routes of large mammals in South-East Asia during the Late Middle Pleistocene: focus on the fossil and extant faunas from Thailand. *Palaeogeography, Palaeoclimatology, Palaeoecology, 168*, 337–358.

Tougard, C., & Montuire, S. (2006). Pleistocene palaeoenvironmental reconstructions and mammalian evolution in South-East Asia: focus on fossil faunas from Thailand. *Quaternary Science Reviews, 25*, 126–141.

Tougard, C., Jaeger, J.-J., Chaimanee, Y., Suteethorn, V., & Triamwichanon, S. (1998). Discovery of a *Homo* sp. tooth associated with a mammalian cave fauna of Late Middle Pleistocene age, Northern Thailand. *Journal of Human Evolution, 35*, 47–54.

Tyler, D. E. (1992). A taxonomy of Javan hominid mandibles. *Acta Anthropologica Sinica, 11*, 292–299.

Tyron, C. A., Logan, M. A. V., Mouralis, D., Kuhn, S., Slimak, L., & Balkan-Atlı, N. (2009). Building a tephrostratigraphic framework for the Paleolithic of Central Anatolia, Turkey. *Journal of Archeological Science, 36*, 637–652.

Urabe, A., Nakaya, H., Muto, T., Katoh, S., Hyodo, M., & Shunrong, X. (2001). Lithostratigraphy and depositional history of the Late Cenozoic hominid-bearing successions in the Yuanmou Basin, southwest China. *Quaternary Science Reviews, 20*, 1671–1681.

van den Bergh, G. D., de Vos, J., & Sondaar, P. Y. (2001). The Late Quaternary palaeogeography of mammal evolution in the Indonesian Archipelago. *Palaeogeography, Palaeoclimatology, Palaeoecology, 171*, 385–408.

Vekua, A. (1995). Die Wirbeltierfauna des Villafranchium vom Dmanisi und ihre biostratigraphische Bedeutung. *Jahrbuch des Römisch Germanischen Zentralmuseums, Mainz, 42*, 77–180.

Vekua, A., Lordkipanidze, D., Rightmire, G. P., Agusti, J., Ferring, R., Maisuradze, G., et al. (2002). A new skull of early *Homo* from Dmanisi, Georgia. *Science, 297*, 85–89.

Verosub, K. L., & Tchernov, E. (1991). Résultats préliminaries de l'étude magnétostratigraphie d'une séquence sédimentaire a industrie humanie en Israël. In E. Bonifay & B. Vandermeersch (Eds.), *Les Premiers Peuplements Humains de l'Europe* (pp. 237–242). Paris: CRNS.

Vishnyatsky, L. B. (1999). The paleolithic of Central Asia. *Journal of World Prehistory, 13*, 69–122.

Vos, J.de., Sondaar, P.Y., Bergh, G.D. van den., Aziz, F. (1994). The *Homo* bearing deposits of Java and its ecological context. *Courier Forschungsinstitut Senckenberg 171*, 129–140.

Vrba, E. S. (1995). The fossil record of African antelopes (Mammalia, Bovidae) in relation to human evolution and paleoclimate. In E. S. Vrba, G. H. Denton, T. C. Partridge, & L. H. Burckle (Eds.), *Paleoclimate and evolution* (pp. 385–424). New Haven/London: Yale University Press.

Walker, A., & Leakey, R. (Eds.). (1993). *The Nariokotome* Homo erectus *skeleton*. Cambridge: Harvard University Press.

Walker, A., & Shipman, P. (1996). *The wisdom of bones: In search of human origins*. London: Weidenfeld & Nicholson.

Wanpo, H., Ciochon, R., Yumin, G., Larick, R., Qiren, F., Schwarcz, H., et al. (1995). Early *Homo* and associated artefacts from Asia. *Nature, 378*, 275–278.

Weinstein-Evron, M., Bar-Oz, G., Zaidner, Y., Tsatskin, A., Druck, D., Porat, N., et al. (2003). Introducing Misliya Cave, Mount Carmel, Israel: a new continuous Lower/Middle Paleolithic sequence in the Levant. *Eurasian Prehistory, 1*, 31–55.

Wheeler, P. E. (1991). The thermoregulatory advantages of hominid bipedalism in open equatorial environments: the contribution of increased convective heat loss and cutaneous evaporative cooling. *Journal of Human Evolution, 21*, 107–115.

Wheeler, P. E. (1992). The thermoregulatory advantages of large body size for hominids foraging in savannah environments. *Journal of Human Evolution, 23*, 351–362.

White, T. D. (1995). African omnivores: global climatic change and Plio-Pleistocene hominids and suids. In E. S. Vrba, G. H. Denton, T. C. Partridge, & L. H. Burckle (Eds.), *Paleoclimate and evolution* (pp. 369–385). New Haven/London: Yale University Press.

Williams, M.A.J., Clarke, M.F. (1995). Quaternary geology and prehistoric environments in the Son and Belan Valleys, North Central India. In S. Wadia, R. Korisettar, & V.S. Kale (Eds.), *Quaternary environments and geoarchaeology of India. Geological society of India memoir 32, Bangalore*, pp. 282–308.

Williams, D. F., Peck, J., Karabanov, E. B., Prokopenko, A. A., Kravchinsky, V., King, J., et al. (1997). Lake Baikal record of continental climate response to orbital insolation during the past 5 million years. *Science, 278*, 1114–1117.

Wood, B. (1984). The origin of *Homo erectus*. *Courier Forschungsinstitut Senckenberg, 69*, 99–111.

Wood, B., & Collard, M. (1999). The human genus. *Science, 284*, 65–71.

Wu, X., & Poirier, F. E. (1995). *Human evolution in China: A metric description of the fossils and a review of the sites*. New York/Oxford: Oxford University Press.

Wu, Y., Cui, Z., Liu, G., Ge, D., Yin, J., Qinghai, Xu, et al. (2001). Quaternary geomorphological evolution of the Kunlun Pass area and uplift of the Qinghai-Xizang (Tibet) Plateau. *Geomorphology, 36*, 203–216.

Xinxhi Wu, X. (2000). Longgupo hominoid mandible belongs to ape. *Acta Anthropologica Sinica, 19*, 1–10.

Yang, S., & Ding, Z. (2006). Winter-spring precipitation as the principal control on predominance of C_3 plants in Central Asia over the last 1.77 Myr: evidence from d13C of loess organic matter in Tajikistan. *Palaeogeography, Palaeoclimatology, Palaeoecology, 235*, 330–339.

Zaim, Y. (2010). Geological evidence for the earliest appearance of hominins in Indonesia. In: J.G. Fleagle, J.J. Shea, F.E. Grine, A.L. Baden, & R.E. Leakey (Eds.), *Out of Africa I: the first hominin colonization of Eurasia* (pp. 97–110). Dordrecht: Springer.

Zheng, H., Powell, C. M., An, Z., Zhou, J., & Dong, G. (2000). Pliocene uplift of the northern Tibetan Plateau. *Geology, 28*, 715–718.

Zhu, R., An, Z., Potts, R., & Hoffman, K. A. (2003). Magnetostratigraphy of early humans in China. *Earth-Science Reviews, 61*, 341–359.

Zhu, R. X., Hoffman, K. A., Potts, R., Deng, C. L., Pan, Y. X., Guo, B., et al. (2001). Earliest presence of humans in northeast Asia. *Nature, 413*, 413–417.

Zhu, R. X., Potts, R., Xie, F., Hoffman, K. A., Deng, C. L., Shi, C. D., et al. (2004). New evidence on the earliest human presence at high northern latitudes in northeast Asia. *Nature, 431*, 559–562.

Chapter 16
Summary and Prospectus

John G. Fleagle and John J. Shea

Abstract The chapters of this volume have addressed many different aspects of our current understanding of the initial dispersal(s) of hominins from Africa and the initial colonization(s) of Eurasia. In each of the contributions the authors have identified how our understanding of early hominin dispersal has changed in recent years and areas in which there are unresolved problems. Although most chapters addressed specific topics or regions of this area of research, there were many overlapping themes that cut across individual contributions. These include the stimulus for the initial hominin dispersal from Africa, the number, timing, the likely routes of dispersal both within Africa and from Africa to and through Eurasia, the identities of the dispersing taxa, and other mammals that might have dispersed along with the hominins. Critical to any understanding of hominin dispersal are issues concerning our ability to identify hominin presence through either fossils or archeological remains, and to obtain accurate dates for this material. Finally, these unresolved issues suggest areas, both geographical and topical, for future research.

Keywords *Homo* • Fauna • Dispersal • Biogeography • Africa • Eurasia

Today, humans are the most successful primate species on earth with a cosmopolitan distribution covering all of the continents except Antarctica. However, the hominin lineage not only originated in Africa, but for the first 4 million or so years of evolution, our hominin ancestors were restricted to that continent. Until relatively recently, it seemed that the initial hominin dispersal to other continents took place approximately 1 million years ago. However, discoveries over the past decade have documented the presence of hominins in Eurasia at the beginning of the Pleistocene nearly 2 million years ago (e.g., Antón and Swisher 2004). New radiometric dates for the fossils from Java (Swisher et al. 1994; Antón and Swisher 2004; Larick et al. 2001; Ciochon 2010; Zaim 2010), new discoveries and dates from Northern China (Hou and Zhao 2010; Potts and Teague 2010; Zhu et al. 2001, 2003, 2004, 2008), and the spectacular collection of ongoing discoveries from Dmanisi in the Republic of Georgia (e.g., Lordkipanidze et al. 2007; Rightmire and Lorkipanidze 2010) all attest to the presence of hominins in Eurasia soon after the initial appearance of the genus *Homo* in Africa (see Grine et al. 2009).

The increasingly solid documentation of hominins in Eurasia around the beginning of the Pleistocene raises a vast array of additional questions. Why did this dispersal take place at 2 million years ago rather than earlier or later? Is the timing of the initial hominin dispersal out of Africa the result of external factors such as climate change, making northern latitudes more hospitable to tropical primates, or geological events enabling intercontinental interchange of faunas? Or, is it the result of internal factors, evolutionary changes within the hominin lineage that made hominins better adapted for dispersal and life in northern latitudes? What route out of Africa did the initial dispersal(s) follow? Is the presence of hominins in western, eastern and southeastern Asia the result of a single dispersal event from Africa or several independent events? Why is there no record of early human occupation in other parts of Eurasia, especially the vast regions of central and southern Asia? What kind of evidence is needed to reliably document an early hominin presence? Was the initial dispersal of hominins into Eurasia part of a larger biogeographic event involving other mammals? What kinds of adaptations made it possible for these African hominins to survive in northern continents? These are some of the many questions addressed by the scholars that attended the Second Stony Brook Human Evolution Symposium and Workshop and authored the preceding chapters. While none of these questions have been satisfactorily answered, the chapters in this volume set the stage for future work.

J.G. Fleagle (✉)
Department of Anatomical Sciences, Stony Brook University Medical Center, Stony Brook, NY 11794-8081, USA
e-mail: john.fleagle@stonybrook.edu

J.J. Shea
Department of Anthropology & Turkana Basin Institute, Stony Brook University, Stony Brook, NY, 11794-4364
e-mail: John.Shea@sunysb.edu

J.G. Fleagle et al. (eds.), *Out of Africa I: The First Hominin Colonization of Eurasia*, Vertebrate Paleobiology and Paleoanthropology,
DOI 10.1007/978-90-481-9036-2_16,

External Versus Internal Stimulus to Hominin Dispersal

Many of the papers discuss the factors that likely contributed to the timing and pattern of the initial hominin dispersal out of Africa into Eurasia. Most emphasize the importance of evolutionary developments in hominin technology and mobility. In the view of Leakey and Werdelin (2010) hominin intercontinental dispersal was only possible after the development of culture in the form of stone tools which enabled hominids to access meat and become part of the carnivore guild rather than occupying the normal prey role of other primates. Compared with other mammals, carnivores are excellent dispersers with large ranges, as evidenced by the genera *Panthera* and *Canis* today. Stone tools and evidence of cut marks on bones are present at 2.6 Ma, and by 1.8 Ma there is evidence of hominins butchering large mammals. Moreover, the increased extinction of carnivores between 2 and 1.5 Ma in eastern Africa may be attributable to the hominin entrance into the carnivore guild. Martínez-Navarro (2010) also emphasizes the role of carnivory as an enabling factor in hominin dispersal and in permitting hominins to survive in a temperate environment. Potts and Teague (2010) identify further indications of increased mobility of hominins in the Late Pliocene and Early Pleistocene, including relatively longer hind limbs than earlier hominins, presumably related to enhanced abilities for terrestrial locomotion. Compared with the earliest, Pliocene sites, archeological sites from the Early Pleistocene are found in a wide range of paleogeographic locations and show use of raw materials from increasingly large distances.

Several authors (including Dennell 2009, 2010; and especially Tchernov 1987, 1992) have emphasized the role of climatic change in the Late Pliocene and Early Pleistocene and some (Tchernov 1987, 1992) have explicitly seen hominin dispersal into the Levant as part of northern extension of African habitats and faunas in conjunction with regional warming. However, Belmaker (2010) finds that although the Early Pleistocene fauna of 'Ubeidiya in Israel contains African elements, the overall character is that of a Mediterranean woodland habitat rather than an African savannah. Thus, she argues, the limited African taxa found at 'Ubeidiya are those that are especially adept at long range dispersal with special features. Martínez-Navarro (2010) argues for the importance of both climatic change and innovations in human evolution in the initial hominin dispersal out of Africa. In his contribution, he documents numerous, presumably independent, episodes of faunal exchange between Africa and Eurasia during the Pliocene and Pleistocene. The fact that only one of these seems to have involved hominins suggests that external factors such as climate or geography alone were not sufficient to account for the initial hominin dispersal. Rather he suggests that changes in the nature of the hominins, specifically a carnivorous diet and social groups enabled the initial hominin dispersal out of Africa and successful colonization of Eurasia.

Lahr (2010) offers a different perspective on the role of climate and geography in early hominin dispersal abilities by examining dispersal *within* Africa. As she emphasizes, a prerequisite for any dispersal into Eurasia is an ability of sub-Saharan taxa to disperse across the Sahara into North Africa. Overall, she finds that North Africa was a rather inhospitable place for early hominins with little evidence of long term occupation of the region before the Middle Pleistocene. Ironically, this suggests that hominins were able to successfully colonize other continents, or perhaps parts of other continents well before they were able to survive in many parts of Africa. However, like Potts and Teague (2010), she argues that it was the diversity of habitats and climatic regimes within Africa, and the challenges they provided that led to the evolution of dispersal abilities that ultimately enabled hominins to successfully inhabit the rest of the world.

The Nature and Number of Early Hominin Dispersal(s)

Several authors address the question of the likely geographical path(s) that hominins followed in their initial dispersal from Africa to Eurasia. Lahr (2010) reviews alternative dispersal routes between North Africa and Eurasia, including (1) the Straits of Gibraltar; (2) the Sicily-Tunisia Strait; and (3) the Sinai Peninsula, and concludes that the Sinai Peninsula is the most likely route based on both geological and faunal evidence, even though the Nile Delta offers a potential obstacle separating that route from most of the African continent. Martínez-Navarro (2010) reaches a similar conclusion in his review of possible dispersal routes. Potts and Teague (2010) consider a dispersal across the Bab-el-Mandeb Straits, but reject this possibility because it would have involved a water crossing in the Early Pleistocene. In discussing further dispersal from western Asia eastward, they note that the Levant is surrounded by numerous barriers, including the Tauros and Zagros Mountains and the deserts of the Arabian Peninsula. They identify three possible routes from the Levant to eastern Asia; (1) a Palearctic dispersal corridor running between mountain systems; (2) a southern route along the coast of Asia; and (3) a "middle" route across the Purana basins of north-central India, an option also discussed by Petraglia (2010). In his discussion of hominin dispersal to southeast Asia, Ciochon (2010) advocates a dispersal across the Bab-el-Mandeb Straits and a coastal route along southern Asia to southeast Asia. In his contribution, Chauhan (2010) compares northern routes of dispersal and possible hominin pathways into and within India, noting the diversity of habitats within that subcontinent.

The striking similarity in dates for the earliest records of hominins in widely scattered parts of Eurasia (Java, 1.8–1.6 Ma; northern China, 1.7–1.66 Ma; Dmanisi, 1.77 Ma) is suggestive of a single, very successful dispersal of hominins at the beginning of the Pleistocene. However, as Dennell (2010) emphasizes, there are vast areas in between with no records of a hominin presence and little indication of a continued hominin presence in any of these regions. We really do not know whether these are the same or multiple dispersals and how they relate to later hominins in the area. As Shea (2010) suggests, the pebble-core technology associated with these sites is not necessarily indicative of a uniform culture or population. Moreover, there are several sources of evidence that suggest the possibility of even earlier dispersals. Dennell (2004, 2009, 2010) has repeatedly advocated an earlier, Pliocene dispersal of hominins into Asia based on a large stone tool recovered *in situ* at Riwat, Pakistan. More indirect, but perhaps more compelling, is the growing evidence, discussed below, that *Homo floresiensis*, the fossil "hobbits" from Flores document the presence on isolated islands in Southeast Asia of a hominin more primitive than *Homo erectus* from either Java, China, or Dmanisi (Jungers et al. 2009). If, as the proponents of this view argue, the Flores hominins cannot be derived from an ancestor with the known morphology of any *Homo erectus* remains from either Africa or Asia, then there must have been a dispersal of a more primitive hominin taxon, presumably earlier than the early Pleistocene dispersal documented by fossils attributed to *Homo erectus*, but one for which we have no direct evidence.

Fellow Travelers

An aspect of the Early Pleistocene hominin dispersal from Africa to Eurasia that has received much discussion and debate is the extent to which it accompanied, or was accompanied by, the dispersal of a consistent set of other mammals, and what this might indicate about the nature of the hominin dispersal in terms of both the habitats and behavior of the hominin dispersers. As discussed above, most of the contributors do not support the idea that the hominin dispersal was part of a wholesale extension of African habitats and faunas into Eurasia (see especially Belmaker 2010; and Potts and Teague 2010). Nevertheless, Martínez-Navarro (2004, 2010) and colleagues (Rook et al. 2004) have repeatedly argued that the initial presence of hominins in many parts of Eurasia, especially Europe, is concordant with the appearance of a small number of other African taxa, specifically, the carnivore *Megantereon whitei*, and the large gramnivorous primate *Theropithecus oswaldi*, and *Hippopotamus antiquus*. While Martínez-Navarro and colleagues have argued for a possible ecological relationship between the hominins and *M. whitei*, in which hominins scavenged the carcasses left by the large "super predator", this scenario is questioned by Lewis and Werdelin (2010). Martínez-Navarro (2010) further suggests that hominins may have preyed on the aquatic hippo, *Hippopotamus antiquus*, which is also found in Early Pleistocene sites in Eurasia.

In the debates over these and many other issues surrounding the initial dispersals of hominins form Africa to Eurasia, the contributors to this volume wrestled with many of the basic questions that confront all paleoanthropologists and paleontologists.

Who's Who in the Fossil Record

Ironically, one of the most surprising of the unresolved questions about the initial hominin dispersal out of Africa concerns the identity of the dispersers. Traditionally, most paleoanthropologists have identified this initial disperser as *Homo erectus*, a taxon that was, after all, initially named from fossils found in Java by Eugene Dubois in the early 1890s. However, there are several reasons to question this scenario. As the sample of fossil hominin remains from Dmanisi has continued to grow, some individuals seem to show similarities to a more primitive member of the genus *Homo* such as *Homo habilis* rather than to *Homo erectus* (Rightmire and Lorkipanidze 2010). This raises the possibility that the first hominin to disperse out was Africa may have been something like *Homo habilis* or an even more primitive taxon, and that *Homo erectus* actually dispersed into Africa from Eurasia (Rightmire and Lorkipanidze 2010; Dennell 2010). This interpretation is consistent with the recent documentation that *Homo habilis* and *H. erectus* were sympatric and synchronic for over a half million years in East Africa (Spoor et al. 2007). Moreover, the recent arguments that the small hominins from the late Pleistocene of Flores preserve a skeletal and dental morphology that is more primitive than any *Homo erectus* fossils also suggest a dispersal to Eastern Asia of a more primitive hominin.

In addition to debates about the taxonomic identity and dispersal history of fossil hominins known from relatively complete fossils, there are many debates about the identity of less complete remains. As recounted by Ciochon (2009, 2010), Hou and Zhao (2010) and Potts and Teague (2010) there are numerous controversies over the identity of various isolated teeth that have been recovered from Plio-Pleistocene sites in China. In many cases, the isolated teeth have been subsequently determined to belong to apes rather than hominins (Ciochon 2009, 2010; also Etler 2009), but others remain difficult to clearly assign to taxon. Likewise, there is a long history of purported fossil hominins from India that have subsequently been discounted (Chauhan 2010).

In an effort to understand Early Pleistocene dispersals between Africa and Eurasia, debates over the identity and phylogenetic affinities of fossils are certainly not limited to hominins. The identity of many other fossil mammals from the Early Pleistocene of Eurasia that have been attributed to African species has been questioned and debated including specimens attributed to *Megantereon whitei* (Martínez-Navarro 2010; Lewis and Werdelin 2010) and vertebrae attributed to *Theropithecus* (Rook et al. 2004; Martínez-Navarro 2010; Patel et al. 2007). Part of this reflects the difficulties of positive identification of fragmentary and non-diagnostic remains. However, there are also intellectual issues. As noted by Martínez-Navarro (2010), it is common for similar fossils from different parts of the world to be given different names because the scholars studying these fossils are not familiar with the fossils from parts of the world in which they have never worked, a difficulty also discussed by Potts and Teague (2010) and Patnaik and Nanda (2010). Only careful documentation of the actual fossil specimens and broad revisions by individual scholars can overcome what may be artificial regional faunas in order to identify hidden dispersal events.

What Are Artifacts and Who Made Them?

If identification of hominins from fossils is beset with difficulties, identification of hominin dispersal events from stone tools and other archeological remains is no less problematic. As discussed by numerous contributors (Chauhan 2010; Dennell 2010; Potts and Teague 2010; Hou and Zhao 2010; Shea 2010), the first priority is deciding if purported stone tools are actually the result of hominin activity or whether the flaking is due to geological processes. This is a critical issue for many of the debated early hominin sites from southern and eastern Asia (e.g., Dennell 2009, 2010; Chauhan 2010; Petraglia 2010; Hou and Zhao 2010; Ciochon 2010). In his review of the purported evidence of early Pleistocene stone tools in India, Chauhan (2010) finds no convincing evidence of an Oldowan assemblage.

As discussed extensively by Shea (2010) and Chauhan (2010), there is no direct relationship between the simplicity of a stone tool and either its age or the hominin species that is likely to have made it. Thus, as Shea emphasizes, while it is undoubtedly true that the earliest stone tools were simple flakes and sharpened pebbles, these types of simple, easily made and useful tools remained a part of the hominin arsenal for over 2 million years. Similar tools are still being made and used today. Thus, sites with simple tools can be of virtually any age, can be the work of any of numerous hominin species and can result from any number of different tasks (i.e., tools made for woodworking in one context may not differ from those made for butchery in a different context). This simple, versatile, pebble-core technology appears to have been the "breakout" technology of early hominin dispersal. Stone tools with more clearly imposed designs, such as the hand axes, picks, cleavers, and other "large cutting tools" appear significantly later in those regions newly-occupied by hominins in Plio-Pleistocene times.

How Old?

Any attempt to identify the first dispersal of hominins out of Africa necessarily depends on accurate dating of positively identified evidence of hominin presence. Thus, while some sites such as Dmanisi (Lordkipanidze et al. 2007) seem securely dated, many others remain the subject of ongoing debates. In some cases ongoing paleontological and geological research continues to refine absolute dates for the biochronology of hominin fossils with a few hundred thousand years (e.g., Zaim 2010; Larick et al. 2001; Spoor et al. 2007), in other cases, such as much of the Indian subcontinent and many parts of China, secure dates are difficult to obtain for most sites (e.g., Petraglia 2010; Chauhan 2010; Dennell 2009, 2010; Hou and Zhao 2010). However, with more thoroughly documented excavations and the development of new dating methods, the chronology of hominin dispersals is becoming more secure everywhere.

Prospectus

In paleoanthropology, as in all branches of science, progress usually leads to more questions. As the chapters in this volume and the summary above illustrate, our current understanding of the initial dispersal of hominins from Africa to Eurasia is both more refined and more complex than it seemed just a few years ago (e.g., Antón and Swisher 2004). This is the result of many active programs of new fieldwork and new analyses. With many active ongoing research programs of fieldwork in many parts of Europe, western Asia, southern Asia, China, and southeast Asia, as well as Africa, there is every reason to think that our understanding of early hominin dispersals will continue to change and become both broader and deeper. In addition to those regions that have already yielded important finds, many areas, such as the vast regions of central and western Asia, and the northern coasts of the Indian Ocean, remain poorly surveyed. Yet these are the very regions often mentioned as likely dispersal routes.

In addition to expanded fieldwork, it is especially critical to have broader international efforts to revise the systematics of many groups of fossil mammals that are already available in museum collections. Without direct comparisons of the fossils from different regions, it is impossible to know which faunal similarities and differences between regions are real and which are just guesswork based on older literature, regional tradition, and/or poorly preserved remains. International conferences such as the Stony Brook Human Evolution Symposia and Workshops can greatly advance this effort by bringing together scientists from many parts of the work to compare material and discuss research activities.

References

Antón, S. C., & Swisher, C. C., III. (2004). Early dispersals of *Homo* from Africa. *Annual Review of Anthropology, 33*, 271–296.

Belmaker, M. (2010). Early Pleistocene faunal connections between Africa and Eurasia: an ecological perspective. In J. G. Fleagle, J. J. Shea, F. E. Grine, A. L. Baden, & R. E. Leakey (Eds.), *Out of Africa I: the first hominin colonization of Eurasia* (pp. 183–205). Dordrecht: Springer.

Chauhan, P. R. (2010). The Indian subcontinent and 'Out of Africa 1'. In J. G. Fleagle, J. J. Shea, F. E. Grine, A. L. Baden, & R. E. Leakey (Eds.), *Out of Africa I: the first hominin colonization of Eurasia* (pp. 145–164). Dordrecht: Springer.

Ciochon, R. L. (2009). The mystery ape of Pleistocene Asia. *Nature, 459*, 910–911.

Ciochon, R. L. (2010). Divorcing hominins from the *Stegodon-Ailuropoda* fauna: new views on the antiquity of hominins in Asia. In J. G. Fleagle, J. J. Shea, F. E. Grine, A. L. Baden, & R. E. Leakey (Eds.), *Out of Africa I: the first hominin colonization of Eurasia* (pp. 111–126). Dordrecht: Springer.

Dennell, R. (2010). "Out of Africa 1": current problems and future prospects. In J. G. Fleagle, J. J. Shea, F. E. Grine, A. L. Baden, & R. E. Leakey (Eds.), *Out of Africa I: the first hominin colonization of Eurasia* (pp. 247–273). Dordrecht: Springer.

Dennell, R. W. (2004). Hominid dispersals and Asian biogeography during the Lower and Early Middle Pleistocene, c. 2.0–0.5 Mya. *Asian Perspectives, 43*, 205–226.

Dennell, R. W. (2009). *The palaeolithic settlement of Asia*. Cambridge: Cambridge University Press.

Etler, D. A. (2009). Mystery ape: other fossils suggest that it's no mystery at all. *Nature, 460*, 684.

Grine, F. F., Smith, H. F., Heesy, C. P., & Smith, E. J. (2009). Phenetic affinities of Plio-Pleistocene *Homo* fossils from South Africa: molar cusp proportions. In F. E. Grine, J. G. Fleagle, & R. E. Leakey (Eds.), *The first humans: origin and early evolution of the genus Homo* (pp. 49–62). Dordrecht: Springer. New York: Springer.

Hou, Y.M., & Zhao, L.X. (2010). New archeological evidence for the earliest hominin presence in China. In J. G. Fleagle, J. J. Shea, F. E. Grine, A. L. Baden, & R. E. Leakey (Eds.), *Out of Africa I: the first hominin colonization of Eurasia* (pp. 87–95). Dordrecht: Springer.

Jungers, W. L., Harcourt-Smith, W. E. H., Wunderlich, R. E., Tocher, M. W., Larson, S. G., Sutikna, T., et al. (2009). The foot of *Homo floresiensis*. *Nature, 459*, 81–84.

Lahr, M. M. (2010). Saharan corridors and their role in the evolutionary geography of 'Out of Africa I'. In J. G. Fleagle, J. J. Shea, F. E. Grine, A. L. Baden, & R.E. Leakey (Eds.), *Out of Africa I: the first hominin colonization of Eurasia* (pp. 27–46). Dordrecht: Springer.

Larick, R., Ciochon, R. L., Zaim, Y., Sudijono, S., Rizal, Y., Aziz, F., et al. (2001). Early Pleistocene $^{40}Ar/^{39}Ar$ ages for Bapang Formation hominins, Central Java, Indonesia. *Proceedings of the National Academy of Sciences of the United States of America, 98*, 4866–4871.

Leakey, M., & Werdelin, L. (2010). Early Pleistocene mammals of Africa: background to dispersal. In J. G. Fleagle, J. J. Shea, F. E. Grine, A. L. Baden, & R. E. Leakey (Eds.), *Out of Africa I: the first hominin colonization of Eurasia* (pp. 3–11). Dordrecht: Springer.

Lewis, M.E., & Werdelin, L. (2010). Carnivore dispersal out of Africa during the Early Pleistocene: relevance for hominins? In J. G. Fleagle, J. J. Shea, F. E. Grine, A. L. Baden, & R. E. Leakey (Eds.), *Out of Africa I: the first hominin colonization of Eurasia* (pp. 13–26). Dordrecht: Springer.

Lordkipanidze, D., Jashashvili, T., Vekua, A., Ponce de Leon, M. S., Zollikofer, C. P. E., Rightmire, G. P., et al. (2007). Postcranial evidence from early *Homo* from Dmanisi, Georgia. *Nature, 449*, 305–310.

Martínez-Navarro, B. (2004). Hippos, pigs, bovids, sabertoothed tigers, monkeys and hominids: Dispersals during Late Pliocene and Early Pleistocene times through the Levantine Corridor. In N. Goren-Inbar & J.D. Speth (Eds.), *Human Paleoecology in the Levantine Corridor* (pp. 37–51). Oxford: Oxbow Books.

Martínez-Navarro, B. (2010). Early Pleistocene faunas of Eurasia and hominin dispersals. In J. G. Fleagle, J. J. Shea, F. E. Grine, A. L. Baden, & R.E. Leakey (Eds.), *Out of Africa I: the first hominin colonization of Eurasia* (pp. 207–224). Dordrecht: Springer.

Patel, B. A., Gilbert, C. C., & Ericson, K. E. (2007). Cercopithecoid cervical vertebral morphology and implications for the presence of *Theropithecus* in early Pleistocene Europe. *Journal of Human Evolution, 52*, 113–129.

Patnaik, R., & Nanda, A.C. (2010). Early Pleistocene mammalian faunas of India and evidence of connections with other parts of the world. In J. G. Fleagle, J. J. Shea, F. E. Grine, A. L. Baden, & R. E. Leakey (Eds.), *Out of Africa I: the first hominin colonization of Eurasia* (pp. 129–143). Dordrecht: Springer.

Petraglia, M.D. (2010). The Early Paleolithic of the Indian subcontinent: hominin colonization, dispersals and occupation history. In J. G. Fleagle, J. J. Shea, F. E. Grine, A. L. Baden, & R. E. Leakey (Eds.), *Out of Africa I: the first hominin colonization of Eurasia* (pp. 165–179). Dordrecht: Springer.

Potts, R., & Teague, R. (2010). Behavioral and environmental background to 'out-of-Africa I' and the arrival of *Homo erectus* in East Asia. In J. G. Fleagle, J. J. Shea, F. E. Grine, A. L. Baden, & R. E. Leakey (Eds.), *Out of Africa I: the first hominin colonization of Eurasia* (pp. 67–85). Dordrecht: Springer.

Rightmire, G.P., & Lorkipanidze, D. (2010). Fossil skulls from Dmanisi: a paleodeme representing earliest *Homo* in Eurasia. In J. G. Fleagle, J. J. Shea, F. E. Grine, A. L. Baden, & R. E. Leakey (Eds.), *Out of Africa I: the first hominin colonization of Eurasia* (pp. 225–243). Dordrecht: Springer.

Rook, L., Martínez-Navarro, B., & Howell, F. C. (2004). Occurrence of *Theropithecus* sp. in the Late Villafranchian of Southern Italy and implication for Early Pleistocene "out of Africa" dispersals. *Journal of Human Evolution, 47*, 267–277.

Shea, J.J. (2010). Stone Age visiting cards revisited: a strategic perspective on the lithic technology of early hominin dispersal. In J. G. Fleagle, J. J. Shea, F. E. Grine, A. L. Baden, & R. E. Leakey (Eds.), *Out of Africa I: the first hominin colonization of Eurasia* (pp. 47–64). Dordrecht: Springer.

Spoor, F., Leakey, M.G., Gathogo, P.N., Brown F.H., Antón, S. C. McDougall I., et al. (2007). Implications of new early *Homo* fossils from Ileret, east of Lake Turkana, Kenya. *Nature 448*, 688–691.

Swisher, C.C., Curtis, G. H., Jacob, T., Getty, A.G., Suprijo, A., & Widiasmoro (1994). Age of the earliest known hominids in Java, Indonesia. *Science 263*, 1118–1121.

Tchernov, E. (1987). The age of the Ubeidiya Formation, an Early Pleistocene hominid site in the Jordan Valley, Israel. *Israel Journal of Earth Sciences, 36*, 3–30.

Tchernov, E. (1992). The Afro-Arabian component in the Levantine mammalian fauna – a short bio-geographical review. *Israel Journal of Zoology, 38*, 155–192.

Zaim, Y. (2010). Geological evidence for the earliest appearance of hominins in Indonesia. In J. G. Fleagle, J. J. Shea, F. E. Grine, A. L. Baden, & R. E. Leakey (Eds.), *Out of Africa I: the first hominin colonization of Eurasia* (pp. 97–110). Dordrecht: Springer.

Zhu, R. X., Hoffman, K. A., Potts, R., Deng, C. L., Pan, Y. X., Guo, B., et al. (2001). Earliest presence of humans in northeast Asia. *Nature, 413*, 413–417.

Zhu, R. X., Potts, R., Pan, Y. X., Yao, H. T., Lü, L. Q., Zhao, X., et al. (2008). Early evidence of the genus *Homo* in East Asia. *Journal of Human Evolution, 55*, 1075–1085.

Zhu, R., An, Z., Potts, R., & Hoffman, K. A. (2003). Magnetostratigraphic dating of early humans in China. *Earth Science Reviews, 61*, 341–359.

Zhu, R., Potts, R., Xie, F., Hoffman, K. A., Deng, C. L., Shi, C. D., et al. (2004). New evidence regarding the earliest human presence at high northern latitudes in northeast Asia. *Nature, 431*, 559–562.

Site Index

Subject Index

I

J

K

Q

R

S

T

MIX
Papier aus verantwortungsvollen Quellen
Paper from responsible sources
FSC® C105338

If you have any concerns about our products,
you can contact us on
ProductSafety@springernature.com

In case Publisher is established outside the EU,
the EU authorized representative is:
Springer Nature Customer Service Center GmbH
Europaplatz 3, 69115 Heidelberg, Germany

Printed by Libri Plureos GmbH
in Hamburg, Germany